Lecture Notes in Electrical Engineering

Volume 259

For further volumes:
http://www.springer.com/series/7818

Shengzhao Long · Balbir S. Dhillon
Editors

Proceedings of the 13th International Conference on Man-Machine-Environment System Engineering

MMESE

Springer

Editors
Shengzhao Long
Astronaut Research and Training Center of China
Beijing
People's Republic of China

Balbir S. Dhillon
University of Ottawa
Ottawa, ON
Canada

ISSN 1876-1100 ISSN 1876-1119 (electronic)
ISBN 978-3-642-38967-2 ISBN 978-3-642-38968-9 (eBook)
DOI 10.1007/978-3-642-38968-9
Springer Heidelberg New York Dordrecht London

Library of Congress Control Number: 2013947373

Printed on acid-free paper

Springer is part of Springer Science+Business Media (www.springer.com)

Xuesen Qian's Sky-high Estimation

Grandness Scientist Xuesen Qian's Sky-high Estimation for the Man-Machine-Environment System Engineering

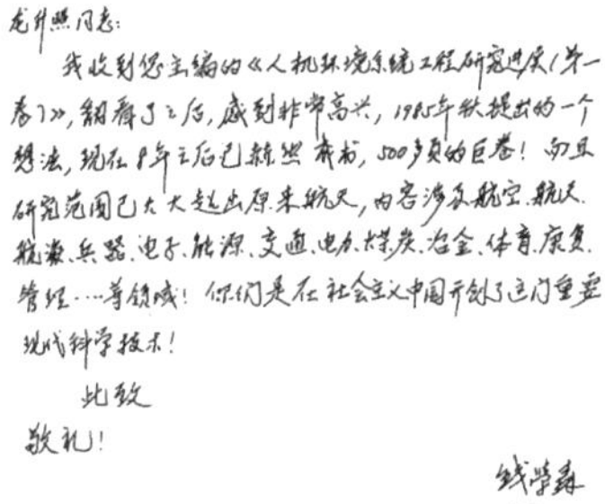

龙升照同志：

我收到您主编的《人机环境系统工程研究进展（第一卷）》，翻看了之后，感到非常高兴，1985年秋提出的一个想法，现在8年之后已蔚然成林，500多页的巨卷！而且研究范围已大大超出原来航天，内容涉及航空、航天、航海、兵器、电子、能源、交通、电力、煤炭、冶金、体育、康复、管理……等领域！你们是在社会主义中国开创了这门重要现代科学技术！

此致

敬礼！

钱学森

1993.10.22

Xuesen Qian's Congratulatory Letter

Grandness Scientist Xuesen Qian's Congratulatory Letter to the 20th Anniversary Commemorative Conference of Man-Machine-Environment System Engineering Foundation

龙升照同志:

你的来信已收到. 欣悉人-机-环境系统工程创立 20 周年纪念大会暨第五届全国人-机-环境系统工程学术会议即将召开，我向你们表示最热烈的祝贺!

20 年来，你们在人-机-环境系统工程这一新兴科学领域进行了积极的开拓和探索，并取得了非常可喜的成绩，我感到由衷的高兴。

希望你们今后再接再励，大力推动人-机-环境系统工程理论及应用的蓬勃发展，为中国乃至世界科学技术的进步作出积极贡献!

祝

工作顺利!

钱学森

2001年6月26日

Preface

In 1981, under the guidance of the great Scientist Xuesen Qian, an integrated frontier science—Man-Machine-Environment System Engineering (MMESE)—came into being in China. Xuesen Qian gave high praise to this emerging science. In a letter to Shengzhao Long, he pointed out, "You are creating this very important modern science and technology in China!" on October 22, 1993.

In the congratulation letter to the commemoration meeting of the 20th anniversary of establishing the MMESE, the great Scientist Xuesen Qian stated, "You have made active development and exploration in this new emerging science of MMESE, and obtained encouraging achievements. I am sincerely pleased and hope you can do even more to make prosper development in the theory and application of MMESE, and *make positive contribution to the progress of science and technology in China, and even in the whole world*" on June 26, 2001.

October 22, which is the day the great Scientist Xuesen Qian gave high praise to MMESE, was determined to be the Foundation Commemoration Day of MMESE by the 2nd conference of the 5th MMESE Committee on October 22, 2010. On this very special day, the great Scientists Xuesen Qian pointed out in a letter to Shengzhao Long, "You are creating this very important modern science and technology in China!" The conference also determined that the annual Conference on MMESE would be held from October 21–25 to cherish the memory of the great contributions that the great Scientist Xuesen Qian had made to MMESE!

The 13th International Conference on MMESE will be held in Yantai on October 21–25 of this year; hence, we will dedicate the *Proceedings of the 13th International Conference on Man-Machine-Environment System Engineering* to our readers.

Proceedings of the 13th International Conference on Man-Machine-Environment System Engineering are the academic showcases of the 13th International Conference on MMESE jointly held by the MMESE Committee of China and Beijing KeCui Academe of MMESE in Yantai. The conference proceedings consisted of 71 more excellent papers selected from more than 400 papers. Due to limitations of space, some excellent papers have been left out, we feel deeply sorry for that. Crudeness in contents and possible incorrectness is inevitable due to the somewhat pressing editing time and we hope you kindly point them out promptly, and your valuable comments and suggestions are also welcome.

Proceedings of the 13th International Conference on Man-Machine-Environment System Engineering will be published by Springer-Verlag, German. Springer-Verlag is also responsible for the related matters on Index to CPCI-S (ISTP), so that the world can know the research quality and development trend of the MMESE theory and application. Therefore, the publication of *Proceedings of the 13th International Conference on Man-Machine-Environment System Engineering* will greatly promote the vigorous development of MMESE in the world, and realize the grand object of "making positive contribution to the progress of science and technology in China, and even in the whole world" proposed by Xuesen Qian.

We would like to express our sincere thanks to Springer-Verlag, German for their full support and help during the publishing process.

Beijing, August 2013 Shengzhao Long

Program and Technical Committee Information

General Chairman

Prof. Shengzhao Long, Astronuat Research and Training Center of China

Program Committee Chairman

Prof. Balbir S. Dhillon, University of Ottawa, Canada

Technical Committee Chairman

Prof. Shangguang Chen, Astronuat Research and Training Center of China

Program and Technical Committee Members

Prof. J. G. Kreifeldt, Tufts University, USA
Prof. Yanping Chen, University of Management and Technology, USA
Prof. Hongfeng Gao, University of California, USA
Prof. Xiangshi Ren, Kochi University of Technology, Japan
Prof. Kinhuat Low, Nanyang Technological University, Singapore
Senior Engineer Baoguo Luan, CH-AUTO Technology Co. Ltd, China
Prof. Baoguo Wang, Beijing Institute of Technology, China
Prof. Baoqing Xia, Weapon Industrial Hygiene Research Institute, China
Prof. Chunhui Wang, Astronuat Research and Training Center of China, China
Prof. Enrong Mao, College of Engineering, China Agricultural University, China
Prof. Fang Jie, China North Vehicle Research Institute, China
Senior Engineer Fong Zhou, AVIC Aerospace Life-support Industries, Ltd, China
Prof. Guangtao Ma, Shenyang Jianzhu University, China
Prof. Guansheng Huang, Beijing Special Vehicle Insititute, China
Prof. Guohua Jiang, Astronuat Research and Training Center of China
Prof. Guohua Shi, Engineering Academy of the Second Artillery, China
Prof. Haiming Wu, Weapon Industrial Hygiene Research Institute, China
Prof. Hongjun Xue, Northwestern Polytechnical University, China
Prof. Honglei Zhang, Air Defense Forces Academy, China
Senior Engineer Jiajun Sun, Petrochina Changqing Oilfield Company, China
Prof. Junfeng Liu, Educational Technology Center of Dalian, China
Prof. Li Ding, Beijing University of Aeronautics and Astronautics, China
Prof. Lijing Wang, Beijing University of Aeronautics and Astronautics, China
Prof. Long Ye, Beijing Jiaotong University, China
Prof. Xiaochao Guo, Institute of Aviation Medicine, Air Force, China
Prof. Xinqin Jin, Harbin University of Science and Technology, China
Prof. Yongqing Hou, China Academy of Space Technology, China
Prof. Yuhong Shen, Quartermaster Research Institute of the General Logistics Department of CPLA, China
Prof. Zhibing Pang, Air Defense Forces Academy, China

Contents

Part VII Research on the Overall Performance of Man-Machine-Environment System

Part VIII Theory and Application Research

Biography of Editor-in-Chief

Prof. Shengzhao Long is the founder of the Man-Machine-Environment System Engineering (MMESE), the Chairman of the MMESE Committee of China, the Chairman of the Beijing KeCui Academy of MMESE, and the Former Director of Ergonomics Lab of Astronuat Research and Training Center of China. In October 1992, he was honored by the National Government Specific Allowance.

He graduated from the Shanghai Science and Technology University in 1965, China. In 1981, directing under famous Scientist Xuesen Qian, he founded the MMESE theory. In 1982, he proposed and developed Human Fuzzy Control Model using fuzzy mathematics. From August of 1986 to August of 1987, he conducted research in Man-Machine System as a visiting scholar at Tufts University, Massachusetts, U.S.A. In 1993, organized MMESE Committee of China. Published "Foundation of theory and application of MMESE" (2004) and "MMESE" (1987). Edited "Proceedings of the 1st–12th Conference on MMESE" (1993–2012)., e-mail: shzhlong@sina.com

Dr. Balbir S. Dhillon is a Professor of Engineering Management in the Department of Mechanical Engineering at the University of Ottawa, Canada. He has served as a Chairman/Director of Mechanical Engineering Department/Engineering Management Program for over 10 years at the same institution. He has published over 345 (i.e., 201 journal + 144 conference proceedings) articles on reliability, safety, engineering management, etc. He is or has been on the editorial boards of nine international scientific journals. In addition, Dr. Dhillon has written 34 books on various aspects of reliability, design, safety, quality, and engineering management published by Wiley (1981), Van Nostrand (1982), Butterworth (1983), Marcel Dekker (1984), Pergamon (1986), etc. His books are being used in over 85 countries and many of them are translated into languages such as German, Russian, and Chinese. He has served as General Chairman of two international conferences on reliability and quality control held in Los Angeles and Paris in 1987.

Prof. Dhillon has served as a consultant to various organizations and bodies and has many years of experience in the industrial sector. At the University of Ottawa, he has been teaching reliability, quality, engineering management, design, and related areas for over 29 years and he has also lectured in over 50 countries, including keynote addresses at various international scientific conferences held in North America, Europe, Asia, and Africa. In March 2004, Dr. Dhillon was a distinguished speaker at the Conf./Workshop on Surgical Errors (sponsored by White House Health and Safety Committee and Pentagon), held at the Capitol Hill (One Constitution Avenue, Washington, D.C.).

Prof. Dhillon attended the University of Wales where he received a BS in Electrical and Electronic Engineering and an MS in Mechanical Engineering. He received a Ph.D. in Industrial Engineering from the University of Windsor., e-mail: dhillon@genie.uottawa.ca

Part I
Research on the Man Character

Chapter 1
Research on Fatigue Risk Management of Airport Staff

Xiaoli Luo, Shan Zhao, Xianlin Zeng and Limin Li

Abstract In order to effectively carry out fatigue risk management of airport staff, the risk factors that caused airport staff fatigue were analyzed from five aspects based on "5 M" model which had been proposed by FAA. The five aspects were men, machine, media, management, and mission. Then, a corresponding hierarchy structure model was established. By using fuzzy analytic hierarchy process and fuzzy comprehensive evaluation method, the fatigue risk factors of airport staff were evaluated. The result showed that the comprehensive evaluation of the fatigue risk factors is medium risk, and the management factor and human factor are the main fatigue risk factors to airport staff. At last, targeted risk control measures were proposed.

Keywords Airport staff · Fatigue · Risk factor · Fuzzy comprehensive evaluation method · Risk management

1.1 Introduction

Fatigue is a complex state characterized by a lack of alertness and reduced mental and physical performance, often accompanied by drowsiness. Fatigue is objectively observed as changes in many aspects of performance [1], including lapses in attention, reduced situational awareness, and reduced motivation.

X. Luo (✉) · S. Zhao · X. Zeng · L. Li
Department of Postgraduate, Civil Aviation Flight University of China,
618307 Guanghan, China
e-mail: crmluo@yahoo.com.cn

S. Zhao
e-mail: zshaohaiziyi@163.com

S. Long and B. S. Dhillon (eds.), *Proceedings of the 13th International Conference on Man-Machine-Environment System Engineering*, Lecture Notes in Electrical Engineering 259, DOI: 10.1007/978-3-642-38968-9_1,

The fatigue phenomenon of the employed persons in civil aviation has affected the flight safety and benefit. In recent years, civil aviation organizations like ICAO, FAA, IATA and some aviation-developed countries like the UK have paid more attention to fatigue management problem of the employed persons in civil aviation and written it into the laws and regulation documents. CAAC also brought it into the important measures that will improve civil aviation safety, and at the safety working conference, CAAC made it as one of the key tasks that must be implemented. Airport security management is one of the main components of aviation safety, so carrying out an effective fatigue management of its staff will help to improve the overall level of aviation safety.

1.2 Identification of Fatigue Risk Factors of Airport Staff and Construction of its Hierarchy Structure Model

In 2008, the “5 M” model was proposed in ATC management manual by FAA, and it has been used more and more widely in the identification of human error risk factor in aviation ever since. The “5 M” refers to men, machine, media, management, and mission, which is shown in Fig. 1.1 [2].

1.2.1 Human Factor

The characteristics of airport round-the-clock service will make human circadian rhythm desynchronized and sleep disturbed. The high working pressure at airport has already become a big threat of mental health and physical health to some staffs of the front line, such as operation center, mechanical engineering department, and security inspection. Age and gender, health status, unhealthy living habits, and bad interpersonal relationships will also affect the staffs’ fatigue-resistant ability.

Fig. 1.1 “5 M” model

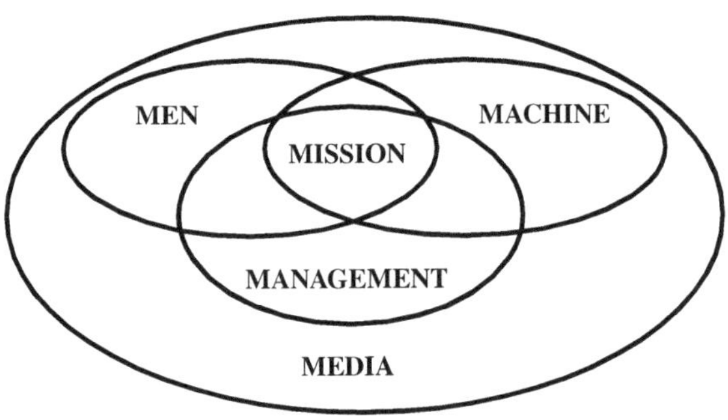

1.2.2 Machine Factor

In some middle-small airports, the monitoring systems need to be improved, aging facilities should be updated, and whenever the emergency equipment must be prepared well [3]. What is more, there are human factor defects in the design of some facilities and equipments, which make the staffs cannot match up with hardware equipments (e.g., monitors, communication systems).

1.2.3 Media Factor

The geographical environment, and temperature, noise, sanitary condition inside the airport all have an impact on its staff. Adverse working atmosphere will make staffs feel stressed, lack enthusiasm, and easily tired of working.

1.2.4 Management Factor

In the airport, supervisory mechanisms are not perfect, supervision of relevant laws and regulations is absent, and effective continuous monitoring is also inadequate. Unreasonable shift systems lead many staffs to the life which is monochrome reversed. In human resource management, the phenomenon of setting position by employee, overstaffed, and uneven makes some frontline staffs are insufficiency, even have oversaturated workload [4].

1.2.5 Mission Factor

As to continuously provide service to passengers and different resident units, staffs have to work long hours and only take short rest, they often under overload work in the day-to-day running state. At the same time, the flight accidents, runway accidents, terminal building accidents, or emergencies will cause irregular situations of airport service [5]; they will double the workload of airport staff within a short period of time.

1.2.6 The Hierarchy Structure Model of Fatigue Risk Factors of Airport Staff

Based on the analysis about factors in “5 M” model, a hierarchy structure model of fatigue risk factors of airport staff can be established, as shown in Fig. 1.2.

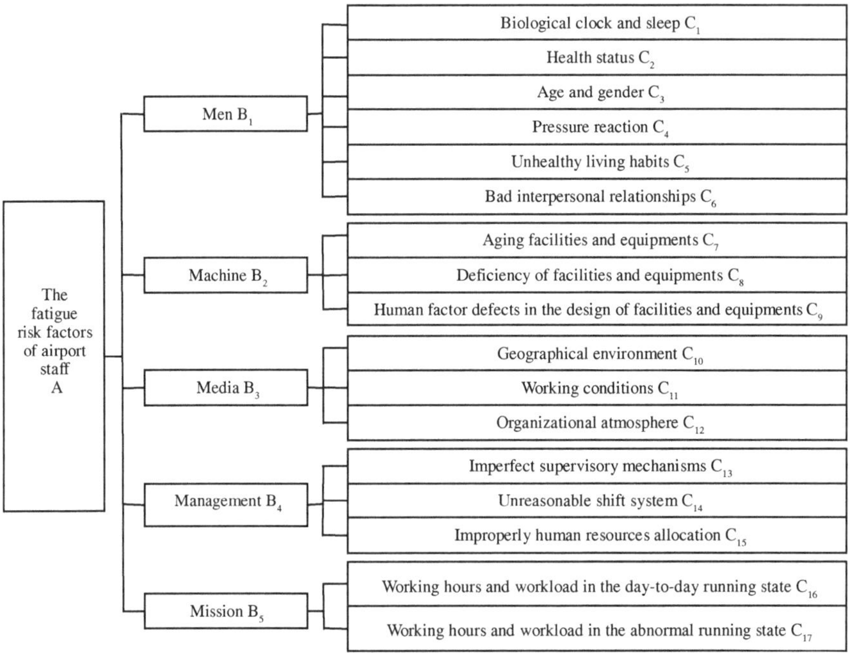

Fig. 1.2 The hierarchy structure model of fatigue risk factors of airport staff

1.3 Fatigue Risk Factors Assessment of Airport Staff

Forty staffs of the frontline and senior management cadres at airport were asked to fill a questionnaire according to the established hierarchy structure model above. The questionnaire was based on fuzzy analytic hierarchy process (FAHP) and fuzzy comprehensive evaluation method in the form of design. For the results of those questionnaires, weights of fatigue risk factors were fixed by using fuzzy analytic hierarchy process, and then, a comprehensive evaluation on these factors through fuzzy comprehensive evaluation method was made.

1.3.1 Weights Calculation of Fatigue Risk Factors of Airport Staff by FAHP

In order to enable the experts to judge the importance of the various factors in quantification, the 0.1–0.9 scaling method was used to construct priority relation matrices. Then, the priority matrices were turned into fuzzy consistent matrices [6]. Due to the limited space, here all the priority relation matrices and fuzzy consistent matrices were not listed.

Power method was adopted to strike weight value of factors in each layer [7]. In the first layer, the weight set consisted of $W_{A-B} = (0.2337\ 0.1745\ 0.1644\ 0.2532\ 0.1743)^{\mathrm{T}}$

In the second layer, the weight set of human factor consisted of $W_{B1\ -\ C1j} = (0.1425\ 0.1922\ 0.1526\ 0.1804\ 0.1610\ 0.1713)^{\mathrm{T}}$

The weight set of machine factor consisted of $W_{B2-C2j} = (0.3092\ 0.3385\ 0.3523)^{\mathrm{T}}$

The weight set of media factor consisted of $W_{B3-C3j} = (0.2655\ 0.3747\ 0.3598)^{\mathrm{T}}$

The weight set of management factor consisted of $W_{B4-C4j} = (0.3171\ 0.3307\ 0.3522)^{\mathrm{T}}$

The weight set of mission factor consisted of $W_{B5-C5j} = (0.5357\ 0.4643)^{\mathrm{T}}$

1.3.2 Evaluation of the Fatigue Risk Factors of Airport Staff by Fuzzy Comprehensive Evaluation Method

1.3.2.1 Establishment of Evaluation Index Set

According to the hierarchy structure model above, evaluation index sets were established: the set of human factor $U = \{u_{11}\ u_{12}\ u_{13}\ u_{14}\ u_{15}\ u_{16}\}$, the set of machine factor $U = \{u_{21}\ u_{22}\ u_{23}\}$, the set of media factor $U = \{u_{31}\ u_{32}\ u_{33}\}$, the set of management factor $U = \{u_{41}\ u_{42}\ u_{43}\}$, and the set of mission factor $U = \{u_{51}\ u_{52}\}$.

1.3.2.2 Establishment of Selected Evaluation Set

Appropriate evaluation level was established to constitute evaluation set, generally access to five levels, namely $V = \{V1, V2, V3, V4, V5\} = \{$very low risk, low risk, medium risk, high risk, very high risk$\}$.

Fuzzy evaluation matrix of human factor

$$R_1 = \begin{bmatrix} 0 & 0.4 & 0.1 & 0.1 & 0.1 \\ 0.1 & 0.3 & 0.5 & 0 & 0.1 \\ 0 & 0 & 0.3 & 0.4 & 0.4 \\ 0 & 0.4 & 0.4 & 0.2 & 0 \\ 0 & 0.2 & 0.2 & 0.4 & 0.2 \\ 0.1 & 0.3 & 0.4 & 0 & 02 \end{bmatrix} \tag{1.1}$$

Fuzzy evaluation matrix of machine factor

$$R_2 = \begin{bmatrix} 0.1 & 0.2 & 0.5 & 0.1 & 0.1 \\ 0.1 & 0.4 & 0.3 & 0.3 & 0 \\ 0.1 & 0.3 & 0.4 & 0.1 & 0.1 \end{bmatrix} \tag{1.2}$$

Fuzzy evaluation matrix of media factor

$$R_3 = \begin{bmatrix} 0 & 0.1 & 0.2 & 0.4 & 0.3 \\ 0 & 0.2 & 0.4 & 0.4 & 0.1 \\ 0 & 0.2 & 0.4 & 0.2 & 0.1 \end{bmatrix} \tag{1.3}$$

Fuzzy evaluation matrix of management factor

$$R_4 = \begin{bmatrix} 0 & 0.1 & 0.5 & 0.4 & 0.1 \\ 0 & 0.3 & 0.5 & 0.1 & 0.1 \\ 0 & 0.5 & 0.4 & 0.1 & 0 \end{bmatrix} \tag{1.4}$$

Fuzzy evaluation matrix of mission factor

$$R_5 = \begin{bmatrix} 0 & 0.3 & 0.3 & 0.2 & 0.2 \\ 0 & 0.5 & 0.1 & 0.3 & 0.1 \end{bmatrix} \tag{1.5}$$

1.3.2.3 The Primary Fuzzy Comprehensive Evaluation

It had been determined by FAHP of each factor weight earlier and then the fuzzy comprehensive evaluation of human factor $B_1 = W_{B1-C1j} \cdot R = (0.0397\ 0.2666\ 0.3731\ 0.1710\ 0.1496)$.

The fuzzy comprehensive evaluation of machine factor $B_2 = W_{B2-C2j} \cdot R = (0.1187\ 0.2878\ 0.4023\ 0.1440\ 0.0473)$.

The fuzzy comprehensive evaluation of media factor $B_3 = W_{B3-C3j} \cdot R = (0\ 0.1953\ 0.3449\ 0.3057\ 0.1540)$.

The fuzzy comprehensive evaluation of management factor $B_4 = W_{B4-C4j} \cdot R = (0\ 0.2932\ 0.4497\ 0.2108\ 0.0463)$.

The fuzzy comprehensive evaluation of mission factor $B_5 = W_{B5-C5j} \cdot R = (0\ 0.3852\ 0.2194\ 0.2475\ 0.1480)$.

1.3.2.4 The Second Class Fuzzy Comprehensive Evaluation

Fuzzy comprehensive evaluation of the fatigue risk factors of airport staff

$$B = W_{A-B} \cdot \begin{bmatrix} B_1 \\ B_2 \\ B_3 \\ B_4 \\ B_5 \end{bmatrix} = (0.0300\ 0.2860\ 0.3662\ 0.2119\ 0.1061)$$

In accordance with the maximum subordination principle, the result of the comprehensive evaluation is medium risk. At the same time, from the risk assessment we can draw a conclusion that management factor and human factor are the main cause of airport staff fatigue, and then followed by machine factor, mission factor and media factor. Management factor is mainly shown as improper human resources allocation and unreasonable shift system. Human factor is mainly shown in health status and pressure reaction. Machine factor is mainly shown as the human factor defects in the design of some facilities, and the shortage of equipments. The mission factor is mainly shown as long working hours and heavy workload in the day-to-day running state of the airport. The media factor is mainly shown in working environment and organization atmosphere.

1.4 Fatigue Risk Control and Prevention of Airport Staff

By the fuzzy comprehensive evaluation method, the general result of the fatigue factors now is medium risk, but with the development of civil aviation, the workload of airport staff will increase along with the flight and passenger flow. If there is no effective control and preventive measures, airport staff fatigue status will be more and more serious, and ultimately, it will transformed into high risk, which will affect the airport safety.

1.4.1 Control and Prevention of Human Risk Factor

The airport staff firstly should keep healthy diet and moderate physical exercise. Then, try to change life value and keeping self-affirmation when under pressure is extremely important. And transposition consideration will help to establish a good interpersonal relationship. Bad habit, like smoking, excessive drinking, should be improved. Establishing a positive work attitude is very helpful for raising working efficiency. Finally, for those people who often work in shifts, understanding human circadian rhythm can help to ensure the quantity and quality of sleep as far as possible.

1.4.2 Control and Prevention of Machine Risk Factor

For device designers, they should focus on the human–machine interface issues and lay stress on human factor principles in the design of monitor and controller devices at the airport. Staffs can take a short break after in front of the machine for a long time to reduce error rates. Then, make an attempt to raise the automation level of machines and equipments and turn to lower-risk devices instead of higher-risk devices [8]. Pay attention to the problem of checking the equipment regularly, reducing repair cycle, and updating old facilities and equipments.

1.4.3 Control and Prevention of Media Risk Factor

It is suggested that the layout of airport workplace should be designed in a reasonable way, and working conditions, operation procedure, and methods should be improved. Do best to keep the airport clean and put an end to dirty and disorder. Smoother lines of communication mechanism should be effectively established, and exchanges between inter-department and department cooperation should be firmly strengthened. The manager is better to mobilize each staff's working enthusiasm, creativity, and initiative, and gradually, it can form complementary team spirit. At the same time, building a proactive airport safety culture with self-features is in significance [9]. At work, airport management leadership and engineering and technical operators of various subordinate bodies should be responsible for airport security layer by layer, to establish a reliable safety production responsibility system.

1.4.4 Control and Prevention of Management Risk Factor

Before the staffs have access to work, the manager should analysis the position in a scientific and reasonable way and rationally divide the duties of each position and clear the work contents and professional requirements of each position [6]. Then, according to the different periods, deploy staff in a flexible way and carry out different post-distributing systems at the peak and valley periods of flights, as well as in slack seasons and busy seasons of airport operation. The work of establishing effective and rational staff assessing, and incentive mechanisms, should be done in time. Do well in pre-job training, in-service training, and annual refresher training. In order to lay down a rational shift system, integrate the airport job and scientific data into the schedule and take human circadian rhythm characteristic fully into account. At last, it is very important to improve safety supervision system and establish real-time monitoring system.

1.4.5 Control and Prevention of Mission Risk Factor

The work time and workload of airport staff in the day-to-day running state and abnormal state are, respectively, the direct external factors which will cause chronic fatigue and acute fatigue. For the high density and high strength work task of airport staff in the day-to-day running state, it is suggested to implement the "double-post system" and strengthen work division and cooperation. For the abnormal running state of airport, it is advised to do well basic airport security work, establish highly efficient emergency management system, improve the running command and rescue system, and accomplish emergency plan, emergency drilling, and training work [10].

1.4.6 Systematic Fatigue Risk Management [1]

From the angle of airport safety management system, the airport organization should carry out up-to-down systematic management. At first, systematic fatigue risk management strategy must be made, and the conception of fatigue risk management system (FRMS) should be introduced. Secondly, a fatigue risk reporting system should be established, and the accident/incident or unsafe events which are entirely or partly caused by staff fatigue can be reported, investigated, and recorded. Again, based on data-driven means, the potential consequences caused by airport staff fatigue must be analyzed, and then, FRMS should be established; it can continuously monitor and manage the fatigue-related safety risk. Finally, the manager should also pay close attention to the feedback of the effectiveness of control measures, continuously monitor the job performance of airport staff after implementing the FRMS, and regularly review the FRMS.

Acknowledgements This work is supported by Grant 60832012 of the National Science Foundation of China, partially supported by Grant MHRDZ200884 of Science and Technology Fund of CAAC.

References

1. U.S. Department of Transportation (2010) Fatigue risk management systems for—aviation safety. FAA, 2010, 8
2. Zhang. Y (2007) Airport Safety Risk Management.JIANGSU AVIATION, (2):4–7
3. Fan L, She L, Ping Y (2002) On a Forewarning Management System for Civil Aviation airport disasters in China. J WUT, 24(6):41–44
4. Y Wu. The Challenges of Human Resources Management in the Airport Industry. Human Resource Management, 2011, (3):48–50
5. Y Li, Zhu T. Introduction to Civil Aviation Airport Ground Services. Beijing: China Civil Aviation Press, 2006:199–203

6. Min Yao (1999) A Practical Fuzzy Analytic Hierarchy Process. Soft Science 1:44–52
7. Y Li. The Matlab Implementation of Fuzzy Analytic Hierarchy Process. Natural Sci, 2009:37–38
8. X.Guimei(2010) Research on Techniques of Risk Management for Airport Runway Accident. Nanjing: Doctor's Degree Thesis of Nanjing University of Aeronautics and Astronautics.
9. Huanting Yu (2007) Discuss on the Safety Culture Construction of Civil Airport. Jiangsu Aviation 2:21–22
10. Yu Yang. The Research of Civil Airport Emergency Management Evaluation and Development Solutions .Guanghan: Master's Degree Thesis of Civil Aviation Flight University of China, 2011
11. Matschnigg Guenther, Graham Nancy, Wykoff Don (2011) Fatigue Risk Management System (FRMS) Implementation Guide for Operators. IATA 7:2–4

Chapter 2
The Foot Sizes of Chinese Male Pilots

Xiaochao Guo, Qingfeng Liu and Yu Bai

Abstract To probe the foot sizes of Chinese male pilots on the bases of databank for GJB4856, statistical analysis was made with indexes of foot length and foot girth in the frames of GB/T3293.1 and GJB1095. Results show that there are 11 foot marks for foot length of 230–280 mm, which cover 72.24 % of the pilot population in GJB4856, and 7 foot types for foot girth of 227–270 mm. It suggests that the most foot sizes of the pilots are distributed in 9 marks of 235–275 and 5 types of 2–4, which cover 65.2 % of the population mentioned above.

Keywords Pilots · Foot sizes · Human dimensions · Anthropometry

2.1 Object

The data of human dimensions of pilots are considered as the fundamental basis to design the cockpit dimension and layout, ejection seat and security access, and personal protection and survival equipment. Generally, the data of the foot dimensions are mainly used in the designing and manufacturing of production such as shoes, socks, the pedal, and so on. *Human dimensions of Chinese male pilot population* (GJB4856) were made in 2000–2001 by actual measurements, providing data of 7 fundamental items and 10 recommended items of the foot [1]. However, it is difficult for such code to make guidance on the research and development of foot articles including shoes and socks as a result of lacking data relation and distribution pattern. The foot dimensions of 16,383 were measured in the national public welfare project—Study on foot type rule of Chinese population in 2002 [2]. The database of the foot type of the Chinese was established. *Chinese*

X. Guo (✉) · Q. Liu · Y. Bai
Institute of Aviation Medicine PLAAF, Beijing 100042, China
e-mail: guoxiaochao1965@163.com

S. Long and B. S. Dhillon (eds.), *Proceedings of the 13th International Conference on Man-Machine-Environment System Engineering*, Lecture Notes in Electrical Engineering 259, DOI: 10.1007/978-3-642-38968-9_2,

last systems (GB/T3293) was formulated after this study to guide the shoe production and promote the development of shoe manufacture in China. Then, what characteristics do pilots have in the foot type distribution as a particular professional group? What should be paid attention to in the research and manufacture of foot articles for pilots? The thesis is intended to make deep exploration and analysis in relation to the foot data of pilots on the basis of the foot dimensions in the database of GJB4856, with the purpose of providing scientific guidance on the utilization of fundamental data.

2.2 Methodologies

2.2.1 Measured Items

Seven foot fundamental items in GJB4856: foot length, foot width, foot girth, toe height, acrotarsium height, medial malleolus height, and girth above malleolus.

2.2.2 Measurement Devices

Foot measuring plate, square gauge, and tape.

The measuring accuracy is 1.0 mm.

2.2.3 Pilot Samples

Totally 1,739 male pilots were measured, aged 21–55 (32.4 ± 6.8), and weighted 48.5–106.0 (68.8 ± 8.5) Kg. They were operating fighter, attacker, transport, bomber, or helicopter aircraft.

2.2.4 Database

The anthropometry database was developed in Visual FoxPro6.0 language. Alarm for error of measures on the site guarantees the quality of the data [3, 4].

2.2.5 Statistics and Analysis

Various statistics and analyses were made with SPSS 16.0 software.

2.3 Results

The measured data on 7 foot dimensions are as in Table 2.1.

The foot types of pilots, with reference to GB/T 3293.1 and GJB1095 [5, 6], on the basis of the foot length, taking 255 mm as the mean value, are classified into 11 types, of which the sample data are as in Table 2.2.

The regression analysis was made on the measured mean values of 9 subsamples, excluding the two subsamples of 230 and 280 because of subsample size, obtaining the 2-D relation between the foot width, foot girth, and the foot mark, as indicated in Fig. 2.1.

Table 2.1 The foot dimensions of Chinese male pilot population (in mm)

Measured items	Description	Mean	Std. Dev.	Percentage				
				1	5	50	95	99
1. Foot length	Max. straight distance between the heel point and toe point, parallel to the line between the heel point to No. II toe point	252.7	9.2	232.4	238	252	268	275
2. Foot width	Straight distance between tibial metatarsus point and fibular metatarsus point	98.5	4.1	89	92	98	105	108
3. Foot girth	The girth length starting from the tibial metatarsus point, along acrotarsium, fibular metatarsus point, and pelma to the starting point	251.9	10.3	230	235	251	270	276
4. Toe height	The vertical distance from the upper edge of the first MPJ to the standing plane	35.6	3.0	28	31	36	40	42
5. Acrotarsium height	The vertical distance from the front lower point of tibia to the standing plane	84.1	5.1	73	76	84	93	97
6. Medial malleolus height	The vertical distance from the medial malleolus point to the standing plane	72.8	5.4	60	64	73	82	84
7. Girth above malleolus	The horizontal girth at the thinnest point above the medial malleolus point	220.2	11.4	196	203	220	240	250

Table 2.2 The data of pilot subsamples in different foot size (in mm)

Foot mark / Measured items		230	235	240	245	250	255	260	265	270	275	280
Foot length	Mean	231.0	235.5	240.4	245.2	250.1	254.9	259.8	264.6	269.6	275.0	279.0
	Std.Dev.	0.9	1.4	1.4	1.4	1.4	1.4	1.4	1.3	1.3	1.4	0.8
	Range	230 ~ 232	233 ~ 237	238 ~ 242	243 ~ 247	248 ~ 252	253 ~ 257	258 ~ 262	263 ~ 267	268 ~ 272	273 ~ 277	278 ~ 280
Foot width	Mean	93.3	94.6	96.0	96.7	97.7	98.9	99.9	101.4	102.6	104.7	104.8
	Std.Dev.	3.7	3.6	3.7	3.5	3.5	3.6	3.2	3.9	3.7	3.0	5.4
	Range	89 ~ 99	87 ~ 107	86 ~ 109	87 ~ 107	86 ~ 108	89 ~ 112	92 ~ 111	89 ~ 112	94 ~ 111	98 ~ 112	99 ~ 112
Foot girth	Mean	241.5	241.3	245.8	247.6	250.3	252.8	254.8	259.7	262.0	266.8	269.0
	Std.Dev.	8.5	8.8	9.1	8.8	8.8	8.9	8.2	9.8	8.2	7.5	7.4
	Range	225 ~ 254	222 ~ 269	223 ~ 271	220 ~ 270	228 ~ 275	230 ~ 280	235 ~ 285	233 ~ 296	246 ~ 280	250 ~ 280	260 ~ 278
Toe height	Mean	34.6	34.4	35.1	35.3	35.1	35.6	35.8	36.4	36.7	36.8	40.3
	Std.Dev.	3.3	3.2	3.0	2.7	3.1	2.8	2.9	2.9	3.3	3.5	2.8
	Range	28 ~ 39	27 ~ 41	27 ~ 42	27 ~ 45	24 ~ 43	28 ~ 44	28 ~ 44	27 ~ 44	29 ~ 43	30 ~ 42	37 ~ 43
Acrotarsium height	Mean	81.7	81.4	81.7	83.5	83.4	84.1	85.3	86.4	87.4	87.2	88.0
	Std.Dev.	6.8	5.5	4.3	4.6	5.3	4.6	5.2	4.9	4.3	5.6	5.8
	Range	75 ~ 101	69 ~ 97	71 ~ 94	70 ~ 96	65 ~ 101	73 ~ 103	73 ~ 104	74 ~ 100	77 ~ 98	78 ~ 96	83 ~ 94
Medial malleolus height	Mean	69.8	70.9	71.6	72.4	72.1	72.5	73.8	74.3	75.7	77.6	77.3
	Std.Dev.	4.4	5.5	5.0	5.5	5.4	4.9	5.0	5.6	5.5	4.9	3.9
	Range	62 ~ 76	54 ~ 80	52 ~ 84	49 ~ 88	52 ~ 85	57 ~ 84	55 ~ 87	60 ~ 87	65 ~ 92	67 ~ 88	75 ~ 83
Girth above malleolus	Mean	208.3	211.9	214.7	216.3	219.2	220.6	223.4	225.9	230.7	233.0	236.8
	Std.Dev.	9.3	10.7	10.7	10.3	10.0	10.1	10.4	11.1	12.1	9.2	10.5
	Range	194 ~ 222	195 ~ 244	186 ~ 241	190 ~ 246	194 ~ 246	190 ~ 265	199 ~ 259	204 ~ 256	206 ~ 265	217 ~ 247	228 ~ 252
Subsample size		14	60	160	281	360	338	245	178	72	23	4

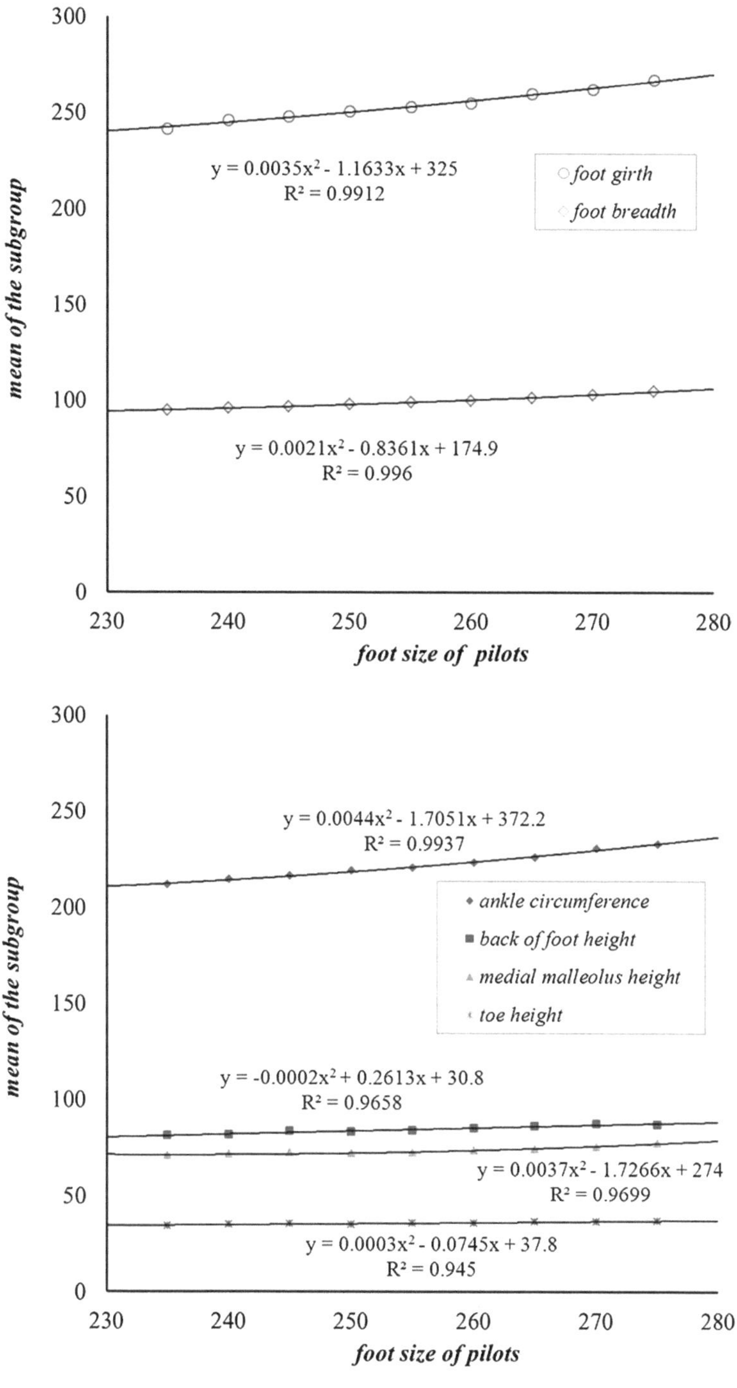

Fig. 2.1 The foot dimensions of pilot subgroups regressed from foot sizes

2.4 Analysis and Discussion

2.4.1 Sizes of Pilots' Foot Types

According to GB/T3293.1, the shoe type system is made on the basis of the two measured values of foot length and width. However, it does not ignore the foot girth completely in different stages of shoe manufacture. According to GJB1095, "mark" is the length sign of the shoe on the basis of the foot length, while "type" is the size sign of the shoe on the basis of the foot girth. According to the classification method on the foot type in GJB1095, pilots' feet can be classified into different types, as indicated in Table 2.3. For 1,739 pilots in the database of GJB4856, the coverage is 72.74 % calculated on 7 types, while the coverage is 46.35 % calculated on 3 types of 3, 3.5, and 4, in the common shoe mark range of 230–280. Therefore, it seems to require further consideration to set up only three types of 3, 3.5, and 4 in the male shoes in GJB1095.

The type 4 are set up as the foot length of 250 mm and the foot girth of 253.5 mm. The shoe size is determined by the calculation on the foot girth difference of different types (type difference) and the foot girth difference of different marks (mark difference) [7]. According to GB/T3293, the mark difference of the foot length is 10 mm and the semi-mark difference is 5 mm; the type difference of the foot girth is 7 mm, and the semi-type difference is 3.5 mm. The regression calculation in Fig. 2.1 shows that the pilot mark difference of the foot length is 9.4–9.8 mm, semi-mark difference is 4–5.4 mm, and mean value is 4.8 mm, which are consistent with the China National Standards basically. However, there are no statistical differences on foot mark 230, 235 and 240 and 245 of two independent subsamples, which seems inconsistent with the stipulations of the type difference of the foot girth of 7 mm and the semi-type difference of 3.5 mm.

Table 2.3 The distribution of pilot's foot size based on GJB 1095 (%)

Mark Type	230	235	240	245	250	255	260	265	270	275	280	Total
1	0.00	0.06	0.12	0.23	0.40	0.63	0.40	0.23	0.23	0.12	0.00	2.42
1.5	0.00	0.12	0.17	0.58	0.92	1.21	0.58	0.81	0.29	0.00	0.00	4.66
2	0.06	0.17	0.35	0.98	2.07	0.63	2.01	0.58	0.52	0.00	0.06	7.42
2.5	0.06	0.29	1.09	2.19	1.44	2.88	2.01	1.61	0.29	0.06	0.00	11.90
3	0.06	0.58	1.61	1.38	3.11	3.11	2.65	0.92	0.92	0.40	0.06	14.78
3.5	0.00	0.75	0.69	3.28	3.68	4.14	2.01	2.30	0.40	0.23	0.06	17.54
4	0.17	0.29	1.67	2.30	3.91	1.73	1.84	1.04	0.86	0.23	0.00	14.03
Total	0.35	2.24	5.69	10.93	15.53	14.32	11.50	7.48	3.51	1.04	0.17	72.74

2.4.2 Possible Characteristics of Particular Professional Group

The pilots may be different in the regional and age distribution from those of other national population for they belong to a particular professional group after occupational selection. It is naturally shown in the particularity of Chinese male pilots' measure samples in GJB4856-2003.

2.4.2.1 China Region

The pilots' samples are roughly distributed, as indicated in Fig. 2.2, according to the birth place. Compared with national census data [8], the pilot ratios of Shandong (14.3 %), Hebei (11.6 %), Jiangsu (10.1 %), and Liaoning (6.5 %) are remarkably higher than those corresponding male population ratios (7.1, 5.3, 5.8, and 3.3 %).

However, the pilot ratios of Guangdong (1.2 %), Guangxi (0.1 %), Yunnan (0.1 %), and Guizhou (0.1 %) are quite lower than those male ratios in the census (6.8, 3.6, 3.5, and 2.9 %).

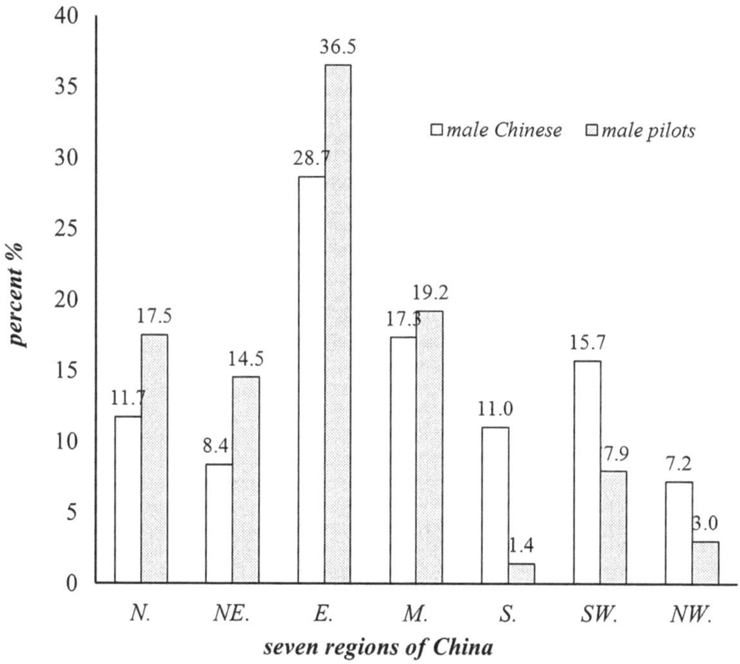

Fig. 2.2 The distribution of male Chinese and pilots in region

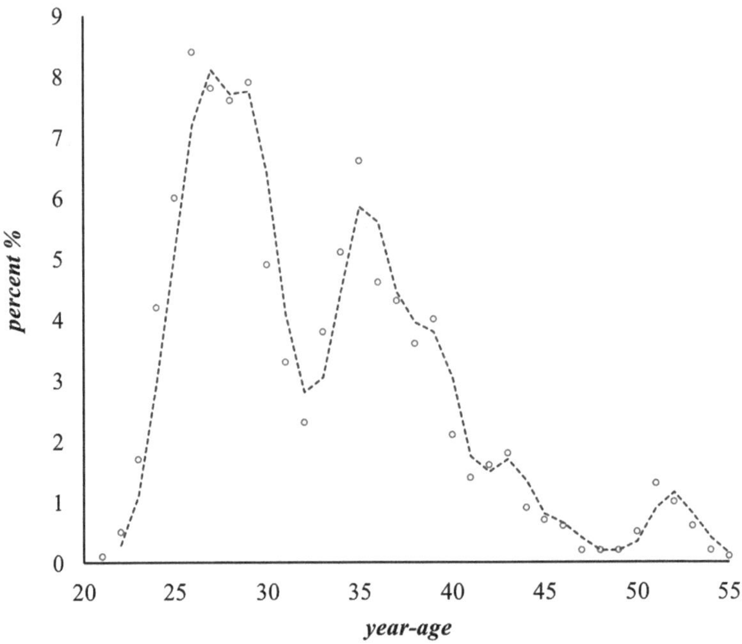

Fig. 2.3 The age distribution of the male pilots

2.4.2.2 Pilot Age

The age distribution of pilot samples is shown in Fig. 2.3, indicating more young-age pilots in the professional growth stage, middle-aged pilots in the professional mature stage, and aged pilots in the professional perfect stage. It also shows that there are more pilots aged 50–55 with transport or bomber aircraft operation.

It is mentioned that the types 1, 1.5, 2, and 2.5 in GJB1095 only for female shoes mostly appear in the pilot group aged 20–40. However, the shoe size distribution of pilots is generally anastomotic with the age distribution in Fig. 2.3 with the correlation coefficient $r = 0.96$. Therefore, it can be inferred that the slim foot types 1, 1.5, 2, and 2.5 exist in pilots of different ages.

2.5 Summaries

With reference to GB/T3293 and GJB1095, 11 foot marks are classified in the foot length of 230–280 mm and 7 types are classified in the foot girth of 200–276 mm, covering 72.74 % of pilots in GJB4856. The foot types of the pilots mainly distribute in nine marks of 235–275 and five types of 2–4, covering 65.2 % of the pilots totally. The occupational selection of pilots may be partly explaining the foot size differences of the pilots with that of common male Chinese population.

References

1. GJB 4856–2003 (2003) Human dimensions of Chinese male pilot population. Military standard Publishing House of the General Armament Department, Beijing
2. GB/T 3293–2007 (2007) Chinese last systems. China Standard Press, Beijing
3. R. Li, H. Wang, M. Wang et al. (2001) Anthropometric error proofing site warning design. Chin J Ergon 7(4):27–30
4. B. Liu, X. Guo, X. Ma (2012) Quality analysis of the third Chinese male pilot anthropometric data. In: The sixth national aerospace medicine academic conference. Chin J Aerosp Med 13(2):123
5. GB/T3293.1-1998 (1998) Shoes sizes. Beijing: China Standard Press, Beijing
6. GJB 1095–1991 (1991) The sizes for military shoes of PLA. Industry of National Defense Publishing House, Beijing
7. Shoes sizes. (2013) In http://baike.baidu.com/view/1356277.htm
8. National Bureau of Statistics of China. (2013) Data of the fifth census. In: http://www.stats.gov.cn/tjsj/ndsj/renkoupucha/2000pucha/pucha.htm

Chapter 3
Effect of Exhaustive Swimming on the Kidney Urinary Concentration Function

Lina Wang and Yang Chen

Abstract This paper aims to study the effect of exhaustive swimming on the kidney urinary concentration function. After 6 weeks of exhaustive swim training, plasma osmolality (Posm), urinary osmolality (Uosm), serum Na^+, K^+ concentration, the expression of kidney aquaporin 2 (AQP_2), and its mRNA are all observed. The results show that 6 weeks of exhaustive swim training will be harmful to the kidney urinary concentration and can deduce the body dehydration. The decrease in the expression of AQP_2mRNA significantly plays an important role in the change in the kidney urinary concentration function. The decrease in renal reabsorption of water maybe the main cause of fatigue resulted from the extreme load.

Keywords Exhaustive swimming · Extreme load · Kidney · Urinary concentration · Function

3.1 Introduction

Important, elusive, and highly efficient special operations require fighters having sound stamina and who can bear extraordinary physical load. It is important and urgent to make study on the effect of extreme load on human functions for it is

Subsidized by: Guangdong Natural Science Foundation Doctor Starting Project Subsidization Project No. 9451050201003592

L. Wang · Y. Chen (✉)
Department of Military Sports Center, The PLA Institute of Physical Education, Guangzhou, China
e-mail: 516729828@qq.com

S. Long and B. S. Dhillon (eds.), *Proceedings of the 13th International Conference on Man-Machine-Environment System Engineering*, Lecture Notes in Electrical Engineering 259, DOI: 10.1007/978-3-642-38968-9_3,

directly related to making the physical training plan, training level, and recovery effect after physical training of fighters. AQP_2 is a new access to study the effect of extreme physical load on the water reabsorption process of kidney and possible abnormality.

3.2 Study Purpose and Methodology

3.2.1 Study Purpose

It is intended to make the study on the effect of the extreme physical load on the urinary concentration function of kidney and to explore possible physiological mechanism, with the model of extreme physical load training by the exhaustive swimming.

3.2.2 Methodology

Twenty male, healthy, adult Wistar rat, with mean weight of 250.58 ± 16.13 g, are divided randomly into control group (Group C) and exhaustive swimming group (Group ES) for comparison training for 6 weeks: ten rats in Group C, commonly fed with any interference and no physical exercise and ten rats in Group ES, trained 6 days each week, swimming once each day. After training for 6 weeks, 5 ml blood from chest vein was obtained and 20 μl serum was absorbed for the test of serum osmotic pressure and Na^+ and K^+ concentrations. 20 μl of urine was obtained from the bladder with the injector for the test of urinary osmotic pressure; 100 mg of tissue block was taken at the same position in the left kidney quickly to extract the total RNA and to make reverse transcription for the test on kidney AQP_2mRNA with polymerase chain reaction (PCR). To set up AQP_2mRNA expression internal control GAPDH is used. To design the premier to be the following:

AQP2	F′CCCTTATAACAACCCTGTGCCC,
	R′AGGTTCCAATGACCAGGACCA;
GAPDH	F′TGCTGAGTATGTCGTGGAG,
	R′GTCTTCTGAGTGGCAGTGAT

Make initial denaturation for 5 min at 94 °C, make denaturation for 15 s at 94 °C, make renaturation for 30 s at 65 °C, make elongation for 30 s at 72 °C, amplify for 35 cycles, and make final elongation for 10 min at 72 °C. Make agarose gel electrophoresis of 2 % (120 V) on RNA product after the action.

Prepare the tissue slice from the same position in the right kidney for AQP_2 expression test.

3.2.3 Data Processing

The microscope is used to observe the positive signal. The measurement is made with LEICA QWIN image processing and analysis system. Three fields are selected randomly in the 40×40 high-power field, to indicate the relative content of protein with the optical density. mRNA data are processed with the gel image processing software, obtaining the gray value.

The analysis is made on the results with SPSS12.0 statistics software. *T* Test is made on two sample mean values for the statistical processing. $P < 0.05$ indicates the remarkable difference between them (Fig. 3.1 and Table 3.1).

3.3 Study Results

3.3.1 Effect of Exhaustive Swim Training on the Posm, Uosm, and Permeability Ratio

Compared with that of rats in Group C, the serum osmotic pressure of rats in Group ES increases much significantly ($p < 0.01$) and the urine osmotic pressure ($p < 0.01$) and the permeability ratio ($P < 0.01$) decrease much significantly (Table 3.2).

3.3.2 Effect of Exhaustive Swim Training on the Serum Ionic Concentration

Compared with that of rats in Group C, the serum Na^+ concentration of rats in Group ES decreases significantly ($P < 0.01$) and the serum K^+ concentration increases significantly ($P < 0.05$) (Table 3.3).

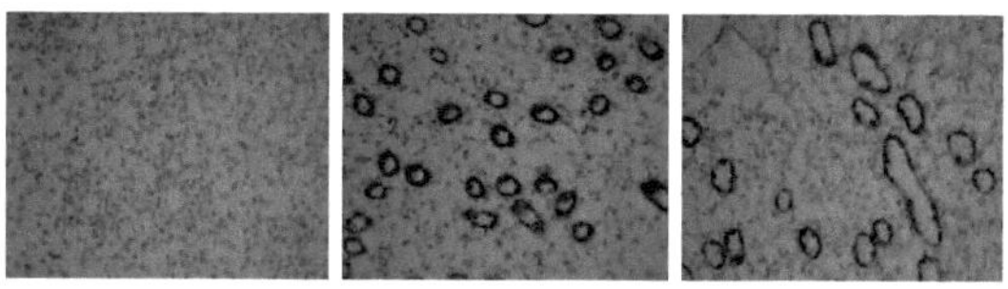

Negative control Control group Exhaustive swim group

Fig. 3.1 The picture of kidney AQP_2 expression after exhaustive swim training

Table 3.1 The data of Posm, Uosm, and permeability ratio after exhaustive swim training

Group	Posm (mOsm/kgH_2O)	Uosm (mOsm/kgH_2O)	Permeability ratio
Group C	306.86 ± 3.6	1192.43 ± 221.94	3.88 ± 0.77
Group ES	$339.14 \pm 20.18^{**}$	$746.86 \pm 75.05^{**}$	$2.17 \pm 0.35^{**}$

** $P < 0.01$

Table 3.2 The data of serum Na^+ and K^+ concentration after exhaustive swim training

Group	Na^+ (mmol/l)	K^+ (mmol/l)
Group C	139.86 ± 1.68	5.74 ± 0.74
Group ES	$136.66 \pm 1.28^{**}$	$6.89 \pm 0.42^{**}$

** $P < 0.01$

Table 3.3 The data of kidney AQP_2 expression after exhaustive swim training

Group	AQP_2 expression optical density
Group C	2882.86 ± 928.35
Group ES	$1945.43 \pm 205.39^{*}$

** $P < 0.05$

3.3.3 Effect of Exhaustive Swim Training on Kidney AQP_2 Expression

The cell has the strengthened coloring of tawny, indicating the remarkable positive action of the immunohistochemical cell. In the kidney of rats in Group ES, AQP_2 immunohistochemical semi-quantitative value decreases, which is significantly less than that of Group C.

3.3.4 Effect of Exhaustive Swim Training on the kidney AQP_2mRNA Expression

Compared with rats in Group C, the AQP_2mRNA expression of rats in Group ES decreases significantly (P < 0.05) (Fig. 3.2 and Table 3.4).

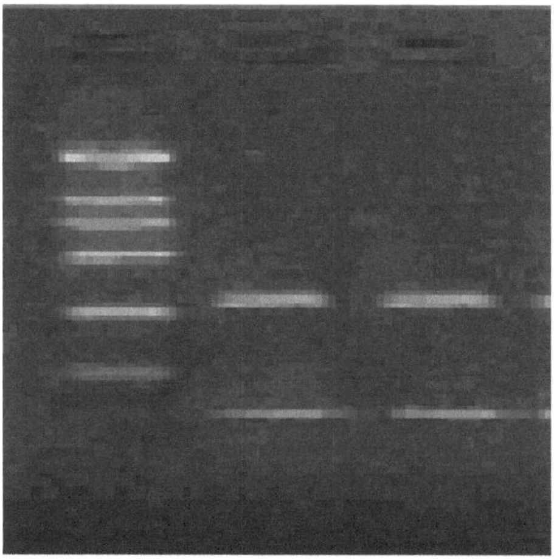

Fig. 3.2 The electrophoresis of kidney AQP_2mRNA expression after exhaustive swim training

Table 3.4 The data of kidney AQP_2mRNA expression after exhaustive swim training

Group	AQP_2mRNA expression gray value
Group C	0.65 ± 0.11
Group ES	$0.49 \pm 0.09^{*}$

* $P < 0.05$

3.4 Conclusions

3.4.1 Exhaustive Swim Training Leads to Decrease in Kidney Urine Concentration Function

The water in the tubular fluid is reabsorbed, and the solute remains in the tubular fluid, resulting in the urine concentration. The urine concentration function of the kidney plays an extremely important part in maintaining the body fluid balance and keeping the body in hydration state and in constant osmotic pressure. The kidney urine concentration capacity is reflected directly with the index of Posm and Uosm, which are also considered generally as the indices to indicate the body hydration state.

The serum osmotic pressure consists of crystal osmotic pressure and colloid osmotic pressure, of which the former mainly comes from the crystal substance dissolved in the serum, especially 80 % of the density of electrolyte comes from Na^+ and Cl^- and the latter comes from that of protein. The serum contains much protein. However, the protein has less molecules with larger molecular weight; thus, the colloid osmotic pressure is less, not larger than 1.5 mOsm/kgH_2O. So, the serum osmotic pressure is considered as an important index of the water–

electrolyte metabolism balance. Related study shows that Posm appears, deepens, and increases with the body dehydration and increases with the increasing percentage of weight loss caused by the dehydration [1], indicating that Posm can reflect effectively the body dehydration level in the emergency dehydration and recovering to the normal hydration state. Thus, it can be used to assist the judgment on the urine concentration function of the kidney.

Uosm mainly reflects the particle number of molecules and ions of solute in the urine of a unit volume, indicating the concentration function of the distal renal tubule. Usom will decrease in case of damage of the distal rental tubule. The urine osmotic pressure between 700 mOsm/kgH_2O and 800 mOsm/kgH_2O can reflect the body dehydration accurately and sensitively [2].

The moderate intensity training for a long time will lead to remarkable increase in Posm, serum Na^+, and K^+ density of rats [3]. Compared with that of athletes before the game, the Uosm pressure of young Taekando athletes starting the training is significantly higher, indicating that the long-time training will lead to the deepening of dehydration, excluding the condition where the athletes limit drinking water to control the weight [4]. High urine volume and low Uosm indicates the rapid development stage of kidney diseases. The higher the urine is, the faster the filtration rate of the glomerulus decreases and the lower Uosm is [5]. It shows that the water reabsorption of the body reduces and the water is losed with Posm increasing and Uosm decreasing.

The study discovers that after the exhaustive swim training, compared with rats in Group C, Posm of rats in Group ES increases significantly ($P < 0.01$) and Uosm ($P < 0.01$) and permeability ratio ($P < 0.01$) decrease significantly. The decreasing Uosm of rats in Group ES is related to the reducing numbers of solute molecules and particles in unit volume of urine. The decrease in urine osmotic pressure of rats in Group ES results from the following two reasons: on the one hand, the amount of solute molecules reduces in the unit volume, possibly resulting from the damaged kidney function, leaving the uric acid in the blood. The increase in β2-microgrobulion content leads to the increase in Uosm. On the other hand, it is because of the increase in solvent molecules in the urine of the unit volume, which possibly results from the decrease in water reabsorption capacity due to the damaged water reabsorption function of renal tubule. With the above-mentioned factors, the decrease in Uosm due to the decrease in water reabsorption capacity is larger than the increase in Uosm due to the increase in protein content. Thus, rats in Group ES have hyposthenuria, damaging the urine concentration function of the kidney. Posm of rats in Group ES is significantly higher than that of rats in Group C ($P < 0.05$), indicating the dehydration.

3.4.2 Exhaustive Swim Training Leads to Increase in Posm, Closely Related to the Serum Electrolyte Disturbance

The increase in Posm is closely related to the serum ionic concentration and water content of plasma. In the study, compared with Group C, serum Na^+ concentration in Group ES decreases significantly ($P < 0.01$). The serum K^+ concentration increases significantly ($P < 0.05$). It shows that exhaustive swim training results in the serum electrolyte disturbance, which is possibly related to the extreme load bore by the body.

The change in ionic concentration in and out of the cell can lead to the change in crystal osmotic pressure of the plasma. Na^+ remains in the cell in a large amount results in the decrease in the concentration of serum sodium, causing the decrease in plasma osmotic pressure. However, the increase in Posm after the exhaustive swim training is possibly related to the loss of water in a large amount in the urine. The water loss is larger than the effect of the decrease in the concentration of serum sodium on Posm, causing the increase in Posm, indicating the possible serious dehydration of rats in Group ES. Besides, the ionic concentration in and out of the cell is closely related to the cytomembrane potential. In the serum electrolyte disturbance, a large amount of K^+ remains in the blood and the K^+ concentration in the cell decreases, which may cause a change in cell polarity, leading to the loss of the polarized state of the cell.

Therefore, the change in Posm and Uosm of rats in Group ES reflects comprehensively the decrease in kidney urine concentration function and water reabsorption capacity of the kidney, resulting in the water loss in the body. The water loss in the body is directly related to the appearance and development of the fatigue because the dehydration can result in the decrease in work capacity of the body, thus affecting the heart rate at rest, degree of fatigue, and perceived exertion [6]. It will cause the increase in central temperature of the body, higher thirsty degree, and decrease in perceived exertion [7]. Other studies show that Posm can result in strengthened sympathetic nerve activities of muscles [8], higher heart rate [9], and restricted sympathetic nerve activities of the kidney, causing the vasodilatation and increasing blood flow [10].

Therefore, in the study, the decrease in kidney urine concentration, electrolyte disturbance, and great dehydration in the body of rats in Group ES is possibly one of the reasons of fatigue of rats in Group ES. The timely water supplement may be good for the elimination of the fatigue and recovering the kidney function of rats in Group ES.

3.4.3 Decrease in AQP_2 Expression is the Major Reason for Decrease in Urine Concentration Function After Exhaustive Swim Training

AQP_2, as a kind of highly conserved protein, can complete the physiological function, including water permeability, which is quite important for the body homeostasis. AQP_2 can allow the reabsorbed water into the cell quickly. Each AQP_2 can allow 3 billion water molecules passing in a second. In case of lack of AQP_2, the water can only pass the hydrophobic lipid bilayer by dripping [11]. AQP_2 mainly expresses in the chief cell connecting renal tubules and collecting duck. AQP_2 plays an important part in the urine concentration and participates in the adjustment of body hydrobalance.

The decrease in AQP_2 expression will lead to the defects of urine concentration. Diabetes insipidus features chiefly the lack of functional AQP_2. Acute and chronic renal failures and hyperuresis after obstruction will have the similar symptom due to the decrease in urine concentration function. The change in AQP_2 plays an important role in the damaged urine concentration function of kidney. In the pathology, the main performance is that the patient with decrease in water reabsorption capacity in the urine has the reduced AQP_2 expression or lack AQP_2 expression and plasmalemma AQP_2 transfer [12].

In the study, the immunohistochemical technology is used to make direct observation on the AQP_2 expression of renal tubules. The result shows that the kidney AQP_2 expression of rats in Group ES is significantly lower than that of rats in Group C ($P < 0.01$). It shows that the kidney reduces the water reabsorption after the exhaustive swim training. The decrease in AQP_2 expression in the exhaustive swim training leads to the reduction in water reabsorption, failing to balance the increasing water consumption in the body exercise, resulting in the water loss finally. The defect of kidney urine concentration function due to the decrease in AQP_2 expression is possibly considered as the direct reason for the body fatigue due to the exhaustive swim straining. Therefore, water permeability function of AQP_2 is the most important part in the adjustment of water reabsorption, affecting the performance of the kidney urine concentration function directly. The above-mentioned result shows that the extreme load training has remarkable effect on the water reabsorption and kidney urine concentration.

AQP_2mRNA expression of rats in Group ES is decreasing significantly than that of rats in Group C ($P < 0.05$), indicating that it is possible for the decreasing AQP_2 expression of the renal tubules in the exhaustive swim training to happen in the transcriptional level.

The decreasing kidney AQP_2 expression can reduce the water reabsorption of the kidney, possibly connected with the accumulation of AQP_2 on the membrane. Related studies show that the reduction in endocytosis blocks can reduce the accumulation of AQP_2 on the membrane. The increase in AQP_2 endocytosis can result in the decrease in water permeability of AQP_2, thus affecting the urine concentration function of the kidney [13]. However, further verification will be made on the study on the accumulation of AQP_2 of the kidney on the plasmalemma.

Acknowledgments It is Guangdong Natural Science Fund Subsidization Project. I appreciate deeply the hard work of all participants in the scientific research of the Project.

References

1. Sharif-Naeini R, Ciura S, Zhang Z, Bourque CW (2008) Contribution of TRPV channels to osmosensory transduction, thirst, and vasopressin release. Kidney Int 73(7):811–815
2. Verbalis JG (2007) How does the brain sense osmolality? J Am Soc Nephrol 18(12):3056–3059
3. O'Neil RG, Heller S (2005) The mechanosensitive nature of TRPV channels. Pflugers Arch 451(1):193–203
4. Woo SK, Lee SD, Na KY, Park WK, Kwon HM (2002) TonEBP/NFAT5 stimulates transcription of HSP70 in response to hypertonicity. Mol Cell Biol 22(16):5753–5760
5. Heo JI, Lee MS, Kim JH, Lee JS, Kim J, Park JB, Lee JY, Han JA, Kim JI (2006) The role of tonicity responsive enhancer sites in the transcriptional regulation of human hsp70-2 in response to hypertonic stress. Exp Mol Med 38(3):295–301
6. Zhou X, Ferraris JD, Burg MB (2006) Mitochondrial reactive oxygen species contribute to high NaCl-induced activation of the transcription factor TonEBP/OREBP. Am J Physiol Renal Physiol 290(5):F1169–F1176
7. Go WY, Liu X, Roti MA, Liu F, Ho SN (2004) NFAT5/TonEBP mutant mice define osmotic stress as a critical feature of the lymphoid microenvironment. Proc Natl Acad Sci U S A 101(29):10673–10678
8. Tsai TT, Guttapalli A, Agrawal A, Albert TJ, Shapiro IM, Risbud MV (2007) MEK/ERK signaling controls osmoregulation of nucleus pulposus cells of the intervertebral disc by transactivation of TonEBP/OREBP. J Bone Miner Res 22(7):965–974
9. Haase VH (2006) Hypoxia-inducible factors in the kidney. Am J Physiol Renal Physiol 291(2):F271–F281
10. Norman JT, Clark IM, Garcia PL (2000) Hypoxia promotes fibrogenesis in human renal fibroblasts Kidney Int 58(6):2351–2366
11. Gade W, Robinson B (2006) A brief survey of aquaporins and their implications for renal physiology. Clin Lab Sci 19(2):70–79
12. Bouley R, Hawthorn G, Russo LM, Lin HY, Ausiello DA, Brown D (2006) Aquaporin 2 (AQP2) and vasopressin type 2 receptor (V2R) endocytosis in kidney epithelial cells: AQP2 is located in 'endocytosis-resistant' membrane domains after vasopressin treatment. Biol Cell 98(4):215–232
13. Yamamoto T, Sasaki S, Fushimi K, Kawasaki K, Yaoita E, Oota K, Hirata Y, Marumo F, Kihara I (1995) Localization and expression of a collecting duct water channel, aquaporin, in hydrated and dehydrated rats. Exp Nephrol 3(3):193–201

Chapter 4
Heart Rate Changes Induced by Rest-Workload Alternating Pattern

Yanjun Li, Wei Chen, Xinming Yu, Hua Chen, Qianqian Liu and Chuang Yu

Abstract This paper specifically distinguishes between external load and interior load and points out the relationship among external load, interior load, and fatigue. Rest-workload alternating pattern is designed as the external load to seek the adapting changes of heart rate (HR). When fatigue level is increasing, the RR intervals decrease (i.e., HR increases) under physical load, which shows the weakened capability of physical reaction to stress. The linear correlation between external load strength and interior load index RR interval is strong, i.e., RR intervals decrease (i.e., HR increases), when external load strength increases.

Keywords Heart rate · External load · Interior load · Rest-workload alternating pattern · Fatigue

4.1 Introduction

The heart rate (HR) means the frequency of the heart beating, which is sensitive to interior and exterior stimulus of the body [1]. The HR increases gradually with the increasing exercise intensity or extending exercise period [1] for which it is influenced by the neuromodulation and humoral coordination. The HR keeps well linear relation with the exercise intensity in the exercise with medium or subultimate intensity [1]. Generally, the physical exercise intensity is monitored by the HR [2], which means the proper HR in exercise is equal to HR at rest + (Max. HR−HR at rest) $\times$ (60–70) % [2]. The maximum HR is generally calculated with the formula of 220-age. However, Ye et al. [3] thought that the formula of $208 - 0.7 \times$ age was more suitable for calculating the maximum HR for the professional athletes.

Y. Li (✉) · W. Chen · X. Yu · H. Chen · Q. Liu · C. Yu
China Astronaut Research and Training Center, Beijing, China
e-mail: yanjunli211@sina.com

S. Long and B. S. Dhillon (eds.), *Proceedings of the 13th International Conference on Man-Machine-Environment System Engineering*, Lecture Notes in Electrical Engineering 259, DOI: 10.1007/978-3-642-38968-9_4,

The HR is the most widely used index for the evaluation of physiological workload at present [4]. The changing HR can reflect the physical and mental adaptable changing and the physiological responds to the external load, which is the comprehensive result of the physical stress and mental stress. In the thesis, the rest-workload alternating pattern is used as the external load to explore the rule on the HR changing during the exercise.

4.2 Theories

4.2.1 Concepts of External Load and Interior Load

The external load [4], also named physical load [1], means the exercise or work that the body needs to complete within a unit period, which is generally expressed with the distance, speed, quality, and time consumption [1]. The external load shows the intensity and quantity of stimulus from the external factors that applied on the body. For example, weight of the snatch or jerk of the weightlifting, work and torque of pedaling are all considered as external loads. During exercise, the external load should be controlled at a proper degree according to the responding of the body.

Interior load [4], also named physiological load [1] or physiological stress, means the physical and mental adaptable changing of the body to external load, which is the adaptive adjustment within the body to the external stimulus. Two types of index are mainly used to monitor the physiological load [1]. The physiological indices include HR, blood pressure, oxygen consumption and breath frequency, and biochemical indices are composed of sarcolactic acid, blood lactic acid, urine lactic acid, testosterone and cortisol, and PH value of saliva. Most biochemical tests belong to the invasive surgery, while the physiological test can be carried out without any invasion. HR can be monitored swiftly, continuously at real time without any interference, realizing the transfer from the interrupted test afterward to the continuous monitoring at real time. For example, the weight of the dumbbell and the lifting speed are considered as external loads, while the increasing HR or blood pressures resulting from the dumbbell lifting are generally considered as the interior loads. The interior load intensity is closely connected with the individual quality, e.g., the athletes capacity to do exercise is remarkably better than that of common people.

4.2.2 Relation Between External Load and Interior Load

The load consists of external load and interior load [4], and both of them contain exercise load and mental load. The studies on exercise load focus on the interior load, i.e., the physiological responding of the body inspired by the exercise stimulus. The stronger of the external stimulus, the higher level of the interior stress [4]. However, such corresponding relation may be different in different exercise modes or individuals [4]. For example, the external load of the same intensity may cause the interior load with different intensity in different individuals.

The external load applied on the body, which leads to the interior load in the body. Therefore, the relation between the external load and the interior load is the cause and effect [4] with the body as the intermediary. With external load applied on the body continuously, the interior load generally increases as the time goes on (such as the decreasing cardiac reserve arising from the increasing HR and the systolic pressure). The fatigue means the negative feedback when the body fails to adapt to the interior load with high intensity, which informs the body to have a rest to avoid the transfer from the functional change to the organic damage. It will result in exercise fatigue when the interior load exceeds the physiological reserve of the body. The relation of the external load, interior load, and fatigue is as follows in Fig. 4.1.

4.3 Methodologies

4.3.1 Data Collection

ECG data of limb leads II from 21 male healthy subjects (aged 25.33 ± 2.92 year) at rest (stand still) and exercising (walking and running) on the treadmill were collected by Medilog AR12 (Oxford Instrument) with time sampling rate of 1,024 Hz, 16 bits A/D transfer, amplitude resolution of 0.3 uV, and magnification times of 1,000. The test process consists of 6 cycles of "rest for 3 min and exercise for 3 min" and stand still for 3 min finally. Thus, the time length of ECG data of

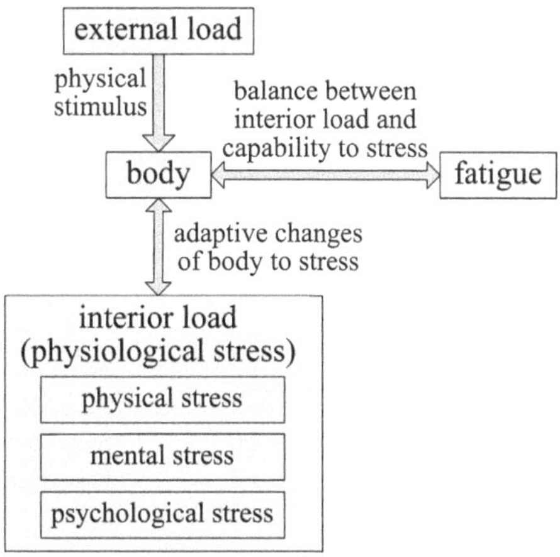

Fig. 4.1 Relations among external load, interior load, and fatigue

each volunteer is 39 min. The test load is shown in Table 4.1, where the letter "S" means the segment, "R" the rest, "W" the workload, and "P" the "rest-workload" period.

4.3.2 Data Processing

Heart beating segmentation is the basis to calculate the HR. In this paper, QRS complex detection is based on the matched filtering and triangle characteristic analysis. Meanwhile, the HR changing tendency chart is used to verify or correct the heart beating that omitted or mistakenly detected. Generally, R wave of the limb lead II is upward. The interval of adjacent peaks of R waves is defined as RR interval (RR(i)) with the unit of the second. The mean value of RR interval of each minute is defined to be $RR_M(i)$, and the mean value of RR_M of each segment is defined as $RR_{SM}(i)$. RR_{SM} of each workload segment is written as $RR_{WM}(i)$, and the minimum value of RR_M as $RR_{SN}(i)$. RR_{SN} of each workload segment is written as $RR_{WN}(i)$. Each segment (3 min) has 3 values of RR_M, of which the difference between the last value and the initial value is defined to be $RR_{SD}(i)$. RR_{SD} of each workload segment is defined to be $RR_{WD}(i)$, and RR_{SD} of each rest segment as $RR_{RD}(i)$.

The differences between the adjacent workload segment (such as difference between *W*2 and *W*1 in Table 4.1) are defined as follows: the difference of RR_{WM} between adjacent workload segments is $DRR_{WM}(i)$ by Formula (4.1), and the difference of RR_{WN} between adjacent workload segment is $DRR_{WN}(i)$ by Formula (4.2).

$$DRR_{WM}(i) = RR_{WM}(i+1) - RR_{WM}(i) \tag{4.1}$$

$$DRR_{WN}(i) = RR_{WN}(i+1) - RR_{WN}(i) \tag{4.2}$$

where $i = 1, 2, 3, \ldots, 5$

Table 4.1 Load imposed order and load level

Segment	Type	Period	State	Speed (km/h)	Time (min)
*S*1	*R*1	*P*1	Stand still	0	3
*S*2	*W*1	*P*1	Slow walking	2.5	3
*S*3	*R*2	*P*2	Stand still	0	3
*S*4	*W*2	*P*2	Fast walking	5	3
*S*5	*R*3	*P*3	Stand still	0	3
*S*6	*W*3	*P*3	Slow running	7	3
*S*7	*R*4	*P*4	Stand still	0	3
*S*8	*W*4	*P*4	Slow running	8	3
*S*9	*R*5	*P*5	Stand still	0	3
*S*10	*W*5	*P*5	Moderate running	9	3
*S*11	*R*6	*P*6	Stand still	0	3
*S*12	*W*6	*P*6	Fast running	10	3
*S*13	*R*7	/	Stand still	0	3

4.3.3 Statistics Analysis

4.3.3.1 Variance Analysis

Generally, the comparison is made between the relative ratio and the unit vector, or the relative difference and the zero vector to reduce the statistic deviation due to the individual difference (different fundamental HR at the rest) actually existing in subjects. In the thesis, the analysis of variance (ANOVA) is made between the statistic difference of each segment of 21 subjects and zero vectors of 21 zero values. The remarkable change is considered to exist in case of remarkable difference level $P < 0.01$.

4.3.3.2 Correlation Analysis

The speed at walking or running is considered as the external load level to define the workload vector L_W during workload segments ($W1$, $W2$,..., $W6$), as in Formula (4.3). The workload vector L_A during the whole process ($S1$, $S2$, ..., $S13$) is define by Formula (4.4). The Formula (4.5) is used to calculate the correlation coefficient of RR_{SM}, RR_{SN}, RR_{SD} during workload segments ($W1$, $W2$, ..., $W6$) and workload vector L_W, and the correlation coefficient of RR_{SM}, RR_{SN}, RR_{SD} in the whole process ($S1$, $S2$, ..., $S13$) and the workload vector L_A, respectively.

$$L_W = [2.5\,5\,7\,8\,9\,10] \tag{4.3}$$

$$L_A = [0\,2.5\,0\,5\,0\,7\,0\,8\,0\,9\,0\,10] \tag{4.4}$$

$$r = C(X,Y) \Big/ \left(\sqrt{D(X)}\sqrt{D(Y)}\right) \tag{4.5}$$

where C means covariance, D for variance, X for RR_{SM}, RR_{SN} or RR_{SD} and Y for L_W or L_A. Correlation coefficient r reflects the linear correlation degree of the two sequences.

4.4 Results

Figure 4.2 shows the parameter curve of one subject that selected randomly from 21 subjects. RR is in the decreasing trend during workload segment and in the increasing trend during rest segment after the workload in Fig. 4.2. Statistics results of the parameters from all 21 subjects are shown in Tables 4.2 and 4.3. Table 4.2 shows that RR is in remarkably decreasing trend (ANOVA, $P < 0.001$) with the increasing workload intensity. Table 4.3 shows that there is strong linear relation between the external workload intensity and RR mean value, which means the larger the external workload intensity is, the smaller the RR interval (the higher of the HR).

4.5 Discussions

4.5.1 Concept Distinction

In most documents at home and abroad, no distinction is made on external load and interior load, in which the terms of load or physiological load refer to interior load specially. In the preliminary study [5], the term of load refers to the exterior load, and stress refers to interior load. Therefore, it is simple and clear to express such terms in analyzing the relation of the load, stress, and fatigue. However, such practice with the load referring to the external load and the stress referring to the interior load [5] is still considered as an individual idea. It is inconsistent with most documents at home and abroad, which is misunderstood easily. To meet the urgent requirements to inherit the existing research achievements at home and abroad and also to make distinction on two different loads that within and beyond the body, it is suggested in this paper that the concepts of external load and interior load (or physiological load) are to be used, of which the former refers to physical stimulus source beyond the body and the latter refers to the stress reaction of different systems, organs and tissues within the body on the external stimulus. On such

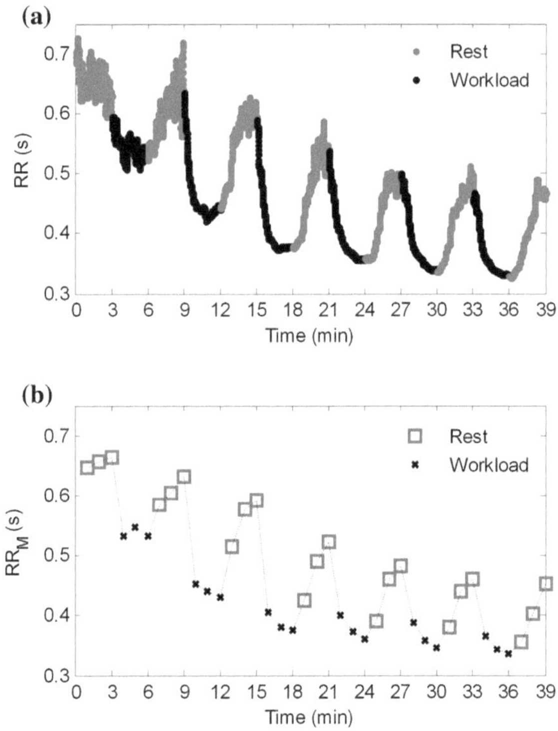

Fig. 4.2 RR curve of a typical sample **a** RR, **b** RR_M

Table 4.2 Indices difference between workload segments

Difference	DRR_{WM}	DRR_{WN}
$W2 - W1$	−0.06 ± 0.03**	−0.05 ± 0.03**
$W3 - W2$	−0.09 ± 0.04**	−0.10 ± 0.04**
$W4 - W3$	−0.03 ± 0.02**	−0.03 ± 0.02**
$W5 - W4$	−0.02 ± 0.01**	−0.02 ± 0.01**
$W6 - W5$	−0.02 ± 0.01**	−0.02 ± 0.01**

Note ANOVA, $^{**}P < 0.001$

Table 4.3 Correlation coefficients between statistical indices and workload level

Index	Workload segment	Rest segment	Whole process
RR_{SM}	−0.979 ± 0.010	−0.917 ± 0.083	−0.738 ± 0.078
RR_{SN}	−0.976 ± 0.014	−0.944 ± 0.055	−0.634 ± 0.081
RR_{SD}	−0.779 ± 0.236	0.704 ± 0.243	−0.696 ± 0.173

basis, the physiological load is consistent with the physiological stress, the exercise load with exercise stress, and mental load with mental stress.

Moreover, there is still no distinction on terms of stress and fatigue in some documents. Stress is the physiological response of the body to the external stimulus, in which the external load is the "cause" of the stress and the physiological adaptive adjustment is the "effect" of the stress. The imbalance between the energy substance supply and oxygen-carrying capacity of the blood and the requirements of the body on the energy substance and oxygen, together with the factors of too slow removal of the metabolic waste can cause the fatigue of the body. The "cause" of fatigue is to supply the useful substance and remove the metabolic waste too slowly. The "effect" of the fatigue is shown in the sore muscles, slow response, and decreasing work efficiency. The external load can be removed instantly, while the interior load and fatigue cannot be removed instantly, which only can be removed in a long process.

Besides it needs avoiding the generalization of the concept of stress. For example, "the HR load" [4] considers the HR value higher than that at rest to be a measuring index on the load intensity. Such generalization of concepts will lead the study about stress to the extreme. Firstly, the HR at the rest is changing actually. The HR variability at rest often contains important information on the balance between sympathetic nerves and vagus nerves. Thus, it is difficult to determine the reference HR for the HR load. Secondly, the stress will exist independently from the stimulus source, which means the body will make stress reaction in case that the HR is higher than the reference HR, even without any change in the external environment.

4.5.2 Evaluation on Exercise Load

Exercise load is closely connected with exercise fatigue. The external load applied on the body and the body will make reaction to exercise stress. Too much exercise stress will result in the fatigue of the body. Therefore, the fatigue arises from exercise stress. The exercise stress will weaken in case of removal of the load, thus, relieving fatigue gradually.

The lactic acid is the best biochemical index to make evaluation on the exercise load, which includes sarcolactic acid, blood lactic acid, and urine lactic acid [1]. The sarcolactic acid can reflect the changing of the lactic acid in the muscles. However, it needs biopsy on muscles. The blood lactic acid analysis requires collecting the blood sample from the finger or ear, which is still an invasive method. The urine lactic acid analysis is still collection analysis afterward, failing to use in the continuous and real time monitoring on the load strength.

Compared with biochemical test, the physiological signals have the advantages of lower invasion and higher accessibility. The accurate blood pressure measurement requires cuff pressure, which is noncontinuous measurement. The continuous blood pressure measurement always requires collecting cardiac signals of two channels, which is not reliable. The HR changes faster than the blood pressure during exercise and recovery segment, which guarantees the sensitivity. The HR is simple, accurate, easy and continuous to monitor. In the experiments of this thesis, the exercise stress increase (accumulated fatigue) along with the external load increases gradually, indicated by the decreasing of RR mean value with the increasing external load.

The rest state and exercise state cannot be distinguished with the HR only. However, the changing trend of the HR can show the rest and the exercise state. In Fig. 4.2, RR is in decreasing trend (HR increase) under the exercise load with the same intensity, showing the accumulative effect of the exercise fatigue. RR is in increasing trend (HR decreases) in the rest segment after each exercise load, showing it is a slow process to eliminate the exercise fatigue and recover the physiological functional reserve. In Table 4.3, RR mean value is in stronger linear relation with the external load intensity, showing the HR changing is in a certain dependency with the external load changing.

Acknowledgments Supported by Advanced Space Medico-Engineering Research Project of China (SJ201006).

References

1. Tang J, Li M (2012) Monitoring and application of physiological load intensity in exercise. J Clin Rehabilitative Tissue Eng Res 16(20):3784–3788
2. Chen C (2006) A pilot analysis on the function of rhythm of the heart monitor law in police's physical fit abnity exercise. J PLA Inst Phys Educ 25(2):117–119

3. Ye W, Ma C (2004) Experimental research on the predicting equation of university students in specialty of physical education. J Shenyang Inst Phys Educ 23(3):382–383
4. Lv Y, Hou J (2002) An analysis on conception of sports load and heart rate controlling in basketball. J Capital Coll Phys Educ 14(4):34–37
5. Li Y, Yan H, Yang X, Wang Z (2010) Study of mental fatigue based on heart rate variability. Chin J Biomed Eng 29(1):1–6

Chapter 5
Method of Formulating the Physical Fitness Standard of Serviceman

Weiming Deng, Wenbin Yang, Dongsheng Ai, Lina Wang and Junhua Li

Abstract Basic physical fitness standard refers to the criteria that all servicemen are supposed to reach, according to the gender and age of individuals. This chapter focuses on developing the method to formulate military basic physical fitness standard. Specifically, sampling survey of male servicemen was made in accordance with eight different age classes, in which push-up, sit-up, 10 m × 5 shuttle run, and 3,000 m long-distance running are involved as the fundamental test items. As for the methodology, this thesis adopts normal distribution theory to determine and calculate how individual serviceman performs in the several test items mentioned previously can reach 70, 75, 80, and 90 %, respectively, which enables the physical fitness test criteria to be more scientific and reliable.

Keywords Physical fitness · Serviceman

The physical fitness standards of servicemen are the basic requirements of the physical and mental capacities of the servicemen to adapt the all-weather, highly competitive, and three-dimensional operation in complicated electromagnetic environment in the future on the basis of the present conditions of troops. Highly technical war operations in complicated electromagnetic environment feature more diversified mission, more complicated and changeable environment, and more miniaturized combat unit [1], making higher requirements on the physical fitness of servicemen.

The physical fitness standards of servicemen consist of the following three parts, namely basic physical fitness, professional fitness, and supplementary physical fitness [2]. Basic physical fitness standard refers to the criteria that all servicemen are supposed to reach, according to the gender and age of individuals. Such basic physical fitness standards should guarantee the physical fitness of servicemen to meet the requirements of normal life and those of military profession as well.

W. Deng (✉) · W. Yang · D. Ai · L. Wang · J. Li
Institute of Physical Educations of PLA, Guangzhou, China
e-mail: mmm333www@126.com

S. Long and B. S. Dhillon (eds.), *Proceedings of the 13th International Conference on Man-Machine-Environment System Engineering*, Lecture Notes in Electrical Engineering 259, DOI: 10.1007/978-3-642-38968-9_5,

The professional standards are listed in different types of general standards, recruits, flight crew (cadets), and astronauts, which may be selected according to the professional requirements and actual conditions by different units. The supplementary fitness standards are optional without any stipulations, which contain any sports events improving the physical fitness of servicemen.

The physical fitness standards of servicemen are formulated with the normal distribution theory in the mathematical statistics on the wide survey on the physical fitness of present servicemen. It is complicated to categorize the troops and physical fitness test standards. Therefore, in the thesis, the discussion is made and only made on the formulation of uniform basic physical fitness test standards, and the data for analysis only involves the part of the survey and the male servicemen only, which is considered as the example.

5.1 Study Subjects and Test Items

5.1.1 Test Subjects

The test was made on the physical fitness of servicemen with the sampling mode of random sampling and classic sampling for several times in certain units of the troops from November 2001 to March 2005, obtaining the first-hand survey data. The mathematical statistical analysis was made on related data [3, 4]. The age structure of the test subjects are as follows in (Table 5.1).

5.1.2 Test Items

The basic physical fitness items consist of push-up (unit: times/2 min), sit-up (unit: times/2 min), 10 m × 5 shuttle run (unit: second), and 3,000 m run (unit: minute second). The test and rules for each item were made in accordance with stipulations of the physical fitness test standards of servicemen.

Table 5.1 The age structure of the test

Age group	Number	Effective number	Effective rate (%)	Composition rate (%)
Under 24	4,470	4,220	94.41	68.01
25–29	1,263	1,196	94.70	19.27
30–34	519	482	92.87	7.77
35–39	187	166	88.77	2.68
40–44	77	67	87.01	1.08
45–49	48	46	95.83	0.74
50–54	20	20	100.00	0.32
55–59	8	8	100.00	0.13
Total	6,592	6,205	94.13	100

5.1.3 Formulation Principle

The random sampling is used to test the physical fitness of servicemen without special training. The results reached by 70 % of the servicemen are taken as the test standards in accordance with the mathematical statistical principle. It is hoped that more than 90 % of the servicemen can reach such standards after the special physical fitness training, with the purpose of improving the achievements of 20 % of the servicemen.

5.2 Statistical Principle

As is well known, results of test items of physical fitness generally comply with the normal distribution principle, of which the density function is as follows

$$f(x) = \frac{1}{\sigma\sqrt{2\pi}} e^{-\frac{(x-\mu)^2}{2\sigma^2}}, \quad (-\infty < x < +\infty)$$

where μ is the mean value, σ the standard deviation. Then, the probability distribution function is

$$\varphi(x) = \frac{1}{\sigma\sqrt{2\pi}} \int_{-\infty}^{x} e^{-\frac{(x-\mu)^2}{2\sigma^2}} \mathrm{d}x, \quad (-\infty < x < +\infty)$$

The corresponding physical fitness test standards can be formulated respectively according to actual conditions with such normal distribution principle. Specific calculation procedures can be referred to in the references [5, 6].

5.3 Analysis on Statistical Results

With the purpose of reducing individual error, the data in the survey test is arranged in order, of which the data of 10 % at the either end is truncated. Only the data of 80 % in the middle is taken for statistical analysis. The standards reached by 70, 75, 80, 85, and 90 % are calculated, respectively, for the comparison and selection.

5.3.1 Standards of Push-up

The push-up is intended for the strength training of servicemen, to develop large muscle groups on the upper limb, including pectoralis major and triceps brachii and to stretch the chest to improve the function of respiratory system as well.

Table 5.2 Standard of push-up (unit: times/2 min)

Age	Sample content	Average value	Standard deviation	70 %	75 %	80 %	85 %	90 %
Under 24	4,220	42.17	8.48	37.73	36.46	35.04	33.39	31.32
25–29	1,196	38.54	9.85	34.23	32.54	31.71	29.35	27.34
30–34	482	36.68	10.15	31.36	29.84	28.13	26.16	23.69
35–39	166	28.65	8.08	24.42	23.21	21.85	20.29	18.32
40–44	67	20.80	5.12	18.11	17.34	16.48	15.49	14.24
45–49	46	17.68	3.41	15.89	15.38	14.81	14.14	13.31
50–54	20	12.90	3.11	11.27	10.81	10.28	9.68	8.92
55–59	8	10.67	1.15	10.06	9.89	9.69	9.47	9.19

The subject requires testees to make preparation pose with two hands on the floor and with distance between slightly wider than shoulder, fingers forward, two arms stretched straightly, two feet close with each other, and the body straight. By hearing the signal of start, the subject is required to bend the arm with the elbow outward and to low the body straightly with the shoulder lower than the elbow joint level and to push-up with two arms and to return to the preparation pose, which is considered an action. The test will be completed in 2 min, during which, the subject may have a rest, but testees are still required to keep the preparation pose. Otherwise, the test will be over. Thus, the number of actions completed in the stipulated period is considered as the final results (Table 5.2).

5.3.2 Standards of Sit-up

The sit-up is intended for training abdominal muscles and rectus abdominis, to improve the strength of abdomen in and forward stretch. It can guarantee the elasticity of muscles, protecting the back and improving the posture.

The subject is required to make preparation pose, sitting on the floor or the mat, with two legs stretched closely and straightly (two feet or ankle joint may be fixed) and two hands crossed under the brain. Hearing the signal of start, the subject is required to lay on the back and bend the trunk forward, keeping the angle of the trunk and leg less than 90 degree, which is considered to be an action. The test will be completed in 2 min, during which, the subject may have a rest, but testees are still required to keep the preparation pose. Otherwise, the test will be over. Thus, the number of actions completed in the stipulated period is considered as the final results (Table 5.3).

Table 5.3 Standard of sit-up (unit: times/2 min)

Age	Sample content	Average value	Standard deviation	70 %	75 %	80 %	85 %	90 %
Under 24	4,220	45.37	6.42	42.00	41.04	39.96	38.72	37.15
25–29	1,196	39.29	5.25	37.18	36.37	35.87	34.51	33.08
30–34	482	35.90	4.75	33.41	32.70	31.90	30.98	29.82
35–39	166	31.67	3.15	30.02	29.55	29.02	28.41	27.64
40–44	67	22.62	4.08	20.49	19.88	19.19	18.40	17.40
45–49	46	17.68	2.10	16.58	16.27	15.91	15.50	14.99
50–54	20	12.60	2.95	11.05	10.61	10.11	9.54	8.82
55–59	8	12	2	10.95	10.65	10.32	9.93	9.44

5.3.3 Standards of 10 m × 5 Shuttle Run

10 m × 5 run is intended for training the sensibility, harmony, and turning speed and capacity. In the training to change the posture quickly and shuttle times, it will improve the harmony and flexibility of the whole body and sensibility of nerves and muscles and it will also improve the cardiovascular function and that of the respiratory system and develop the muscles and speed quality of lower limb.

The marker lines (starting line and finishing line) are set up with two parallel lines 10 m away from each other. Hearing the signal of start, the subject is required to run from the starting line and touch the finishing line with the foot and return to the starting line with the foot touching it, which is considered to be an action. It is required to complete 5 actions. The subject is required to touch the line again in case of failing to touch it. Otherwise, it will be considered a foul, and no result will be recorded. The subject in foul may be tested one more time. The time from the signal of starting to the time when the subject completes the stipulated actions will be considered as the final result, accurate to the second (Table 5.4).

Table 5.4 Standard of shutter run for 10 m × 5 (unit: s)

Age	Sample content	Average value	Standard deviation	70 %	75 %	80 %	85 %	90 %
Under 24	4,220	26″76	0″62	27″08	27″17	27″27	27″39	27″54
25–29	1,196	27″08	0″83	27″39	27″62	27″74	27″90	28″09
30–34	482	28″27	0″98	28″79	28″96	29″09	29″30	29″42
35–39	166	30″17	1″16	30″68	30″94	31″19	31″33	31″67
40–44	67	32″13	1″27	32″84	32″97	33″17	33″50	33″80
45–49	46	35″65	2″33	37″06	37″28	37″47	37″75	38″41
50–54	20	41″18	3″24	43″05	43″65	44″10	44″48	45″23
55–59	8	45″15	5″36	48″12	48″71	49″33	50″36	51″02

Table 5.5 Standard of 3,000 m run (unit: min, s)

Age	Sample content	Average value	Standard deviation	70 %	75 %	80 %	85 %	90 %
Under 24	4,220	12′37″	0′59″	13′08″	13′17″	13′27″	13′39″	13′53″
25–29	1,196	13′53″	0′53″	14′01″	14′23″	14′39″	14′58″	15′19″
30–34	482	14′32″	0′48″	14′57″	15′40″	15′12″	15′21″	15′33″
35–39	166	14′57″	1′00″	15′29″	15′38″	15′49″	16′00″	16′15″
40–44	67	15′56″	1′09	16′32″	16′43″	16′54″	17′08″	17′25″
45–49	46	17′10″	1′04″	17′44″	17′54″	18′05′	18′17″	18′33″
50–54	20	18′11″	1′46″	19′07″	19′23″	19′41″	20′01″	20′27″
55–59	8	20′35″	0′53″	21′03″	21′11″	21′20″	21′30″	21′43″

5.3.4 Standards of 3000 m Run

A 3000 m run is mainly intended for training the aerobic endurance of servicemen, to improve the function of cardiovascular system, muscle, and cardiopulmonary endurance and to develop the volitional quality, which is considered as one of the main means to improve the physical fitness. After a period of running, the subject will feel extremities of sports, including dyspnea, Chest distress, weak limbs and decreasing speed. At the time, the subject shall continue running with perseverance. At the time, the subject shall continue running with perseverance. The subject is required to deepen the breath, make proper adjustment in the speed, and the extremity will be relieved. The so-called Second wind will appear. The subject will feel better remarkably after a period of continuous running. Therefore, 3,000 m run will be one of major items to improve the muscle and cardiopulmonary endurance and develop the volitional quality (Table 5.5).

5.4 Conclusion

1. In this essay, the discussion is made on the method to formulate the basic physical fitness test standards for the servicemen. The sampling survey is made on the present male servicemen of 8 different age groups and the standards of different levels for four basic physical fitness items, including push-up, sit-up, 10 m × 5 shuttle run, and 3,000 m run is formulated.
2. In order to guarantee the scientific test on the physical fitness of servicemen, the study must be made on the scientific method to formulate the physical fitness standards. In the thesis, the scientific physical fitness standards are formulated with the normal distribution principle in the sports statistics, solving the practical problems.

References

1. Cao Z, Wu R et al (2006) Informationized joint operations. The People's Liberation Army Press, Beijing
2. Sun X et al (2009) Guide of physical fitness for serviceman. The People's Liberation Army Press, Beijing
3. Lu W (2008) SPSS for Windows about statistic analysis. Publishing House of Electronics Industry, Beijing
4. Sun Q, Hao W, Hong H (2010) Sports measurement and evaluation. Higher Education Press
5. Chen J et al (2006) Sports statistics People's Sports Publishing House of China, Beijing
6. Ma G, Shen S (2009) Sports statistics and application of SPSS. Jingling university press, Jiling

Chapter 6
Fatigue Analysis of Load Carriage Marching

Yuhong Shen and Chenming Li

Abstract To research the relationship between fatigue of shoulder, waist, back, and whole body, load carriage simulation experiments were carried out. The relationship between fatigue of whole body and heart rate changes was also studied. The results showed the following: (1) The slopes of the fatigue curve of whole body and shoulder were consistent. The fatigue of shoulder was an important factor to whole body fatigue. (2) At the beginning period of experiment, heart rate increased significantly. However, when fatigue achieved moderate levels, heart rate increased slowly. (3) When the subjects reached fatigue limit, the experiments were stopped, and the heart rate maintained at 130–140 bpm, which consistent with the provisions of GJB1136-92.

Keywords Load carriage · Fatigue · Heart rate

6.1 Introduction

The fatigue is quite common in the march with load carriage for a long time. Generally, the fatigue will appear on the whole body or part of it due to too much load, unscientific carriage means, and improper load distribution, which restricts the maintenance and development of the physical fitness seriously. According to Washington Times, too much field equipment for a single soldier of the US troops caused too many cases of soldiers muscle strain or diseases on bones in Iraq and Afghanistan, impacting the military mission. The statistics of the US military shows that 257,000 cases of leg strain and knee damage due to too much field equipment in 2007, 10,000 cases larger than that of 2006 approximately [1, 2].

Y. Shen (✉) · C. Li
The Quartermaster Research Institute of the General Logistics Department, CPLA, Beijing 100010, China
e-mail: yuhong810@yahoo.com.cn

S. Long and B. S. Dhillon (eds.), *Proceedings of the 13th International Conference on Man-Machine-Environment System Engineering*, Lecture Notes in Electrical Engineering 259, DOI: 10.1007/978-3-642-38968-9_6,

At present, the standard field equipment of each soldier of US Marine Corps consists of the body armor, weapons of rifle, ammunition, drinking water, food, and communication equipment, totaling 44–61 kg approximately. The load carriage of a single soldier of the US army for a mission of three days was as much as 59–68 kg approximately in Afghanistan. The load of a single soldier of the US army in the march covered 75 % of his weight approximately. The muscle and bone damages increased with the increasing combat load [3, 4]. It is quite significant to make discussion on the fatigue rules and evaluation index, with the purpose of studying the measures to fight against the fatigue and making scientific guidance on the load carriage, thus, reducing the damages due to the fatigue.

6.2 Experiment Method

6.2.1 Subjects

The subjects are three male healthy adults, aged 23 year ± 1, with height of 175.3 cm ± 4.93 and weight of 70.67 kg ± 9.02, making sure no remarkable difference in basic physiological parameters. The subjects all received the education above the university, guaranteeing the stronger capacity to understand and cooperate with the experiment.

6.2.2 Experiment Conditions

The environmental temperature in the laboratory is controlled at 20 ± 2 °C, with the wind speed less than 0.5 m/s, the march speed of 5 km/h, the ground slope of 0° and the carried load of 25.5 kg. All experiments are made in the morning to guarantee the comparison of the experimental data.

6.2.3 Measurement Parameters

The experiment is mainly intended to measure the heart rate and subjective fatigue. The heart rate table of Polar is used to measure the heart rate, to record the heart rate changing in the whole process. The subjective inquiry is made on the subjective fatigue, to record the fatigue of the shoulder, waist, back, and the whole body.

6.2.4 Experiment Terminal Conditions

To avoid unexpected damages in the experiment, the experiment will end in any of the following conditions:

(1) the heart rate exceeding 90 % HRmax (HRmax is predicted to be 220—age);
(2) the subject feeling hard to continue the experiment, with the symptoms as instable walking, pale face or dizziness, stuffiness, fluster, and sick.

6.3 Experiment Results

6.3.1 Subjective Fatigue

Borg table is used to evaluate the fatigue of the subjects, with zero indicating no fatigue, 10 extreme fatigue in Table 6.1 [5, 6].

The curve of the fatigue at the shoulder, waist, back, and the whole body of three subjects is as follows in Figs. 6.1, 6.2, and 6.3.

The changing trend of the subjective fatigue of the subjects shows that the fatigue at the shoulder is close to that of the whole body. The fatigue at the shoulder is stronger than that at the waist and back. The fatigue at the back is the slightest for the knapsack is made in supporting structure. The body bears little force due to a certain gap between the muscle at the back and the bullet proof vest, which can be shown in the following force analysis at the back. The linear regression is made on the subjective fatigue data of three subjects, obtaining the equation of the fatigue at the whole body, shoulder, waist, and back with the changing time:

$$\text{Fatigue at whole body} = 0.132 \times t(\text{min}) + 0.769 \tag{6.1}$$

Table 6.1 The Borg scale of fatigue level

Marks	Fatigue descriptions
0	No fatigue
1	Rather slight
2	Weak (slightly)
3	Temperate
4	Slightly strong
5	Strong
6	Moderately strong
7	Very strong
8	Quite strong
9	Super strong
10	Extreme

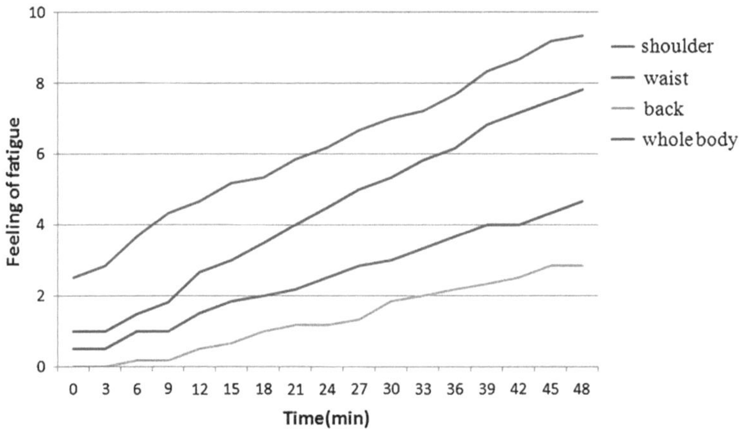

Fig. 6.1 Fatigue curve of 1# subject

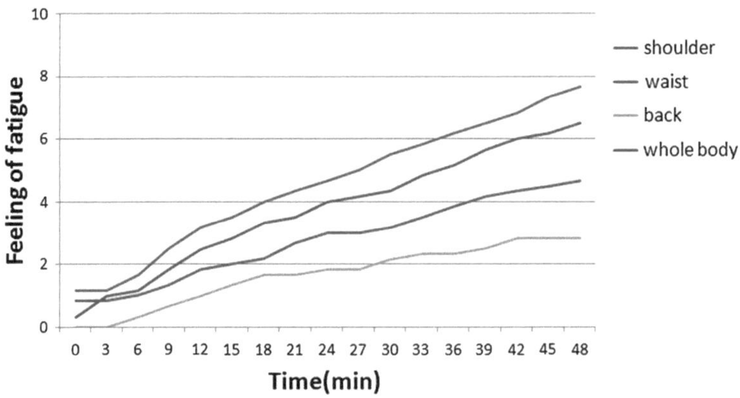

Fig. 6.2 Fatigue curve of 2# subject

$$\text{Fatigue at shoulder } = 0.133 \times t \text{ (min) } + 1.574 \tag{6.2}$$

$$\text{Fatigue at waist } = 0.097 \times t \text{ (min) } + 0.444 \tag{6.3}$$

$$\text{Fatigue at back } = 0.062 \times t \text{ (min) } - 0.094 \tag{6.4}$$

The regression equation shows that the slope of the fatigue at the whole body is basically consistent with that of the fatigue at the shoulder. In the initial stage, the fatigue at the shoulder is more obvious than that of other parts.

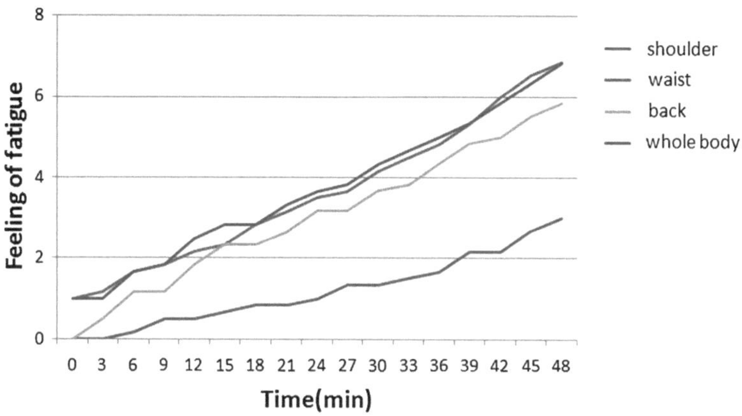

Fig. 6.3 Fatigue curve of 3# subject

6.3.2 Heart Rate

In the experiment, the heart rate increases gently. Figure 6.4 shows the heart rate changing of 1# subject in the experiment.

In the figure, the heart rate curve can be divided into four parts generally. The first part is the rest stage (2 min after the start) where the subjects have no load at rest, with relatively stable heart rate of 76 bpm ± 5.5 approximately. The second part is the preparation stage (2–9 min) where the subjects wear bullet proof vest and knapsack with the remarkable waving heart rate. The mean value of heart rate reaches 88.7 bpm. The third part is the initial stage of the experiment (from the starting of exercise on the running platform to 25 min) where the heart rate of the

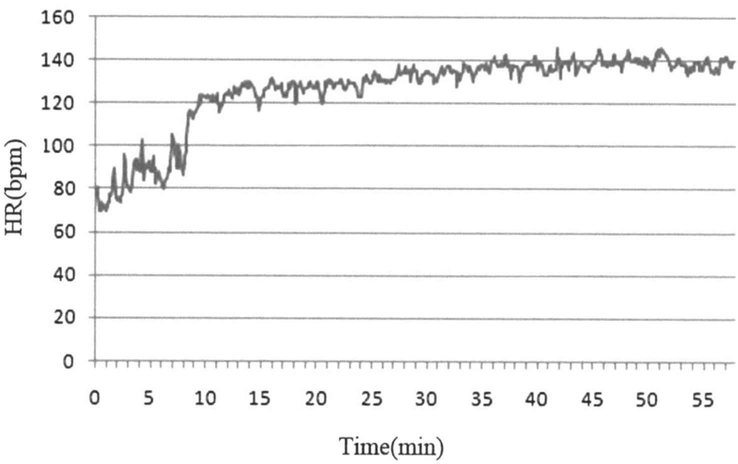

Fig. 6.4 Heart rate curve of 1# subject

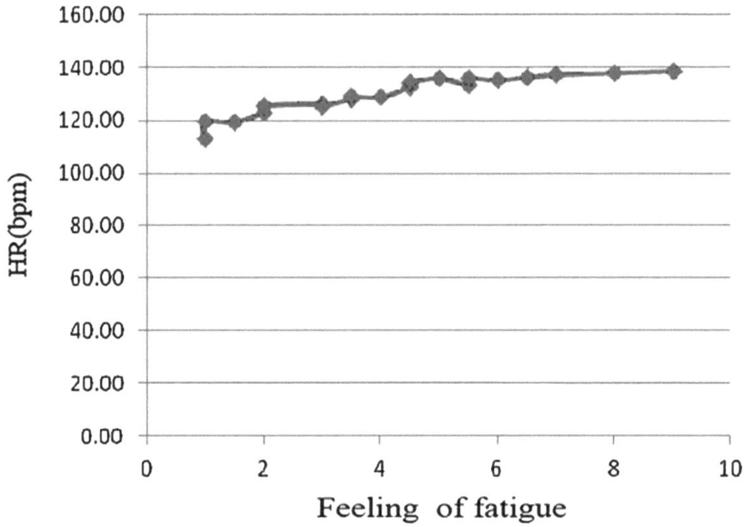

Fig. 6.5 Scatter plot of fatigue and heart rate

Table 6.2 Subjects' heart rate at the end of experiments

Subjects	Average heart rate
1	132.4
2	128.4
3	135.1

subjects is increasing and the subjects feel gradually stronger fatigue subjectively. The subjects are in the transition stage from the slight fatigue to fatigue when whey moves to 25 min (the fatigue of the whole body is 4.5 at 24 min). The fourth part is the stage of stable heart rate where the subjects have basically stable heart rate after 25 min. The subjects feel increasing fatigue. However, the heart rate remains 137.14 bpm $\pm$ 3.57.

Figure 6.5 is the scatter plot of fatigue and heart rate, showing the heart rate is increasing remarkably with the subjective fatigue mark of 0–5. However, the heart rate changes gently with the subjective fatigue mark exceeding 5 and remains at 125–140 bpm until the end of the experiment. The mean values of heart rate of three subjects in the last 3 min at the end of the experiment are as follows in Table 6.2.

6.4 Discussions

The heart rate trend chart shows that in the initial stage before going to the running platform, the heart rate of the subject increases by 10 bpm approximately when added the load to, indicating the increasing load will lead to the increasing heart rate and energy consumption when added the load to. The heart rate of subjects remains 130–140 bpm at the end of the experiment, indicating the subjects have reached the endurance extremity in accordance with *Military Physical Work Intensity Classification* (GJB1136-92), which is consistent with the subjective evaluation result of the fatigue.

The scatter plot of heart rate and fatigue shows that when the subjective evaluation mark of fatigue is less than 5, the body is in the adaption stage where the cardio-pulmonary function is adjusted rapidly to adapt the changing intensity of the exercise, to meet the requirements of metabolic quantity of the body. Thus, the heart rate is in remarkable increasing trend, increasing by 30–40 bpm than that of the initial stage of the experiment. When the subjective evaluation mark exceeding 5, the cardio-pulmonary function has adapted to the requirements of metabolic quantity and the heart rate is in a gentle changing trend, increasing by no more than 15 bpm, compared with that of the stage with the subjective evaluation mark of 5. The heart rate maintains at 130–140 bpm at the end of the experiment, when the heart rate and the fatigue evaluation mark not in linear changing trend. The average heart rate of three subjects at the last 3 min of the experiment is listed in Table 6.1, where the data show that even the same subject has not the same heart rate at the end of the experiment or at the stage when the subjective fatigue reaching the maximum value. For example, 2# subject has the minimum heart rate of 128 bpm and the maximum heart rate of 135 bpm at the end of the experiment, indicating the single index of heart rate cannot make evaluation on the subjective fatigue of subjects [7].

References

1. Kudryk IA (2008) A biomechanical analysis of a specialized load carriage technique and the development of an assistive load carriage device. Queen's University
2. Anderson AM, Meador KA, McClure LR (2007) A biomechanical analysis of anterior load carriage. Ergonomics 50:2104–2117
3. Stevenson JM, Bossi LL, Bryant JT (2004) A suite of objective biomechanical measurement tools for personal load carriage system assessment. Ergonomics 47:1160–1179
4. Stevenson JM, Bryant TJ, Reid SA (2004) Development and assessment of the Canadian personal load carriage system using objective biomechanical measures. Ergonomics 47:1255–1271
5. Martin J, Hooper R (2000) Military load carriage: a novel method of interface pressure analysis. Meeting on "Soldier Mobility: Innovations in Load Carriage System Design and Evaluation"
6. Bryant JT, Stevenson JM, Bossi LL (2004) Optimizing load carriage systems. Ergon Des 4:12–17
7. Lloyd R, Parr B, Davies S (2009) Subjective perceptions of load carriage on the head and back in Xhosa women. Appl Ergon 9:1–8

Chapter 7
Pilot Error and Error Recovery

Xiaoyan Zhang and Hongjun Xue

Abstract Three pilot error mechanisms have been presented from the analysis of aviation accidents and the limit of pilot performance. The three mechanisms are speed limited, learned carelessness, and cognitive locked which is all validated through experiments. The essential causes for the three errors are the capability limit, human cognitive inertia, and the limited cognitive resource. The recovery suggestions have already been presented from the cockpit design and pilot training. For the cockpit HCI design, the persistent time for the information should be 500 ms at least to give pilot enough time to react; if the pilot has to execute multitasks at the same time, the information should be designed in ladder to avoid unreasonable attention resource allocation; for the pilot training, the responsibility of the job should be enhanced and the study of rules and procedures should be stricter to avoid carelessness learned.

Keywords Pilot error · Error recovery · Speed limited · Learned carelessness · Cognitive locked · Cockpit design · Pilot training

7.1 Introduction

The statistic results of aviation accidents in recent years are shown in Fig. 7.1, which shows nearly 70 % causes of all aviation accidents is the crew. According to the survey conclusion of the aviation accidents, the major causes of the accidents are cognition and environment. The environment factors includes mainly the rough

X. Zhang (✉) · H. Xue
School of Aeronautics, Northwestern Polytechnical University, Xi'an 710072 Shanxi, China
e-mail: zxyliuyan@163.com

H. Xue
e-mail: xuehj@nwpu.edu

S. Long and B. S. Dhillon (eds.), *Proceedings of the 13th International Conference on Man-Machine-Environment System Engineering*, Lecture Notes in Electrical Engineering 259, DOI: 10.1007/978-3-642-38968-9_7,

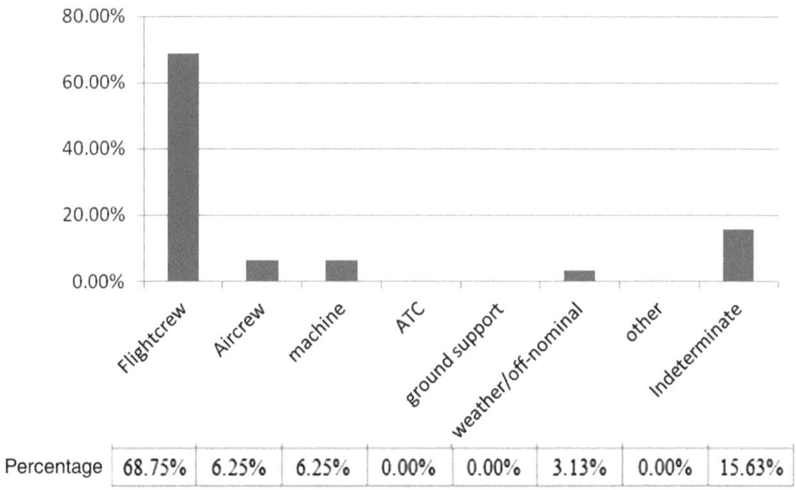

Fig. 7.1 Statistic analysis of commercial aviation accidents during 1999–2008

HCI design [1], automatic operation hard to understand [2] and adverse weather conditions [3] while the cognition factors includes poor situation awareness [4], disobeying operation procedures [5], and coordination and communication of crew [6], mainly related with the pilot. Foreign scholars established some cognition models of pilots [7], including ACT-R, AIR-MIDAS, and A-SA to simulate the decision error of pilots. The models can make simulation on the mistakes in selecting the taxi track for landing, obtaining ideal results. Such teams established the cognition capacity models of pilots; however, they were devoted to improving the simulation capacity of the people [8] and failed to connect the mechanism of error and actual aviation accidents; thus, such models had no significant guidance for the aviation security. The thesis, on the basis of survey and study on the present aviation accidents, presents three mechanisms of the pilot error, namely speed limited, learned carelessness, and cognitive locked. The experiments have been made to verify such three mechanisms. Then, the thesis makes suggestions on recovering such three errors in the cockpit HCI design, operation program design, and trainings on pilots.

7.2 Mechanism of Pilot Error

The pilot error means an action or omission of the pilot, resulting in disobedience of the will of crew or requirements of actual situation, including the regulations, rules, and standard operation procedures [9]. The pilot error mainly results from awareness of situation, fatigue, work load, HCI design level, organizational culture, and external environment, etc. In the thesis, the discussion is made on the three mechanisms of the pilot error from the limit of the performance, namely speed limited, learned carelessness, and cognitive locked.

1. Speed limited

According to the speed limited, people have the inherent limit in the performance and cannot make correct reaction in a certain speed limited. Especially in adverse and emergent operation situations, the aircraft is in particular condition. It is difficult to leave enough time for the pilot to make cognition and reaction, thus, resulting in the error.

2. Learned carelessness

If the pilot failing to operate according to the procedures in a certain mission does not result in the accident or any accident sign, he will have the tendency to simply the procedures in the future, causing the accident finally.

3. Cognitive locked

According to cognitive locked, the cognitive source of the pilot is limited [9]. Therefore, the pilot will make unreasonable attention resource allocation, even given enough time in executing multitasks, thus, causing the error.

In the following part of the thesis, the experiment is made to verify the three mechanisms of the pilot error. The discussion is made on the speed limited in the operation, the tendency to simply the operation procedures and limited cognition resources.

7.3 Error Mechanism Verification Experiment and Analysis

7.3.1 Speed Limited Experiment

1. Subjects

12 students on school, including 3 girl students. Their right hands are dominant. The cognition level is restricted to be the faster the better on the basis of correctness. The subjects are in normal spirit without any abnormal conditions as fatigue in the experiment.

2. Experiment design

The experiment is designed as one with a single variable of the appearing time of the stimulus, 800, 500, 400, 350, 300, 275, 250, and 200 ms, each of which the subject is required to be tested for 20 times. In the experiment, on the screen will appear red and green warning lights to indicate different warning levels. The subjects are required to click different buttons according to the warning light of different colors. The computer will make warning sound and record the reaction time and accuracy rate of the subjects. The subjects are required to participate in the formal experiment after being skilled at the experiment.

3. Experiment result and analysis

The experiment results show that the error rate of the subjects increases in nearly linear mode as the present time decreases. In the present time of the warning information of 200 ms, the error rate reaches 90 %. No reaction errors are 195 in the total 216 errors, covering 90.3 %. In the present time of the warning information of 800 ms, the error rate reduces to 2.9 %, of which only one no reaction error. It is a complete cognition control process for the subject to make judgment on the information and corresponding reaction. If the present time is not enough for the process, the subject will make errors of no reaction or the subconscious error.

Besides, the experiment results show that the reaction time of the subjects decreases in a nearly linear mode with the shortening present time. In the stimulus present time of 200 ms, the correct reaction time is only 163.5 ms. However, the studies of anthropologists show that the limit of the reaction time is 0.2 s. It shows that the time pressure will inspire the potential to make reaction beyond the capacity limit. However, such state beyond the capacity limit will not last for a long period. The people will feel frustration and tend to give up in case of the reaction frequency failing catching up with that of the present stimulus (Figs. 7.2 and 7.3).

The experiment of speed limited is made when the subjects are quite familiar with the operation procedures. The results show that the people have capacity limit in cognition or operation. In case of the reaction time less than 400 ms, the error rate will be more than 30 %. In case of the reaction time more than 800 ms, the error rate of the subjects tends to zero.

7.3.2 Learned Carelessness Experiment

1. Subjects

Same with the above-mentioned experiment.

2. Experiment design

Besides the judgment on the warning information, the subjects are required to check the flight altimeter beside the warning information and to press different buttons in case of abnormal conditions on the altimeter. In the experiment, the altimeter is designed to be normal altitude range, interspersed with abnormal altitude range. The record is made on the reaction time and accuracy rate of the subjects. The whole experiment is required to be made for 20 times. The subjects are required to participate in the formal experiment after being skilled at the experiment.

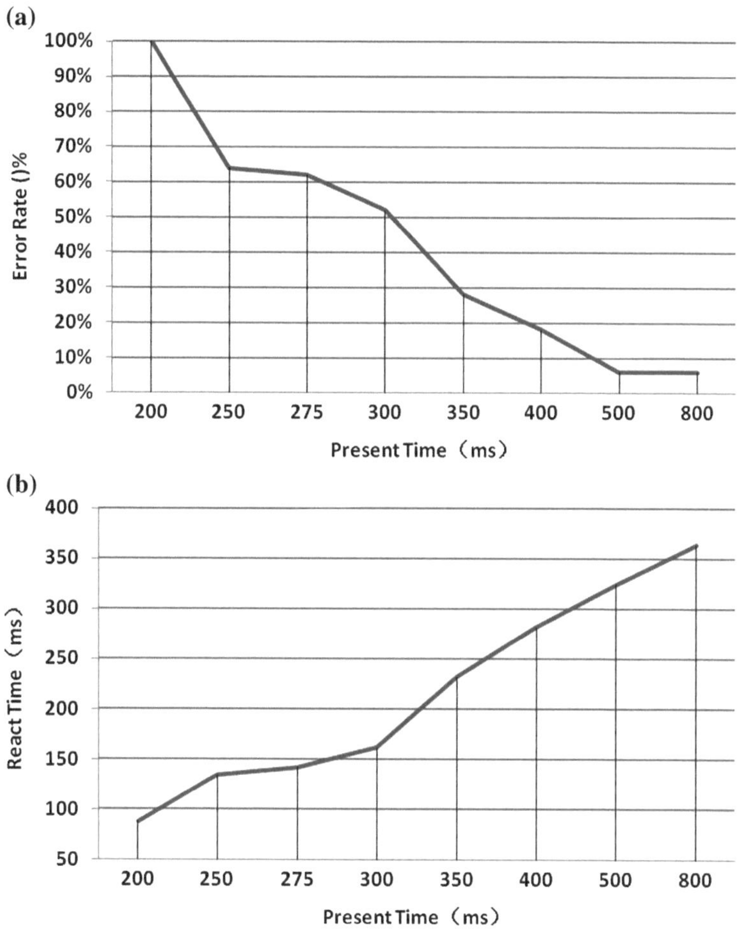

Fig. 7.2 Result of speed limited. **a** Relation between information present time and error rate. **b** Relation between information present time and reaction time

3. Experiment results and analysis

In the experiment, the flight altimeter is set up in the normal range for the first 5 times on purpose, to observe whether the subjects have the tendency to ignore examining the altimeter. The experiment results show that the reaction time of the subjects is longer at the beginning of the experiment, indicating the subjects check the altimeter according to the requirements of the procedures. However, finding the altimeter in normal condition for several times successively, the subjects will ignore the check on the altimeter and simply the operation procedures. The reaction time will be shortened to 744 ms on average and the error rate will increase to 6.25 %. In 240 experiments, nearly all errors appear in the inflection points set up in the experiment, that is, the time when the altimeter changes from normal condition to abnormal condition. It shows that the people will have the

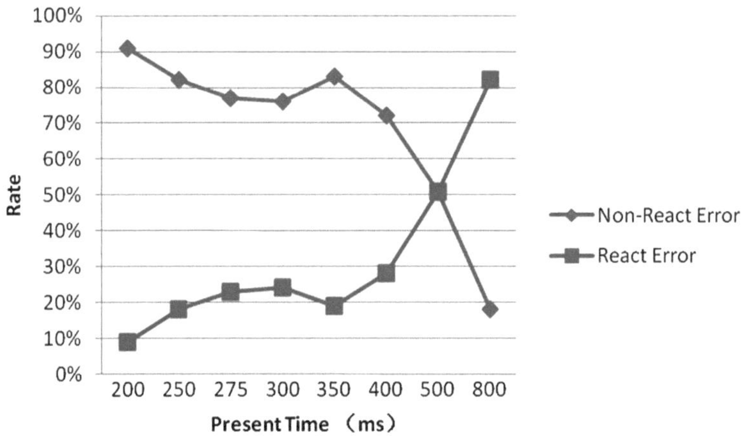

Fig. 7.3 Different errors and information present time

cognitive inertia. Finding no check on the altimeter does not result in the accident or accident signs informed in advance, the subjects will tend to ignore checking the altimeter. Such proves the existence of learned carelessness, which is the error mechanism due to the simplification of the operation procedures of the pilots.

7.3.3 Cognitive Locked Experiment

1. Subjects

   Same with above-mentioned experiment.

2. Experiment design

   Besides the judgment on the warning information, the subjects are required to check whether there is any prompt information at the lower right corner of the screen. The subjects are required to press corresponding buttons in case of any other prompt information. The recorded is made on the reaction time and accuracy rate of the subjects. The whole experiment is required to be made for 20 times. The subjects are required to participate in the formal experiment after being skilled at the experiment.

3. Experiment results and analysis

   In the experiment, the subjects are required to execute two tasks simultaneously on purpose, to make judgment on different warning information levels and to notice whether there is any abnormal condition at the lower right corner and to make corresponding reaction. The reaction time of the subjects is shorter than that of the subjects in the learned carelessness. However, the error rate increases remarkably. The average reaction time is 496 ms and the error rate is 12.9 %.

The analysis on the errors of the subjects shows that the subjects make errors on the reaction to the abnormal conditions at the lower right corner of the screen. Besides, the interview to the subjects shows that the subjects tend to focus on the judgment on the warning information level, thus, resulting in ignoring the judgment on the abnormal conditions at the lower right corner of the screen. It shows the people have limited cognition resources and will make unreasonable attention allocation in multitasks. The subjects will ignore the important information and make the error, of which the rate is higher than that of learned carelessness and which is hard to recover.

7.4 Conclusion

The results of three experiments verify three error mechanisms of pilots, namely speed limited, learned carelessness, and cognitive locked. Detailed analysis is also made on the causes of three error mechanisms in the thesis. It requires making corresponding recovery measures for the errors of pilots with the purpose of improving the aviation security. In the thesis, some suggestions are made on the recovery of three errors in the cockpit HCI design and the pilot training:

1. The speed limited shows the capacity limit is one of sources of errors. Consideration is required to make in the cockpit HCI design. The information needs being presented no less than 500 ms and more than one warning measure is made in case of any fault to avoid the pilot ignoring the warning information;
2. The learned carelessness shows the people have the thinking and operation inertia. Therefore, the pilot training will focus on the responsibility of job and mastering rules and operation procedures. In the cockpit HCI design, the reasonable warning will be made in case of any pilot failing to operate in accordance with the operation procedures;
3. The cognitive locked shows that the people have the limited cognition resources. Therefore, in the cockpit HCI design, the attention shall be paid to keeping the stimulus presented in gradient and trying not to require the pilot to execute multitasks in the same time. Otherwise, there should be warning device to interrupt the cognitive locked of the pilot. In the pilot training, the training on the capacity to execute multitasks should be strengthened because it is impossible to avoid the pilot executing multitasks simultaneously in the flight, especially in adverse and emergent conditions.

The thesis presents three error mechanisms of pilots in the capacity limit of pilots and suggests corresponding recovery measures. However, the factors are complicated to affect the aviation security and the serious accidents or disasters are generally from various factors as the environment, crew and cockpit HCI design simultaneously. Therefore, the further study will be made on the complex factors affecting the pilot errors in the adverse and emergent conditions to improve the aviation security.

References

1. Degani A, Shafto M, Kirlik A (1999) Modes in human-machine systems: review, classification and application. Int J Aviat Psychol 9:125–138
2. Olson WA, Sarter NB (2000) Automation management strategies: pilot preferences and operational experiences. Int J Aviat Psychol 10:327–341
3. Wiegman DA, Goh J (2001) Pilots' decisions to continue visual flight rules (VFR) flight into adverse weather: effects of distance traveled and flight experience (Tech. Rep. No. ARL-01-11/FAA-01-3). Aviation Research Laboratory, Savoy: University of Illinois
4. Endsley MR, Smolensky MW (1998) Situation awareness in air traffic control: the big picture. In: Smolensky MW, Stein ES (eds) Human factors in air traffic control. Academic, San Diego, CA, pp 115–154
5. Bisantz AM, Pritchett AR (2003) Measuring judgement in complex, dynamics environments: a lens model analysis of collision detection behavior. Hum Factors 45:266–280
6. Fousee HC, Helmreich RL (1988) Group interaction and flight crew performance. In: Wiener EL, Nagel DC (eds) Human factors in aviation. Academic, San Diego, CA, pp 189–277
7. Foyle DC, Hooey BL (2007) Human performance modeling in aviation. CRC Press, Florida, U.S.A, 2007
8. Byrne MD, Kirlik A (2005) Using computational cognitive modeling to diagnose possible sources of aviation error. Int J Aviat Psychol 15(2):135–155
9. Reason J (1990) Human error. Cambridge University Press, Cambridge

Chapter 8
Research on Evaluation of Operation Command Capacity for Growing Cadre Students Position Training

Rongzhi Yang, Hai Chang, Xianghe Jing, Xiaodong Zhang and Jie Xing

Abstract Evaluation of operation command capacity for growing cadre students position training is the process of measuring, analyzing, and comparing by the scientific index system and the effective method. On the basis of analyzing, the problem in evaluation of operation command capacity for growing cadre students position training, the significance, principle, method, and process of evaluation of operation command capacity for growing cadre students position training is proposed. It provides a clear aim for students, a reliable method for academy to evaluate operation command capacity for growing cadre students, and a reliable gist for ameliorating train project and optimizing course setting.

Keywords Growing cadre students · Position training · Operational command capacity · Evaluation

8.1 Introduction

After the academy education for 4 years, the growing cadre position training students (hereinafter referred to as growing cadre students) will receive the position training for one year in the position education colleges and universities. They have relatively solid foundation on political thought, military quality and science and culture, a certain capacity on organization and command, cooperation and innovative study, making better foundation for the position training [1]. The growing cadre students will have various teaching contents in the training, with how to improve the operation command capacity of the squad. To improve the operation command capacity of students, the colleges and universities shall,

R. Yang (✉) · H. Chang · X. Jing · X. Zhang · J. Xing
Air Defense Forces Academy, Zhengzhou, China
e-mail: skyguard999@126.com

S. Long and B. S. Dhillon (eds.), *Proceedings of the 13th International Conference on Man-Machine-Environment System Engineering*, Lecture Notes in Electrical Engineering 259, DOI: 10.1007/978-3-642-38968-9_8,

besides the operation command capacity, strengthen the teaching, learning, management, and guarding and also make effort on the evaluation, to prompt the reform and improvement with the evaluation. The scientific evaluation is made to judge the operation command capacity of the growing cadre students. In the education, the talent training program will be improved and course setting will be optimized to improve the teaching and to provide the service for improving the operation command capacity of the growing cadre students greatly.

8.2 Analysis on Existing Problems in the Operation Command Capacity Cultivation of the Growing Cadre Students

8.2.1 Unstable Though of Preparing for the War Weakens the Urgent Sense of Improving Operation Command Capacity

The disputes of territorial land and sea between China and surrounding countries are increasing with more insecurity factors. Therefore, the military of China is faced with serious challenges of new missions. However, some growing cadre students fail to realize the urgent conditions. They do not have the stable thought of preparing for the war. They consider "to win the war" only as a slogan and do not have strong urgency and power to improve operation command capacity.

8.2.2 Inadequate Teaching Time Restricts the Development of the Operation Command Capacity

The strongly practical operational command requires the growing cadre students to improve the operation command capacity by the practice. As Engels said, it is quite difficult to be a real military expert without hearing the bullet whistle and smelling the smoke of the powder in the battle field [2]. In the present training on the growing cadre students in the position education colleges and universities of troops in China, the practical teaching process is complete. However, the total teaching and training time is not adequate. The students are not trained deeply and the people are quite few who have practical experience. Such condition results in the following problems: the students master the military theory but lack the test in the practice; the organization and command capacity of the students lack the test in the practice as well; the psychological quality of the students lacks the training in the practice.

8.2.3 *Low Information Quality Affects the Improvement of the Operation Command Capacity of Students*

Recent partial wars show that the information confrontation lasts the whole process of the organization and command in the informational wars. With the fast development of the information construction in troops of China, the command information system will be widely used in the operation command. In the present colleges and universities, the knowledge structure of the students has been improved. However, there are still outstanding problems including low information quality and inadequate capacity of organization and command of the information war [3]. The low information quality of students results from two essential factors: firstly, inadequate attention is paid to improving the information quality of students. No definite target has been set up in the information quality in the training program. The information command system is not used fully in various teaching and training and the conventional operation command mode is still in use in the teaching and training; secondly, the students lack adequate awareness on the information war and do not struggle to improve the information quality. The students lack the persistence to practice and develop the information quality and the practical action to master the information knowledge in the integrity of the learning and training, especially the new equipment.

8.3 Significance of Evaluation on the Operation Command Capacity of the Growing Cadre Students

8.3.1 *It will Benefit for the Target of Position Training*

The commanding officer is mainly responsible for the leading soldiers for the war with the main target to inspire the people play the maximum part and utilize the maximum efficiency of weapons, developing the strongest fighting capacity with the best integrity of the people and weapon, thus, wining the partial war in the information condition. The operation command capacity is the essential capacity of the growing cadre students. The definite stipulations should be made on the operation command capacity necessary for the growing cadre students, solving the basic target of the growing cadre students on how to command the operation. Only the training target on the operation command capacity of the students is definite, can the training in colleges and universities target strongly.

8.3.2 *It will Benefit for the Learning Target*

The learning is the foundation to improve the capacity. However, to improve the learning effect, the learning target must be set up. Only the focus and difficult points and acting point are made clear can the learning is made with definite target, achieving more with less efforts. The definite requirements on the operation command capacity standards establish the definite direction of the growing cadre students to improve the capacity in the learning and to develop the capacity in the practice. That means the specific requirements on the operation command capacity will be the focus of the learning and acting point of improvement for the growing cadre students in the future.

8.3.3 It will Benefit for the Normalization of Test Evaluation

The foundation and premise to judge whether the students meet the training target is to make correct test and accurate evaluation on the operation command capacity of the growing cadre students. Without the test evaluation, the training target of the growing cadre students on the operation command capacity cannot be implemented. The previous training on the operation command capacity of the growing cadre students had poor effect and achievements, of which one important cause is the lack of standards to make test and evaluation on the operation command capacity of the growing cadre students; thus, it is impossible to make accurate and effective test and evaluation on the capacity. The study is made on the composition, standards, and basic requirements on the operation command capacity of the growing cadre students, which solves the standards of the test and evaluation in a better mode and makes the foundation for the gradual establishment of the comprehensive capacity test and evaluation system for the growing cadre students.

8.3.4 It will Benefit for Deepening the Reform on the Teaching and Training

The teaching and training in colleges and universities are not changeless but changing according to the actual conditions of military technical revolution and reform of troops training. The colleges and universities shall make reform on the teaching and training according to the training target and method. The position education colleges and universities are the main fields to train the growing cadre students, of which the main direction to the reform on the teaching and training is to train the organization and command capacity. The colleges and universities will, by making evaluation on the operation command capacity of the students, systemize the experience, find existing problems, and put forward instructions and suggestions to solve the problems, providing the service for the reform on the teaching and training contents and improvement on training method and means and also prompting the reform on position training for further development [3].

8.4 Principles of Evaluation of the Operation Command Capacity of Students

8.4.1 The Principle of Capacity Development Rule

The growing cadre students will develop their operation command capacity according to the basic rule of making solid military foundation, mastering basic skills, commanding in the practice and practicing in the command, which includes the following five aspects: firstly, common awareness on the operation command is the basic premise of the development of the operation command capacity; secondly, mature operation and training system is the important basis of the development of the operation command capacity; thirdly, definite training target and scientific course system is the major support of the development of the operation command capacity; fourthly, complete training system is the reliable guarantee of the development of the operation command capacity; fifthly, solid military knowledge and rich military practice is the strong foundation of the development of the operation command capacity [4].

8.4.2 The Principle of Meeting the Actual Background of the War

The operation command is one military action with strong practice, which can only be developed and improved in the practice of the operation command. The training on the operation command of the growing cadre students in colleges and universities is generally made under the nearly actual background in different parts of stage, comprehensive, and live ammunition trainings. The evaluation on the operation command capacity of the students should be made in the practice, especially in the informatization meeting the actual background of the war, which has more actual evaluation effect [5].

8.4.3 The Principle of Prompting Reform with Evaluation

The evaluation on the operation command capacity of the students will be made in the whole process of the teaching, where the students are under observation and evaluation. The students can understand the faults in time and make improvement according to the feedback information. The teachers can make special training on students on the basis of the evaluation on the operation command capacity of different students. Existing problems can be discovered in the evaluation and made feedback to teachers and students, who can made corresponding adjustment and modifying the procedure or direction of the reform on the teaching and training, obtaining more ideal teaching effect.

8.5 Method and Procedure of Evaluation of Operation Command Capacity of Students

8.5.1 To Formulate Evaluation Index System

To formulate scientific evaluation index system is the key to make evaluation on the operation command capacity of the growing cadre students.

Firstly, to set up the evaluation factors and weight the previous evaluation on the operation command capacity of the students is usually a subjective judgment without quantitative standards. The evaluators would make remarks on the basis of subjective impression, and it was impossible to avoid unfairness. To make sure the more scientific evaluation process and more just evaluation result, we, on the basis of teaching and training experience for many years and integrity of questionnaire on 100 of officers of the regiment and battalion levels, divide the operation command capacity of the growing cadre students into 7 factors, namely, military theory quality, planning capacity, decision making capacity, organization capacity, command procedure, expression capacity, and volitional quality with the reference weight, respectively, of 0.1, 0.2, 0.1, 0.2, 0.2, 0.1, and 0.1.

Secondly, to definite essential contents of evaluation factors, the study shows that the military theory quality mainly contains modern military theory, operation policies, principles and strategies and tactics; the planning capacity contains accurate understanding of the upper intension, receiving operations, troops distribution, and apparatus and material preparation; the decision making capacity contains mastering conditions of both parties of the war, analysis on officers and soldiers quality, making decision on major direction and action; the command procedure contains normalized procedures and treatment on emergency; the organization capacity contains the management and control of troops, operation organization and analysis and summarization; the expression capacity contains accurate delivery of the upper instruction and requirement, simple and clear order and mobilization to the point; the volitional capacity contains daring to sacrifice, fearing no hardship, calmness, and sound psychological quality.

Thirdly, to formulate evaluation standards, the quality index reflects the general contents of the person to be evaluated. To make up for the disadvantages, the index of the quality index is specialized to keep the standards specific and reduce the possibility of personal judgment and more definite conclusion. Several clear, specific, and observable criteria are made definite for each capacity. The evaluation conclusion will be made in four different levels of excellent, good, passing, and failing. Only the levels of excellent and passing are presented in the specific evaluation standards. Those between the excellent and passing are considered as the level of good and those failing to meet the passing standards are considered to be the level of failing.

8.5.2 To Organize and Execute Evaluation

Firstly, in the diversified evaluation method, the marks are made to classify the levels of the operation command capacity of the students. The classification is made with the methods of theoretical test, simulate command, and operation command evaluation in the field. The evaluation is made on the operation command capacity of the students with a series of teaching and training, including professional training, stage training, comprehensive drill, and live ammunition trainings.

Secondly, in the diversified evaluators, the elevator means the individual or team to participate in the organization and execution of the evaluation and make the value judgment on the object to be evaluated according to a certain standards. The training of the operation command capacity of the students is interactive in the whole training program, involving various people in the teaching system as students, squadron cadres, teachers of the operation command, and professional teachers. Such people can made observation and evaluation on specific activities of the students.

Different evaluators involve aspects of the students in the learning life, featuring multi-aspect and diversification, avoiding the lopsided view due to the separate evaluation by a certain evaluator. Diversified evaluators can reflect the real overall operation command of the students and show the leadership potential of the students. Besides, the students can be evaluators, too. Thus, the students can have deeper understanding on the requirements on the operation command quality and understand the development direction more clearly.

8.5.3 Feedback of the Evaluation Results

The quality evaluation is intended to promote the improvement of the quality, to make the students understand themselves, to display their advantages and improve their disadvantages, and to make timely adjustment in case of any error, thus, guiding the development of the students. The evaluation of the operation command capacity aims to develop such capacity by continuous and gradual practice. The training on the operation command capacity of the students is a cycling development of study, training, evaluation, feedback and supplement and training again. In the interaction of learning and evaluation, the operation command capacity is improved continuously, meeting the requirements of an excellent primary officer finally. In the cycling, the time feedback of the evaluation results is the motivation of the training on the operation command capacity and the interaction of learning and evaluation. Such mechanism can guarantee the operation command capacity of the students in continuous development to the predicted target with increasing improvement.

References

1. Zou P (2011) Shulin reading of military position training. Military Science Publishing House, Beijing P94–P95
2. Guo J, He X, Song L (2010) Theory innovate and practice of armed forces capability. Gold Shield Publishing House, Beijing P269–313
3. Xu W, Ma H, Gao R (2009) Research on Training of the new compere for diversification military affairs. Tide Publishing House, Beijing P425–428
4. Li Y, Feng Z, Yang F (2007) Research on campaign command with informationization. Tide Publishing House, Beijing P637–640
5. Chai Y (2004) Transformation of military training and education. PLA Publishing House, Beijing P186–192

Chapter 9
Countermeasures to Improve the Role Quality of a College Course Designer

Hua Li, Guoxiang Tian, Jianhe Wang and Zhiwen Yuan

Abstract The role quality is an important factor in the college course designer's psychological quality structure, to research the countermeasures on improving the role quality of a college course designer can promote the personality developing and potential improving, achieving the higher quality of a college course design. The role quality of a college course designer mainly include those abilities such as college course designer's forethought and harmonizing ability, sociability, collaborative consciousness and collaborative ability, perform ability, etc. This thesis makes analysis on the two sorts of college course designers and proposes the countermeasures that can improve the role qualities.

Keywords College · Course designer · Role quality · Improvement

9.1 Introduction

The psychological quality is the core of the overall quality of the designers of the college courses and the role quality is one of the psychological qualities of them. The role quality is the capacity to understand the role code and to master the role activities, which is the key factor of the psychological quality structure and can be used for evaluation. The study on the role quality of the designers of college courses is to promote the personality improvement, maintenance sound work state, and potential development of the college course designers; thus, to improve the design quality of the courses and education quality of colleges and training talents with development in all-round way.

H. Li (✉) · G. Tian · J. Wang · Z. Yuan
Air Defense Forces Academy, Zhenzhou 450052, China
e-mail: fkb_mmes@163.com

S. Long and B. S. Dhillon (eds.), *Proceedings of the 13th International Conference on Man-Machine-Environment System Engineering*, Lecture Notes in Electrical Engineering 259, DOI: 10.1007/978-3-642-38968-9_9,

The role quality means the awareness, feeling, volition, and evaluation of the role in fulfilling rights and obligations in the socialization [1]. It guides the people to make definite the responsibilities and obligations, emotional direction, ambitions and beliefs, and the timing, scene, and method of self-adjustment with observing professional ethics and virtues as the foundation, completing the role mission as the target and harmonious interpersonal relationship as the guarantee.

The role quality of the designers of the college courses has irreplaceable guidance for the designers to make definite the role mission, make correct role positioning, enter the role state rapidly, discover the role characteristics, and develop role acting capacity. Meanwhile, it also has quite important evaluation effect for the designers to establish the motivation of reasonable requirements, to set up the confidence, master and control feelings accurately, to solve the interpersonal dispute, and to make correct behavior attribution and understanding themselves. The role quality of the designers is mainly displayed in the planning and cooperation capacity, scalability, cooperation awareness and capacity, and performance ability, of which the performance ability is the core.

It should be made clear that the subject of the study on the role quality of the designers of the courses is the community of two types of people [2], designers of the courses, mainly including teaching and management staff and teachers of colleges and participators of the course design, which is the student. Two types of people are working together in the course design. Thus, the improvement strategy is to be studied according to different features.

9.2 Role Misunderstanding of the Designers of the College Courses

The role is the product of the socialization, a unity of the responsibilities, rights and interests in the social life, displaying with the role capacity, including role awareness, role expectancy, role attitude, and role play [3]. How related people to understand the status and part of themselves and other people in the course design has influence on the recognition of the designers on their responsibilities and their feeling and volition to design the course, thus, affecting the design effect of the course. Two major groups, namely the designers and participators, are significantly different in the role awareness, resulting in role misunderstanding, including inaccurate role positioning, and ignoring the role part in the course design.

The wrong positioning on the role of the designers of the course can be expressed with two terms of generalization and minimization. The designers of the college course have the common concept on the role and position for a long time, thinking that they have no direct role in the design. They always say: "We are only the executor to complete what the upper requires and the personal role won't display in the course design." They fail to realize the importance of the role of the designers and their responsibilities and missions. They also ignore the role expectancy of the students, which is considered the minimization of the role awareness of the designers. Meanwhile, due to the academic authority of the designers and bureaucracy nature of the management staff existing inveterately in the course design, the designers, considering their academic and administrative authority absolute, decide everything in the actual course design, which is considered as the generalization. The designers of the course fail to make accurate positioning, directly shown in the lack of the democracy and interaction and development in the course design.

The role and part of the students as the participators in the course design is not recognized for a long time, which results from the ignorance on the role by the colleges and teachers and unawareness of the students on the important role and ignorance their role status in the course design. On the one hand, the supremacy of the administration authority in the management, the academic authority in the course, and teachers in teaching imprison the ideas of the designers of the college course; thus, the students have not been accepted in the participation in the course design. On the other hand, the students will start their learning according to the previously arranged course program after admission. They need not worry about what to learn, how to learn, or even what level to learn. They just attend the class on schedule, focus on the speech and write down key points, guaranteeing a high mark in the examination, which is quite common in the colleges. The students are accustomed to accepting the existing conclusion, disobedience rules in the management, having no definite ambition and not being outstanding. They do not have the wish and make any effort to express themselves due to lack of subject consciousness. They fail to find the independent themselves, not even to express themselves. The students are ignoring their rights and obligations to participate in the course design.

9.3 To Improve the Role Quality of the College Course Designers

9.3.1 To Strengthen the Planning Capacity of the Course Designers

Different roles have various orientations of the thought [4]: it is necessary for the college course designers to establish the planning thought on the basis of the nature, status, and idea of the college education and particular requirements of the course development. They need to develop the necessary thought habits and quality of the researchers of the college teaching. They also need to develop thought characteristic with the distinctive individuality. Such thought shall be fully overall, harmonious, systematic, integrative, and forward looking.

9.3.2 To Strengthen the Coordination Capacity of the Course Designers

The course design is an overall mission, involving in various aspects. Therefore, it requires the course designers to make timely and reasonable allocation of the limited course resources and to make scientific arrangement, guaranteeing the orderly process and maximum efficiency. The coordination capacity means mainly the scientific arrangement on the manpower, materials, and financial resources, the overall mastering of the process of the course design and also the coordination in various conflicts and issues.

9.3.3 To Strengthen the Cooperation Capacity of the Course Designers

It is the cooperation and trust in the course designers group that results in taking advantages of collective wisdom. For such purpose, it requires that the course designers shall have adequate toleration to be responsible for the individual action, not to show off the achievements, not to dodge the responsibility, not to cover any mistakes and to share the achievements and honors.

9.3.4 To Strengthen the Responsibility Awareness of the Course Designers

The course designers shall keep the mission and role responsibility in the mind and shall recognize that they are the designers and executors of the blueprint to train talents of high quality in the college. They shall, remembering the role expectancy from the students, make improvement continuously, to meet the expectancy and requirement on the course designers.

9.4 To Improve the Role Quality of the College Course Participators

9.4.1 To Strengthen the Sense of Participation of the Students

The students are indispensable participators in the course design. The people-oriented and students-oriented education idea will not come true if the students are dissociated from the course design and development, the core of the college education as passive audience in absolute obedience without any right to participate and to express themselves. The students are allowed to participate in the course design and development actively, which is consistent with the education values and expectancy indicated in the replacement of the original teaching plan and course outline with the talent training program and course standards. Only the full concern and promotion are made to the students to participate in the course design and development actively and innovatively, can the supremacy in the course of the colleges and teachers for a long time be changed, realizing democracy in the college course design, guaranteeing the high energy of the college course, and reaching benefiting for the students.

9.4.2 To Strengthen the Share Capacity of the Students

Learning needs mean the desire for learning. The share of learning needs means accepting the others to express their requirements and also allowing themselves to accept the requirement of the others. It also means considering themselves and others equally and seeking for the access of common development. The students will expect the course to meet their desire for learning and to learn what they need with the best effect, which is the key point of the course design. The students need to learn how to share the learning need, to express what they are interested in and what course present mode is popular and what are their learning targets and experience and to participate in the course design, guaranteeing the effective and targeted course design.

9.4.3 To Strengthen the Sense of Participation of the Students

Whether they are allowed to participate is equally important with whether they will participate in the course design. It will be no effect in case of the students having no desire to participate. Whether the student are active, serious, and desirable for the participation in the course design depends on the mastering and understanding of the students on the role, the attitude, and idea of the colleges on the students' participating in the course design and the understanding of the students on the important part of the course design. To inspire the desire of the students to participate in the course design, we can obtain the valuable opinions and suggestions of the students and inspire the students to be quite interested in the course to learn deeply the content, nature of the course spontaneously, making sure the quality of learning the course. It also can make the students feel the recognition of their status of the subjects, obtaining the sense of achievements on supporting the college course construction.

9.4.4 To Strengthen the Effective Expression Capacity of the Students

The effect of the students on the course design depends on the present mode and expression level of their opinions and suggestions in a great level. The students shall learn how to systemize the thinking and requirements on the course, making well organized, definite standpoint and feasible opinions and suggestions and express with proper modes in writing or orally. Such effective expression means the students shall not close off and not refuse to accept the feedback. The students shall express themselves decently, properly, definitely, fully, and equally. With such expression, the students can attract the other concern, express their feelings, and express even amateur opinion without worry about the criticism. They can be inspired from the information of the others, leaving open impression to others.

References

1. Alsubel.etc. Education psychology—the cognitive view. Yu Xingnan etc. translated. People's Education Press, Beijing
2. Huang G, Cai Q (2005) Course design—Theory and practice. Normal university Press, Nanjing
3. Lu J, Wei Q, Li Q (2004) Psychology—the basic theory and education application (revision). Shanghai People's Press, Shanghai
4. Tanner D, Tanner L (1995) Curriculum development: theory into practice (3rd ed) Merrill, New York

Chapter 10
Improved Air Defense Command Personnel Training Quality Several Ponders

Bingjun Zhang, Rongzhi Yang, Peng Han, Meng Kang, Jianjun Li, Tie Wang and Yan Sun

Abstract Since ancient times, Chinese people have laid sufficient importance to teachers and moral education. Teachers are the guides of ideology, culturists of socialism successors, discriminators of good and evil, right and wrong, beauty and ugliness, practitioners of promoting traditional Chinese virtue, executants of training students "morality foremost." Noble teachers' morality and behavior are the vivid, direct, and meaningful education mode. We cannot cultivate high-quality talents without noble ethics of teachers; teachers' construction is weightier than Mount Tai. So, strengthening teachers' morality construction plays an important and profound significance. The paper mainly analyzes the present situation and puts forward basic countermeasures from basic connotation of teachers' morality construction.

Keywords Teachers' morality · Construction · Understanding and thinking

B. Zhang (✉) · R. Yang · P. Han · M. Kang · J. Li · T. Wang · Y. Sun
Air Defense Forces Academy, Zhengzhou, China
e-mail: Sophia_awia@sina.com

R. Yang
e-mail: marsyangrongzhi@163.com

P. Han
e-mail: hanpengmine@163.com

M. Kang
e-mail: francais365@hotmail.com

J. Li
e-mail: salut_lijianjun@163.com

T. Wang
e-mail: monsieurwt@163.com

Y. Sun
e-mail: fengfengsunny@hotmail.com

S. Long and B. S. Dhillon (eds.), *Proceedings of the 13th International Conference on Man–Machine–Environment System Engineering*, Lecture Notes in Electrical Engineering 259, DOI: 10.1007/978-3-642-38968-9_10,

10.1 Introduction

Chinese people have been respecting teachers since ancient times. Teachers not only teach knowledge to students but also show good behaviors to students. Their virtues and behaviors are strictly normalized. In the *Book of Rite*s, only strict teachers can be respected. It shows that teachers must improve their virtues and behaviors, and then, they will be respected socially. The people will respect them for what they teach. Hu Jintao, General Secretary of CPC, said at the symposium of national outstanding teacher representatives on August 31, 2007, that noble teachers' morality and behavior are the vivid, direct, and meaningful education mode. Therefore, it is greatly significant to strengthen the study on teachers' morality construction.

Teachers' morality is understood as the moral trait that teachers must have and the code of practice that they must follow. The moral trait could be further interpreted as love of their career and care for students, being diligent and studious, innovative and progressive, and showing indifference to fame and wealth but always aspiring for more. They will also be innovative and ambitious. Teachers will follow the following behavior standards: civil moral standards, professional moral standards of teachers, and related behavior standards formulated by the education administrative sectors and colleges and universities.

Teachers' morality construction refers to the relevant measures taken or positive activities conducted by governmental administrative sectors or university/college education management sectors to improve teachers' moral trait and better observe rules for behavior. Educational administrative organs of various governmental levels and different universities/colleges might have emphasized on different points. However, the education and promotion of teachers' morality, improved systems, assessment management, and series of activities they organize are all considered as the morality construction.

10.2 Principles to Strengthen Teachers' Morality Construction

The principle means the code or standards to say or to act. The teachers' morality construction will be made according to the following basic principles: cultivation oriented, morality education first, to improve the quality, and to develop innovatively [1].

10.2.1 The Cultivation-Oriented Principle

The cultivation-oriented principle is the core of the education development of the Party and the basic mission of colleges and universities, which is the first principle for the teachers' morality construction. The teachers' morality construction is strengthened for the cultivation. The fundamental target of such principle is to cultivate qualified talents developing morally, intellectually, and physically with the proper treatment of teachers' morality construction and cultivation with the virtue.

10.2.2 The Principle of Morality Education First

Comrade Hu Jintao expressed at the national work conference on strengthening and improving ideological and political education that the education in colleges and universities should focus on the cultivation and the morality education was the most important in the moral, intellectual, physical, and aesthetic education. In the education of colleges and universities, the moral education is the requirement to cultivate talents of high quality. The socialism education under the direction of the Communist Party of China will pay more attention on the moral education, not only educating students to learn how to behave, to act, and to learn but also to establish great Communist ambition.

10.2.3 The Principle of Improving Quality

The target of colleges and universities is to improve the quality of the students, mainly including the following three aspects: firstly to improve the quality of the administrators of colleges and universities, mainly including the planning, organization, and implementation capacity; secondly, to improve the quality of teachers, focusing on the ideological and political quality, teaching and research capacity and professional virtue; thirdly, to improve the quality of students, including ideological and political quality, science and culture quality, professional quality and physical and mental quality as well.

10.2.4 The Principle of Innovative Development

The innovative development is the requirements of the Party on the whole society and the expectancy to the education in colleges and universities, which is an important principle to strengthen the teachers' morality construction. The workers of the education in colleges and universities should consider the innovative development as the motivation of the education and creating the environment for the innovative development as the important content. The education will cultivate a quality of leading talents of the first class in the education, teaching, and research, thus promoting the deep development of the teachers' morality construction.

10.3 Analysis on Present Conditions

The Party and the government have been stressing emphasis on the teachers' morality education in colleges and universities for a long time. A series of guiding, directional and policy decisions and opinions on the teachers' morality construction in different historical stages make foundation for the better development of the teachers' morality construction. Education administrative sectors of different levels and colleges and universities also take the teachers' morality construction as the primary task and take various measures to strengthen the teachers' morality construction. On the one hand, they actively refine their rules to perfect teachers' professional moral trait and academic moral trait. Various assessments will be made on the teachers' morality, and the long-term mechanism has been established on the teachers' morality construction. On the other hand, they attach great importance to normal morality education practices. Various education administration sectors and colleges and universities stress emphasis on the key part of the teachers' morality education. As a result, the teachers' morality has been improved. Teachers are respected socially for their virtue in cultivation and devotion on the teaching. Teachers of colleges and universities enjoy an unprecedentedly academic atmosphere, obtaining great achievements in the research. A sound situation has been developed basically, where different points of view are like various flowers in bloom and more and more talents stand out in the prosperous academic circle.

However, we still have to recognize consciously the problems in the teachers' morality construction which can never be ignored: Firstly, some teachers do not care for students enough and fail to set an example for students. A survey by Tencent in 2008 showed that caring for students (22.7 %) and setting an example for students (21.4 %) is considered as one of most important basic virtue qualities of the teacher. In the great earthquake in Wenchuan, Fan Meizhong ran away from the classroom without caring for the students, which was a most concern in 2008, showing that the people expected too much from the teacher on their virtue. The teachers' morality should be much higher than the common one. A teacher should set an example for students on his speech and behavior, which will be imitated by the students. The morality can make up for the defect on the knowledge. However, the knowledge cannot make up for the defect on the morality. The influence of the loss of the teachers' morality is much larger than that of the loss of the knowledge. Secondly, some teachers work on the teaching without high positivity and sense of responsibility [2], mainly including inadequate understanding on the cultivation, unscientific course design, carelessly prepared teaching plan, no new collected data, unspecific class design, inadequate learning on teaching subject, and failure to implement trial teaching, thus resulting in insufficient preparation and impacting the teaching effect greatly. Thirdly, the teaching in the class is made without new content, flexible method, and inadequate innovation: The teaching content of some teachers is too old without any new ideas. The teaching is made with a boring method, mainly reading the book to the students; especially the teacher fails to inspire the positivity of students, resulting in poor teaching effect. Fourthly, there is no innovative research and false research achievements, mainly including that some teachers do not complete the academic research innovatively but copy other achievement, or require someone to write the thesis for them or make signature on other people's thesis, violating the professional morality standards. Some teachers practice fraud in the authentication of research achievements, project assessment, and inspection of degree authorization and evaluation in colleges and universities. Some teachers even copy other people's scientific research achievement or teaching reform achievement and change the author's name in the professional title conferring.

10.4 Measures to Strengthen Teachers' Morality Construction

Teachers are the leading force to cultivate students. We stress the emphasis on the teachers' morality construction with the purpose of keeping teachers to work without correct attitude and to improve the teaching and research capacity and to be devoted to the teaching and cultivation [3]. Therefore, the following measures shall be made, according to the author.

10.4.1 To Unify the Ideological Understanding

To strengthen the teacher's morality construction is one indispensable procedure to cultivate talents of high quality. Therefore, the colleges and universities should unify the understanding, considering it in the strategic point of view of the construction and development of colleges and universities.

Firstly, to strengthen the teachers' morality construction is basic to the education of colleges and universities. The core of the basic mission in the education of colleges and universities is to cultivate what type of talent. The talents cultivated in colleges and universities should adhere to the four fundamental principles of the Party, establish the outlook of honor and disgrace with eight honors and eight disgraces as the major content, which is the political direction of the talent cultivation in colleges and universities. Therefore, the effort should be taken in colleges and universities to strengthen the teacher's morality construction, struggling for cultivating a team of teachers qualified politically and behaving well in the morality.

Secondly, to strengthen the teachers' morality construction is an effective measure to carry out the idea of moral education first. Hu, General Secretary of CPC, said in the report of the National Congress of CPC that the education should develop, adhering to the principle of the moral education is the first in the cultivation. To strengthen the teachers' morality construction means strengthening the excellent virtue qualities of teachers to educate, move, and influence the students, realizing the target of setting an example for students. Therefore, such practice is an effective measure to carry out the principle of the moral education first.

Thirdly, to strengthen the teachers' morality construction is in urgent demand to cultivate talents of high quality. To strengthen the teachers' morality construction, the first target is to improve the moral education level of the team of teachers, making sure the cultivated talents of highly political quality. The practice shows that the teachers' high consciousness in the politics and ideology is the strong foundation of teachers devoted to the teaching, making effort in research and improving the quality. Only in this way can more and more talents of high quality be cultivated.

10.4.2 To Set Up Sound Mechanism

To strengthen the teachers' morality construction and realize the development target, colleges and universities should establish sound operation mechanism, guaranteed with strong organization, policies, and systems, making sure the scientific development of the teachers' morality construction along a healthy road.

First, efforts should be made to strengthen the leadership. Emphasizing the organizing and leading of teachers' morality construction is the primary task to establish a good mechanism. Colleges and universities should, on the basis of the actual condition of the unit, set up the leading group of the teachers' morality construction with the secretary of Committee of the Party of colleges and universities as the group leader, and vice leader of colleges and universities in charge of ideological politics as the vice leader, consisting of leaders from related business sectors and primary colleges or department. The leading group can establish different teams, including organization team, promotion team, scientific innovation team, and quality cultivation team on actual conditions.

Second, efforts should be made to perfect the system. Recovery of teachers' morality cannot be realized merely depending upon individual teacher's self development. Morality construction must have perfected rules and regulations as a guarantee. Therefore, the following systems should be formulated and improved: system to bring the teachers' morality construction into the construction of CPC, education evaluation system, focusing on teaching quality and virtue of teachers, scientific and reasonable evaluation system on the teachers' morality construction and the teachers evaluation method and index system to normalize the policy direction and to guarantee the teachers' morality construction in the system.

Third, efforts should be made on competition and stimulation. A set of methods to encourage competition and stimulation should be stipulated to reward the good and fine the bad. Thus, the positivity, activity, and innovation of teachers and students will be promoted. The effective stimulus mechanism should be established and improved, providing preferential policies to the teachers with excellent achievements in teaching and cultivation, such as to authorize major science research project, to recommend to various academic committees, academic societies, evaluation committees, to send to related colleges and universities for further education, to arrange for foreign investigation, study, and visit, and to provide the chance of professional title promotion and technical level adjustment.

10.4.3 To Create Sound Atmosphere

To strengthen the construction of teachers' team is the essential task of colleges and universities in the new historical era and also the major measure to cultivate talents of high quality.

First, diversified morality education should be carried out. 1. Ideological education should be stressed by highlighting basic theories of Marxism, particularly by highlighting the important thoughts of "three representatives" and scientific outlook on development. Teachers will establish solid communist and socialist believes, advancing with the Party under the leadership of the Party and implementing the instructions and requirements of the Party. 2. To stress the emphasis on the professional ambition education of teachers, colleges and universities should make professional ambition education on teachers in different aspects to assist them to be devoted in the teaching. The new teachers should take an oath, just like that of joining the Party. They will start the glorious mission of cultivation with high enthusiasm and selfless dedication. 3. To stress emphasis on the ideological education activities, colleges and universities should make targeted ideological education activities in proper time on the basis of the actual conditions of the unit, including targeted education with the integrity of promotion, evaluation, major festivals, and important events, making sure the teachers with clear mind and in stable mood at any time and in any conditions for the teaching and education with due diligence.

Second, training talented persons should be set as the ultimate goal. 1. A correct attitude toward work is a must. Teachers should go into the midst of their students and turn an open ear to students' voices. They should collect various materials and learn new knowledge, new theory and new technology so as to make full preparations for each lecture. All possible problems should be solved before the class, making sure the overall teaching preparation. 2. To improve the teaching capacity, such capacities mainly include teaching design capacity, class organization capacity, vivid expression capacity, capacity to inspire student to think, and organization capacity of teaching after class. 3. To guide students to make scientific innovation, teachers should make the study on the innovative education on the basis of the teaching, guiding the students studying problems arising out of the research and learning. Teachers should make innovative scientific research on the basis of the frontier problems in the subject, guiding the students participating in the innovative research. Teachers should also make academic research on the basis of the important and difficult points in the social science, guiding the students improving their academic level.

Third, virtues of indifference to fame and wealth should be advocated. Teachers with morality should deal with fame and wealth correctly. The study shows that the traditional virtue of teachers being indifferent to the fame and wealth is one of glorious and rare virtue of a teacher. How can a teacher be indifferent to the fame and wealth? On the first hand, to improve the basic qualities of teachers with different methods and to educate and guide teachers to treat the fame and wealth correctly and fulfill their task in a thoroughgoing manner. If a senior teacher works with a status of lecturer and wins wide recognition from students for his excellent performance, he is actually considered as a professor in the mind of student, indicating the teacher's correct treatment of the relation of the individual and the fame and wealth. On the other hand, the organization of the Party and leaders of different levels of colleges and universities should make scientific and orderly evaluation on the fame and wealth in the just, fair and open principle, really conferring the fame and wealth to the teachers with outstanding achievements in the education and teaching. Meanwhile, it should be forbidden for some teachers to obtain the fame and wealth with dishonest means as private relationship or bribery.

10.4.4 To Cultivate Qualified Talents

To strengthen the teachers' morality construction is intended for cultivating talents of high quality. Colleges and universities should specialize the talent cultivation target and take effective measures for cultivating talents of high quality.

First, perfecting teaching contents while focusing on the development of students' comprehensive qualities. The proportion of the moral education and humanity course will be increased. Many means and methods will be taken to strengthen the moral education and humanity education. Emphasis will be stressed on teaching new knowledge, theory, and technology with increased the hi-tech content; especially, the latest achievements in modern science and technology and social development will be introduced to the students as the most effective knowledge to learn the essence and rule of things. The integrity of professional education and general education, scientific education and humanity education, and common education and individual education will be realized with effort, to promote the transfer of knowledge to the capacity and quality, improving the overall qualities of students.

Second, perfecting teaching methodology while focusing on the enhancement of teaching efficiency. In the teaching method, attention should be paid to tendency emphasizing intelligence, teaching, knowledge, memory training, heritage, and deductive reasoning but ignoring moral education, learning, capacity cultivation, thinking training, innovation, and inductive analysis. It will adhering to the principles of integrity of teaching and learning, interaction of teaching and learning, teaching and learning promoting each other, to integrate the dominating role of teachers and subjectivity of students, intelligence education and moral education, knowledge teaching and capacity cultivation and quality improvement, sticking to effective traditional teaching method, and referring to modern teaching method.

Third, perfecting teaching management system while focusing on the reform of means of exam. Therefore, the examination standards and methods will be established to meet the requirements of the quality education, targeted at problems of focusing on memory, knowledge, and skill examination in the previous practice. The talent basic quality standards should be explored and formulated actively to meet the requirements of the social development, which is considered as the basic foundation of the examination. The examination content should be determined according to the basic quality standards, making sure more innovative content and more overall examination. The organization manner of the examination should be reformed in a further manner, making sure overall and accurate reflect of the requirements of the quality education.

10.4.5 To Create Sound Environment

Advancing the development of school ethos, teaching attitude and study style by means of creating the education environment featured by *Education first, moral education ahead* is the major initiative to strengthen teachers' morality construction.

First, advancing the development of school ethos. Institutions on various levels should foster such ethos as rigorousness, diligence, innovativeness and excellence. The school spirit of the first class is displayed in the heart and soul service for the teaching, teachers and students, devotion to the teaching and continuous innovation and fast implementation and higher work quality. The CPC committee of colleges and universities should focus on the policy guidance, considering the construction of the school spirit to be quite important. Leaders and service sectors of different levels should take effort in the planning, service for teaching and implementation, creating sound environment for teachers and students to work, study, research, and make innovation.

Second, advancing the development of teaching attitude. First-rate teaching attitude is understood as charismatic personality and influential academic competence, new educational idea and noble professional ethics, as well as skillful teaching techniques [4]. Teachers should set a good example for students in serious working, strict teaching, and cultivation. Teachers should take measures to strengthen the communication with students, to organize students to make study, to improve the teaching with advanced teaching means and method.

Third, advancing the development of study style. First-rate study style is understood as positive study attitude, explicit study objectives, innovative study method and good study results. Colleges and universities should take measures to promote the positivity of students, to organize the students for learning after class and related discussion activities. The students should take an examination with correct attitude and obey regulations of examination.

References

1. Chenning (2008) Construction of teachers' morality: analysis and construction of multiple perspectives. Capital Normal University Press
2. Tan C-B (2009) Toward the new morality—research ethics present situation and the teacher's professional moral construction. Beijing Normal University Press
3. Puweizhong (2009) University teacher's ethics construction theory and practice. Beijing Institute of Technology Press, Military Science Press, Beijing
4. Wang X (2013) Teaching style to teach arts preliminary: on the morality construction of Renmin University of China. History of the communist party of China publishing house

Chapter 11
Modeling and Analysis on Quayside Crane Loading/Unloading Based on Event Sequence Diagram

Yang Pan, Chengji Liang and Huiqiang Zheng

Abstract To analyze the features of information processing in a man-machine-environment system of quayside crane loading/unloading, a model is established and analyzed. First, a meta-operation of loading/unloading is defined, which specifies the beginning and the ending of a cycle loading or unloading a container. Then, according to the controlling operating of each step in a meta-operation, an event sequence diagram (ESD) model is built to describe information processing as the operator is regarded as an information processor. Then, a Gantt chart is drawn with the independent events of ESD model as abscissa and the input information and the output response operations as ordinate. Then, the features of input information and output response are analyzed separately.

Keywords Quayside crane (QC) · Event sequence diagram (ESD) · Load/unload · Gantt chart · Information processing

11.1 Introduction

Quayside crane (QC), as heavy quayside equipment, works with containers of dozens of tons. Therefore, the operation performance of the QC operator is quite important to the security and benefit of the ports. Operator's operation of loading/ unloading containers is a typical one person one machine manual performance.

Y. Pan (✉) · H. Zheng
College of Mechanical Engineering, Tongji University, Shanghai, China
e-mail: pany@shmtu.edu.cn

C. Liang
Scientific Research Academy, Shanghai Maritime University, Shanghai, China

Y. Pan
College of Logistic Engineering, Shanghai Maritime University, Shanghai, China

S. Long and B. S. Dhillon (eds.), *Proceedings of the 13th International Conference on Man-Machine-Environment System Engineering*, Lecture Notes in Electrical Engineering 259, DOI: 10.1007/978-3-642-38968-9_11,

The operation based on the physical rule of the system will be made in a strict time sequence. The surrounding environment of the operation is changing gradually. Thus, it is necessary for QC operator to make repeated testing and processing the information of the environment and equipment state.

With the increasing size of containers and increasing loading and unloading pressure of ports, the loading and unloading of containers is becoming a concern. Garrido [1] made study on the loading and unloading of containers with QC in the view point of the overall operation efficiency of ports with the theory of queuing. Cheng et al. [2] made fault tree analysis and modeling on QC operation. Zhong et al. [3] set up the system dynamics model with the equipments as the object to study the anti-sway control in the QC loading and unloading operation, optimizing the lifting track.

Event sequence diagram (ESD) method is a kind of graphic modeling tool to make description on correlated time series. ESD is widely used in different fields in foreign countries [4]. In the nuclear industry, ESD is used as the documents management tool and the qualitative aids of the event tree structure. Stutzke et al. [5] proposed to use ESD to improve the comprehension of operators on the accident scene. ESD used in such fields is called function event sequence diagram. ESD is used as the quantitative study tool in studies in the chemical industry and stage mission.

The loading and unloading operation procedure is taken as the main study object in the thesis. ESD is used to set up the functional ESD model to the previously defined meta-operation cycle, describing the operation procedure and scene in details. Gantt chart is drawn on the man-machine-environment input information of each independent event in ESD model and output response operation. The analysis is made on the information characteristics, information processing process, and the features of output response operation.

11.2 Modeling of Loading/Unloading Containers Based on ESD

11.2.1 Definition of Meta-Operation of QC Loading/Unloading Operation

QC operator works 12 h each shift, in which the operation is highly repeated in case of no accidents and equipment faults in reasonable truck dispatching and scientific distribution in the storage yard. Taking the unloading as an example, a container is lifted from the container vessel to the land and then is put on the truck. Figure 11.1 shows that the a spreader without load moves from Point A to Point B and moves downward to Point C to pick up the container. The spreader then will return to Point B along the same route and moves to Point A and then moves downward to Point D to land the container on the truck. The spreader will return to

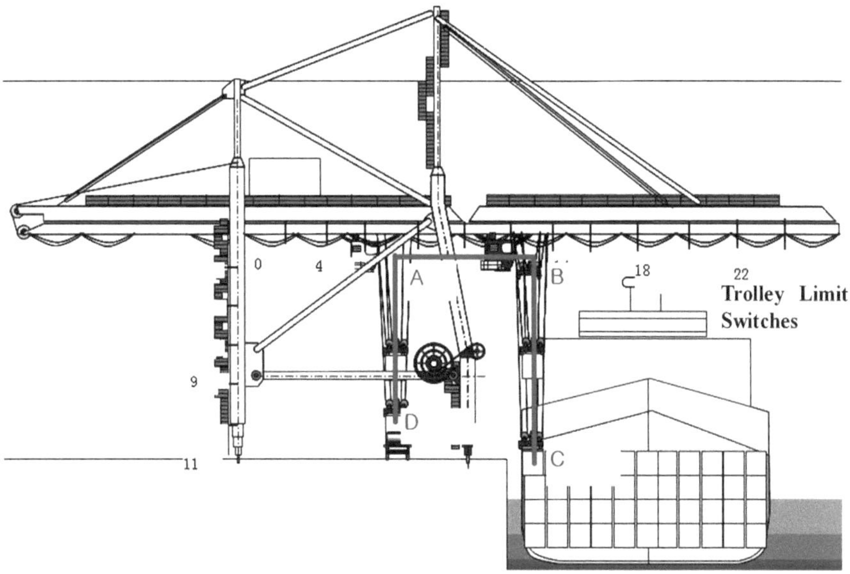

Fig. 11.1 Unloading recycling scheme

Point A along the same route for the next operation. That is: A → B→C → B→A → D→A.

QC operation: to run the trolley → to operate hoist → to confirm the choice of spreader consistent with the length of container → to aim the spreader at the container → to confirm the twistlock locked → to lift the container → to run the trolley → to move the container to the position → to aim at the truck → to confirm the landing of the container → to confirm the twistlock open. For the convenience of study on the basis of the operation data of the hoist recorded by PLC, the next time after the container landing and the twistlock opening is considered as the starting point of the next operation. The time when the container lands steadily and locking signal goes out as the end point of the operation. The lifting operation is made along the track similar to "∧". The modeling is made on such operation cycle in the thesis.

11.2.2 ESD Basic Framework

ESD is a kind of graphic modeling tool to make description on correlated time series. ESD is defined with a six-tuple array of {E, C, G, P, CB, DR}, indicating events, conditions, gates, process parameter set, conditions of boundary, and association rules.

E stands for the events, referring to all observable physical phenomena.

Table 11.1 Types of events, conditions, and gates in an ESD and their iconic representation

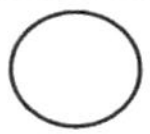	Initiating box	The first event in an ESD which initiates a sequence of events terminating in an end state
	End state	It is the terminating point of an ESD scenario. Each scenario starts with an initiating event and ends in an end state. An ESD can have multiple end states
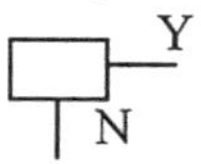	Pivotal event	An event which has two outcomes typically a "Yes" and a "No", corresponding to event occurrence and non-occurrence
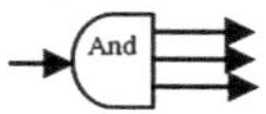	Output AND gate	Multiple concurrent independent processes can be modeled using this gate. This gate has one input and multiple outputs
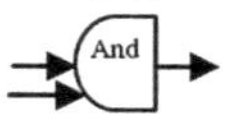	Input AND gate	Situation where scenarios cannot progress until a certain set of independent scenarios have occurred (synchronizations) can be described using the input AND gate. This gate has multiple inputs and a single output
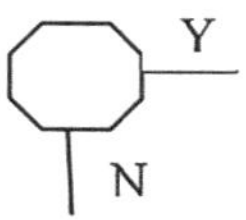	Physical variable switch	They are based on deterministic process variable limits and represent a condition. They lead two outcomes depending on whether the condition is met or not

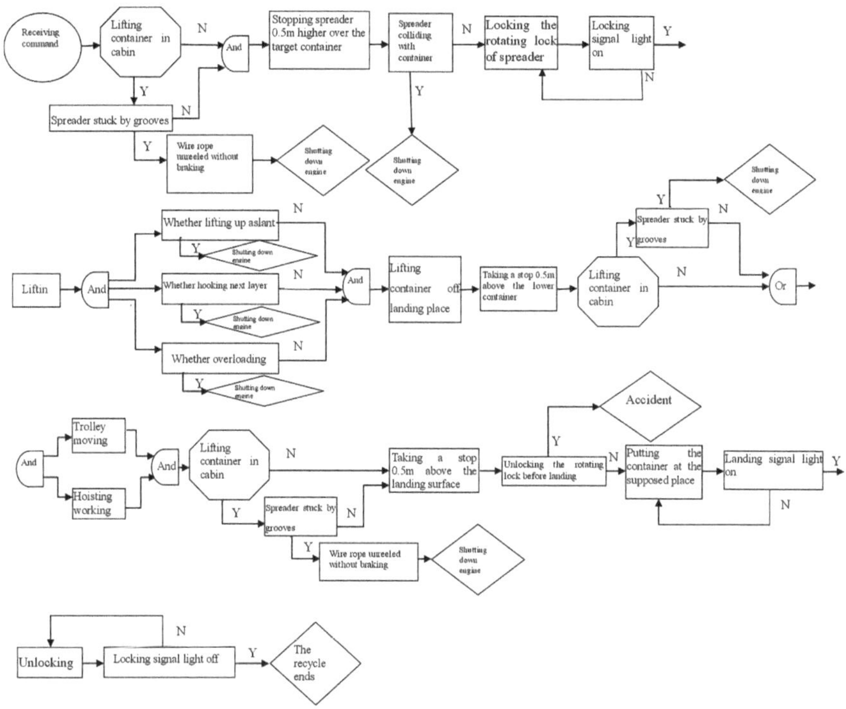

Fig. 11.2 Framework of ESD model of meta-operation

C stands for the conditions. The event scene of the system will develop in different paths on whether the condition is satisfied or not.

G stands for the gates, including AND gates and OR gates for modeling on single input multiple outputs or multiple inputs single output.

P stands for the process parameter, consisting of the events and physical process variables influencing the system.

CB stands for the constraint/boundary conditions.

DR stands for the association rules [6] (Table 11.1).

11.2.3 Modeling on Operation Decision Based on ESD

The receiving of loading/unloading command is considered as the initial event. The stage requiring the operator to make judgment or operation response in the meta-operation of the loading/unloading are considered as independent events to deduce the event sequence. ESD model is set up on meta-operation of the loading/unloading procedure with the initial event as the starting point, as indicated in Fig. 11.2.

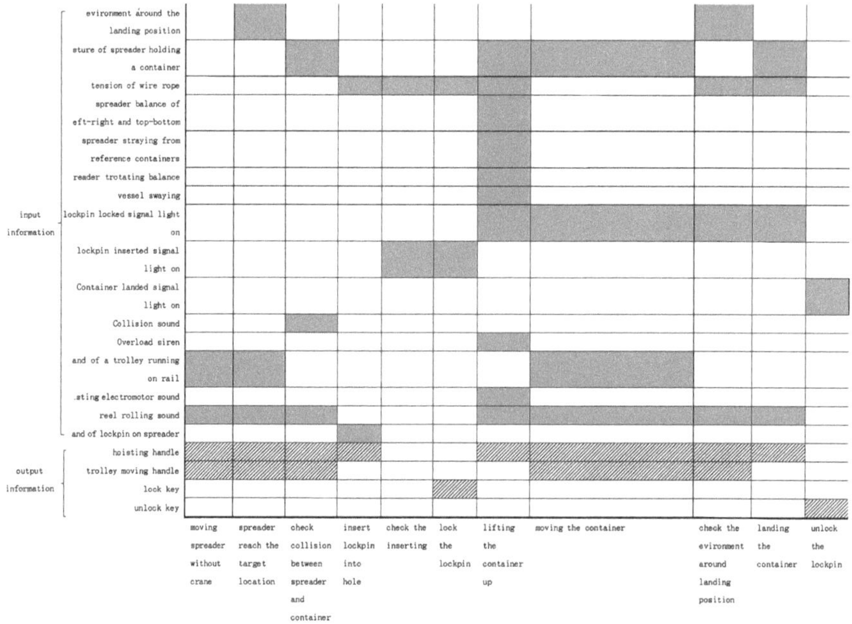

Fig. 11.3 Gantt chart of information processing of loading/unloading

11.3 Information Processing and Analysis on Loading/Unloading Operation by QC Operator

11.3.1 Gantt Chart on the Information Processing of the Loading/Unloading Operation Based on ESD

In the loading/unloading process, QC operator is considered as a processor, who can collect signals with visual and auditory organs, make identification and judgment on external environment, output correct information on the basis of

Table 11.2 Features of input information from machine and environment

Input information	Vision information	Provided from the field of view	Environment around landing position
			Posture of spreader holding a container
			Tension of wire rope
		Provided by equipment	Lock pin locked signal light on
			Lock pin inserted signal light on
			Container landed signal light on
	Auditory information	Provided by equipment	Collision sound
			Overload siren

Table 11.3 Features of response operation of human

Output operation	Executed by left hand	Controlling span	Hoisting handle
	Executed by right hand	Controlling span	Trolley moving handle
		Triggered by rotating key 90°	Lock key
			Unlock key

operation memory, and make correct operation control [7]. The analysis is made on the information detected, judged, and processed on each stage and corresponding action selection, taking independent event of ESD model as abscissa. Gantt chart is drawn on information processing of loading/unloading operation by QC operator, as indicated in Fig. 11.3. The event list and time scale are used to indicate the environment information, activity sequence, and duration of operation output.

11.3.2 Analysis on Features of Input Information Provided by Machines and Environment

In the system of man-machine-environment of loading/unloading operation, external factors affecting the decision and response selection of the operator are passed to the operator by visual and auditory organs, of which on the basis, the operator will make decision. Specific features are indicated in Table 11.2.

The auditory signal, as early warning signal, indicates the abnormal condition. The operator receives such signal passively. Therefore, it is least possible to make false alarm or miss on such information in case of the equipment is normal. The response operation is only to shut off QC as soon as possible in case of receiving such signals. The response requirement for such information is to make quick response as soon as possible.

The visual information is examined actively by the operator. The information of the equipment signal is quite clear, for which it is all right for the operator to make correct judgment. However, QC operator works high altitude with average 20–30 m high above the ground. To make correct judgment on visual field information is considered as a skill requiring the training for a long time to develop.

11.3.3 Analysis on Features of Response Operation

Information decoded by perception will be transferred into the understanding of situation with working memory. It will trigger a selection of action to realize the target response. Features of response operation in the loading and unloading operation are listed in Table 11.3.

QC operator needs to complete the vertical lifting control and horizontal translation control during the loading/unloading operation, for which the operation is considered as a system of multiple input and output.

As indicated by the abscissa in Fig. 11.3, the lifting up, checking the environment around landing position, and landing the container are stages in which more information are processed. The practice has proved that more accidents happen in those stages. Therefore, the safety operation rules are to move slowly at the starting and ending points and fast in the middle part.

As indicated by the ordinate in Fig. 11.3, handling the lifting stick and trolley stick are operations lasting for the longest time and are all range responses. The response timing and controlling accuracy depend on the capacity, experience, physical and psychological states of the operator. Optimum control does exist in such operation.

11.4 Conclusion

1. In the thesis, the definition is made on the meta-operation of the loading/unloading operation. The subject is made definite in the study on the loading/unloading operation. A simple graphic ESD model is set up to describe the process and state of the information flow in the meta-operation cycle.
2. A Gantt chart is drawn with the independent events of ESD model as abscissa and the information detected, judged, and processed by the operator and response actions in each stage as ordinate.
3. Analysis is made on features of the input information and output response operation in the man-machine-environment.

Acknowledgments This project was subsidized by National Natural Science Foundation (71071093), Shanghai Natural Science Foundation (10ZR1413300), and Shanghai Education Commission Innovation Foundation (11YZ136).

References

1. Garrido RA et al (2002) Modeling the internal transport system in a container port. Mar Transp Port Oper Marine Transp 1782:84–91
2. Chen L, Liang C (2006) Fault tree analysis on container falling by port crane operation. J Shanghai Maritime Univ 27(2):25–28
3. Zhong B (2007) Critical technique for crane's fast contraposition based on load's anti-sway control. Mechanical Electronic Engineering, Southwest Jiaotong University
4. Tan L, Wang W, Guo B (2007) Supply chain modeling and analysis based on event sequence diagram method. Ind Eng Manage 6:18–22
5. Stutzke M et al (1911) Accident sequence analysis task procedure, R.E. Ginna Nuclear Power Plant PRA Project, Procedure TQAP-2118-1.1

6. Swaminathan S et al (1999) The event sequence diagram framework for dynamic probabilistic risk assessment. Reliab Eng Syst Saf 63:73–90
7. Wickens CD, Hollands JG (1999) Engineering psychology and human performance, 3rd edn. Prentice Hall, New Jersey

Chapter 12
Study on the Relationship Between Sitting Time Tolerance and Body Pressure Distribution

Xingwei Wang, Qiran Pu and Heqing Liu

Abstract By using body pressure measure system, the pressure distribution data and tolerable sitting time of 12 subjects in three different seats were recorded. The paper analyzes the relationship between sitting time tolerance and pressure distribution index and the effect of time on sitting comfort, provides a theoretical basis for reasonably optimizing design of cushion and evaluation of sitting comfort from the view of the human engineering. Experimental results show that the softer the cushion is, the longer tolerance sitting time is. And softer cushion will improve sitting comfort. The quantitative relationship between sitting time tolerance and body pressure distribution index was determined.

Keywords Body pressure distribution · Sitting time tolerance · Sitting comfort

12.1 Introduction

The seat is quite important in the daily life, of which the fitting design cannot only support body weight reasonably but also increase the sitting comfort. Hertzberg [1] thought the comfort and discomfort were two different conscious states. The comfort means there is no discomfort. Therefore, whether the subjects have the discomfort is considered as the fundamental index to evaluate the sitting comfort of a seat. However, consideration needs to be taken on the factor of time in terms of comfort or discomfort. For example, it is comfortable to sit on a certain seat for

X. Wang (✉) · H. Liu
Institute of Aviation Medicine, 28 Fucheng Road, Haidian District, 100142 Beijing, China
e-mail: xingweiwang2013@sina.cn

Q. Pu
School of Biological Science and Medical Engineering, BUAA, 100190 Beijing, China
e-mail: pqrdoris2@yeah.net

S. Long and B. S. Dhillon (eds.), *Proceedings of the 13th International Conference on Man-Machine-Environment System Engineering*, Lecture Notes in Electrical Engineering 259, DOI: 10.1007/978-3-642-38968-9_12,

1 h while uncomfortable to sit for 2 h continuously. The judgment of the user on sitting comfort is generally obtained by the subjective survey [2]. However, such method will take too much time, and the individual difference is usually significant. For such reason, the thesis is expected to make judgment on the subjective comfort of the seat with the objective evaluation method. The body pressure distribution means the pressure distribution of the body on the contact surface with the seat. Most body weight passes to the seat surface by the bulge parts of hip/back and the muscles. The body pressure is closely connected with related parameters of the seat design and also correlates to the subjective evaluation on the discomfort. Therefore, it is popular to make prediction on the seat comfort by the pressure distribution [3]. The study on the pressure distribution has achieved greatly. In most present studies, mathematic model or neural network model about the body pressure distribution and the subjective comfort are made to evaluate the seat comfort [4], but the effect of the time factor is not considered.

In different environment, the static sitting posture is kept for quite different periods. For example, it is no more than 1 h approximately on the student seat or bus seat. However, it is much longer on the driver seat or office seats, even as much as 5–8 h. Therefore, consideration is required to take on the requirements of different seats on the static sitting time. Different seats should be designed according to different environment requirements to improve the seat comfort. In this thesis, analysis is made on the relationship between different pressure distributions and the static tolerable sitting time to evaluate the seat comfort, and the relationship is quantitatively described.

12.2 Experiment Method

12.2.1 Subjects

Twelve male volunteers from college (aged 25 ± 3.3, stature= 172.4 ± 4.1 cm, mass= 67.6 ± 8.4 kg) participated the experiment. They do not take part in any intense physical activities in 24 h ahead of the experiment and do not feel the muscle fatigue.

12.2.2 Test Content and Sitting Posture Requirements

The experiment is intended to test the pressure distribution of the hip on three different cushions. The subjects are required to sit naturally backward eudipleurally on the back of the seat with knee bending 110°, feet on the floor flatwise, two shoulders related and two hands on the lap. The specific procedures are as follows:

1. To introduce the purpose and method to the subjects before the experiment, making sure they know about the experiment to reduce the possible error due to

misunderstanding on the procedure, record their individual information, and measure their stature and weight.
2. To measure body pressure distribution data with body pressure measure system (BPMS) of Tekscan Inc., USA; to spread the pressure sensing cushion on the seat; to require the subjects to keep the sitting posture after adjustment to the most comfortable position; then record the pressure distribution for 2 min with the sampling rate of 8 f/s after the relative stability;
3. To require the subjects to remain the sitting posture until serious pains of the body or part of the body which is unbearable; then end the experiment and record the tolerable sitting time of the subjects;

The seat is equipped with three kinds of cushion of different hardness. The seat has moderately hard back and sound elasticity with the seat surface height difference (surface height—body sitting height—shoe thickness) of 0 mm and depth difference (surface depth—body seat depth) of 0 mm, the inclination angle of the seat back is 107°, and the seat surface is 0°.

12.3 Pressure Distribution Data Processing

12.3.1 Preconditioning

The edge of the pressure cushion on the seat in the experiment can unavoidably make some interference pressure values, affecting the experiment data. So, they should be removed manually. The preconditioning results are indicated in Fig. 12.1.

12.3.2 Pressure Distribution Feature Index

The physical parameter to describe the body pressure distribution on the seat must have definite physical meaning, making sure the sound relevance of the subjective and objective evaluation results. Therefore, the evaluation index should indicate

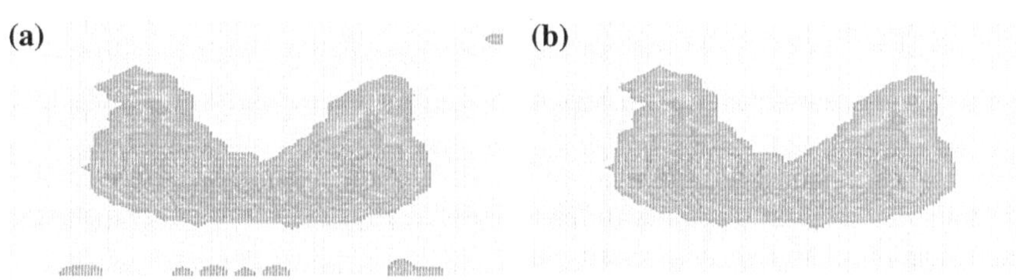

Fig. 12.1 Preconditioning results. **a** Before Preconditioning. **b** After Preconditioning

seat features in different aspects, including pattern, hardness and hardness distribution [5]. Parameters as total pressure, maximum pressure, average pressure, contact area, and maximum pressure gradient [6] are used in the thesis.

Total pressure means the total pressure on all measuring points, indicating the rigidity and pattern of the seat, greatly affected by the body weight and hip coverage;

Maximum pressure means the maximum pressure value of all measuring points, indicating the seat rigidity in the physical features of the seat. It should be at the ischium nodule in the body; otherwise, the hip pressure is distributed unreasonably.

Average pressure means the mathematic average value of the pressure of all pressure points. The average pressure is affected by the seat rigidity directly. However, the seats of the same material are quite different due to different surface pattern and pressure points.

Contact area means the total contact area between the hip and the cushion, which is related with the seat rigidity and pattern. The contact area is related with the rigidity and pattern of the seat, which is larger when the cushion covers the hip well.

Maximum pressure gradient: The pressure gradient means the changing rate of the pressure in a certain direction. The maximum pressure gradient means the maximum value of pressure gradients in all directions in a certain section area, indicating the seat rigidity and pattern. The cushion with large rigidity has a large maximum pressure gradient. The cushion of reasonable pattern has a small maximum pressure gradient, which feels more comfortable.

The statistic calculation is made on various pressure distribution indexes to obtain the relationship between them and the static tolerance sitting time, determining the effectiveness and using range of the pressure distribution index. Excel and SPSS software are used in the statistic analysis.

12.4 Experiment Results and Discussion

12.4.1 Pressure Distribution Data

The statistic analysis is made on the experiment data to obtain average value and standard deviation of each item of 12 subjects. The results are shown in Table 12.1.

The experiment errors due to repeated experiments are shown in Table 12.2. The comparison of the data in Tables 12.1 and 12.2 shows that the experiment error due to individual difference is quite larger than that of repeated experiments. It means the subjects are quite different individually, indicating the experiment results can only be used in a certain range.

Table 12.1 Pressure distribution data

Distribution index	Cushion 1	Cushion 2	Cushion 3
Total pressure (N/cm^2)	5929 ± 1309	5661.8 ± 901.4	5975 ± 937.5
Max. pressure (N/cm^2)	5.2 ± 0.73	3.4 ± 0.51	1.51 ± 0.29
Average pressure (N/cm^2)	0.60 ± 0.09	0.544 ± 0.083	0.4 ± 0.05
Contact area (cm^2)	852.4 ± 131.37	924.25 ± 96.68	1283.9 ± 102.6
Max. pressure gradient 1	3.07 ± 0.76	1.88 ± 0.35	0.809 ± 0.17
Max. pressure gradient 2	2.56 ± 0.59	1.53 ± 0.25	0.66 ± 0.17
Max. pressure gradient 3	1.84 ± 0.44	1.11 ± 0.2	0.46 ± 0.14
Time (min)	100 ± 12	132 ± 18	230 ± 16

Note ① The hardness of cushion 1, 2, and 3 is changing from the largest to the smallest. ② Max. pressure gradient 1, 2, and 3 refers to that of the area of 0.9 × 0.9 cm, 1.8 × 1.8 cm, and 3.6 × 3.6 cm

12.4.2 Pressure Distribution Data Analysis

12.4.2.1 Total Pressure

Figure 12.2 shows that the total pressure between the hip and the cushion does not change much with the hardness of the cushion changing from the largest to the smallest. The rigidity of the cushion does affect the total pressure. However, the overall total pressure is determined by the body weight.

12.4.2.2 Maximum Pressure

Figure 12.3 shows that the maximum pressure between the hip and the cushion decreases with the hardness of the cushion changing from the largest to the smallest. The harder cushion will make greater pressure on the ischium nodule in the body.

Table 12.2 Standard deviation of repeated experiment

Distribution index	Cushion 1	Cushion 2	Cushion 3
Total pressure (N/cm^2)	98.6	105.4	104.7
Max. pressure (N/cm^2)	0.16	0.18	0.15
Average pressure (N/cm^2)	0.03	0.05	0.02
Contact area (cm^2)	18.5	20.5	14.7
Max. pressure gradient 1	0.15	0.09	0.03
Max. pressure gradient 2	0.11	0.06	0.04
Max. pressure gradient 3	0.13	0.07	0.04

Note ① The hardness of cushion 1, 2, and 3 is changing from the largest to the smallest. ② All pressure indexes are standard deviation of 3 repeated experiments. ③ Max. pressure gradient 1, 2, and 3 refers to that of the area of 0.9 × 0.9 cm, 1.8 × 1.8 cm, and 3.6 × 3.6 cm

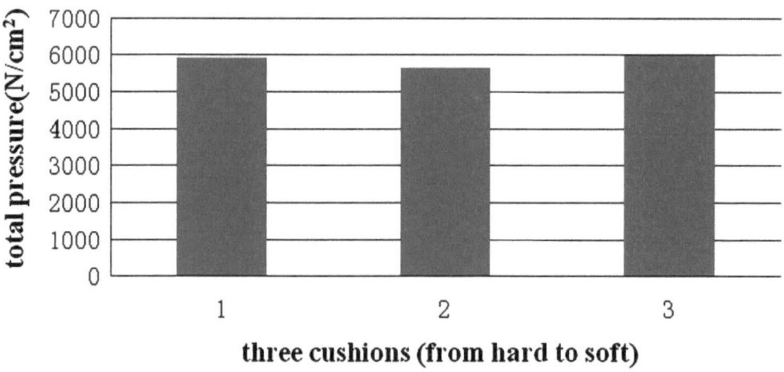

Fig. 12.2 Total pressure of three different cushions

12.4.2.3 Average Pressure

Figure 12.4 shows that the average pressure between the hip and the cushion decreases with the hardness of the cushion changing from the largest to the smallest. The softer the cushion is, the better it covers the hip. It is easier to change according to the hip pattern. The hip will have more even pressure, resulting in less average pressure.

12.4.2.4 Contact Area

Figure 12.5 shows that the contact area increases with the hardness of the cushion changing from the largest to the smallest. The softer the cushion is, the deeper the hip sinks in the cushion, resulting larger contact area with the cushion.

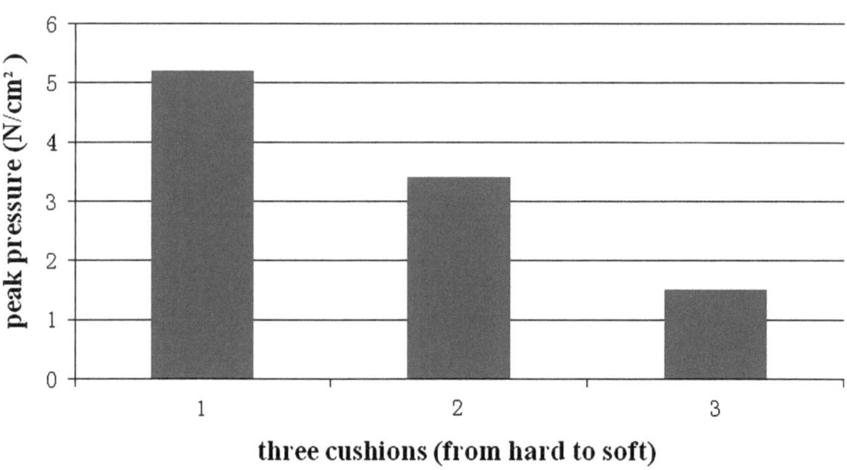

Fig. 12.3 The maximum pressure of three different cushions

12.4.2.5 Maximum Pressure Gradient

Figure 12.6 shows that the maximum pressure gradient between the hip and the cushion decreases with the hardness of the cushion changing from the largest to the smallest. Besides, the maximum pressure gradient decreases with the increasing area. The softer the cushion is, the better it covers the hip. The hip will have more even pressure, resulting in lower maximum pressure gradient.

12.4.3 Relationship Between the Pressure Distribution Data and Static Tolerable Sitting Time

Table 12.3 shows that the total pressure is in least correlation with the tolerance sitting time. The maximum pressure and average pressure are in high negative correlation with the tolerance sitting time, while the contact area is in high positive correlation with the tolerance sitting time. The maximum pressure gradients in different areas are different in the correlation with the tolerable sitting time. The data show that the maximum pressure gradient in the area of 1.8 × 1.8 cm is in strongest negative correlation with the tolerance sitting time. The maximum pressure gradient in such area is selected.

The total pressure is in little correlation with the tolerable sitting time; therefore, the total pressure is not selected to evaluate the seat comfort in the thesis. Figure 12.7 shows the maximum pressure, average pressure, contact area, and maximum pressure gradient are in linear relationship with the tolerance sitting time, which can be used to predict the corresponding tolerance sitting time of different pressure distribution.

For example, the tolerance sitting time of the cushion with the maximum pressure of 5 N/cm^2 is no longer than 100 min, that of the cushion with the maximum pressure of 3 N/cm^2 is no longer than 150 min and that of the cushion with the maximum pressure of 1.5 N/cm^2 is no longer than 240 min. The tolerance

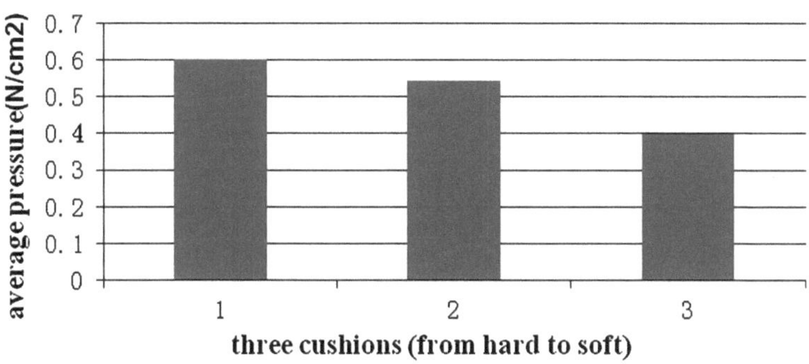

Fig. 12.4 Average pressure of three different cushions

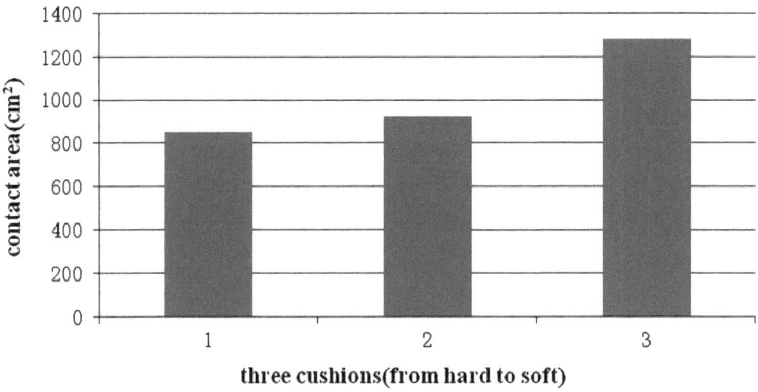

Fig. 12.5 Contact area of three different cushions

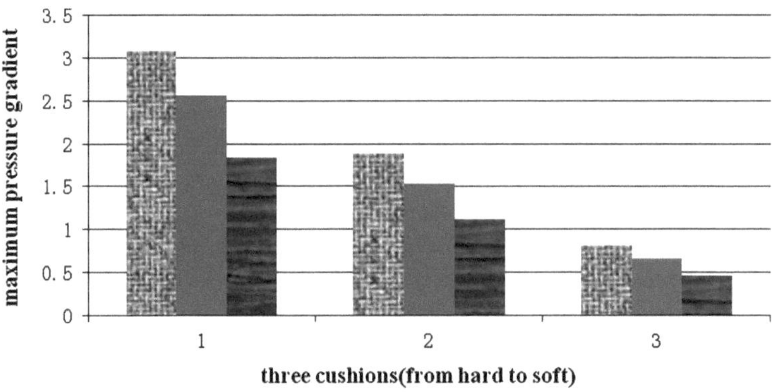

Fig. 12.6 The maximum pressure gradient of different cushions. *Note* Max. pressure gradient 1, 2, and 3 refers to that of the area of 0.9 × 0.9 cm, 1.8 × 1.8 cm, and 3.6 × 3.6 cm

Table 12.3 The correlation between tolerable sitting time and pressure distribution data

Correlation coefficient	Tolerable sitting time
Total pressure	0.063
Max. pressure	−0.904
Average pressure	−0.752
Contact area	0.882
Max. pressure gradient 1	−0.749
Max. pressure gradient 2	−0.85
Max. pressure gradient 3	−0.78

Note Max. pressure gradient 1, 2, and 3 refers to that of the area of 0.9 × 0.9 cm, 1.8 × 1.8 cm, and 3.6 × 3.6 cm

sitting time of the cushion with the average pressure of 0.6 N/cm^2 is no longer than 100 min, that of the cushion with the average pressure of 0.5 N/cm^2 is no longer than 150 min, and that of the cushion with the average pressure of 0.4 N/cm^2 is no

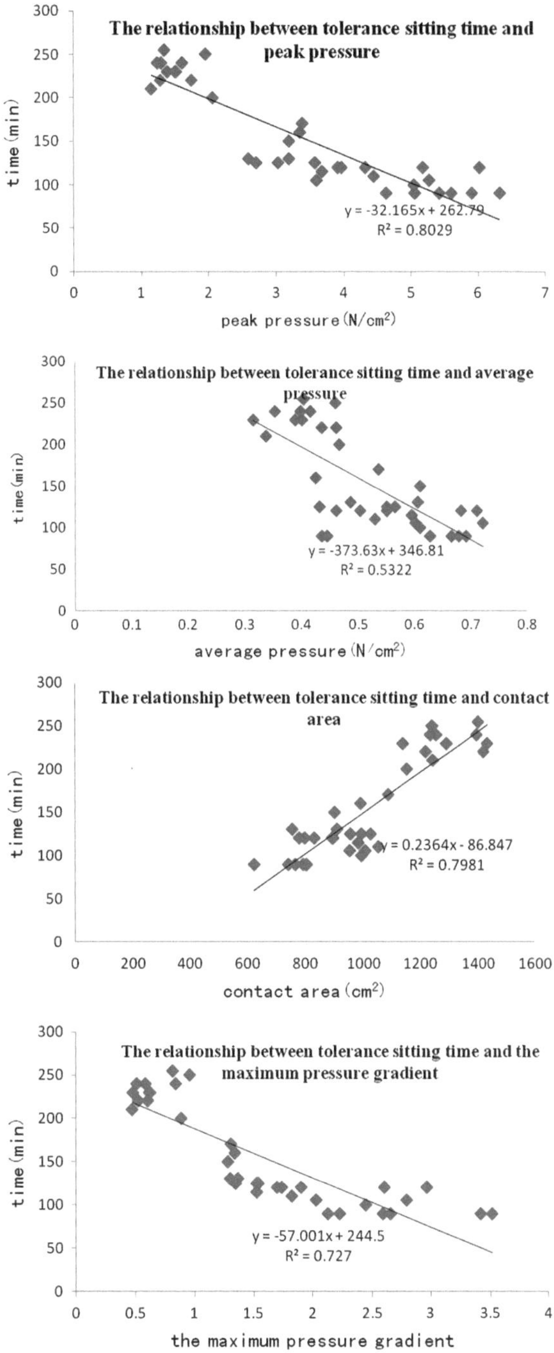

Fig. 12.7 The relationship between tolerable sitting time and pressure distribution data

longer than 240 min. The tolerance sitting time of the cushion with the maximum pressure gradient of 2.5 is no longer than 100 min, that of the cushion with the maximum pressure gradient of 1.5 is no longer than 150 min, and that of the cushion with the maximum pressure gradient of 0.5 is no longer than 240 min.

12.5 Conclusions

In the cushions selected in the thesis, the softer the cushion is, the longer the tolerable sitting time is, and the better the seat comfort is. The static seat comfort can be evaluated with the maximum pressure, average pressure, contact area, and maximum pressure gradient. The results can provide quantitative pressure distribution for design of seats with requirements of different tolerable sitting time and postures.

References

1. Hertzberg HTE (1998) Annotated bibliography of applied physical anthropology in human engineering Wright-Patterson air force base, OH Aero-Medical Library
2. Annett J (2002) Subjective rating scales in ergonomics reply. Ergonomics 45:1042–1046
3. Verver M, Hoof J (2004) A seat sensitivity study on vertical vibrations and seat pressure distributions using numerical models. SAE 1:2142
4. Kolich M (2004) Predicting automobile seat comfort using a neural network. Int J Ind Ergon 33(4):285–293
5. Zhang E, Jun H, Jian L et al (2005). Research on the biomechanical properties of the man-machine interface determination of mechanical system. Mechanical Design 22:188–190
6. Ming X, Qunsheng X (1997). Body pressure distribution index. China Mechanical Engineering, 8(1):65–68

Chapter 13
Modeling and Application Virtual Human for Ergonomics Evaluation of Armored Vehicle

Haiyan Niu, Menghong Li and Yu Yang

Abstract For the lack of evaluating means for research on ergonomics of armored vehicles in the armored vehicle design, based on anthropometry, human models were customized according to those data on Jack software. Virtual models of armored vehicle's cabins were created that based on "virtual model-unit-motion object" structure. The method of ergonomics evaluation based on virtual human changed the conventional mode of armored vehicle development. It would make armored vehicle design and ergonomics evaluation synchronously. Ergonomics problems would be found and solved before the armored vehicle had been manufactured. It would be an effective way of saving cost and shortening development period.

Keywords Armored vehicle · Virtual human · Ergonomics · Human modeling

13.1 Introduction

In the traditional ergonomic analysis, the observation is made on the real experiment person to operation the appointed equipment with observation method, measure method, experiment method, and survey method. The analysis and evaluation are made on existing problems in ergonomics and improving measures are put forward for such problems according to the principle of ergonomics. Such method is real and reliable, but it is afterward evaluation with longer period and more cost. Ergonomic engineering simulation means the analysis and evaluation on the man-machine system with methods of data processing and image display and result analysis on the basis of the integrity of the ergonomics theory and

H. Niu (✉) · M. Li · Y. Yang
Beijing Special Vehicle Institute, Beijing 100072, China
e-mail: niuzjs@126.com

S. Long and B. S. Dhillon (eds.), *Proceedings of the 13th International Conference on Man-Machine-Environment System Engineering*, Lecture Notes in Electrical Engineering 259, DOI: 10.1007/978-3-642-38968-9_13,

computer image simulation technology. Basically, the visual simulation and evaluation are made on the man-machine system in the virtual scene of the virtual human operating the virtual equipment with the 3D virtual machine replacing the real machine and the virtual human replacing the real human. The analysis and evaluation in advance can be made in the program design stage with the ergonomic simulation. The serious defect in the design can be discovered, and the improvement measure can be proposed to provide technical support for the ergonomic analysis and evaluation in the program discussion stages.

In the model reconstruction method of virtual human in the thesis, the virtual human is used in the 3D model of vehicles for ergonomic virtual evaluation, to provide the support for the ergonomic analysis and evaluation in the type argumentation and program discussion stages.

13.2 Ergonomic Virtual Human Profiles

13.2.1 Ergonomic Virtual Human

With the development of the virtual human technology, it is increasingly widely used in the ergonomic analysis, including project design, manufacture, maintenance, and training as the HCI design of the aircraft, cockpit layout, equipment maintenance analysis, and virtual training. Besides the real sense, it is more important for the virtual human in the engineering to be accurate, to meet the requirements on the operability evaluation and working space analysis. The accuracy means the human mode complying with the data standards of the anthropometry and fitting physical model of kinematics as well.

A complete virtual human model consists of many models to express the human motion, physiological structure, and action, including body geometry model, body motion model, and physiological and psychological fatigue model. The body model for the ergonomic analysis shall meet the following basic requirements:

1. The body model must comply with the accurate and effective data of the anthropometry. The data can be managed with the database system to meet the requirement on ergonomic analysis.
2. In the human modeling, the body is required to be divided into different segments reasonably according to the study requirement, and then, the simplification is made on the bone shape and joint contact surface. Meanwhile, it must guarantee the correction interaction between the body and different segments.
3. The computer stimulation and information processing functions are used to show the motion and operation of the body model effectively. It is required to have higher human–computer interaction and operation on the posture and motion of the body model. The comparison is made with the response data of the real human, to improve the ergonomic analysis capacity.

4. It is required to have function of ergonomic analysis, including viewshed analysis, accessible range analysis, collision detection, force analysis, and fatigue analysis, to provide the sound interface to the user, making sure the direct and simple operation.

13.2.2 Jack Software

Jack software is a kind of commercial software widely used in the human modeling at present. It was researched and developed by University of Pennsylvania in 1995 and commercialized by Siemens Industry Software Co., Ltd. It has main functions as follows: (1) strong joint graphic modeling system. It has a complete hierarchical data base, complete joint database, motion constraint, interference detection, and real-time motion and dynamic simulation; (2) complete human models. It has a dimension model of anthropometry based on the database, hi-fi complicated joint biomechanical model, hand joint and spine joint models, a database containing several automatic capture and handshape and walking along appointed access, head–eye coordination, and balance functions; (3) real-time browsing environment, including interactive browsing, multiple window, light and viewing angle, texture and static images; (4) complete animation system to produce animation and output documents in video format; (5) strong expansion function, including macrolanguage, dynamic modeling system, customized menu, embedded Lisp and Tcl programming system; (6) complete VR system, which can be in total immersion, supports stereoscopic glasses, fast or helmet mounted display, data glove, whole-body trackmotion (whole-body sensor) and simulation posture input equipment [3].

In Jack software system, the virtual human model and ergonomic analysis functions are two major parts, which are considered as the soul of Jack software. Jack system, as the real-time visual simulation system, can introduce CAD 3D models, create simulation environment, to introduce 3D body model with biomechanical features. The system can assign the operation for the virtual human and obtain related ergonomic analysis information by simulation analysis on the actions of the virtual human to complete the operation.

13.3 Jack Virtual Human Description Method

13.3.1 Peabody Model Structure

Jack has the strongest function to create virtual human in a virtual environment with proper biomechanical, anthropometry, and ergonomic features. Peabody model structure is used in human modeling in Jack, which is a widely used joint graphic modeling method. In the human model, the skeleton modeling is mainly

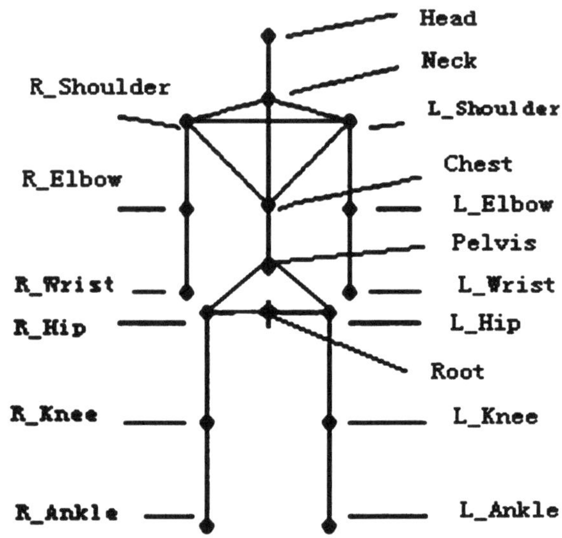

Fig. 13.1 Simplified human skeleton model

used to define the motion part of the human model. Each bone of the body can be modeled, and the relative motion relation between bones can be figured out. However, it is unnecessary to create so fine human model for most geometric analysis. Generally, the human model is created according to the body dimensions and simple joint rotating motion, as indicated in Fig. 13.1 [1, 2].

The joint graph is one with connected joints, just like simplified skeleton. In Jack system, Peabody data structure contains a model description language and interface. So, it is convenient to select and create the joint graph. Such data structure provides the information of each segment dimension and joint angle and also provides a highly effective calculation, storage, and geometry information access mechanism.

13.3.2 Jack Virtual Human Description

The body data of Jack software come from US military human survey in 1988. The human model consists of 69 parts, spine of 68 sections and 17 segments, hand of 16 segments, and shoulder/clavicle joint and 135 degree of freedom. The human model is made up of 10 major parts as head, neck, trunk, hip, upper arm, forearm, hand, thigh, crus, and foot, each of which consists of several segments. The segment is a rigid body described with the surface of polyhedron, to describe the geometry shape of a certain part of the body. The joint feature point is used to connect segments. For example, in Jack system, upper trunk consists of thoracic vertebra and lumbar vertebra, of which the former contains 12 segments from T1 to T12 and the latter contains 5 segments from L1 to L5, totaling 17 segments. The segment has three properties: pattern, feature point, and segment dimension.

1. Segment pattern

The basic pattern of the segment is the polygon surface set. The vertex coordinates of each polygon are defined with a normalized local coordinate system. Generally, X indicates the depth direction, Y the width direction, Z the length direction, of which the value range of X and Y is $[-1, 1]$ and that of Z is $[0, 1]$.

2. Feature points of segment

The feature points are distributed in the surface or the inside of the segment to mark the measuring points of anthropometry or joint connection points. Features points consist of joint feature points, feature points of anthropometry, and surface feature points. Generally, speaking features points of anthropometry and joint feature points are distributed on the segment surface in coincidence. The surface feature points are used to describe vertexes in direction of up and down, left and right, and front and back of the segment. Each segment has proximal point, a joint connection point with the previous segment and distal point, a joint connection point with the next segment. The normalized local coordinate system is relative one, so the positions of feature points of the segment are relatively fixed.

3. Segment dimension

The segment dimension determines the actual shape, described with the length, width, and depth/thickness.

13.4 Remodeling of Virtual Crew of Armored Vehicle

The virtual human for the ergonomic analysis should have necessary visual real sense. However, the accuracy of the model is particularly important. When Jack software is used for the ergonomic analysis of armored vehicles in China, it is necessary to make reconstruction of the virtual human model inbuilt in Jack software due to great difference between Chinese people and Western people. Therefore, it is a must to make remodeling with anthropometry data of armored vehicle crew to make sure the effectiveness and reliability.

13.4.1 To Generate Crew Model with Anthropometry Data

In Jack software, the measuring data are related with the dimensions of several segments, which are accumulated with the offset of these feature point pairs inside the segment on the appointed axial coordinate. In other cases, several anthropometry data are related with a segment dimension. Therefore, the mapping relation is considered to exist between the anthropometry data and segment dimensions. The calculation of 3D dimension of each part of the body with the anthropometry data can be considered to solve the linear equation.

In the human modeling, the segment dimension, as the zoom factor of the segment pattern, controls the actual shape of the segment. The coordinate of feature points will change accordingly. The anthropometry data can be used to make inverse calculation of the dimensions of parts of human model by the modeling of each segment in the local coordinate system. If the segment dimension zooms, the coordinate of features points will change accordingly. The remodeling of Jack human model is actually a zoom process, of which the coefficient is to calculate the scale of segment dimensions of the anthropometry data to those of Jack standard model [4, 5].

In calculating the zoom scaling of human model, the simple constraint relation to solve the dimensions of some segments, and then, the known segment dimensions are used to solve other dimensions. There are many constraint relations between the anthropometry data and segment dimension, between different segments, which can be divided into the following conditions generally:

1. If a segment dimension is in direct correspondence to an anthropometry data, the segment dimension is equivalent to the anthropometry data, as total head height and hand width.
2. If anthropometry data are determined with a segment dimension to be solved, the segment dimension can be solved with the known related segment dimensions, the scale between the axial distance of known segment feature point pair and the segment dimension, as the length of thigh/crus.
3. A segment dimension is in dependency with other segment dimensions. Some segment dimensions are not connected with the anthropometry data for the ergonomic analysis, so it is in dependency with other segment dimensions to guarantee the smooth connection of different segments, as hip thickness in consistence with that of the upper trunk.
4. If several segment dimensions are in correlation with multiple anthropometry data, such segment dimensions will be calculated by solving the system of linear equations, as the dimensions of three segments (neck length, upper trunk length, and hip length) are determined (body stature, shoulder height, and sitting height).

The above-mentioned method can be used to calculate dimensions of 10 parts of head, neck, trunk, hip, upper arm, forearm, hand, thigh, crus, and foot. The zooming can be made to Jack standard model on such dimensions, obtaining Nos. 1, 5, 10, 50, 90, 95, and 99 typical percentage human models based on the anthropometry data. The error of anthropometry data of Nos. 5, 50, and 95 % to created human model should not be larger than 2 %.

13.4.2 Generation Method of Human Model with the Parameter of Stature

There are as many as 26 anthropometry data closely related with the virtual human. The operation is complicated for too many parameters. To generate the human model rapidly, the stature is generally considered to be the basic parameter to measure or calculate dimensions of parts of the body. The stature can be used to derivate other measuring dimensions, and then, the segment dimensions of the human model can be obtained, thus generating human models with any stature. The above-mentioned method proves applicable by the verification with the body dimension measuring data of the armored vehicle crew.

13.4.3 To Create Human Model with Particular Percentage

Generally, the particular percentages are used in the actual ergonomic analysis. However, they are not provided directly in the measuring data of body dimensions of the crew. The anthropometry data are used to get the fitting function of the body stature to the percentage. The body stature of any percentage from 1 to 100 % can be calculated from such functions. The body stature to the typical percentage is drawn, as indicated in Fig. 13.2. Figure 13.2 shows that 7 % are not distributed evenly. The section fitting is used to construct the fitting functions of 1–10 % with the values of 1, 5, and 10 %, that of 10–90 % with the values of 10, 50, and 90 %, and that of 90–99 % with the values of 90, 95, and 99 %. The fitting function of quadratic polynomial can be created with known values of three points. The quadratic fitting function can be obtained with the above-mentioned method as follows:

$$y = \begin{cases} -0.0583x^2 + 1.2750x + 154.1833 & (1 \le x < 10) \\ 0.0001x^2 + 0.1819x + 159.2719 & (10 \le x < 90) \\ 0.0583x^2 - 10.2917x + 630.15 & (90 \le x < 99) \end{cases}$$

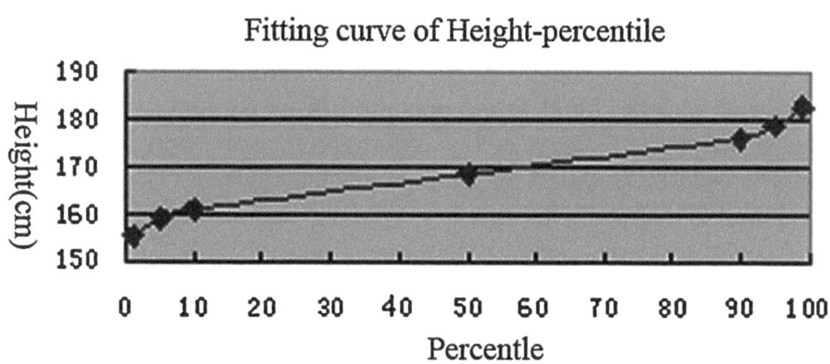

Fig. 13.2 Function *curve* of stature and percent

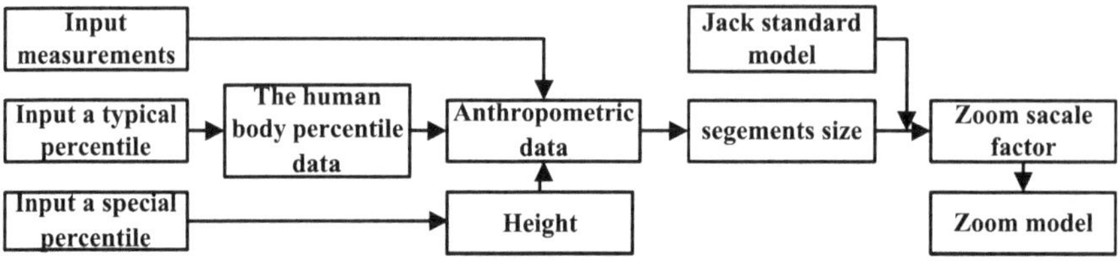

Fig. 13.3 Human body remodeling flow

The fitting function curve is drawn according to the fitting function, as indicated in Fig. 13.3. The figure shows that the fitting function of the body stature to the percentage can describe well the changing of the body stature with the percentage.

Figure 13.3 shows the brief flow chart of human modeling. The dimensions of various parts of the human model can be calculated based on a series of anthropometry data with input typical percentage, particular percentage and parameters of stature. The zooming is made on Jack standard model to get the required human model.

13.5 Application Examples

Jack ergonomic analysis tool has the following functions: view, force and torque analysis, comfort analysis, posture prediction, accessible range, fatigue and recovery, manual operation limit analysis, metabolism energy consumption, NIOSH lifting analysis, Ovako posture analysis, predefined time evaluation, quick upper arm analysis, and static force prediction. The visibility of display device in the cockpit of the armored vehicle and the reachability of the operation device are taken as examples.

13.5.1 Visibility Analysis

The display parts in the cockpit of the armored vehicle mainly include display instrument and warning light. In the operation of the system, the display instrument makes feedback of the machine information to the operator. The display instrument is used most widely in the man-machine system. The display instruction is required to guarantee the operator rapid and accurate reading, avoiding east fatigue. According to the design requirements and the visual features, the view distance and visual area (horizontal and vertical) of the installation position are taken as the analysis principle of the display instrument. The normal visual distance of the people is 460–710 mm. Considering the internal space of the cockpit of the armored vehicle, the visual distance of the display instrument shall be no less than 380 mm and not larger than 800 mm. It is best to be 560–610 mm. The instrument should be installed in the front visual field of the operator, of which the important should be no larger than 40° in the visual angle. The common instrument should be installed within the visual angle of 30°. It is better to keep the instrument at the same level of the eye, within the visual angle of 39°–41°.

The visual come model of the virtual human is made with the visibility analysis function of Jack with the visual distance of 560 mm and visual angle of 40°. Visual humans at 5, 50, and 95 % are created successively. Suppose the human keeping the normal sitting posture, turning the head to the instrument panel for horizontal view with the eyeball still. The visual cone model is used to make visibility analysis on the instrument panel in the cockpit of the armored vehicle. The range covered by the visual cone model is considered as the best visual range, which is the proper area for the display devices. Some range out of the visual cone is difficult to be observed and the display devices should be avoided within such range. Figure 13.4a shows the range that can be observed in the instrument panel. It shows that the indicator bar and instrument bar are within the best visual range of the human at 5 %. The visual cone model of the human at 50 % can only cover the indicator bar while that of the human at 95 % can only cover partial indicator bar.

13.5.2 Reachability Analysis

The operation instrument is required to be operated with the hand or foot of the virtual human, as switch, button, operating leveler, and gear shift. The operation instrument should be arranged according to the basic dimension of the operator in

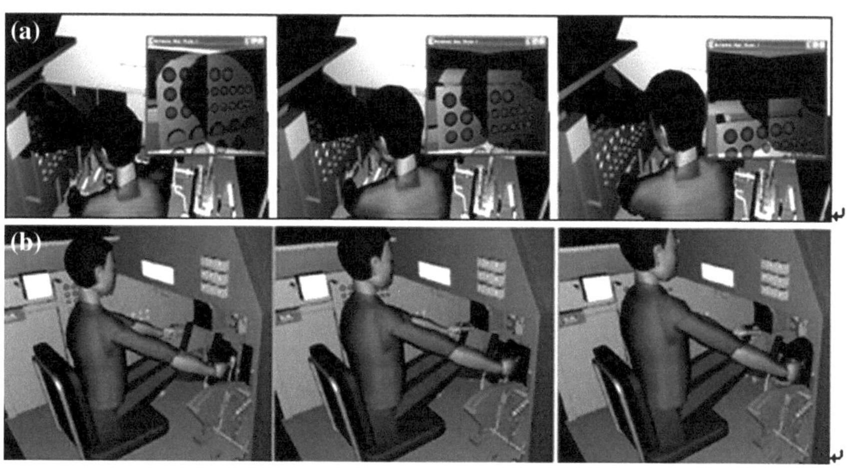

Fig. 13.4 Application. **a** P5, P50, P95 human model instrument visibility analysis sketch. **b** P5, P50, P95 human model gear shift reachability analysis sketch

the working posture and the limbs activity range of the crew. The common operation instrument should be arranged in the best operation range. For the operator of the armored vehicle, the common operation instrument includes gear shift, steering leveler, gas pedal, brake pedal, and clutch. In Fig. 13.4b, the reachability analysis tool of Jack is used to make analysis on the left steering leveler and gear shift in the cockpit of the armored vehicle, respectively.

13.6 Conclusions

The ergonomic analysis can be made in the discussion and program design stages of the armored vehicle with the established crew model in the 3D model of the vehicle by the ergonomic analysis function of Jack software, avoiding serious defect in the application process.

References

1. Norman I Badler. Virtual humans for animation, and simulation. Nonrigid and Articulated Motion Workshop, 1997. Proceedings IEEE Published. 1997:28–36
2. Norman I Badler. Cary B Phillips, Bonnie L Webber. Simulation Human: Computer Graphics, Amimation and Control. Oxford University Press. 1999, 3
3. A Bruderlin, T W Calvert, Goal-directed dynamic animation of human walking. Computer Graphics. 1999, 23(3):233–242
4. Li Yan, Wang wei, Lu xiao-jun, A method of virtual human modeling based on anthropometry. System simulation transaction, 2003, Vol.15, 210–212
5. Wang wei,Li yan, He han-gen, A scaling method of virtual human modeling based on GB anthropometry. computer simulation, 2006, 23(7), 219–222

Part II
Research on the Machine Character

Chapter 14
Car Collision Compatibility Research and Improvement Measures

Guirong Zhang

Abstract Car collision is now the traffic accident with higher incidence and more death. In the process of collision between large trucks and cars, the cars are likely to drill into the trucks in the front or in the back and be compacted on the side due to the high frame of the trucks. What's more, the quality of trucks is much larger than small cars, which will lead to the problem of compatibility.

Keywords EV · Safety · Test method

14.1 Introduction

Along with the economical development, cars not only bring convenient for people but also cause a lot of traffic accidents, posing extremely great threat to people's life and property. In vehicle collision accident, with the two cars' collision of poor quality increases, the accident's mortality rate is also rise; in the accident, two cars' mass differences exceeds more than 200 kg account for about 40 %. When two cars' quality are large difference happen collision accident, light quality vehicles obvious at a disadvantage appear the risk of death is extremely high, and this is because of the collision two cars caused the incompatibility. Therefore, the compatibility of vehicle collisions focuses on research and development and design of auto.

G. Zhang (✉)
Shandong Jiaotong University, Jinan, 250023 Shandong, China
e-mail: zgr6933@163.com

S. Long and B. S. Dhillon (eds.), *Proceedings of the 13th International Conference on Man-Machine-Environment System Engineering*, Lecture Notes in Electrical Engineering 259, DOI: 10.1007/978-3-642-38968-9_14,

14.2 Car Collision Compatibility's Concept

At present, car collision compatibility is automotive safety's new relatively concept, and in accordance with the safety, car traffic accident is divided into active and passive safety and security. The initiative safety of auto is designing the structure of the car effectively and reasonably, the initiative to prevent and avoid collision of power; while passive security refers to the collisions in cars and collision, through the body energy-absorbing structure and vehicle protection system to effectively protect the safety car occupant, items, damage and loss minimized. And car collision compatibility is passive security category, and it studies how to protect crew's safety after a car-to-car collision.

We can define car collision compatibility as follows: In car-to-car (especially in the car with different types of vehicle collision between) traffic accident, car should not only can protect themselves in-car crew member's safety, but also can protect the safety of other crew members, make the crew of the both collision vehicles under equal damage, and achieve the collision accident in personnel and property damage minimization, so we call the compatibility of the collision is good.

Instead, if a car in the collision can protect themselves very well, but the car has obvious offensive and give other occupant vehicle caused great damage, so namely the incompatibility of auto crash, this kind of circumstance should be avoided as far as possible.

The economical vehicles because of light quality trailers, cheap, are a very suitable for China's national conditions of economical family by using models. But in collision accident, this kind of car obviously unfavorable position, namely occurrence collisions are incompatible.

14.3 The Theoretical Basis of the Collision Compatibility

In the car-to-car collision process, the different mass, body stiffness, and front body structure of auto were considered influence three main factors of compatibility. When between different quality vehicles collisions, according to the formula (14.1) listed in the law of conservation of momentum that the change in velocity before vehicle collision with each other is proportional to the vehicle quality, formulas (14.2) and (14.3) marked lighter cars in the quality of the collision was bigger impact; body stiffness different may also affect car of collision energy absorption.

$$m_1 v_1 - m_2 v_2 = (m_1 + m_2) v \tag{14.1}$$

$$\Delta v_1 = v_p \frac{m_2}{m_1 + m_2} \tag{14.2}$$

$$\Delta v_2 = v_p \frac{m_1}{m_1 + m_2} \tag{14.3}$$

As diameter forcibly, the difference of body collision parts stiffness may also affect vehicles deformation and of collision energy absorption. Light vehicle usually has low stiffness and smaller contact area in two collisions, so it will cause high casualty rate by car collision as well as generate larger deformation. While the heavy vehicle has high stiffness and larger contact area, the deformation is smaller. So considering the front of the influence of stiffness in collision compatibility of body design, light quality car should be appropriately increased front and the quality of the stiffness should be appropriately reduced large vehicles; through this method, we can reduce vehicle to the other front stiffness of damage and improve vehicle collision compatibility. If either the body front bumper or the former longeron width is totally different, it can make the energy absorption thing do not play its full role. The collision between car and truck is a typical example. When collisions happen, the frame of trucks is higher, so cars are likely to bore into the truck from the front or back, causing "underride". The energy-absorbing components can not play its role and most of the energy is directly absorbed by crew cabin so that it brings tremendous damage to the crew.

In the collision between varies size of the vehicle, the mass ratio of the vehicles will determine the speed of change before and after vehicle collision, so light quality vehicles in the crash will be hurt more seriously; but the different collision space frame of two cars will also affect collision energy absorption. In addition, the different body front structure will lead to the front of between longer on happened malposition in collision that happened before the collapse is not fully longitudinal contraction deformation, which reduces the engine compartment energy source. Therefore, to realize the car in collision compatibility, at the same time to solve the three aspects of the problem is very important and difficult.

14.4 The Collision of Compatibility Safety Regulations

In 1994, European formulated to prevent vehicle bore into the truck chassis of the ECE-R93 regulations, specific provisions as shown in Fig. 14.1, in the van on the lower part of the front frame installation FUP to FUP geometrical shape of the specific provisions as follows:

1. The biggest FUP lower surface from ground clearance cannot exceed 400 mm.
2. FUP structure height of not less than 120 mm.

To ensure the performance of FUP, according to ECE-R93 requirements, we do the following test:

1. As shown in Fig. 14.1, *P*1, *P*2, and *P*3, we are, respectively, to impose 80, 160, and 80 kN loads, each applied load to maintain no less than 0.2 S.

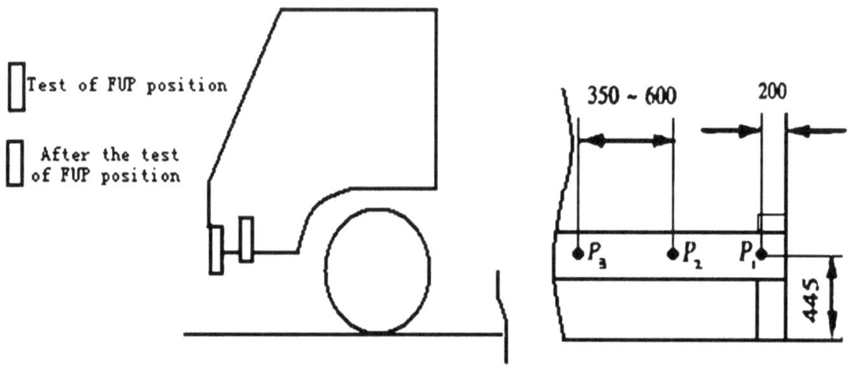

Fig. 14.1 ECE-R93 outlined map

2. The order of loading is followed behind: firstly, load 80kN on *P*1 point, then load 160kN on *P*2 point. If the structure is consistent with the cross-sectional shape does not change, then do not load on *P*3 point; if it changes, load 80kN on *P*3 point.

After the experiment is over, ECE-R93 ruling FUP performance requirements are as follows:

1. FUP front surface move back to the distance not more than 400 mm.
2. The lower part of the biggest FUP surface from ground clearance does not exceed 450 mm.

ECE-R93 regulations of FUP's main function are to prevent vehicle bore into the truck chassis below and the stiffness of protection device and collision energy absorption performance are not provided. ECE-R93 regulations have the biggest problem if FUP stiffness is too big that it will cause passenger car's deformation too large and most of the energy of the collision is absorption by the passenger cars. So, when in reference to ECE-R93 regulations FUP design, we should consider reducing as much as possible in order to improve the stiffness of the FUP truck and passenger car collision compatibility. At present, there is no domestic front collision van compatibility of the evaluation index in our country, and reference may be made to the provisions of the ECE-R93 related research and design.

14.5 The Basic Characteristics of the Safety Car Body

In car design, in order to improve the safety performance, body collision, the auto companies and research institutions have made a test to major parts of the body and the simulation analysis crash in order to ensure that the development of the new model can meet the requirements of the safety regulations collision. General safety body at least have the following three features: (1) We should always keep

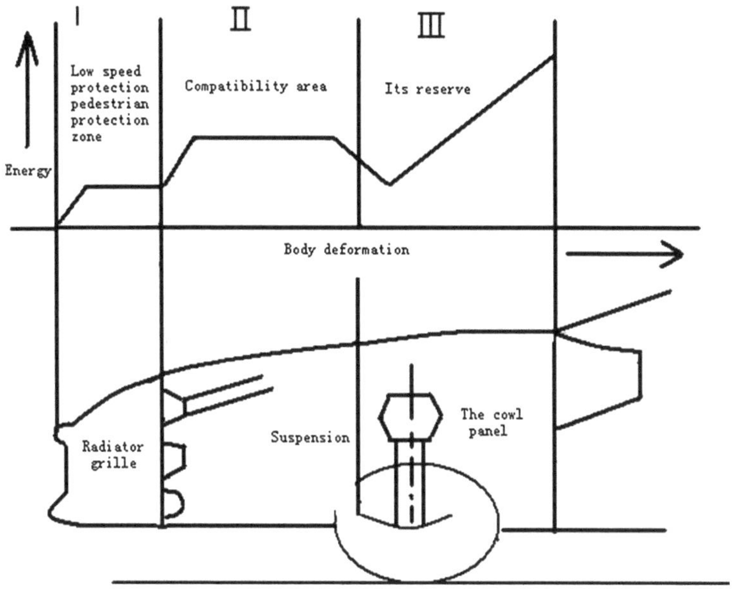

Fig. 14.2 The front ideal deformation characteristics

the room at the deformation process in their integrity and small deformation; (2) be able to ensure crew and automobile internal components of the relative velocity between have lower value; (3) car body structure should have reasonable segmentation collision characteristics. Therefore, the structure of the front car's ideal deformation characteristics is shown in Fig. 14.2. The first section is about soft segments. It can protect the safety of the weak during the collision, such as pedestrians, cyclists, motorcyclists, etc, also can avoid collision in the low speed of vehicle system safe operation of the damage. The section II is used to meet the requirements of collision compatibility; that is, the section of the design can make different quality of the two cars collided; in two cars hit, the area can create the best energy distribution and make the two cars can be effective crew protection. The section III is for self protection when collide with big vehicles or fixed quality objects.

14.6 Improve the Car Collision Compatibility of the Improvement Measures

1. A new three beam structure

To improve the car collision compatibility, the important thing is to improve energy consumption of collision area- the engine compartment. In order to absorb collision energy and reducing speed control vehicle, the traditional body structure design is often on the engine compartment two sides decorate two root keys. In this way, if the collisions of two cars' body width of overlap in smaller or because the two cars' bumper highly different that lead to dislocation; at this time, former longitudinal cannot fully happen collapse and make the other side shrinkage deformation of the cabin crew vehicle's deformation increase.

In addition, the traditional body basically is to rely on longitudinal absorb energy, so the strength of the longitudinal is more than other parts of the engine compartment to high; therefore, the strength of the components inside the engine will become is uneven. Body structure design not only can prevent front each other between the longitudinal misalignment but also can make the engine the strength of the cabin, even make not only rely on the collision absorption of energy and other components of the former longitudinal but can also assist in energy absorption, eventually improve the engine compartment of energy consumption.

New type of three-beam structure is that below beam *A*, beam *B*, and former longitudinal *C*, as shown in Fig. 14.3. In the small amount of overlapping collision, underneath beam can prevent the other vehicles to crew member of the cabin through and the upper part of the engine compartment beam can increase the energy absorption that the former is the major keys collision energy-absorbing component.

The blow beam located in the upside of the front vehicle wheel cover, extending to the front, and with former side member and side panel of beam in connected. So, when a collision of two cars to avoid the dislocation through happens, the two cars overlap smaller or bumper height of different happen.

2. Strengthening the front and the rear of a bull steel beam

Because our country are not to publish relevant laws and regulations on the installation of bumper to provisions and design at present, it makes a lot of manufacturer drill a loophole in the law and leave out the rear of a bull steel beam. After the accident happened through the survey, they found that many cars have no bull and the steel of the bull in the front is very thin so that when in car of the case tracing caudal it

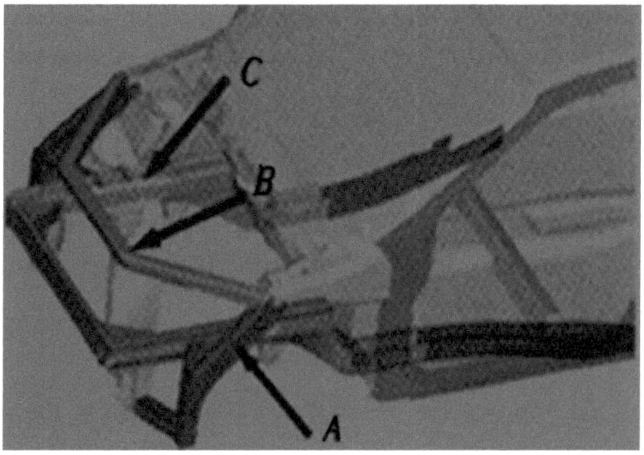

Fig. 14.3 New body structure schematic drawing

is very dangerous. In this, the manufacturers should proposed body structure design in the front and the rear of the car that also strengthen the bull steel beam.

14.7 Improve Commercial Vehicles Collision Compatibility's Improvement Measures

14.7.1 Large Parts of the Characteristics of the Car Accident

Large parts of the car accident characteristics as follows:

a. In large parts of the crew of the vehicles in the collision, accident casualty rate is very low.
b. The highest frequency of traffic accident is the crash between commercial vehicles and cars. In this kind of accident, the victims are generally the crew in the car.
c. In large parts of the car and car collision accident occurrence frequency, highest is behind collision form, but behind collision accident often only cause minor injuries.
d. Large parts of the car and the car is a head-on collision form have caused the deaths and serious injuries frequency of the highest collision form.

14.7.2 Improve Commercial Vehicles Collision Compatibility of the Improvement Measures

1. In the front down, we set reasonable stiffness and a value of protection device

In a head-on collision because of the more violent collisions and the quality of the car is less than the quality of large commercial vehicles, if large commercial vehicles are not equipped with front protection device, the deadly dived into collision will cut off the car and a cover of the death of the crew of the car, if the protection devices are installed that they will hinder the car in the collision bore into the large commercial vehicles below. But the current laws ECE-R93 only limit the protection device and stiffness of the lower limit, so if the protection device design is not reasonable, too "strong" the lower part of the protection device will bring collision compatibility, but when collision that the quality of the vehicles will be small in the vulnerable, so if we want to solve the head-on collision compatibility between the large commercial vehicles and cars that we should carry on the lower part of the protective stiffness and thorough research of the collision of the compatibility; from the point of view, the quality of the stiffness of the big vehicles should be little some that in collision as far as possible many absorption collision energy. The quality of the stiffness of the small vehicles soprano some, in order to reduce the damage degree in collision. So for large parts of the car before protection device should have reasonable stiffness value.

2. In the bottom of the protection device should have a reasonable stiffness of the values

And compared with the front protective device on the blow of large commercial vehicles, bottom protection device the current laws ECE-R58, FMVSS 203 is only designed the lower stiffness of the protective frame but it is the big rigid so protection device is able to meet the requirements of the current regulations, that is, the lower the protection device with large commercial vehicles in behind is also exist compatibility, but because behind collision is not as violent as a head-on collision, so in the behind collision accident the death ratio is not high, but in serious injury accident proportion is high by reasonable design of bottom protection device and in guarantee that stop bore into the collision and let the bottom protection device as often as possible and it can improve energy absorption between collision car and commercial vehicles to ensure the safety of the passengers.

Behind collision in our traffic accident is higher, and this may mortality and current GB11567-1994 of commercial vehicles in the bottom of the provisions of the protection device is not reasonable relevant. The regulations in 1998 in 34 item were strong inspection, but its provisions implementation after the commercial vehicles the protective ground clearance is achieved for 700 mm, so that it cannot effectively prevent most cars, motorcycles roll into behind commercial vehicles in the accident. And GB11567-1994 bottom protection device to the strength of the calculated by using the bending beam section examination modulus method of 20 cm^3 to determine, in the implementation of no maneuverability and there are also not have comprehensive assessment bottom protection device of the strength of the protective performance.

3. It should have the side of the appropriate protection, bottom protective regulations and standards

In our country, the mixed traffic mode, bicycles, motorcycles, and large commercial vehicles in suburban highway drive, so that it is very important to take necessary protective measures on the side of commercial vehicles, and this paper puts forward reasonable lateral protection that should be in the bottom of protection device, regulations, and standards. Because of these commercial vehicles, protective measures made for commercial vehicles are paid by owners and in the protection of the accident was the other's safety, so compared with the security measures the same with car, it is difficult to collect commercial vehicles or parts of the owner of the car active equipment with protection device so that it is only through these regulations and compulsory standards in the way of new factory equipment these protection device are forced to make a improvement in the purpose of passive security commercial vehicles.

14.8 Conclusion

Because do not consider the design thought of compatibility, manufacturer is: for the small vehicles, it should have front stiffness small, while for large vehicles, manufacturers often increase the stiffness of the front collision way to improve security. In the current positive bias collision test evaluation system, auto makers can only by improving the body stiffness of the measures to improve the front of the vehicle collision so that the collision of security safety competition will cause the deterioration of the collision compatibility. In the car to commercial vehicles of the collision of the phenomenon, in commercial vehicles of the lower side, the former should be based on the height of the car design and installation of drill touch these barriers devices to solve problems that rolling and drill touch tracing caudal problem. As a result of the present car collision compatibility of research methods are not existing national laws and regulations and if the regulations of impact from compatibility, as if reasonable norm production enterprise will improve the car body compatibility to meet the requirements of the regulations to reduce road traffic accidents.

References

1. Wang X, Li H etc (1997) Modern automobile safety. People's Traffic Press, Beijing
2. Zhong Z, Zhang W (2003) Auto crash security technology
3. Lei Y, Yan B, Cheng K (2004) In the collision compatibility research. Vehicle Technology (1)
4. Liu J, Li X (2008) Automotive safety and regulations. People's Traffic Press, Beijing (3)
5. Cheng Y, Zhu X (2002) Large bills car and the characteristics of the passive safety improvements. Vehicle Technology, (5)
6. Xu H, He B (2000) Road traffic accidents analysis and reappear. Officer education publishing house, Beijing

Chapter 15
Study on Methods of Electric Vehicle Safety Test

Guirong Zhang

Abstract In the process of charging and driving, EV accidents may occur, such as collision, rollover, and other dangerous situations, and those would cause traverse of power system, extrusion, short circuit, cracking, leakage, thermal shock, explosion, and combustion conditions, which may result in electric vehicle occupant's mechanical injury, electrical injury, chemical injury, burn injuries, and battery explosion injury, etc. This may lead to greater bursts of accidents and secondary damage. By the research of electric vehicle safety test methods, we can provide the establishment and improvement of safety and technical standards of electric vehicles with the necessary basis.

Keywords EV · Safety · Test method

15.1 Introduction

Electric vehicles as the twenty-first century green cars, with huge market prospects, in the future will account for a large automotive market position. With in-depth study of electric vehicles and the acceleration of the industrialization process, study of the safety of electric vehicles and the resulting number of special requirements of electric vehicles are few. Therefore, the analysis and study of electric vehicle safety, establish and improve safety and technical standards of electric vehicles through the study of electric vehicle safety test methods, will be imperative.

G. Zhang (✉)
Shandong Jiaotong University, Jinan 250023 Shandong, China
e-mail: zgr6933@163.com

S. Long and B. S. Dhillon (eds.), *Proceedings of the 13th International Conference on Man–Machine–Environment System Engineering*, Lecture Notes in Electrical Engineering 259, DOI: 10.1007/978-3-642-38968-9_15,

15.2 Potential Security Risks of Electric Vehicles

Potential security risks of Electric Vehicles are the following:

1. Electrical injury: the voltage of electric vehicles is much higher than the safety voltage that the human body can withstand; in the event of leakage, electric shock, and other accidents, the crew will be subject to more injury. o riction, extrusion, and so on are likely to make the original iAuto parts during operation of the mutual collision occurs, insulated wire insulation damage occurs, there terminals overlap with the surrounding metal. These may cause leakage and short circuit, which may cause the body to the danger of electric shock. In addition, the spark from short circuit may ignite surrounding combustibles.
2. Risk of battery internal pressure: generally, the battery's internal pressure stays at normal levels; however, when the battery is used incorrectly or when subjected to mechanical shock, the pressure of the battery may exceed the limits of the battery container which it can withstand, then an explosion occurs.
3. Risk of combustion and explosion: electric vehicles in the process charging or running for various reasons may cause burning or explosion.
4. Risk of electrolyte splash: strong acid and strong alkaline electrolyte material generally, the general condition of electric vehicle battery electrolyte leak and will not be a big splash. However, manufacturing defects could cause violent oscillation and collision of the battery cells or rupture of battery case, leading to electrolyte leakage, and splashing will be a lot to burn crew.
5. Battery heat damage on the human body: during charging and working process, the battery will generate heat, which mainly comes from three aspects: (1) heat effect of electrochemical reaction, (2) current through the ohmic resistance of the battery, (3) heat generated when the current flows through the polarization resistance. Some of the highest battery temperature can reach 500 °C, so that proper arrangement of the location of the battery and the thermal isolation between battery and the passenger can prevent heat damage caused by the battery on the occupant.

15.3 Electric Vehicle Battery Safety Experiment and Battery Safety Management System

15.3.1 Electric Vehicle Battery Safety Experiment

According to national technical safety law, the indicators of battery's safety performance are resistant to overcharge, over discharge, short circuit, high temperature, drilling test, and corrosion resistance tests.

15.3.1.1 Test for Resistance to Overcharge and Over Discharge Capacity

Burst, endangering personal safety. This is not allowed in the battery design, generally excessive accumulation of gas inside the battery can be avoid by Negative electrode Excess, and add anti-pole material to the posi Overcharge, over discharge will make the gas accumulate excessively in the sealed battery, as a result of that the pressure will increase, If it is beyond the design limit, the battery will tive electrode Implementation of anti-pole protection.

Test the battery's capacity of overcharge, over discharge, generally according to the method recommended in GB/Z18333-2001. If have the continuous charging test, that is at (20 ± 5) °C, using constant voltage, control the starting current is less than or equal to I_1 (A) currents, when a battery reaches the charge termination voltage of the first (up to 4.2 V), the battery should can automatically stop charging, continuous operation five times, charging protection device can always work well Another example is over discharge or over-charging method. Firstly, at (20 ± 5) °C, make discharge current be I_3 (A) first, when a battery reaches the discharge termination voltage (2.52 V),then use the charger that the plant provided to charge until it reaches the charge termination voltage (4.2 V). Secondly at (20 ± 5) °C discharge with the I_3 (A) until the battery voltage is zero, then the battery discharge with the I_3 (A) at (20 ± 5) °C, until a battery voltage reaches 0.5 V. According to the standards, after the above two test, the battery can not have the phenomenon of leakage, gas discharge, explosion, fire, obvious deformation and other abnormal (Fig. 15.1).

As a family car used in the winter when the outside temperature lower than 0 °C, hypothermia is a high current start car battery's basic performance requirements. Figure 15.2 shows the high-power lithium-ion battery's 8-Ah low-rate characteristics. At −10 °C in the environment for 24 h or more batteries, 5 C discharge rate, still able to release electrical energy than 3 Ah.

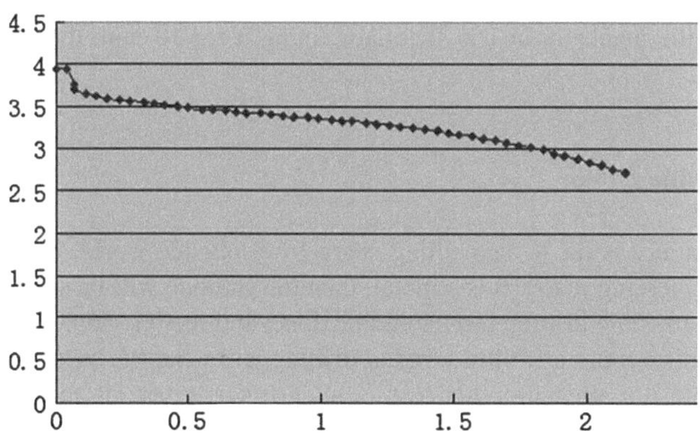

Fig. 15.1 High-power batteries 8 Ah 20 C under 50 % SOC discharge performance (> 75 % DOD)

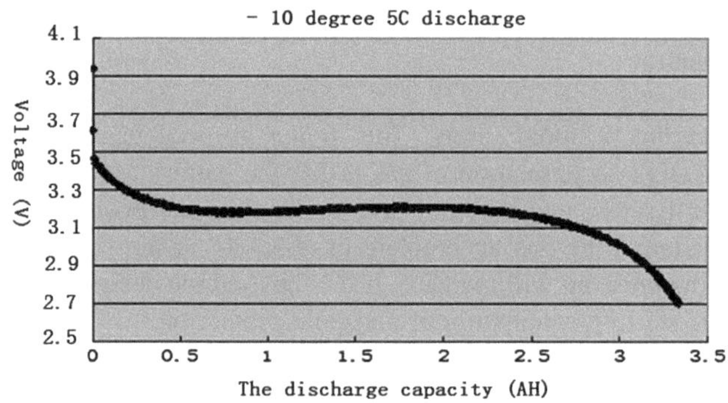

Fig. 15.2 The battery's discharge characteristics at very low temperatures

15.3.1.2 Short-Circuit Test

In the short-circuit test, the circuit may be injection, leakage, and so on. Protective measures should normally be done. Common test conditions are as follows: the battery is fully charged, at room temperature, short circuit the battery poles 1 h. Allow for leak, but the battery cannot fire or explode.

15.3.1.3 High Temperature Test

It is not allowed to throw the battery into fire, and the battery is provided with the proper storage and temperature conditions. Normally, the temperature in high-temperature test can be divided into high and low temperatures in two stages. High-temperature test is testing it in fire; low temperature test is testing it at the temperature of 100–200 °C. Low-temperature test can be done in two ways: First, the test battery is fully charged, then put in boiling water (100 °C), maintained for 2 h, and the battery should not leak and explode; Second, put a fully charged battery into the incubator at 150 °C, maintaining it for 10 min, the battery should not explode.

15.3.1.4 Drilling Test

When the battery is hit by something sharp from outside world, it may be punctured; if the piercing material is a metal, then the cathode will be short circuit, and that is dangerous. So drilling test should be done, and the bit should be conductive, and test conditions are as follows: radial drilling, allowing the battery leaks, fever, but not explosion. Because this test is a destructive test, it must have safety measures and equipment to ensure safety.

15.3.1.5 Mechanical Performance Test

Leakage, explosions, did not produce significant de Commonly mechanical Performance Test includes crash tests and vibration tests. According to the GB/T18333.1-2001, at (20 ± 5) °C, the battery was dropped from 1.0 m height to the board, and drop test was made twice in one direction to observe the battery electrolyte formation defects.

According to China's regulations, the electric road vehicles, anti-vibration test battery can be carried out by three steps:

1. Use the charger provided by the manufacturer to charge the battery with the required method, until the battery is fully charged;
2. Install the fully charged battery in the vibration test stand and make the battery discharge at I_3 (A);
3. Let the battery vibrate at the frequency of 30–35 Hz; maximum vibration acceleration is 30 m/s^2, vibrating for 2 h and no abnormal changes in the battery discharge voltage, leakage, and so on.

Simulation of the Road Vibration Testing of Lead–Acid Batteries

The surface of the battery was observed, and no electrolyte lead was found. In the test, on the good and the uneven road surface, simulation speed was 40 km/h, on cross-country road is 20 km/h, test for 30 min. From the experimental curves on the good roads, it can be seen that the battery voltage has almost no change. In the uneven road and cross-country road, the battery voltage change is 0.4 V, and the frequency of voltage change on cross-country road would higher than the uneven road. During the test, the temperature rise in the battery cannot be felt. Obskage, as for the single cell test, and good ventilation conditions, consider the actual situation is far more complex simulation conditions, and other confounding factors, lead–acid battery electric vehicles on the uneven road and cross-country road, the battery voltage will have a greater Fluctuations, poor ventilation, and battery working long hours and the temperature will inevitably rise, which will affect the electric vehicle's reliability and safety.

Crash Test for Lead–Acid Batteries

The test was a simulation of electric vehicle collision case, to test battery system outlooks and the voltage changes, with reference to SAE J1766: 1998 electric vehicles and hybrid electric vehicle battery system and the overall impact test procedures recommended by relevant national standards. Installed the battery in the jumbo to determine the impact speed of jumbo 50 km/h.

15.3.1.6 Test for Resistance to Corrosion

The test generally contains electrochemical corrosion tests which are salt spray test method. In the test, the battery is exposed to the test box, and then the spray test solution is injected into the box. Fine mist fell on the battery surface evenly. Test solution is 5 % salt solution, which does not exceed the total solid content of 20 mg; pH value of 6.5–7.2 in the test temperature should be kept. Battery should stay in the salt spray box for 48 h; after the test, battery capacity should be significantly reduced, the shell should not have many metal parts rust, may not have rust pitting holes; leakage and explosion are not allowed.

15.3.2 Battery Management System

In the use of electric vehicle batteries, the battery should work at a reasonable voltage, current, and temperature Therefore, the use of electric vehicle batteries needs to be managed effectively. It is particularly important for the nickel-metal hydride batteries and lithium-ion batteries. Poor management not only may significantly shorten the battery life but also may cause fire and other serious security incidents. Battery pack safety management means that when the battery pack's voltage, current, and temperature are out of security state, to give timely warning and emergency treatment like short circuit and so on.

15.3.2.1 Battery Heat Management

Battery pack heat management is working to ensure that the battery pack is at the appropriate temperature range of the entire system, including battery boxes, fans, heat transfer media, monitoring equipment, and other components. Battery pack heat management is working to ensure that the battery pack is at the appropriate temperature range of the entire system, including battery boxes, fans, heat transfer media, monitoring equipment, and other components. Battery pack thermal management system's main functions: battery temperature measurement and control accuracy; when the battery temperature is too high for effective cooling and ventilation; when the low temperature, rapid heating, to ensure that the battery pack to work properly; harmful gases that were in effect Ventilation; to ensure uniform distribution of temperature, the battery pack.

15.3.2.2 Battery State Test

Battery management system should arm and make the appropriate management in the following situations:

1. The total voltage of the battery: when the battery voltage less than 200 V display voltage. When the battery is charging, stop charging if the total voltage is higher than 280 V.
2. Single voltage: battery voltage to determine monomer is divided into two parts
   a. Cell battery voltage range of the normal work is [10, 14 V]; beyond that, voltage range needs to make the appropriate prompt.
   b. Battery voltage difference between the cells. If the voltage difference is more than 0.5 V prompt, pointed out the cell with error voltage.
3. Current judgments: discharge current is too large, the battery pack and electric vehicles are very dangerous. The size of the discharge current determined by the load and the scope of safe discharge from the battery pack may be safe discharge current. On overcurrent, the battery management system will directly trigger the relay, cutting off the connection with car batteries to ensure safety.
4. Remaining capacity judgment: lead–acid battery SOC is less than 30 %, and timely reminder to the driver battery is low.
5. Temperature judgment: this mainly refers to the battery surface temperature. Operating temperature range of different battery is not the same. Lead–acid battery operating temperature is −40 to 60 °C. When the temperature is too low to alert the driver, the battery cannot work. Battery temperature beyond the normal range usually indicate the battery is bad.

State test is carried out at any time; accurate and timely detection of a direct impact on the state of the battery pack and the safety of electric vehicles is very important.

15.4 Safety Tests of the Electric Vehicle

15.4.1 Water Resistance Test

15.4.1.1 Simulation of Cleaning Test

Simulation of the normal cleaning conditions of electric vehicles do not include the use of high-pressure water washing and the special underbody cleaning, which should be provided in detail by vehicle manufacturers in the conditions of these special cleaning in the user manuals. What simulation of cleaning test involved the safety of electric vehicles is the boundary line of hazardous area, that is the seal between the two parts. For example, the trap can open parts of the glass ring, the first column of the border, and sealing ring of lights. Using hose nozzle and clean water in the test, water to all of the boundary line in all possible directions, which flow should be in rate of 12.5 L/min, 0.1 m/s speed and the distance of nozzle and the boundary line is 3 m.

15.4.1.2 Simulation of Rainfall Experiment

The experiment is simulated as follows: when parts leading to passenger compartment, cargo compartment and motor compartment are in the condition of opening, suddenly heavy rain (example: thunderstorm) began to fall. If the B-level voltage equipment is shielded and exposure to water is avoided, then independent test can replace vehicle test by equivalent.

What the important regional of electric car in the simulation of rainfall experiment are those accessible and opened parts that can be opened.

Experimental use of artificial rainfall in 1PX3 nozzle GB4208-1993 (Fig. 15.3): use clean water, flow 10 l/min, water on the open part of all surfaces as far as through the nozzle regular movement, time is 5 min.

15.4.1.3 Simulation of Wading Experiment

This test is to simulate electric car through the flood situation in the streets or puddles. Car should be in 10-cm-deep pool, to 20 km/h driving speed of 500 m, approximately, 1.5 min. If the pool length is less than 500 m, it needs several times and the total time (including time outside the pool) should be less than 10 min.

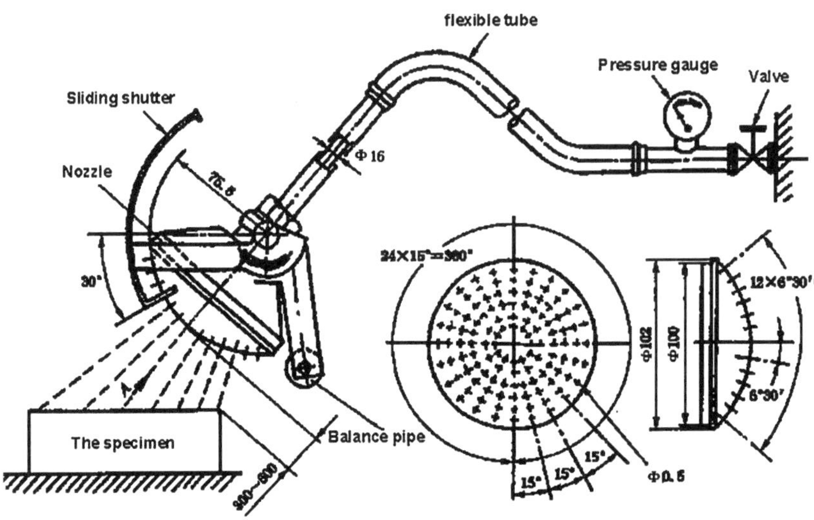

Fig. 15.3 Simulated rain test nozzle

15.4.2 Collision Experiment

For the vehicles in which burn occurs when the battery explosion when squeezed, impacted, and punctured, alternative tests should be carried out first. Use something that has the same size and weight as the battery to replace the battery. To determine whether the battery will be squeezed in the collision, the collision acceleration to withstand puncture and its size. Standards are in accordance with CMVDR294 regulations. Evaluation: battery runout, the extent of damage to the battery, the electrolyte leakage of harmful gas emissions, vehicle deformation, and the damage indicators of the crew.

15.4.3 Overcharge Over Discharge Test

Put electric vehicle at room temperature (20 ± 5 °C) environment, overcharge: constant current mode to charge single battery until the voltage is 5 V; over-discharge: the discharge of constant current mode to single battery until the voltage is 0 V. Evaluation: the hydrogen concentration, electrolyte leakage, whether it causes accidents, and the accident caused.

15.4.4 Vehicle Insulation Test

There should be an insulation resistance between electric vehicle's electric chassis and each circuit and other circuits The test vehicle should be at (5 ± 2) °C for 8 h before the next stage; it should be at (23 ± 5) °C; humidity 90 %; pressure of 86–106 kPa in the following measure time for 8 h.

Between the following live part using appropriate measuring instruments (for example megohm meter) to measure the resistance: (1) electric power systems and vehicle electrical chassis; (2) power systems and auxiliary circuits.

15.5 Conclusions

Electric vehicles should be in accordance with the simulated vehicle washing, heavy rains, and the situation when wading to have safety test safety performance of electric vehicles in the insulation of these dangerous conditions is directly related to electric vehicle driving safety, occupant safety, and may lead to greater security incidents, and personal injury. Therefore, in China, during the process of research and development and the promotion of electric vehicles, the safety performance of electric vehicles in dangerous conditions of the test and the ability to test and the test conditions are also the important base construction projects and basic tasks.

References

1. Cao L, Yang J (2008) Safety analysis of electric vehicles. Hunan University, Changsha
2. Zhu Z (2007) Electric vehicles. Yellow River Conservancy Press, China, 3
3. Zhang C, Wu Z (2008) Electric vehicle safety issues. Electric Vehicle Technology (4), 4
4. Xie Q, Zhong Z, Zhang W (2008) Diamond-shaped electric vehicle crash safety simulation. Hunan University, Changsha
5. Hu Y, Huang M (2010) Characteristics of multi-functional electric car collision. Wuhan University of Technology, Wuhan
6. National Standards Commission (2001) Electric vehicle safety part 1. GB/T 18384.1 GB/T 18384.1-2001
7. National Standards Commission (2001) Electric vehicle safety part 2. GB/T 18384.1 GB/T 18384.2-2001
8. National Standards Commission (2001) Electric vehicle safety part 3. GB/T 18384.1 GB/T 18384.3-2001
9. National Standards Commission (2001) Nickel metal hydride batteries for electric vehicles. GB/T 18332.1-2001

Chapter 16
Improvement on Jiangxi Digital Seismological Network

Cuixiang Zhong

Abstract Scientific research shows that a tectonic earthquake is generally caused by underground water erosion or volcanic eruption instead of inter-continental plate collision. According to this new law, it is easy to explain many past great earthquakes around the world and predict some future possible earthquake foci by observing landforms. These research results are important for economic protection against earthquakes and reducing disasters. Especially, according to this new discovery, many digital seismograph networks in the world can be improved. Through on-the-spot investigation or observing topographical maps, one can find that many areas in Jiangxi Province of China lack underground water, and the possibility of devastating earthquakes occurring in these areas is vey low. Hence, the observation on these areas can be lessened. But, some areas such as Jiujiang, Poyang, and Huichang are seismic risk areas, and the observation on these areas should be enhanced.

Keywords Tectonic earthquake · Root cause · Landform · Topographical maps · Digital seismological network

16.1 Introduction

Earthquakes are the most common and dreaded natural disasters. However, before most earthquakes really occurred, earthquake monitoring institutions could not predict their occurrences [1]. In fact, due to the complexity of earthquake prediction, seismologists have not been able to accurately predict the occurrences of earthquakes, and they have not even given thorough explanation for the cause of

C. Zhong (✉)
Software College, Jiangxi Normal University, Nanchang 330022, China
e-mail: cuixiang_zhong@sohu.com

S. Long and B. S. Dhillon (eds.), *Proceedings of the 13th International Conference on Man-Machine-Environment System Engineering*, Lecture Notes in Electrical Engineering 259, DOI: 10.1007/978-3-642-38968-9_16,

earthquakes [2]. As for some great earthquakes around the world, although some experts have given tentative explanations for their causes, their opinions vary greatly [3, 4]. Since the existing explanations for the cause of earthquakes are not thorough enough to convince other people or to be used to predict earthquake foci, the authors of this chapter, based on existing research results, made a scientific analysis on the geographical environments and regional geological tectonic features of some notable earthquakes over the world, revealed the root cause of tectonic earthquakes, and discovered the law of tectonic earthquake occurrence—that is, a tectonic earthquake is generally caused by underground water erosion instead of inter-continental plate collision. According to this new law, it is easy to explain many past great earthquakes around the world and predict some future possible earthquake foci by observing landforms, that is, great earthquakes occur easily in coastal areas or areas retaining large amounts of water, but not in areas lacking water. So, these research results are important for economic protection against earthquakes and reducing disasters. Especially, according to this new explanation, many digital seismograph networks in the world can be improved. China Digital Seismological Observation System and one of its subnets—Jiangxi digital seismological network—can also be improved.

In order to understand how to optimize a digital seismograph network, we should first study the root cause of earthquakes and then use the law of earthquake occurrence to guide the improvement in digital seismograph networks.

16.2 Root Cause of Earthquakes

Earthquakes are classified as naturally occurring earthquakes and artificial earthquakes [2]. Naturally occurring earthquakes can be further divided into three categories: (1) tectonic earthquakes; (2) volcanic earthquakes; and (3) impacting earthquake (the impact of large meteorites on the ground). Most of the world's earthquakes are tectonic earthquakes, accounting for 90 % of the total number of earthquakes, and most of these are shallow earthquakes, which affect a wide area, cause great destruction to ground and buildings, and usually cause a great loss of life and property. Hence, the cause of tectonic earthquakes and the earthquake focus mechanisms have been studied more deeply. In fact, in the last century, scientists had already recognized that general earthquakes are closely related to the construction processes causing widespread deformation of earth surface. These deformations also created mountains, rift valleys, ocean ridges, and sea trenches. Geologists inferred that massive fault dislocation or abrupt rock collision is the cause of violent earthquakes. Their inference immediately became a credible assumption, so the cause of most earthquakes was discovered [1]. However, what is the natural force that makes the rocks beneath the earth's surface break, dislocate, or subside abruptly? Geologists have not clearly given the answer. In fact, a colorless, odorless, tasteless, transparent, permeable, changeable, and erosive agent—water—is the real force that makes the earth's surface soften, break, subside, dislocate, or collide. Under the constant erosion of water, some parts of the Earth's lithosphere become more and more thin, and some places of these parts were even worn out by stagnant water, making water enter the mantle and finally causing volcanic earthquakes.

According to the above law, we can see that a big river, lake, or ocean with constant flowing water or stagnant water must make a wide area of earth's surface soften or break. Especially under the high pressure of big and high mountains or ocean ridges, it is easy for the earth's surface to subside, dislocate, or collide. This is the reason why earthquakes occur easily in costal areas or inland basin. In addition, big and high mountains that can catch rain clouds help generate heavy rainfall, which results in mountains' softening, limestone caves, landslides, and mountain rocks' rupture or collapse to cause earthquakes. This is the reason why earthquakes also occur at the foot of big and high inland mountains, where inland basin retains large amounts of water constantly. According to this explanation, it is easy to explain many world earthquakes such as the 1906 and 1989 San Francisco earthquakes [5] in the USA, the 1976 Tangshan earthquake [6], 2008 Wenchuan earthquake [7], and 2005 Jiujiang earthquake [8] in China.

On November 26, 2005, a moderate earthquake known as the Jiujiang earthquake (Supplementary Fig. 16.1) that was measured at 5.7 M [9] and occurred between Jiujiang and Ruichang in China's Jiangxi province, killing dozens of people and damaging many houses. According to a survey from Earthquake Administration of China [8], the chief cause of this earthquake was the difference between the Jiuling uplift and the Poyanghu depression. It is thus clear that they have not given the root cause of the earthquake.

In order to find out the root cause of the earthquake, one should analyze the geographical environment and tectonic settings of this earthquake. In fact, the epicenter was between Jiujiang and Ruichang, which is the lowest lying place in Jiangxi Province. It is located near the Yangtze River and is covered by many other rivers or lakes such as Poyang Lake, which is the biggest freshwater lake in China. Hence, the area is rich in surface water and groundwater. These waters erode and soften the earth's surface in this area severely and even penetrate the rock layer to cause volcanic earthquake, especially under the great pressure produced by the nearby high mountains such as the Jiuling Mountains, making the earth's surface break or cave in easily and finally causing the earthquake. But,

Fig. 16.1 Deep pools in field produced by Jiujiang-Ruichang earthquake

since the mountains near Jiujiang and Ruichang are lower than Wenchuan's mountains and the water surrounding Jiujiang and Ruichang is less than that in seas, the Jiujiang earthquake was only a medium earthquake.

According to the above research on some typical earthquakes, we can conclude that a tectonic earthquake is generally not caused by any collision between two continental plates, but usually results from a local movement of the earth's crust. Usually due to mountain bodies, ocean ridges, or the earth's surface suffering from long-lasting underground erosion by water or violent impact of volcanic eruption, the earth's crust will finally break, subside, or slip to cause earthquakes. Since coastal areas are rich in water, the earth's crust in these areas is eroded more severely by water, so earthquakes occur more frequently in these areas. Inland areas join tightly with continental plate and are eroded less severely by underground water than coastal areas, so earthquakes occur less frequently or severely in these areas. In an area with many high mountains catching much vapor and rain, the waters flowing down from the high mountains may form big rivers, and if the river valleys at the foot of mountains retain large amounts of water for a long time, the river valleys may form a strong earthquake belt. The places far from big rivers or short of water have fewer or no earthquakes. Hence, we can predict some future possible earthquake foci from the general configuration of the earth's surface. If we focus our attention on these places to monitor seismic activity, we can effectively predict earthquakes and reduce the damage of earthquake disasters.

16.3 The Present Condition and Improvement in Jiangxi Digital Seismological Network

Supported by the central and local governments, China Earthquake Administration began to establish "China Digital Seismological Observation System" [4] in 1996. According to the principle of uniform distribution of seismic stations and at the same time ensuring intensive observation in some key administrative areas, the observation system was designed to consist national and regional monitoring, as well as mobile seismograph networks [10]. The system began operation at the end of 2000 and was basically completed at the end of 2007. The National Digital Seismograph Network (CDSN) is equipped with 152 seismic stations (Supplementary Fig. 16.2). All of the seismic stations perform 24-bit data acquisition, and the waveform data are synchronously transmitted to the Center of National Digital Seismograph Network via a satellite network. There are altogether 31 regional digital seismograph networks in China, which have 685 digital seismic stations performing 16-bit data acquisition. The waveform data are synchronously transmitted to the centers of local seismograph networks. The Mobile Digital Seismograph Network has 800 portable digital seismographs, which are of exactly the same type as those in the regional digital seismograph networks. Between 1999 and 2001, the Capital-Circle Digital Seismograph Network (covering Beijing

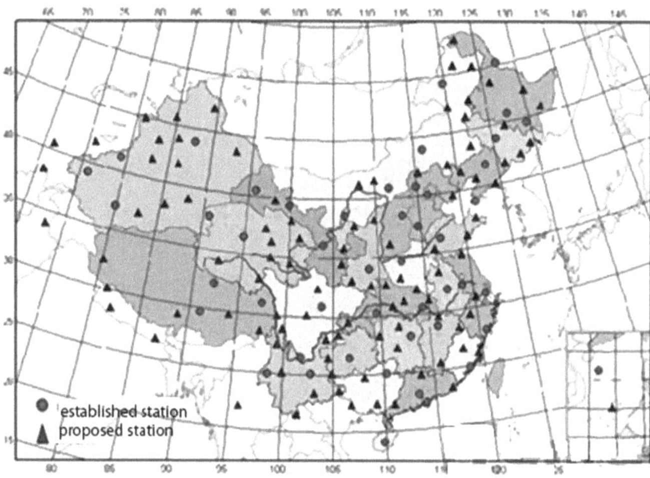

Fig. 16.2 Distribution of CDSN

Municipality, Tianjin Municipality, and Hebei Province) for real-time data transmission was established, which has 107 seismic stations performing 24-bit data acquisition. The national, regional, and Capital-Circle Digital Seismograph Networks have gone into full operation since 2002 and have yielded a great deal of observational data. The seismological observation system in China has experienced rapid development since 2003: CEA has accomplished analog-to-digital conversion of all seismological networks, thus, full digitization of seismological observation in China has been realized.

An important part of the China Digital Seismological Observation System is 31 regional digital seismograph networks, which are mainly used to monitor the areas of frequent seismicity, key economic regions of high background seismicity or population accumulation areas of high background seismicity. These regional digital seismograph networks were also designed according to the principle of uniform distribution of seismic stations. For example, Jiangxi digital seismological network completed in 2007 consists of 24 substations and 1 network center. These substations are, respectively, located at Jiujiang city, Duchang county, Yongxiu county, Xiushui county, Jingdezhen city, Shangrao city, Yugan county, Jinxian county, Yichun city, Gaoan city, Fengcheng city, Nancheng county, Le'an county, Ji'an city, Wan'an County, Jinggangshan city, Ganzhou city, Huichang county, Shicheng county, Dayu county, Anyuan county, Xunwu county, and Longnan county. The network center is located at Nanchang city. The aperture of the network is 500 km long from east to west, and 600 km wide from south to north. The distribution and data transmission of the network are shown in Fig. 16.3. All substations are equipped with EDAS-24IP type seismic data acquisition device, except Huichang substation and Shangrao substation equipped with EDAS-24L6 type seismic data acquisition device. In addition, Huichang national substation is equipped with a JCZ-1 type super-broadband seismograph, the other three national

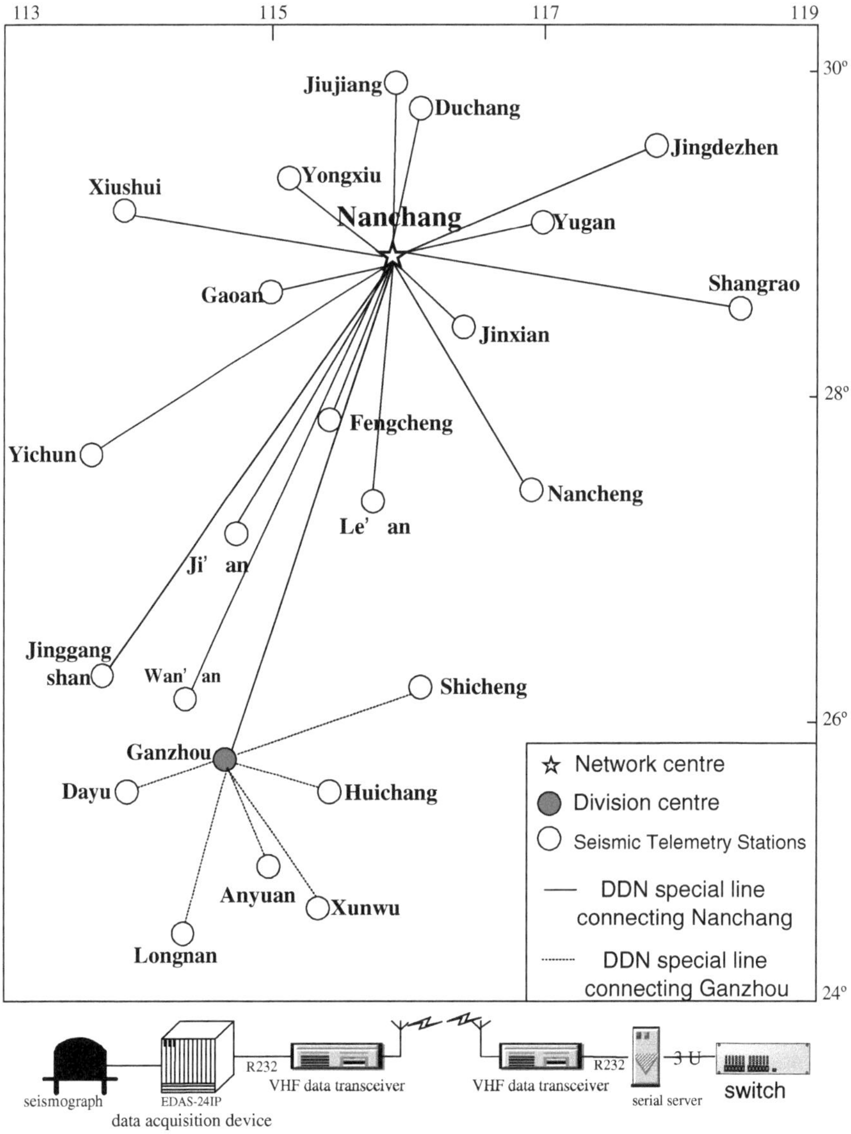

Fig. 16.3 Distribution of Jiangxi seismological network

substations located at Nanchang, Jiujiang, and Shangrao are equipped with CTS-1 type broadband seismographs, four other substations are equipped with BBVS-60 type broadband seismographs, and 16 substations are equipped with KS2000 M type broadband seismographs. All seismic data acquisition devices' sampling frequency was set to 100 Hz, and they all use SDH firber optic link to convey information to the network centre of the provincial seismic telemetry station. The network centre uses JOPENS, a data processing system for seismic station network centre, to collect, store, transfer, and process the seismic observation data. Since the data processing system receives multiple links' data simultaneously, the efficiency of this system is usually low. Hence, it is urgent to improve this system.

In fact, according to the principle of tectonic earthquake occurrence, we can optimize Jiangxi digital seismological network to improve its processing efficiency of seismic observation data. Through on-the-spot investigation or observing topographical maps, we can find that the terrain of Jiangxi Province slopes from south to north, forming a big basin called Poyang Lake in the northern part of Jiangxi. Therefore, the rain or water received by this land can flow down into the basin and then enter Yangtze River. So, most areas of Jiangxi Province do not retain large amounts of water, except some basins in Huichang, Jiujiang, or Shangrao, even Poyang Lake would dry in winter. Therefore, many areas, such as Longnan, Xunwu, Anyuan, Dayu, Shicheng, Jinggangshan, Ji'an, Le'an, Fengcheng, Yichun, Gaoan, Jingdezhen, and Xiushui, lack underground water, and the possibility of devastating earthquakes occurring in these areas is very low. Hence, the observation on these areas can be lessened, for example, reducing the acquisition, storage, and processing of seismic data from these areas. But, some areas such as Jiujiang, Poyang, and Huichang retaining large amounts of water constantly are seismic risk area, and the observation on these areas should be enhanced.

16.4 Conclusions

A scientific analysis on the geographical environments and regional geological tectonic features of some notable earthquakes over the world revealed the root cause of tectonic earthquakes—that is, a tectonic earthquake is generally caused by long-term erosion of underground water or volcanic impact instead of intercontinental plate collision. According to this law, it is easy to explain many past great earthquakes around the world and predict some future possible earthquake foci by observing landforms or topographical maps. The place covered with larger bodies of water such as large rivers, lakes, or seas has a possibility of earthquake occurrence, and the place retaining no much water is seismic safe area. Through on-the-spot investigation or observing topographical maps, we can find that many areas in Jiangxi province lack underground water, and the possibility of devastating earthquakes occurring in these areas is very low. Hence, the observation on these areas can be lessened. But, some areas such as Jiujiang, Poyang, and Huichang have some basins retaining large amounts of water constantly; these places are seismic risk areas, and the observation on these areas should be enhanced.

Generally, a national or international digital seismograph network designed according to the principle of uniform distribution of seismic stations can be improved similarly. This research can be developed into a new theory called macroseismology.

References

1. Bolt BA (1993) Earthquakes and geological discovery, scientific American library. W.H. Freeman, New York
2. Geller RJ, Kagan YY (1997) Earthquake cannot be predicted. Science 275:1616–1617
3. Wang Y et al (2011) The genetic mechanism of Wenchuan earthquake. J Mt Sci 8:336–344
4. Parker RN et al (2011) Mass wasting triggered by the 2008 Wenchuan earthquake is greater than orogenic growth. Nat Geosci 4:449–452
5. Wald DJ, Kanamori H, Helmberger DV, Heaton TH (1993) Source study of the 1906 San Francisco earthquake. Bull Seismol Soc Am 83(4):981–1019
6. Liu Q et al (2007) Seismogenic tectonic environment of 1976 great Tangshan earthquake: results given by dense seismic array observations. Earth Sci Front 14(6):205–213
7. Yang S et al (2006) A review of the research on complex erosion by wind and water. J Geog Sci 16(2):231–241
8. Dun W et al (2007) On seismogenic and its background tectonics of Jiujiang-Ruichang seismic sequence. J Geodesy Geodyn 27(6)
9. Li CY, Zeng XF, Zhang JX (2008) The tectonic settings and seismogenic tectonics of the M5.7 Jiujiang earthquake in 2005, Jiangxi Province, China. Sci Chin (Ser D:Earth Sci) 5:640–653
10. Liu R et al (2008) Construction and development of China digital seismological observation network. Acta Seismol Sinica 21(5):535–541

Chapter 17
Electronic Control Technologies for Safety of Modern Automobiles and Their Development Trend

Qing Gao

Abstract With the rapid development of electronic information technology, more and more electronic information technology was introduced to car, and the intelligent safety devices and electronic control technology became the key to the modern car safety technology. This chapter describes a variety of modern car safety electronic control technology and its development trends. The development of advanced vehicle safety technology contains the prevention of security technology, accident prevention techniques to reduce harmful levels of conflict between the techniques, and post-crash technology to prevent the expansion of the four aspects of disaster. The road to complete the automatic identification, automatic driving, and automatic detection of vehicles running set all kinds of security control in a car, making the car a highly intelligent safety car, and safety car is the future direction of development.

Keywords Security · Electronic control technology · Safety car

17.1 Vehicle Safety and Social Environment

In developed countries on car transport, car safety hazard has become a serious threat to social development issues. As automotive safety electronics technology used in the car is on the gradual introduction, coupled with enforcement of traffic laws, the accident rates of view tended to decrease. In some developed countries, such as the United States, Britain, France, Germany, and Japan, automobile safety electronics technology are widely applied in cars, as well as that intoxicated driving is severely forbidden, and vehicle crews are obligated to wear safety belts; the overall traffic accident rates has shown a decreasing trend [1].

Q. Gao (✉)
Shandong Jiaotong University, Jinan, China
e-mail: qing917@126.com

S. Long and B. S. Dhillon (eds.), *Proceedings of the 13th International Conference on Man–Machine–Environment System Engineering*, Lecture Notes in Electrical Engineering 259, DOI: 10.1007/978-3-642-38968-9_17,

17.2 The Type of Car Electronic Control Technology

Based on before and after traffic accidents, vehicle safety is classified into two kinds as active safety and passive safety. Active safety is taken in the traffic safety measures before they occur, especially when in the imminent danger, the driver manipulates the steering wheel to avoid or emergency braking to avoid traffic accidents. Passive safety is to minimize the extent of directly affected to passengers, and to ensure the safety of pedestrians when a car accident occurs. When the accident occurred, the safety devices and systems, such as electronic control airbag system and seat belts prevent the expansion of disasters, including the prevention of fire and the crew can expand rapidly from a car accident.

17.3 Development of Advanced Safety Vehicle

Advanced safety vehicle (ASV) is the application of sensor in the car to check around the car or the road traffic environment, the application of information and communication processing technology to prevent of traffic accidents, or to develop the automatic control or automatic steering technology of decreasing collision damage extent. Electronic technology in automotive applications leads to the development of the vehicle to a higher intelligence, manipulated by the driver of the car to "man-machine–electric", the integration of structure, in order to prevent two aspects as security (active safety) and the conflict safety (passive safety) for further improving vehicle safety.

The development of ASV safety technology can be divided into four major areas: (1) preventive safety technology; (2) accident prevention techniques; (3) degree of conflict of harm reduction techniques; and (4) prevent disasters after the crash to expand technology [2].

These systems set preventing accidents as the main purpose. For example, drowsy driving and other alarm systems, not only issue a warning to motorists when risk increases, but also use automatic braking and automatic steering automatically to prevent accidents.

17.4 Car Safety Electronic Control Technology

17.4.1 Anti-lock Brake System Electronic Control

Electronically controlled anti-lock brake system (ABS) can ensure the wheels to brake with the state of slipping and rolling to obtain the desired braking effect, shorten the braking distance, and brake to keep the vehicle driving safety.

The wheel that has a rate range of 10–30 can obtain the maximum braking force. ABS systems installed in the wheel by wheel speed sensors to detect wheel speed signals and send the signals to the ABS ECU, ECU for operational analysis to determine the wheel slip ratio, and control the brake pressure regulator (actuator) continue to boost and maintain blood pressure, so that the wheel slip rate remained at 10–30 % range, to prevent the wheels from locking, so get the best braking effect [3].

17.4.2 Electronic Controlled Anti-skid Drive System

Electronically controlled anti-skid drive system is also known as traction control (TRC). It can adjust the wheel braking force and at the same time control the engine power output to prevent the vehicle at the start, acceleration, or when driving on slippery road from driving wheel skidding. ABS is the implementation of anti-lock brakes to prevent wheel slip ratio becoming too high and increase the braking effect; TRC is a non-slip transfer control, transferring roller to prevent the high rate of car and drive traffic, and increase lateral stability; it is only designed for drive wheels.

Electronic-controlled anti-skid driving system monitors driving speed with wheel-mounted speed sensor and sends the speed signal to TRC's ECU. When the ECU calculates the slip rate of more than 30 %, which controls the TRC, hydraulic actuator is used to gain the appropriate wheel brake. Meanwhile, ECU also controls the auxiliary throttle, so that appropriate auxiliary throttle is closed, thereby reducing the engine's power output, so that slip rates are kept in the best range.

17.4.3 Electronic Controlled Airbag System (SRS)

In order to reduce the driver's injury caused by large inertial force from car head-on collision, the modern car equipped with air bags in the steering wheel center and some cars in front of the co-driver panel airbags are also installed. During a head-on collision accident, when the impact force (collision deceleration), which is detected by the collision sensor of the airbag controlling system, exceeds set value, the airbag ECU immediately switches on the electronic detonator circuit, enclosed by gas-forming agent in the inflatable element, thus sets off ignition media in the detonator, which ignites flame powder and gas-generating agent, and the airbag is inflated and expands rapidly within 0.03 seconds, quickly forming a buffer bag in the midst of the passenger and vehicle makeup [4]. Damping by throttling the exhaust air bags absorb the kinetic energy of the crew, so that the second impact—fierce front passenger head and chest on the steering wheel and windshield of the second impact be mitigated, in order to achieve the purpose of protecting the crew.

17.4.4 Cruise Control System

Automobile cruise control system (CCS) is essentially an automatic constant speed controlling system to maintain the constant speed required by the driver after the cruise switch is turned on and there is no need to step on the gas pedal to automatically maintain the speed of the vehicle traveling at constant speed [5]. Using cruise control when driving on the highway for a long time, the driver would not go to control the throttle pedal, reducing fatigue, thus improving the safety of car driving. While also reducing the unnecessary speed changes, it can save fuel. Meanwhile, the CCS also ensures the driver's operating priority; when the vehicle speed exceeds the predefined speed range and other conditions, cruise control will automatically stop working in order to ensure traffic safety.

17.4.5 Electronic Controlled Active Suspension System

Ride comfort, handling and stability, and so on through the proposed vehicle suspension system are more complex and often have conflicting requirements; these requirements also change with the vehicle driving conditions and road conditions. The active suspension is based on driving conditions, turning signals from the steering sensor, speed signals detected by the vehicle speed sensor, height control sensors to detect the change in vehicle height in ride to determine the road conditions, throttle position sensor that detects the vehicle speed, or acceleration brake switch detection brake signal. These signals are sent to the suspension ECU, suspension ECU according to suspension height control actuators and control valves, at any time of the suspension stiffness, shock absorber damping force, and the body to adjust the height and position so that the car ride comfort, stability handling and ride comfort are always at its best.

17.4.6 Electronic Control Power Steering System

In order to improve the light steering, as well as the safety of high-speed driving, modern cars use electronically controlled hydraulic power steering system. Under different driving speed, the speed signal from the speed sensor, through the power steering ECU controls the solenoid valve, and the driver can have a different steering feel, so as to achieve the purpose of driving safety. The electronic speed control device controls the oil pressure of the hydraulic oil pressure reactor based on the speed of the car to let the driver feel different steering force.

When turning at low speed or parking at the time of low or zero speed, ECU received a week speed signal, that is, to transport a high current to solenoid valve, solenoid valve is opened, hydraulic fluid in the reaction chamber through the solenoid valve flows back to tank. Therefore, there is a low pressure in hydraulic reaction chamber, this time to give the driver a smaller steering feeling. At this point, the driver simply puts a smaller steering wheel force.

When the car is at high speed, ECU receives the high-speed signals in the speed sensor, ECU decreases current that is delivered to the solenoid valve, the solenoid valve opening is reduced to closure, and there is a large fluid pressure in the hydraulic piston chamber. The hydraulic reaction force is transmitted to the steering wheel, and the driver requires a larger force to turn the steering wheel to alert the driver to concentrate to improve driving safety.

17.4.7 Electronic Control Rearview Mirror

Electronic exterior mirror control system can operate mirror switch controlled by the ECU to adjust the exterior mirrors to upper, lower, left, and right positions. The adjusted position of the mirror can be stored by operating the driving position reset and store switch and the tilt and telescopic ECU and adjust back to the original position when need.

Mirror switch's various position signals are sent to the exterior mirror ECU, and the exterior mirror ECU controls the corresponding motor work after receiving switch position signals, then running the motor drive the mirror up and down, or turning left and right. Up and down or left and right rotating rearview mirror positions can be sent as feedback to the exterior mirror ECU by the position sensor installed in the motor to monitor the work of the mirror motor.

17.4.8 Fatigue Driver Warning System

The technology is by monitoring the driver's drowsiness level (decreased consciousness), not only the to make the appropriate drivers aware of their sleepy state, but also to stop the car automatically when the driver cannot drive due to sleepiness to reduce traffic accidents. The technology features include the following two parts:

Functionality of detecting level of driver's drowsiness: there are two ways to detect driver's drowsiness: the first is the detection of vehicle operating conditions; and the second is to detect the driver's physical condition. The first method uses the steering wheel angle sensor, video camera, yawn frequency detectors, and other instruments. The second method is used to monitor the activities of the driver's eye by pulse sensors, cameras, and other similar sensors.

17.4.8.1 Driver's Physical Condition Detection

Drowsy driving prevention features: computer deals with signals that are detected by various drowsy sensors, and when the results exceed the normal threshold, it can be determined that the driver is in a sleepy state [6]. In addition, ways such as displaying some of the display information, alert sound vibrating driver seat, or releasing mint flavor so that the driver had regained conscious are also used. If the warnings are ineffective, steering system will automatically control the direction and open the hazard warning lights to warn passing vehicles, while the car automatically brakes and the vehicle stops.

17.4.8.2 Detecting Vehicle Operating Conditions

The technology of detecting vehicle operating conditions is the operational status of equipment to monitor traffic through the route, and the computer can identify the tag on driveway [7]. If the sleepy driver is about to put the car in the ditch, detect equipment will send signals to the computer, and computer-controlled alarms will sound to wake up the driver.

17.5 The Trend of Safety Control Technology Development

In recent years, as the rapid development of electronic information technology, more and more electronic information technologies are introduced to car industry, and car safety devices are becoming intelligent, with higher automation level and reliability, as well as ever-improving performances. Microcomputer pens the door to the car for quickly setting up multiple computers. In the future, a car will be equipped with several computers that can automatically identify the road, automatically drive and detect running vehicles, and set all kinds of security control in a car, so that a highly intelligent vehicle safety in car can be achieved.

References

1. Qian Y, Hu N (2008) Modern automotive safety technology. Shanghai Jiaotong University Press, Shanghai
2. Wang X, Li H (2001) Modern car safety. China Communications Press, Beijing
3. Zhang GR (2003) Automotive positive numerical analysis of the security. Jilin University, Changchun
4. Liu J, Zhou MW, Zhao H (1997) Vehicle crash safety protection structure. Southwest Jiaotong University, Jiaotong, (6)

5. Kay Melbourne MM, Wolf JA (1987) Chen inspirational translation. Modern motor structure. China Communications Press, Beijing
6. Li X (2005) Automotive safety and regulations. China Communications Press, Beijing
7. Zhengbao L, Zhong Z (1999) Automotive passive safety in the study of problems and solutions. Hunan University, Changsha, (1)

Chapter 18
Construction of the Special Vehicle Dynamic Simulation System Based on Virtools

Xue Shi, Sijuan Zheng, Zhongliang Wei, Fang Xie, Liang Ling and Qiufang Wang

Abstract In this paper, the dynamic simulation system for special vehicles is established rapidly on the basis of the vehicle dynamic study with the combined mode of PRO/E, 3DS MAX, and Virtools, realizing the highly reusing of the model. Engineering design and simulation was carried out at the same time, so the efficiency was improved and the design cycle was shortened. Some main contents for this thesis are as the follows: (1) imports data from PRO/E; (2) creates the 3D virtual scene with 3DSMAX; (3) processes and interacts on this model with Virtools Dev. It is showed by practice that it can make the developments of virtual simulation system more simplification, and can improve the efficiency, the facility, and the interaction, etc. At last, a summary and a prospect are made in this thesis.

Keywords Special vehicle · Virtual reality (VR) · Building block · Virtools

18.1 Introduction

Virtual reality (VR) is a kind of advanced human–computer interface technique to create 3D simulation environment based on perspective information with the assistance of the computer. The user can enter the environment to make action control, realizing the interaction of the user and the virtual environment. At present, VR is widely used in military simulation, view simulation, virtual tour, aircraft and automobile manufacture, and science visualization [1].

With the development of the technology, many techniques and software can be used to construct the virtual environment, including 3DS MAX, VRML, and VEGA. However, the tool to create virtual environment on the text based on

X. Shi (✉) · S. Zheng · Z. Wei · F. Xie · L. Ling · Q. Wang
China North Vehicle Research Institute, Beijing, China
e-mail: xuezi911@qq.com

S. Long and B. S. Dhillon (eds.), *Proceedings of the 13th International Conference on Man-Machine-Environment System Engineering*, Lecture Notes in Electrical Engineering 259, DOI: 10.1007/978-3-642-38968-9_18,

VRML cannot provide visual environment. It is necessary for the user to have a certain background knowledge. The scene creator must create the complicated scene from the basic geometric figures provided by VRML. The user must be familiar to Java language and VRML animation interaction programming. For those reasons, it is more difficult to use actually and it is hard to realize the purpose [2]. Therefore, the thesis presents a development method based on 3D and Virtools, where PRO/E software is used to create 3D vehicle model, 3DS MAX is used to construct application scene in 3D virtual environment, Virtools technique is used as the platform to transfer the 3D model into a simulation platform for overall view, and real-time interaction controlled with the keyboard and mouse. Such method with reality and interaction is simple and easy to use, which is widely used in the vehicle design.

18.2 Virtools Dev Software

Virtools Dev real-time 3D editing software can be used by the developer to make integrity of 3D model and building blocks (BB) visually. The developer can assign the BB to the proper object with Drag & Drop and can decide the processing order of the BBs with the flowing chart and finally create a complete interactive virtual environment gradually.

BB is the visual expression of the software functions with an input and output at least. Generally, it has the function of parameter input and output and it may also have a target parameter to determine the object affected by the action.

Virtools Physics Pack for Dev integrates top physical property engines of Havok, making it convenient for the user of Virtools to create 3D interactive scene. Virtools Physics Pack contains 29 new BBs, providing to the user various physical properties as gravity, friction, elasticity, physical limit between objects, buoyancy, field of force, and dynamic physical property of the vehicle. It greatly shortens the creating time of the user and reduces complicated and lengthy object dynamic creation process of engineers and algorithm writing of the program designers.

The vehicle system based on physical dynamics can be created easily with Physics car system. Some basic dynamic concepts are required to be introduced generally.

18.3 Basic Vehicle Dynamic Concepts

The vehicle running on the ground has two major resistance forces, namely the air resistance $R1$ and wheel rotating resistance force $R2$. The air resistance formula is $R1 = (1/2)\rho\, V^2 S_p C_d$, where ρ is the air density, V vehicle speed, S_p the area directly to the front of V, and C_d the resistance coefficient related with the vehicle shape [1]. The wheel rotation resistance is not the friction but the resistance of the tire deformation of the running wheel. Theoretically, it is a quantity hard to calculate accurately. It is a function with various complicated factors, as deformation of the wheel and road surface, pressure of the wheel on the contact surface of the ground, wheel elasticity and material of the ground, roughness of the wheel and the ground. Therefore, in the actual programming, a formula based on the experience is used as $R2 = \mathrm{Cr}w$, from which the rotation resistance of each wheel can be obtained. In the formula, w means the weight of the wheel and the bearing partial. Cr means the rotation resistance coefficient acted on W simply [2]. The tire manufacturer usually provides the design value of such coefficient. Knowing how to calculate the resistance of the vehicle, it is easy to calculate the power of the engine to get the required acceleration to fight against the resistance.

18.4 Geometry Modeling

18.4.1 Environment Modeling

Firstly, 3DS MAX software is used to make polygon modeling, as indicated in Fig. 18.1. A scene model is created with reference to a certain actual environment, including asphalt pavement, road block, grassland slope, predicted collision object, and assigns the model with necessary material.

18.4.2 Model Processing

To store the 3D vehicle body model created with PRO/E software in .igs format and then to introduce it into 3DS MAX for processing. In the introduction and derivation of the model, there always are inverse image (inverse normal), damaged

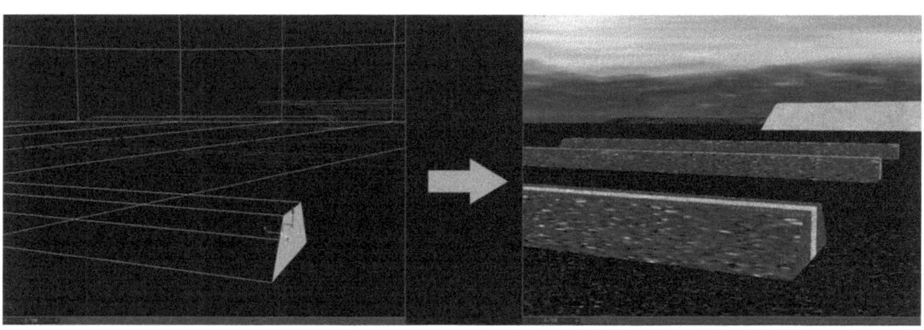

Fig. 18.1 Polygon modeling

Fig. 18.2 Texture processing

image, or lost image. In such case, partial model is required to be modified in 3DS Max. Any image damaged too seriously is required to make remodeling.

Meanwhile, the mapping can be used to replace the complicated model (tire model in Fig. 18.2 is the simple geometry map after processing). For example, the complicate bump facet on the tire surface can be removed and the mapping can be used for texture setting, as indicated in Fig. 18.2.

18.5 Physical Modeling

18.5.1 To Create Major Model

The processed model is loaded into Virtools Dev scene environment. The major model of Physics car needs a vehicle body and four wheels. In Virtools Dev, the coordinate unit is 1 Unit = 1 m. Therefore, it is better to set the coordinate unit to be 1 Unit = 1 m in 3DS MAX modeling. Attention should be paid to the position and direction of the pivot of the vehicle body and wheels. In 3DS MAX, the coordinate system has Y axis frontward and Z axis upward while in Virtools Dev, the coordinate system has Z axis frontward and Y axis upward. The pivots of the vehicle body and four wheels should be set at the center of the object, respectively [3].

18.5.2 Model Naming

In Physics car BB, the major models are required to name according to the programming rules. Otherwise, the system cannot call for them. It is suggested to name the four wheels according to the following rules:

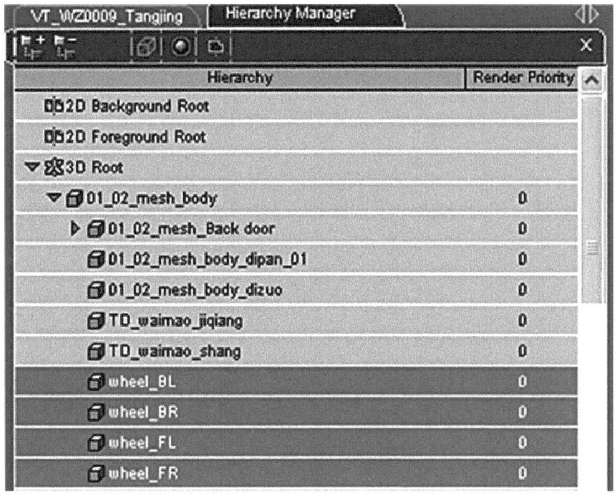

Fig. 18.3 Hierarchy manager

The name of the front right wheel contains "FR";
The name of the front left wheel contains "FL";
The name of the back right wheel contains "BR";
The name of the back left wheel contains "BL".

To open hierarchy manager and add the four wheels into the sub-hierarchy of the vehicle body, establishing membership, as indicated in Fig. 18.3 [4].

18.5.3 Code and Parameter Setting

18.5.3.1 Code Editing

In the logic editor interface of Virtools Dev, the named models can be output into files in .nmo format in Virtools with the export function of 3D MAX. To start Virtools Dev software and click the menu of Resources–Import File to introduce the .nmo file and to add script into the scene. To make it a part of physical environment with the physical initialization module of Physicalize BB. To add script to the vehicle body with Physics Car BB, as indicated in Fig. 18.4.

18.5.3.2 Interactive Action Setting

The moving forward, backward and steering left and right and brake and input of the speeder of Physics Car can be controlled with Switch on key BB. According to the general driving habit, four direction keys are set to control the vehicle to move forward and backward, left and right. The space bar is set to control the brake and the speeder can be selected according to the actual requirements [5].

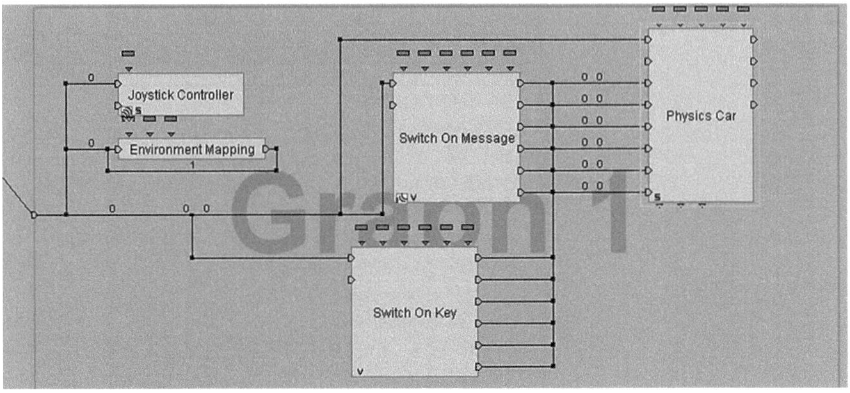

Fig. 18.4 Physics car BB

18.5.3.3 Parameter Setting

Now the vehicle still cannot move, for it is controlled with three arrays of body parameter, wheel-suspension parameter and engine-steering parameter, of which the raw number, column number, and name must meet the requirements of Virtools strictly. With reference to the dynamics formula in above discussion, the parameters of body, wheel and engine of Physics Car are set, which are controlled with the array. The parameters of Physics Car are adjusted for the present state of the model properly. Whether the vehicle can move and whether the vehicle simulation has the accurate and real effect depends on these three groups of data. The specific introduction will be made on such three arrays [6].

(1) Body parameter: It is an array of 1 row and 15 columns, mainly controlling the property of the vehicle body. It mainly depends on whether the parameters of three arrays are adjusted properly;

- Body mass: Mass (ton);
- Body friction: The actual friction is equal to such vale multiply the friction coefficient of the ground;
- Body elasticity: The larger the value is, the larger the elasticity is;
- Body speed damp: It can be considered as the air resistance;
- Body rotation damp: It is a 3D vector, keeping the vehicle body stable;
- Body rotation inertia: It is a 3D vector, indicating the inertia of the vehicle when steering;
- Shift mass center: It is a 3D vector, indicating the offset of the body center to the center of four wheels, of which the value of *Y* decides the center of gravity. The low center of gravity can avoid the overturning in the sharp steering.

(2) Wheel-suspension parameters (front or back wheels):

- Wheel mass: It indicates the mass of wheels. The total mass of four wheels is better to be the mass of the body, without considering the actual conditions, to improve the elasticity of the vehicle;
- Wheel friction: The actual friction is equal to such vale multiply the friction coefficient of the ground. The vehicle even can climb up the wall in case of too much value of it while the too small value leads to wheelslip;
- Wheel elasticity: It indicates the elasticity of the wheel;
- Wheel speed damp: It should be rather small;
- Wheel rotation damp: The vehicle cannot run with high speed in case of the value of it being too large;
- Wheel rotation inertia: The vehicle will be unstable in case of the value of it too small while the vehicle body will be affected in steering in case of the value of it too large;
- Suspension constant: The larger the value is, the farther the vehicle will be away from the ground. Such constant is related with the weight.

(3) Engine-steering parameters: such array of 1 row and 15 columns, mainly controlling-related properties of the steering wheel and vehicle engine. Parameters to control the property of the lengthwise column:

- Max steering: It is the maximum steering angle in case of the low speed;
- Max speed steering: It is the maximum steering angle in case of the high speed;
- Steering velocity: The steering velocity of the front wheel;
- Engine power: It is the power of the engine in the maximum speed;
- Min engine rpm: Rotating speed at the lowest gear;
- Max engine rpm: Rotating speed at the highest gear;
- Axle torque ratio: It indicates the engine power distributed to the back wheel;
- Max speed (km/h): It indicates the maximum speed;
- Rear brake deceleration: It indicates the speed deceleration in the rear brake.

A camera is required to add into provide a comfortable angle for the observer. The camera, as the eye of the observer, plays quite important role in strengthening the whole scene demonstration effect, to adjust the camera to a proper angle and to add script code, to add the logic control in the code, and to set the interactive action to moving with the vehicle body. Thus, the view angle of the observer will focus on the vehicle body, obtaining predicted effect, as indicated in Fig. 18.5.

18.5.4 To Create Environment Light

To strengthen the sense of reality of the virtual environment, it is required to add the lights in the environment scene. To be exact, there are no fixed requirements in the

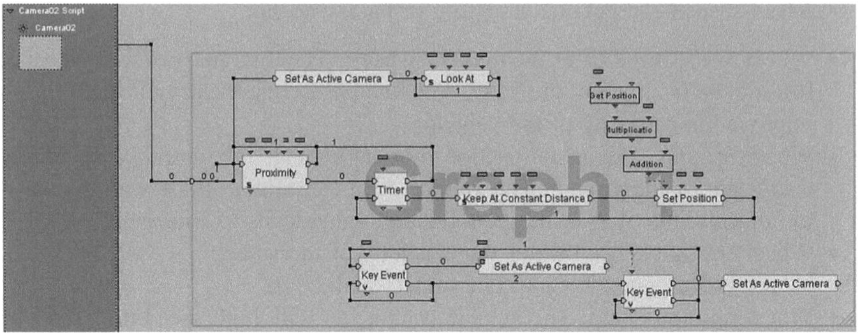

Fig. 18.5 Camera setting

light setting. The general rules will be followed that to set the major light and then to set several supporting lights. Finally, the light parameters are adjusted according to the scene size, space height, and range to light, obtaining the final effect.

18.6 Conclusion

In the thesis, the integrated mode of PRO/E, 3DS MAX, and Virtools is used for successful and rapid construction of the dynamics simulation platform for special vehicles. It highly reuses the model for the simultaneous work of the design and simulation, improving the design efficiency and shortening the design period. Meanwhile, in the dynamics simulation platform in the thesis, the vehicle can be controlled with the keyboard and mouse. Furthermore, the interactive equipment of human factor engineering is added into control, such as driving simulator, space position tracker, and 3D mouse [7]. The further study is required to make on how to drive such external equipment, to establish more perfect virtual simulation platform in the future. The simulation platform has been verified in the immersion environment with Cadwall.

References

1. Dingfang C, Yabo L et al (2002) Virtual design. China Machine Press, Beijing
2. Xu Y (2012) The virtual interaction design instance of virtools. Beijing University of Aeronautics and Astronautics press, Beijing
3. The help documentation of Virtools Dev3.0
4. Stathler W (2004) Beginning math and physics for game programmers. New Riders Games, Indianapolis
5. Zherong W (2008) The design of 3D VR. Wen Kui Zi Xun. Taiwan. 7
6. Mingkun L. The design of 3d games. Sichuan publishing group, Sichuan
7. Xu Y (2009) Virtools designer of 3d games. DianNaoBao Electronic Audio and Video Publishing Group, Beijing

Chapter 19
Study on the Application of RFID in the Visible Military Logistics

Jinshan Chen

Abstract The modern military logistics is more exact in the time, space, and quality for military supplies. Application of RFID technology speeds up modern military logistics visualization and intelligence. This chapter proposes an approach to solve the efficiency and quality problems in the military logistics, through the construction of the military logistics information system based on RFID technology. In this chapter, the current application situation of RFID technology in the military logistics is presented, some tentative plans for the planning of the military logistics information system based on RFID technology and some suggestions on the application of RFID in the military logistics are proposed.

Keywords Military logistics · Military supplies security · RFID

19.1 Introduction

The modern war is typical war with high consumption of the military materials, which makes higher requirements for the military logistics. With the remarkable increase in military security difficulty and strength, the military logistics is more and more important to the war effect. A single soldier consumed 6 kg materials in average in the World War I. However, it increased to 500 kg in Iraq War. The military material consumption increased by 100 times approximately in just over 80 years. It shows that the modern war is the competition of the supporting security in a larger degree.

How to accept, examine and distribute rapidly the military materials especially those transported to the destination with containers, is a problem that needs to be

J. Chen (✉)
PLA Institute of Special Operation, Guangzhou 510502, China
e-mail: js63@163.com

S. Long and B. S. Dhillon (eds.), *Proceedings of the 13th International Conference on Man-Machine-Environment System Engineering*, Lecture Notes in Electrical Engineering 259, DOI: 10.1007/978-3-642-38968-9_19,

solved immediately in the military logistics security. In the Persian Gulf War of 1991, US military sent 48,000 containers approximately to Saudi Arabia. However, 28,000 containers had to be opened, recorded, repacked, and transported again due to unclear signs on them, seriously impacting the transportation efficiency. After the War, there were still 8,000 unopened containers and 250,000 air pallets containing unknown goods, resulting in serious waste of the security materials [1]. According to the estimation afterward by the US military, if RFID technique had been used to track the supporting materials, it would have saved US$2 billion for the Department of Defense. After the Persian Gulf War, US military made the whole property visualization plan to solve serious problems of materials in application, transportation, and distribution. It is considered as one technical access in *Joint Vision 2020* of the US military. It is possible for the US military to realize the supporting materials visualization with the application of RFID technique, which provides the convenient and flexible solution to automatically obtain visualization information on military materials in storage, transportation, and utilization.

RFID technique was also verified in the Iraq War. The US military accurately tracked 40,000 containers sent to the Gulf by the Department of Defense with the RFID chips of SAVI on the containers, realizing the whole track on the personnel, equipment, and materials. The whole visualization in the material supply management improved the efficiency of the military material security greatly. The application of RFID made the supporting supply of the US military transfer from the storage supply to distribution supply. Compared with the Gulf War, the shipment quantity was reduced by 87 %, airlift by 88.6 %, strategic supporting equipment mobilization quantity by 89 %, and the battle material storage quantity by 75 % in the Iraq War. Such a new mode saved billions of expenditure for the US Department of Defense.

19.2 RFID Techniques

RFID is the integrity of the modern radio frequency technique and IC card, which is an automatic identification technique without contact. It can work in various adverse environments with radio frequency signal to make automatic identification on the target object and obtain related data, requiring no manual work.

A complete RFID system consists of a reader/writer, electronic tag, and data management system. The electronic tag is made up of the antenna and RFID chip. Each chip has a unique electronic product code (EPC) to describe the materials to which the electronic tag is attached. The reader/writer can read and write the information in the electronic tag and decode before transmission to the central computer system for processing.

RFID system works in the following procedure: the induction radio wave is used for the non-contact two-way communication between RFID tags and the reader/writer. When the electronic tag enters the working area of the antenna of the reader/writer, it can receive the radio frequency signal from the reader/writer. The electronic tag can send out the product information stored in the chip with the energy from the induction current (passive tag). Some electronic tag can send out a signal of a certain frequency actively to deliver the product information (active tag). The reader/writer will read the information and decode and then send the information to the background computer system for related data processing.

19.3 RFID Technique's Application in the Visualized Military Logistics

19.3.1 RFID Technique Application Profile in the Military Logistics

RF satellite chip of SAVI has been used with great success in the military logistics by the USA since 1994 [2], assisting the US military to realize the following targets:

1. Full visualization of the military logistics supplying chain;
2. Elimination of over storage in warehouse (repeated application for extra materials);
3. Dynamically obtaining the real time data of the military logistics supplying chain and realizing rapid and accurate management;
4. To speed up the delivery of supporting materials from the factory to the fox-hole, realizing accurate point–point delivery control;
5. Elimination of personal interference, improving the efficiency of the military logistics security greatly.

At present, RFID technique is widely used in the military logistics security by the US military, such as particular object searching system, visualization system in transportation, soldier's electronic medical record, physiological state detector and military quilts, and clothes distribution management device. The US military reformed the traditional logistics track mode greatly with RFID technique, obtaining ideal effect in the withdrawal from Europe, Somali peacekeeping action, war in Afghanistan, and the Iraq War. RFID technique greatly reduced the average time of the US military materials supplying greatly.

Besides the US military, Israel military started to attempt to use RFID technique to manage and store military materials, track the delivery trucks, tanks, and the transportation process from the military materials supporting center to the frontier, track and manage properties transportation, and manage goods delivery and assets of supplying chain in 2005. At the beginning of 2006, Australian military began to use SAVI program to participate in the military joint action in Iraq with the USA and UK. At present, many developed countries attempted to use RFID technique to make dynamic track of the flow and supply of military materials.

19.3.2 Visualization of Military Logistics System Structure

With reference to successful examples of foreign militaries, especially the US military in the military logistics management, the military communication technique is used to establish the modern military logistics system architecture (Fig. 19.1) of the comprehensive Internet technology and radio communication technology based on RFID. The model consists of supplier, military materials warehouse, military logistics control center, Beidou navigation satellite system to monitor the transportation vehicles, military Internet, and RFID identification system. RFID technique and related components (GIS server, product database) are used in every part of the military logistics: from materials provided by the supplier to the military, to the materials stored in warehouse, and in the transportation, RFID electronic tags are attached to the materials. The storage information is continuously updated, and each process will be recorded in RFID electronic tags, which guarantees the visualization of the whole military logistics. If the electronic tag recognizer is also installed on the transportation vehicles, it is possible to make examination on the materials' security state at any time. Another important premise to realize the visualization of materials security is the information, recorded in RFID electronic tags, will be passed to the control center in real time. Beidou navigation satellite system is required to pass the information involving vehicle position and driving state accurately to the control center by the military radio communication network. Therefore, the commander will control or adjust the marching vehicles at any time, guaranteeing the military supplies security in the war especially.

19.4 Problems in RFID Application at Present

19.4.1 Information Security

In the modern war with informatization, it is a great challenge for the military to guarantee the information security. For RFID system, if there is no reliable security mechanism, there will be no information security. The radio signal of RFID system is spread and received in an open mode, and the radio wave cannot identify the enemy. We can use RFID to send the information. However, the enemy can also use RFID to obtain the data or even learn the specific position and destination of the equipment and materials, which is a serious security danger for the military action [3]. The radiation power, direction, and frequency spectrum band can be controlled, and information can also be encrypted to guarantee the system security. However, it is quite easy to attack any radio system. The security threats of RFID system in actual application are mainly from the following three parts: firstly, communication between the tag and the radio frequency reader; secondly, communication between the radio frequency reader and the back-end

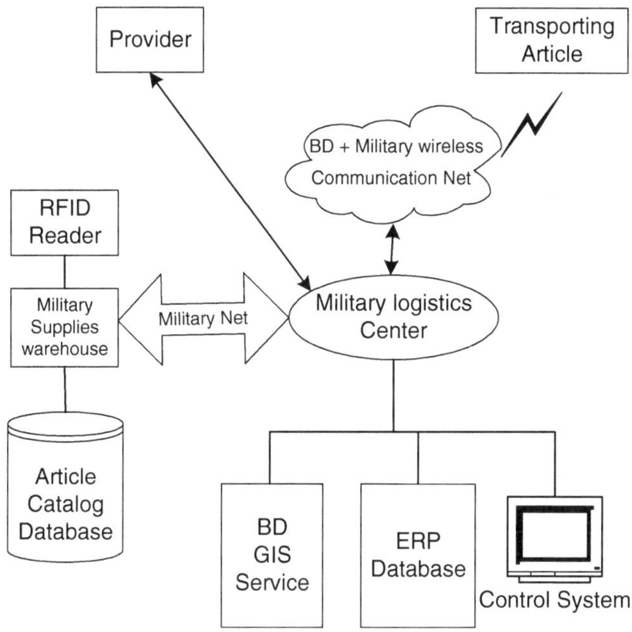

Fig. 19.1 Military logistics system architecture

computer system; thirdly, information transmission between back-end databases. The weakness of RFID system is the communication between the radio frequency tag sending materials information and the radio frequency reader due to the open communication of the radio signal. The tag encryption technique can improve the security of RFID system effectively. However, only a few tags have such function at present. Such tags are not widely used in the military logistics due to high cost. Therefore, it is possible to make encryption on the tag information with traditional code communication in the military of China.

19.4.2 Anti-interference in Complicated Electromagnetic Environment

The information war features complicated electromagnetic environment, compared with traditional battlefield factors. In the relatively limited battle space, various information systems and electronic equipment gather together, producing interference and self-interference of electromagnetic radiation. Besides, there are electromagnetic fighting means of both parties. All such complicated electromagnetic environment affects RFID system greatly. In the actual electromagnetic environment of the battlefield, there are many interference sources that hard to be predicted, including our various communication equipments, electronic interference and attack from the enemy, and unpredicted electromagnetic interference from civil electrical equipment and natural electromagnetic radiation. The electromagnetic radiation is crossing and overlapping in the space, time, frequency spectrum, and power, which is hard to control due to constant change. It is difficult to realize the effect of the electromagnetic radiation on the normal work of the equipment if no on the site. It is a huge challenge in the reliable operation for RFID system with small radiation power of miniwatt/watt in front of the communication equipment with large power of dozens of or hundreds of watt [4]. The anti-interference of the complicated electromagnetic environment is a problem that must be solved in the application of RFID system in the military logistics. It is practical to solve such problem using multiple-frequency tag to replace the single-frequency one. The penetration capacity of the low frequency and the identification distance and anti-interference functions of the high frequency can improve the anti-interference of RFID system significantly.

19.4.3 Application in Sites of Particular Goods

Military materials include many flammables and explosives as various oils, chemicals, bullets, explosives, and electrical fuses. It is forbidden to use radio communication equipment in the gas station. Similarly, it is also a strict security challenge for the super frequency and high-frequency radiation frequency reader to be used in the storage places and warehouse of flammables and explosives. There are not many studies on such aspect at present, so there are no authorized and reliable study results. Thus, it is impossible to make conclusion whether RFID technique can be used in the military logistics involving large quantities of dangerous goods.

19.4.4 Problems Arising from Adverse Natural Environment

The military logistics will face much harsher natural environment. The destination of military equipment and materials is incomparable from that of the general civil logistics. Natural factors as temperature, moisture, salinity–alkalinity, sunshine, and rains and fogs are the challenge to RFID system containing most electronic equipment. Too high or low temperature will lead to abnormal operation of RFID reading equipment and tags. The low temperature will cause the rapid consumption of the battery of the active tags, which fails to meet the requirements of the nominal service time greatly, making it difficult to use and manage the tags. To be used in the military logistics, the RFID reading equipment and tags must have higher work capacity in the high and low temperature and must bear the corrosion and damages from various harsh environments.

19.5 Conclusion

The modern logistics management stresses emphasis on the accurate supply, and thus, it is called as the supply based on the requirement. The military logistics is more accurate because of higher requirements on the time, space, and quality of the logistics. The introduction of RFID logistics technique is the key to promote the accurate military logistics. It is an inexorable trend to establish visual and intelligent military logistics with the informatization of the military of China. Despite various problems and challenges at present, RFID technique, with the development of RFID and increasing improvement in the military logistics system, will be important means to improve the military supplying chain management, to reduce the supporting cost and to strengthen the military power.

Acknowledgments Subsidized by: Guangdong Science and Technology Program Subsidization Project (2010B031000001).

References

1. Yuan Haibo (2009) Research on project of military logistics MIS visualization based on RFID. Shandong University Master's Thesis, Jinan, pp 3–21
2. Shi Y, Gu B, Wang J (2009) Research on the simulation models of military logistics. Logistics Sci Tech 8:88–89
3. Xu C, Guo R, Guo J (2009) Research on military logistics communication system based on the Beidou satellite. Logistics Tech 28(9):147–149
4. Yu B, Zhou X, Rongmin L, Wang L (2011) Research on problems and countermeasures of applying RFID to military logistics. J Military Transp Univ 13(3):72–75

Chapter 20
Research on the File Encryption System Based on Minifilter Driver

Jinshan Chen and Jianwei Ye

Abstract The analysis is made on the security in the information process in this paper. In order to avoid system breakdown under the process by traditional development of file encryption and decryption system, a new approach based on minifilter is proposed. Using a client/server mode and ingenious design for automatically embedding the file encryption tag, the model implements the encryption and decryption functions about the office document under the Windows XP environment. The tests for three different size files have been done with DiskBench for reading and writing hard disk drive. The experiment results show that the approach is reliable, robust, and effective.

Keywords Minifilter driver · File system · Transparent encryption

20.1 Introduction

With the informatization and computer popular application, various internal files including confidential bid documents, reports, table, and design drawings are stored in the computer in the format of the electronic file. To avoid illegal disclosure of such files, various methods appear, such as static encryption and dynamic encryption. In the first method, the user can encrypt the files using password, e.g., in Word, Excel and WinRAR software of Microsoft. However, it is a traditional passive and static encryption method with low efficiency. Besides, the password may be forgotten and the encryption algorithm has the fatal trap and is easy to crack. The representative of the file or file folder active encryption product is EFS encryption system of the operation system above Windows NT. It is active and transparent encryption method,

J. Chen (✉) · J. Ye
PLA Institute of Special Operation, Guangzhou 510502, China
e-mail: js63@163.com

S. Long and B. S. Dhillon (eds.), *Proceedings of the 13th International Conference on Man-Machine-Environment System Engineering*, Lecture Notes in Electrical Engineering 259, DOI: 10.1007/978-3-642-38968-9_20,

with which the user can encrypt the file or file folder. With such product, the operator is required to make active identification which files are confidential and which files need no encryption. However, such files will be decrypted automatically in case of leaving the computer. One representative of the disk encryption products is Bit-Locker of Windows Vista, a product to protect the files with encryption on the logic disk or the full disk volume with the advantage of bottom drive control by Microsoft. Other products including products of PointSec of Sweden, SafeBoot of Netherlands, and a type of hard disk of Seagate of the USA can make real-time dynamic encryption on disks. However, such products have various bugs for it cannot make encryption on the system disk. It is obvious that such methods rely much on the user and application program. All files are in decrypted state in case the user environment is loaded, which fails to avoid the active disclosure by the internal user.

The internal network security and information disclosure cases show that such means as access control, equipment monitoring, and security audit cannot solve the information security completely in the root as long as plain text files are stored in the hard disk. An effective method is to make direct access control and encryption [1] on files in the source terminal and the information media. With the mature of the file system filter driver, to make transparent encryption with such technique on important files has been an important technical means to protect the files. The file protection system based on the file transparent encryption/decryption technique is the concern of the industry. Lin [2] and Li [3] proposed the solution. The thesis presents a file transparent encryption/decryption solution based on Windows file system minifilter model.

20.2 Transparent Encryption/Decryption File System Based on Minifilter Driver

The transparent encryption/decryption based on minifilter driver is a kind of dynamic encryption/decryption. Compared with the static encryption/decryption, the dynamic encryption/decryption means to decrypt as to read the file and to encrypt as to write the file. The transparent means the file encryption/decryption filter, for which the user needs no interference. The transparent encryption/decryption has no impact on the normal operation. The user does not feel any difference and the existence of the encryption/decryption. Neither changes the using habit of the user and the electronic format, nor bring about any trouble to the user after loading the system. The protected electronic files will be stored, used, and managed as the ordinary electronic files. In using the encrypted electronic files, the protected electronic file can be used as the ordinary one in case of the user owning the legal identification and authorized permission. In case the user owns no adequate permission or illegal identification, the system will forbid the user using the protected electronic files automatically. The file will be invalid automatically in case of leaving the using environment; thus, the sensitive information is protected in no disclosure.

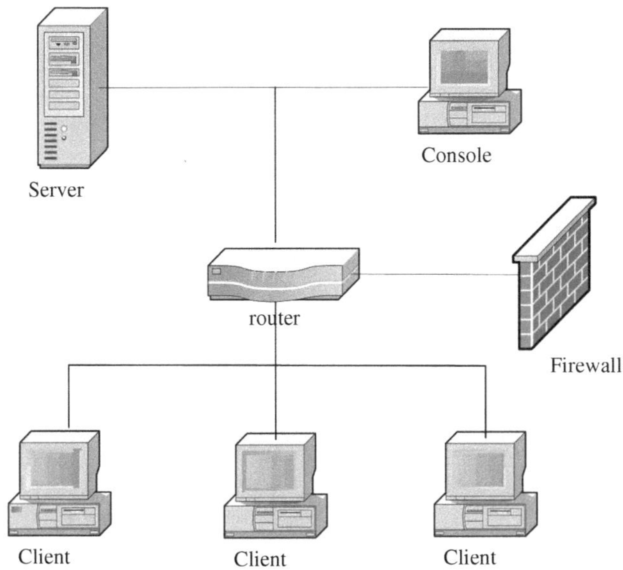

Fig. 20.1 System structure

20.2.1 Structure of Transparent Encryption/Decryption File System Based on Minifilter Driver

The system of C/S structure consists of clients' computer, console, and server (Fig. 20.1).

The server is mainly used to manage the user information, and the console is the control interface of the server. The service terminal software mainly has functions of strategy formulation, distribution and assignment, and identity authentication.

The clients' computer is used to make access control and transparent encryption on the files according to the user's level and strategy control on the file tag. Clients' terminal software can be divided into two parts: The first is the application program responsible for the communication with the service terminal and interaction with the same bottom filter driver and the second is the bottom filter driver to make final encryption/decryption operation on the files.

20.2.2 Principle of File System Filter Driver

The kernel mode driver of Windows NT is in hierarchical structure, namely file system driver, medium driver, and disk equipment driver. The file system driver receives the operation requirements as to open, create, read, write, and close files on disk. Such requirements are generally generated in the user process and sent to

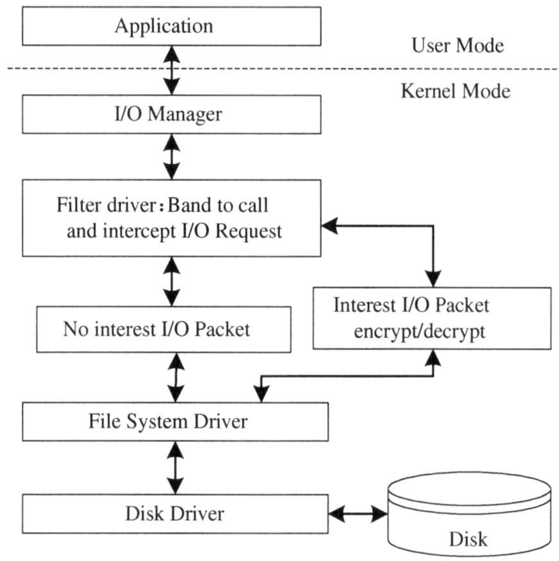

Fig. 20.2 Filter driver principle

the file system by I/O subsystem manager. The filter driver, belonging to the medium driver, is attached to other drivers to intercept the requirements to the bottom drive equipment. Compared with the common hierarchical structure, its device object is anonymous, which is hitched to the device object created by a certain bottom driver for operation [4].

The file system filter driver is attached to the file system. The I/O manager of Windows NT constructs IRP according to the file operation request of the user and sends it to the file system driver. The file system driver transfers the operation of the file system into the operation of the disk driver and to call the disk driver (Fig. 20.2).

20.3 Key Techniques to Realize Minifilter Driver File System

20.3.1 Transparent Encryption/Decryption Technique Based on Minifilter Driver

Filter manager is the core of the minifilter driver. In the traditional filter driver model, it requires to process various IRP and to make judgment on each intercepted IRP package. It also requires constructing IRP memory space, initializing IRP and IRP correlation in IRP construction. Any carelessness will lead to the breakdown of the system. The minifilter model avoids unnecessary trouble of the traditional filter driver model. A third developer can write minifilter driver, which is much easier to develop than the traditional one.

The filter driver is installed with Windows. However, it will act in case of minifilter driver is loaded. Minifilter driver guarantees highly that its example is always loaded in the corresponding position and it decides how to call minifilter driver to process I/O in which order within the filter manager.

The callback mechanism is used by minifilter driver to make definite I/O operation to be filtered. It is indirectly bounded to the file system stack by making registration to the filter manager. Minifilter driver can filter IRP I/O operation, Fast I/O, and FSFilter Callback. It can register a Pre-oper callback routine and a Post-oper callback routine for each I/O operation to be filtered. When an I/O operation is to be made, the filter manager will call the Pre-oper callback routine registered by each minifilter driver. When such callback routine is back, the filter manager will call the Post-oper callback routine registered by the next minifilter driver.

The callback function is an array to define the corresponding operation of the appointed requirement. The code for the callback function array of the system file filter driver is as follows:

```
CONST FLT_OPERATION_REGISTRATION
Callbacks[] = {
{IRP_MJ_CREATE, 0, PreCreate, PostCreate},
{IRP_MJ_READ, 0, PreRead, PostRead},
{IRP_MJ_WRITE, 0, PreWrite, PostWrite},
{IRP_MJ_CLEANUP, 0, PreCleanup, NULL},
{IRP_MJ_OPERATION_END}
};
```

20.3.2 File Encryption Identifier Technique

To make dynamic transparent encryption/decryption of files by the file system filter driver, it is required to make identification between the encrypted files and non-encrypted files. The file identification can be solved with the following programs. One is to place the encryption identifier beyond the file; for example, to create an overall file to store and manage the file encryption identifiers or to create a subsidiary file to the encryption file as the identifier. The other is to place the encryption identifier within the file at the beginning or the end of the file.

The encryption identifier is placed beyond the file, resulting in two problems, including how to keep simultaneous and it is easy to lose or become invalid due to accidental falsification. However, the encryption identifier is placed within the file, which can be guaranteed to be universal, reliable, and permanently valid. If the encryption identifier is placed at the end of the file, it is necessary to make simultaneous movement in case of expanding or reducing the file. If such action is not completed for some reasons, including the power down, the file will be damaged or the encryption identifier is lost. If the encryption identifier is placed in the beginning of the file, it only requires once writing. It does not require writing again in case of expanding or reducing the file. In software like Word, it is possible to make dozens or hundreds of operations to change the file size in saving the file. Therefore, it is most efficient to place the encryption identifier at the beginning of the file.

In the thesis, the file encryption identifier is placed at the beginning of the file. The thesis proposes a method based on the minifilter driver to build in the encryption identifier at the beginning of the file. With the method, a hidden file heading of 4 kB is added to the file and an offset is made on all files in the encryption. Such method has an advantage that it is hardly to increase the system resource consumption. The specific procedure is as follows: When the file filter driver detects the confidential procedure attempts to open a newly created or an existed file, I/O manager will send IRP_MJ_CREATE IRP package to the file system driver at first. Therefore, it is possible to intercept the IRP_MJ_CREATE operation in the kernel with the minifilter driver, and to make judgment whether the file to be opened is newly created or existed one. If it is a newly created file, it will apply for a new secret key and write the information of the secret key, encryption algorithm, and security degree into a new file identifier. It will make encryption on the file identifier and build the encrypted file identifier into the new file. If the file is an existed one, it will extract the encryption identifier of the file and put the file identifier into the encryption list.

The file heading is required to hide all secret procedures. Therefore, an offset is needed to make all file operations. The following is the read offset code to hide the file heading:

```
Void cfIrpReadPre (PIRP irp, PIO_STACK_LOCATION irpsp)
{
PLARGE_ INTEGER offset;
PFCB fcb = (PFCB) irpsp-> FileObject-> FsContext;
Offset = &irpsp-> Parameters. Read. ByteOffset;
If (offset-> LowPart == FILE_USE_FILE_POINTER_POSITION && offset HighPart == −1)
{ASSERT (FALSE);}
Offset -> QuadPart += CF_FILE_HEADER_SIZE;
//decryption data
For (i = 0; i < length; i ++)
{To call encryption/decryption algorithm;}
}
```

The write requirement is similar. However, the sentence of obtaining offset is modified into

```
Offset = &irpsp -> Parameters. Write. ByteOffset;
```

20.4 System Result

The system operation environment is Windows XP sp3, WDK7600. After loading and unloading kernel filter driver module, the execution test results are as follows, as indicated in Figs. 20.3 and 20.4.

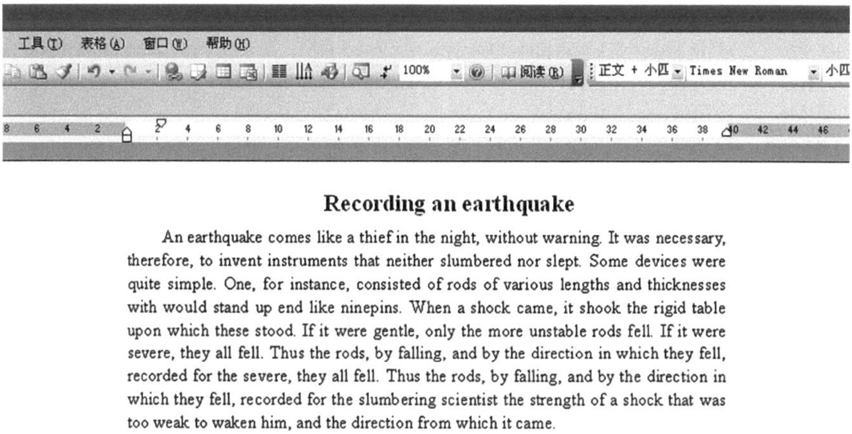

Fig. 20.3 Diagram of opening a document by authenticated users

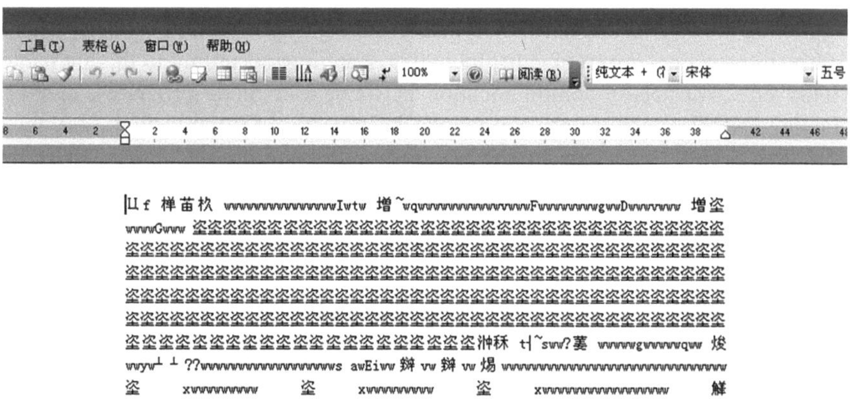

Fig. 20.4 Diagram of opening a document by non-authenticated users

The environment to test the effect of the file encryption system on the computer performance is Windows XP sp3, Core E6300 processor, 2G RAM, and 7200r SATA hard disk. The hard disk speed test software DiskBench is used to make 10 tests under the same condition to the file of the same size under the same environment, of which the average speed is indicated in Table 20.1.

The test result shows that the read/write speed is reduced slightly when the file transparent encryption/decryption system starts. However, such speed loss is acceptable, compared with the security.

Table 20.1 Test time of file read/write

File size	File type	Encryption/decryption closed (s)	Encryption/decryption open (s)
346 KB	Word	0.11	0.13
4.85 MB	Word	0.56	0.63
18.63 MB	Word	1.82	2.01

20.5 Conclusion

The thesis designs and realizes a file transparent encryption/decryption system based on the minifilter driver model provided by Microsoft. The strategic dynamic configuration mode in the system provides wider space for the information management of the user. The automatic embedding the encryption identifier realizes the transparent transmission of the internal information. It provides an effective solution to the serious information security condition at present. The actual test proves it is a stable, reliable, and effective system. Compared with the traditional filter driver model, the development of the minifilter driver mode has the following advantages: firstly, less work load, simpler but more reliable filter driver; secondly, dynamic load and unload, bound and unbound; thirdly, to bound a reasonably fixed position in the filter stack; fourthly, quick, clean, and reliable context management; fifthly, to provide a group of calls, including searching according to the file name, highly effective access, communication between user mode programs, and I/O queuing; sixthly, supporting non-loopback I/O, the I/O application sent by a minifilter is easy to be seen by the lower minifilter and the file system under the stack, avoiding the reentrant; seventhly, it can only filter-interested operation. However, in traditional filter model, it must be attached to each operation interface to pass the operation to a lower hierarchy.

Acknowledgments Subsidized by: Guangdong Science and Technology Program Subsidization Project (2010B031000001).

References

1. Mao R (2008) Research and implementation of transparent encryption file system model. Dalian University of Technology Master's Thesis, Dalian
2. Lin H (2009) The study and implement of a encryption/decryption system based on the windows file system filter manager. Zhejiang University of Technology Master's Thesis, Hangzhou
3. Li JZ, Ling J (2011) Application of minifilter driver in terminal of document security. Comput Eng Des 32(7):2245–2248
4. Sun YY, Zheng KG (2010) File monitoring system based on minifilter. J Comput Appl 30(11):3115–3117

Chapter 21
Research on the Method for Simulating 3D Dynamic Process in Two-Dimensional Space and its Application in Developing Audio and Video Electronic Textbook

Fuxiang Liu

Abstract If we use computer to simulate simple form of three-dimensional dynamic process that is virtual or very difficult to achieve in reality directly in the two-dimensional space, we will reduce the development difficulty of simulation process. This paper introduces the simulation methods and key technologies by some relevant examples in "Audio and Video Electronic Textbook of Mechanical Drawing." The examples include the formation process of gyration curved surface, the process of an object being cut by a plane, the decomposition process of a combined object. The Flash animation software is mainly used in the simulation process, and AutoCAD software and Photoshop software are the tools used for preparing materials. The results show that simple form of three-dimensional dynamic process can be simulated very realistically in the two-dimensional space, not using the tools of three-dimensional animation.

Keywords Two-dimensional space · Three-dimensional dynamic process · Simulation · Audio and video electronic textbook

21.1 Introduction

Some contents of audio and video electronic textbooks will involve in a certain 3D dynamic process or a virtual 3D dynamic process which is difficult to realize in reality, including the formation process of gyration curved surface, the process of an object being cut by a plane, the decomposition and reorganization process of a combined object, the flattening process of six principal views of engineering

F. Liu (✉)
College of Architectural Engineering, Dalian University, Dalian 116622, China
e-mail: dldxlfx@163.com

S. Long and B. S. Dhillon (eds.), *Proceedings of the 13th International Conference on Man–Machine–Environment System Engineering*, Lecture Notes in Electrical Engineering 259, DOI: 10.1007/978-3-642-38968-9_21,

drawings. It is impossible to show the above-mentioned processes with words and figures accurately. Therefore, it is necessary to make simulation on the dynamic processes to deepen the understandings and impressions of learners on concepts related to above-mentioned processes. In thesis, the study is made, with Flash as principal tool and AutoCAD and Photoshop as accessory tools, on methods and key techniques to make simulation of the 3D dynamic processes in 2D space.

21.2 Dynamic Simulation of the Formation of Gyration Curved Surface [1]

The formation of the gyration curved surface is a virtual process, described as a generatrix rotates around a gyration axis for a cycle. The formation of spherical surface is taken as example to explain how to make simulation on the computer. The simulation process consists of two parts of animation, of which the first part is to simulate the semi-circle generatrix rotating around the vertical axis from the left to the right and the second part is to simulate the sphere that appears gradually with the rotating semi-circle generatrix.

21.2.1 Simulation of Semi-Circle Generatrix Rotation

Flash can only be used to make 2D animation. So the rotating generatrix around the vertical axis in the space is only be simulated by the continuous deformation of the semi-circle generatrix in the plane with Flash. In the actual simulation, it is difficult to simulate the continuous deformation of the semi-circle generatrix from the left semi-circle to the right one in Flash. Such part of the animation is divided into two stages for such purpose, of which the first stage is to simulate the deformation of the generatrix from the left semi-circle to a line and the second stage is to simulate the deformation of the generatrix from the line to the right semi-circle. In the 360° rotation of the generatrix around the axis, only in the front 180° the generatrix can be visible. Therefore, only the rotation process of generatrix in the front 180° is required to make in the simulation animation, as indicated in Fig. 21.1a.

21.2.2 Simulation of Gradually Appearing Sphere

Firstly, a sphere is drawn in AutoCAD and then it is copied to Photoshop using the button of Prt. Sc on the key board. In Photoshop, the magic wand is used to select and to cancel the background beyond the sphere. Then, the sphere is stored as PNG

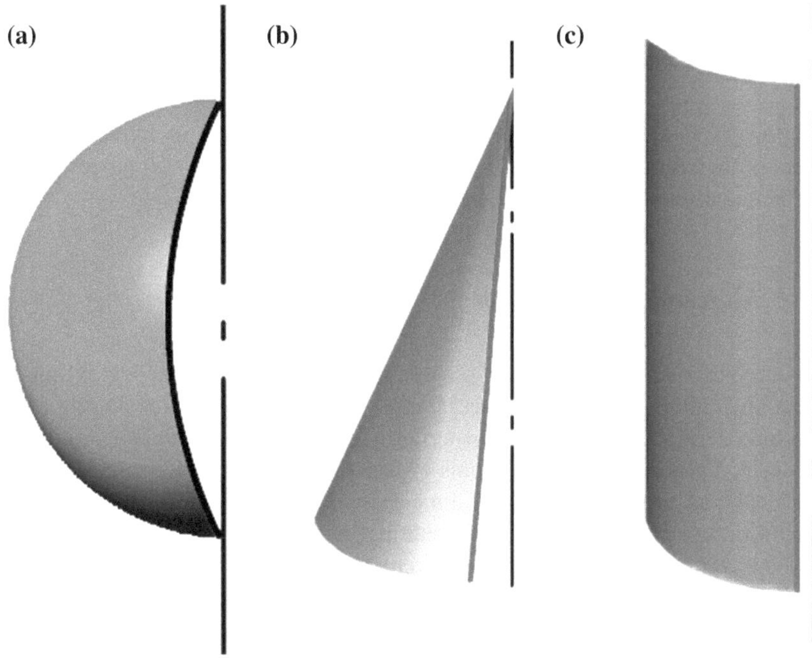

Fig. 21.1 The formation of gyration curved surfaces. **a** *Spherical* surface; **b** *conical* surface; and **c** *cylindrical* surface

images with the transparent background, which is introduced into Flash. Similar to the two stages of animation to simulate the semi-circle generatrix rotation, the process to simulate the sphere appearing gradually is divided into two stages. The first part of the mask animation is to simulate the left sphere appearing with the semi-circle generatrix rotating from the left to the middle. The second part is to simulate the right sphere appearing with the semi-circle generatrix rotating from the middle to the right. The mask layer of the first part is a part of deformation animation from a rectangle with the right side cut to a complete rectangle. The shape of the cut sphere is completely same with that of the semi-circle generatrix. The masked layer of the first part is the above-mentioned static sphere image introduced into Flash. Similar to the first part of the mask animation, the mask layer of the second part is the deformation animation from a complete rectangle to one with a semi-circle projected in the right side. The masked layer is the same static sphere image. In the layer setting of Flash, the animation to stimulate the semi-circle generatrix rotation is above the animation to stimulate the sphere appearing gradually with the starting frames of two parts in alignment. The simulation can be made on the formation of cylindrical and conical surfaces, as indicated in Fig. 21.1b and c.

21.3 Dynamic Simulation of an Object Being Cut by a Plane [1, 2]

Such process is also a virtual one, which may be involved in introducing the concept of the section in the drawing course. The simulation of such process in 2D space is made mainly with the mask animation in Flash.

21.3.1 Materials Preparation

The materials to simulate such cut process mainly include a complete 3D model without any cutting, two parts of the 3D model after cutting and the cutting plane in the space. All above-mentioned material figures should be created in AutoCAD in advance. Various 3D modeling tools in AutoCAD are used to create 3D model of the object according to the given dimensions and make copies of such model. The cutting tool is used to decide the cutting position and to cut the copied 3D model into two parts according to requirements. The space figure of the cutting plane is drawn according to the above-mentioned cutting position. Then, 3D dynamic viewer of AutoCAD is used to adjust the angle of materials in the screen, in the cutting animation the materials will be observed along this angle. Then, the button of Prt. Sc is used to copy these materials into Photoshop to eliminate the backgroud. Then, all images are introduced in Flash for further using.

21.3.2 Simulation Animation

In the animation of the cutting simulation, it is required to combine the image of the object behind the cutting plane with that in front of the cutting plane, making a complete object. However, in the actual animation with Flash, there is a thin line at the joint, just like a seam; however, accurately the material images are placed in alignment. Therefore, the whole object seems like a combination of two parts before the cutting, reducing the reality of the cutting simulation. The key of the simulation animation is how to realize the seamless cutting, to guarantee that the object looks complete without seam before the cutting plane is reached. The mask animation technique of Flash is used to solve such problem after the study. Five image layers are set up from the bottom to the top in Flash: The first is the ordinary layer to place the static image of the complete; the second is the masked layer to place the static image of the part of the object behind the cutting plane after the cutting, of which the position is in total alignment with corresponding part of images in ordinary layer; the third is also the masked layer to place the moving animation of the cutting plane from the top to the bottom; the fourth layer is still the masked layer to place the static image of the part of the object in front of the cutting plane, making sure combining a complete object with the image in the second layer, in alignment with corresponding part of the image in the first layer; the fifth layer is the mask layer with the mask image covering the whole object to be cut with the bottom edge in alignment with that of the cutting plane and moving totally simultaneously with the cutting plane.

The whole simulation process can be described as follows: In the place where the cutting plane reach the object is cut into two parts, the cutting trace can be seen on the surface of object. In the place where the cutting plane does not reach, the complete object beyond the cutting plane can be seen without any similar cutting seam on the surface, as indicated in Fig. 21.2. The dynamic simulation is quite real on the whole cutting process. Specific examples on the dynamic simulation of the object being cut can be referred to in Part 2 of Chap. 5 in Ref. [1].

21.4 Dynamic Simulation of Decomposition of the Combined Object

It is hard to make 3D dynamic simulation of the decomposition process of the combined object in the reality. Such dynamic process will be involved in the three-plane projection drawing in the drawing course. The combination of bearing seat is

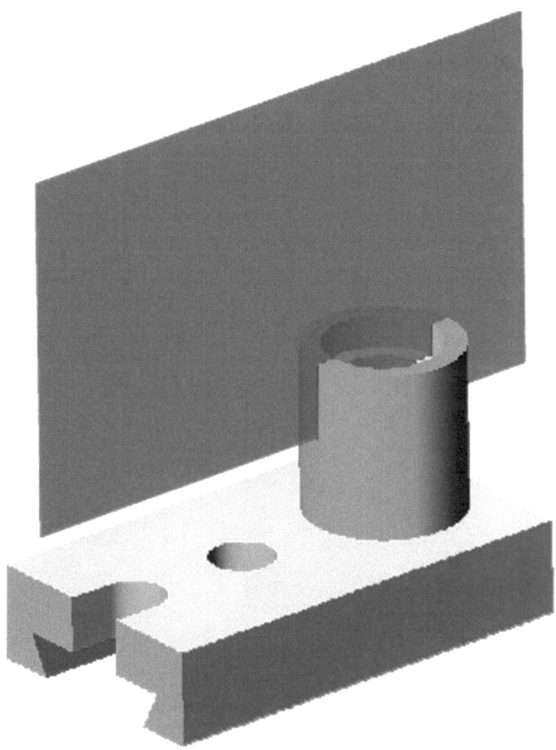

Fig. 21.2 The object is cut by a plane

taken as an example to introduce the main methods and key techniques to be used in the simulation, as indicated in Fig. 21.3.

The bearing seat consists of four parts of baseplate, support plate, rib plate and bearing sleeve. The 3D models of the above parts are created in AutoCAD and copied into Photoshop for further processing. The images are introduced into Flash. Although the images of these parts look vertical, they are still 2D images without the property of 3D images. For example, the shearing sleeve image can be placed simply on the semi-circle position of the supporting plate image to make sure the shearing sleeve can only block the back part of the semi-circle of the supporting plate without covering the front part. It requires creating the images of unblocked and blocked parts of the supporting plate with the initial material images of the supporting plate. The image layer function of Flash is used to place the blocked image at the bottom and to place the unblocked image at the top with images of two layers in accurate combination, making a complete image of the supporting plate. However, in the actual animation with Flash, there is a line like a seam at the joint of two images; however, accurately the images are combined, reducing the reality of images. One of the key techniques to make dynamic simulation is how to make seamless combination of the blocked image and unblocked image. In the experimental study with Flash, the mask layer and masked layer functions are used to solve the problem successfully. A complete image of the supporting plate is placed in the bottom layer of Flash (images of the slab plate and shearing sleeve are placed in the middle layers). A mask layer and a masked layer are created as above. The complete image of the supporting plate is still placed in the masked layer and in full alignment with the image of the supporting plate in the bottom layer. The mask figure in the mask layer is the unblocked part of the supporting plate. The joint action of the mask layer and the masked layer is

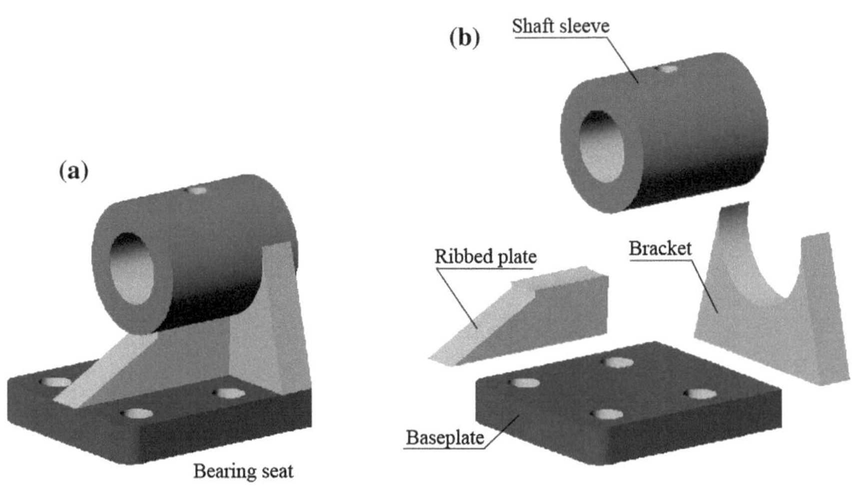

Fig. 21.3 The decomposition of a combined object. **a** Before decomposition; **b** after decomposition

that the top layer displays the image of unblocked part of the supporting plate and the bottom layer displays the image of blocked part of the supporting plate, thus realizing the seamless combination of the two parts of the images.

21.5 Dynamic Simulation of the Flattening Process of Six Basic Views in the Engineering Drawing

Technical drawing standards of China stipulate that the six surfaces of the regular hexahedron are considered as basic projection planes. The views of the object in the regular hexahedron on six basic projection planes are considered as six basic views. The flattening of six basic projection planes results in six basic views on the same plane. The arrangement positions and related reasons of six basic views on the plane are directly related to the flattening method and process of six basic projection planes, which the beginners must understand and memorize. The dynamic simulation of the flattening of six basic views is quite effective to strengthen such understanding and memory. It is difficult to make flattening of six basic views in reality, mainly because the object in the regular hexahedron must be kept suspended in the air. The main method and key techniques to simulate the flattening process in 2D space are as follows:

If the six planes of the regular hexahedron are not transparent, the object inside and projections on six planes will not be visible. It is difficult for the beginners to understand such condition. Therefore, the planes at the left, the front, and the top are set in the transparent state in Flash. In such case, the object inside and the projection figures on six planes of regular hexahedron can be seen out of the hexahedron. The flattening of the regular hexahedron is the plane rotating in the space or the composite motion of the rotation and translation. The simulation is made by the continuous deformation in 2D space of parallelograms indicating each plane and projection figures inside, as indicated in Fig. 21.4. Each continuous deformation is a part of motion animation. However, in the actual operation of the animation, Chinese characters in the projection figures on the planes of the regular hexahedron cannot make continuous deformation with the projection figures, but just become inclined in a certain angle.

The following method is used to solve the problem successfully after the experimental study with Flash. Select all frames of the animation which show the continuous deformation process and transfer each frame into key frame, thus, the motion animation is changed into the animation frame by frame. Separate Chinese characters twice in each key frame using tools of Flash, as a result, Chinese characters are changed into fill patterns. In such condition, Chinese characters can deform continuously with the projection figure. The example of the above-mentioned dynamic simulation can be referred to in related contents in Part 1 of Chap. 5 in Ref. [1].

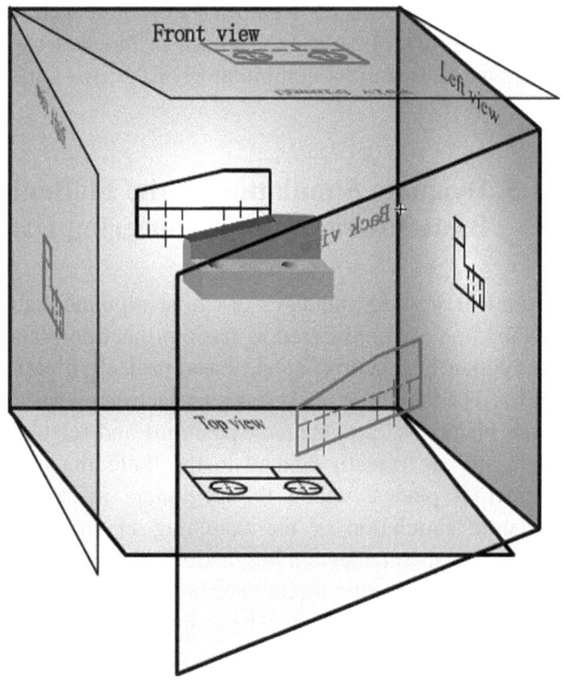

Fig. 21.4 The flattening of six basic views

21.6 Conclusion

In the thesis, the introduction is made, with integrity of related examples in Audio and Video Electronic Textbook of Mechanical Drawing, on main methods to simulate 3D dynamic process in 2D space and key techniques to solve various particular problems in the actual simulation process. It is an approximate simulation. However, the practice proves that the effect of such simulation is quite real in some simple 3D dynamic processes.

References

1. Liu F (ed)(2012) Audio and video electronic textbook of mechanical drawing. Dongbei University Audio and Video Press, Shenyang, ISBN 978-7-88830-039-2 ISRC-CN-Q03-12-0010-5/V.G
2. Xu J, Qi Y, Han Q (eds) (2007) Mechanical drawing. Tianjin University Press, , 10, 1st edn, Tianjin

Chapter 22
Kernel-Based Enhanced Maximum Margin Criterion Algorithm for High-Dimensional Feature Extraction

Chan Zhang and Haifeng Hu

Abstract A new kernel discriminant analysis algorithm, called kernel-based enhanced maximum margin criterion (KEMMC), is presented for extracting features from high-dimensional data space. In EMMC, the local property is taken into account so that the data points of neighboring classes can be mapped far away. Moreover, the regularized technique is employed to deal with small sample size problem. It is extended to a nonlinear form by mapping the input space to a high-dimensional feature space that can make the mapped features linearly separable. Extensive experiments demonstrate the effectiveness of the proposed algorithm.

Keywords Feature extraction · Kernel-based enhanced maximum margin criterion (KEMMC) · Maximum margin criterion (MMC) · Small sample size problem

22.1 Introduction

In the last ten years, with the development of the computer technology, the digital data, and its storage, analysis and access have been in explosive growth. Most of such data are multi-media data, including the texts, images, and video data with quite high dimension. However, it is nearly impossible to learn the data with high dimension in the statistical pattern recognition. The effective method to such problem is to make the dimension reduction [1, 2].

C. Zhang (✉)
PLA Institute of Physical Education, 510502 Guangzhou, China
e-mail: zhangjsh@139.com

H. Hu
School of Information Science and Technology, Sun Yat-sen University, 510275 Guangzhou, China

S. Long and B. S. Dhillon (eds.), *Proceedings of the 13th International Conference on Man-Machine-Environment System Engineering*, Lecture Notes in Electrical Engineering 259, DOI: 10.1007/978-3-642-38968-9_22,

The popular dimension reduction algorithm at present is the linear discriminant analysis (LDA) [1], which attempts to find a linear transformation to maximize between-class scatter matrix S_B and to minimize within-class scatter matrix S_W. However, it requires solving the inverse matrix of S_W with such method. It is generally unrealizable, because S_W is usually a singular matrix due to quite high dimension of the input data in the actual application. Some methods are present for such problem. The common method is PCA + LDA [2], with which to reduce dimension with PCA, thus, making S_W a nonsingular matrix before the application of LDA. However, the dimension reduction with PCA can cause the loss of the discrimination information. Another method is Direct LDA (DLDA) [3], which can eliminate the nullspace of S_B, causing the loss of the discrimination information in the nullspace. TNLDR (Temporal Nonlinear Dimensionality Reduction) also made some improvement [4].

Li et al. proposed a feature extraction method based on maximum margin criterion (MMC) [5], on which Lu et al. made expansion [6]. In MMC, it does not require solution of the inverse matrix of S_W, which is the advantage. Meanwhile, the dimension reduction on the data of high dimension can be made directly in the initial data space, avoiding the loss of important discrimination information. However, there is noise interference in the calculation of the within-class distance. Most importantly, MMC performance may be worsening rapidly in case of small sample size of the training sample.

In the thesis, a new method to reduce the dimension is present, called enhanced maximum margin criterion (EMMC). EMMC can maximize between-class average distance. Besides, the consideration is made on local structure information of the data distribution. EMMC is only a linear method. Therefore, EMMC based on kernel (KEMMC) is proposed with which the input data can be mapped to high-dimensional feature space, and the data are distributed linearly in the space. So KEMMC has better discrimination performance.

22.2 Proposed Method

22.2.1 Formula

Let $x_i \in \mathbb{R}^N (i = 1, 2, \ldots, n)$ be N-dimension sample data, $y_i \in \{1, 2, \ldots, c\}$ the class ID of x_i and c class number.

The linear dimension reduction is intended to find a $N \times r$ transmission matrix w, making the data z_i after the dimension reduction be:

$$z_i = w^T x_i \tag{22.1}$$

Here, $z_i \in \mathbb{R}^r (1 \le r \le N)$ is the low-dimension value to $\boldsymbol{x}_i$. r is the data dimension after the reduction.

22.2.2 MMC

MMC [5] can be expressed to be:

$$J = \frac{1}{2}\sum_{i=1}^{c}\sum_{j=1}^{c} p_i p_j d(i,j) \tag{22.2}$$

where, p_i is the probability of the class $i, d(i,j)$ the between-class distance of i and j, defined to be:

$$d(i,j) = d(\boldsymbol{m}^{(i)}, \boldsymbol{m}^{(j)}) - s(i) - s(j) \tag{22.3}$$

where, $\boldsymbol{m}^{(i)}$ and $\boldsymbol{m}^{(j)}$ is the mean vector of i and j $s(i)$ is the distribution value of the class i, which can be estimated with the total variance $\mathrm{tr}(S_i)$.

The above formula shows that J is the total between-class margin, which will increase in case of the sample distance of the same class becoming smaller and the data distance of the different class becoming larger. It shows that MMC has the similar effect with the traditional LDA. However, it does not require solving the inverse matrix of S_W with MMC.

There are two problems of MMC in the actual application. Firstly, the covariance matrix of each sample class may be error and unstable due to insufficient training samples. As a result, the estimated value of $s(i)$ may be unstable in the height. Secondly, MMC can maintain the global structure of the data in the dimension reduction. However, the local structure is more important in the actual application. Therefore, an enhanced MMC method is proposed for such problems.

22.2.3 Regularized Single Class Covariance Matrix

With the small size sample, the single within-class distribution matrix $\boldsymbol{S}_\ell$ may have extreme deformation. The regularization technique is the common solution.

$$\tilde{\boldsymbol{S}}_l = \boldsymbol{S}_l + \gamma \boldsymbol{I} \tag{22.4}$$

where γ is regularization parameter within the range of [0, 1]. I is the unit matrix.

Another regularization program is:

$$\tilde{\boldsymbol{S}}_l(\alpha) = \boldsymbol{S}_l + \gamma \cdot c \cdot \boldsymbol{S}_W \tag{22.5}$$

where $\boldsymbol{S}_W$ is the within-class average distribution matrix.

22.2.4 Distance Quantity Between Different Classes

Similar to LPP [7], a calculation method on between-class distance considering data structure information is redefined:

$$\tilde{d}(\boldsymbol{m}^{(i)}, \boldsymbol{m}^{(j)}) = \boldsymbol{A}_{i,j}\left\|\boldsymbol{m}^{(i)} - \boldsymbol{m}^{(j)}\right\|^2 \tag{22.6}$$

where, $\|\cdot\|$ is Euclid modulus of $\mathbb{R}^N$, $\boldsymbol{A}_{ij}$ is similar matrix.

$$''\boldsymbol{A}_{ij}'' = \exp\left(-\left\|\boldsymbol{m}^{(i)} - \boldsymbol{m}^{(j)}\right\|^2 \Big/ \tau\right) \tag{22.7}$$

where, τ is positive.

22.2.5 EMMC

With the newly defined within-class distribution matrix and between-class distance quantity, the following EMMC can be obtained:

$$\begin{aligned} \tilde{J} &= \frac{1}{2}\sum_{i=1}^{c}\sum_{j=1}^{c} p_i p_j (\tilde{d}(\boldsymbol{m}^{(i)}, \boldsymbol{m}^{(j)}) - \mathrm{tr}(\tilde{S}_i) - \mathrm{tr}(\tilde{S}_j)) \\ &= \frac{1}{2}\sum_{i=1}^{c}\sum_{j=1}^{c} p_i p_j \tilde{d}(\boldsymbol{m}^{(i)}, \boldsymbol{m}^{(j)}) \\ &\quad - \frac{1}{2}\sum_{i=1}^{c}\sum_{j=1}^{c} p_i p_j \left(\mathrm{tr}(\tilde{S}_i) + \mathrm{tr}(\tilde{S}_j)\right) \\ &= \tilde{J}_1 - \tilde{J}_2 \end{aligned} \tag{22.8}$$

where

$$\tilde{J}_1 = \frac{1}{2}\sum_{i=1}^{c}\sum_{j=1}^{c} p_i p_j \tilde{d}(\boldsymbol{m}^{(i)}, \boldsymbol{m}^{(j)})$$

while

$$\tilde{J}_2 = \frac{1}{2}\sum_{i=1}^{c}\sum_{j=1}^{c} p_i p_j \left(\mathrm{tr}(\tilde{\mathrm{S}}_i) + \mathrm{tr}(\tilde{\mathrm{S}}_j)\right)$$

From Formula (22.6), $\tilde{J}_1$ can be expressed into:

$$\begin{aligned}\tilde{J}_1 &= \frac{1}{2}\sum_{i=1}^{c}\sum_{j=1}^{c} p_i p_j \tilde{d}(\boldsymbol{m}^{(i)}, \boldsymbol{m}^{(j)}) \\ &= \frac{1}{2}\sum_{i=1}^{c}\sum_{j=1}^{c} p_i p_j (\boldsymbol{m}^{(i)} - \boldsymbol{m}^{(j)})^T (\boldsymbol{m}^{(i)} - \boldsymbol{m}^{(j)}) \boldsymbol{A}_{ij} \\ &= \sum_{i=1}^{c}\sum_{j=1}^{c} p_i p_j (\boldsymbol{m}^{(i)})^T \boldsymbol{A}_{ij} \boldsymbol{m}^{(i)} \\ &\quad - \sum_{i=1}^{c}\sum_{j=1}^{c} p_i p_j (\boldsymbol{m}^{(i)})^T \boldsymbol{A}_{ij} \boldsymbol{m}^{(j)}\end{aligned} \tag{22.9}$$

$\tilde{J}_1$ and $\tilde{J}_2$ can be rewritten into:

$$\begin{aligned}\tilde{J}_1 &= \sum_{i=1}^{c}\sum_{j=1}^{c} (p_i \boldsymbol{m}_i)^T \boldsymbol{B}_{ij} (p_i \boldsymbol{m}_i) - \sum_{i=1}^{c}\sum_{j=1}^{c} (p_i \boldsymbol{m}_i)^T \boldsymbol{A}_{ij} (p_j \boldsymbol{m}_j) \\ &= \frac{1}{n^2}\sum_{i=1}^{c}\sum_{j=1}^{c} (\boldsymbol{X}^{(i)} \boldsymbol{1}_{n_i}^T)^T \boldsymbol{P}_{ij} (\boldsymbol{X}^{(j)} \boldsymbol{1}_{n_j}^T) \\ &\quad - \frac{1}{n^2}\sum_{i=1}^{c}\sum_{j=1}^{c} (\boldsymbol{X}^{(i)} \boldsymbol{1}_{n_i}^T)^T \boldsymbol{A}_{ij} (\boldsymbol{X}^{(j)} \boldsymbol{1}_{n_j}^T) \quad \text{and} \\ &= \frac{1}{n^2}\operatorname{tr}\left(\sum_{i=1}^{c}\sum_{j=1}^{c} (\boldsymbol{X}^{(i)} \boldsymbol{1}_{n_i}^T) \gamma_{ij} (\boldsymbol{X}^{(j)} \boldsymbol{1}_{n_j}^T)^T\right) \\ &= \frac{1}{n^2}\operatorname{tr}(\boldsymbol{X}\boldsymbol{L}\boldsymbol{\gamma}\boldsymbol{L}^T\boldsymbol{X}^T) \\ \tilde{J}_2 &= \frac{1}{n}(1 + \gamma \cdot c) \cdot \operatorname{tr}(\boldsymbol{X}\boldsymbol{R}\boldsymbol{X}^T)\end{aligned}$$

where, $\boldsymbol{X} = [\boldsymbol{X}^{(1)}\ \boldsymbol{X}^{(2)} \cdots \boldsymbol{X}^{(c)}], \boldsymbol{\gamma} = \boldsymbol{P} - \boldsymbol{A}$ and $\boldsymbol{P}$ is $c \times c$ matrix.

$$\mathbf{p}_{ij} = \begin{cases} \sum_{j=1}^{c} \mathbf{B}_{ij} & \text{if } i = j \\ 0 & \text{otherwise} \end{cases}$$

$$\mathbf{B}_{ij} = \mathbf{A}_{ij} \cdot \frac{p_j}{p_i}$$

Meanwhile, R is $c \times c$ matrix: $\boldsymbol{R} = \boldsymbol{I} - \boldsymbol{T}$ and

$$\boldsymbol{T}_{ij} = \begin{cases} 1/n_l & \text{if } y_i = y_j \\ 0 & \text{otherwise} \end{cases}.$$

Therefore, Formula (22.8) can be expressed finally into:

$$\begin{aligned}\tilde{J} &= \tilde{J}_1 - \tilde{J}_2 \\ &= \frac{1}{n^2}\mathrm{tr}\left(\mathbf{X}\mathbf{L}\gamma\mathbf{L}^T\mathbf{X}^T\right) - \frac{1}{n}(1+\gamma\cdot c)\mathrm{tr}\left(\mathbf{X}\mathbf{R}\mathbf{X}^T\right) \\ &= \mathrm{tr}\left(\mathbf{X}\mathbf{H}\mathbf{X}^T\right)\end{aligned}$$

where, $\boldsymbol{H} = \frac{1}{n^2}\boldsymbol{L}\gamma\boldsymbol{L}^T - \frac{1}{n}(1+\gamma\cdot c)\boldsymbol{R}$

Till now, EMMC feature vector can be obtained with the following optimization problem:

$$\begin{aligned}&\arg\max_{\boldsymbol{w}} \boldsymbol{w}^l\boldsymbol{X}\boldsymbol{H}\boldsymbol{X}^l\boldsymbol{w} \\ &\quad \mathrm{s.t.}\ \boldsymbol{w}^T\boldsymbol{w} = 1\end{aligned} \tag{22.10}$$

The Lagrange constant method can be used to solve the above optimization problem. Then:

$$\boldsymbol{X}\boldsymbol{H}\boldsymbol{X}^T\boldsymbol{w} = \boldsymbol{\lambda}\boldsymbol{w} \tag{22.11}$$

The above formula shows that feature vector w is the feature vector of $\boldsymbol{X}\boldsymbol{H}\boldsymbol{X}^T$ correlated with maximum feature value. Generally, the transformation matrix w is a matrix of the first r maximum feature vectors of $\boldsymbol{X}\boldsymbol{H}\boldsymbol{X}^T$. H is a symmetric matrix. So w is an orthogonal matrix.

22.2.6 EMMC Based on Kernel

EMMC is linear, so it may be invalid in case of the input data in nonlinear distribution. Therefore, EMMC is expanded to the nonlinear condition.

A known nonlinear map of $\boldsymbol{\Phi}$ can be mapped to higher dimension feature space $\mathcal{F}$ after inputting the data space of $\mathbb{R}^N$.

$$\begin{aligned}\boldsymbol{\Phi} &: \mathbb{R}^N \to \mathcal{F} \\ \boldsymbol{x} &\to \boldsymbol{\Phi}(\boldsymbol{x})\end{aligned} \tag{22.12}$$

For a selected $\boldsymbol{\Phi}$, an inner product can be decided in $\mathcal{F}$, making a regenerated kernel Hilbert Space [8]

$$\langle \boldsymbol{\Phi}(\boldsymbol{x}), \boldsymbol{\Phi}(\boldsymbol{y})\rangle = k(\boldsymbol{x}, \boldsymbol{y}) \tag{22.13}$$

$k(\cdot,\cdot)$ is a positive semi-definite kernel function.

Let $\boldsymbol{v}$ the mapping function in the feature space, MMC of F can be transformed into:

$$\boldsymbol{v}_{\mathrm{opt}} = \arg\max \boldsymbol{v}^T\boldsymbol{\Phi}(\boldsymbol{X})\boldsymbol{H}\boldsymbol{\Phi}^T(\boldsymbol{X})\boldsymbol{v} \tag{22.14}$$

Because of $\boldsymbol{v} \in \mathcal{F}$ in the training sample in the feature space [8].

$$\boldsymbol{v} = \sum_{i=1}^{n} \theta_i \boldsymbol{\Phi}(\boldsymbol{x}_i) \tag{22.15}$$

To define $n \times n$ Gram matrix G $\boldsymbol{G}_{ij} = \boldsymbol{\Phi}(\boldsymbol{x}_i)^T \boldsymbol{\Phi}(\boldsymbol{x}_j) = \langle \boldsymbol{\Phi}(\boldsymbol{x}_i), \boldsymbol{\Phi}(\boldsymbol{x}_j) \rangle = k(\boldsymbol{x}_i, \boldsymbol{x}_j)$

With integrity of (22.14), optimization (22.15) equivalent to:

$$\boldsymbol{\theta}_{\text{opt}} = \arg\max \boldsymbol{\theta}^T \boldsymbol{GHG\theta} \tag{22.16}$$

The above formula can be transformed into the following problem to solve the eigenvalue:

$$\boldsymbol{GHG\theta} = \boldsymbol{\lambda\theta} \tag{22.17}$$

where $\boldsymbol{\theta} = [\theta_1\, \theta_2\, \cdots\, \theta_n]^T$ provides the mapping function $\boldsymbol{v}$ in the feature space.

Let $\{\boldsymbol{\theta}_1, \cdots, \boldsymbol{\theta}_{r-1}\}$ be the first $r-1$ feature vectors in Formula (22.17) on nonzero eigenvalue. The transformation matrix $\Theta = [\boldsymbol{\theta}_1, \ldots, \boldsymbol{\theta}_{r-1}]$ is a $n \times (r-1)$ one. Any input sample of $\boldsymbol{x} \in \mathbb{R}^N$ can be projected into $r-1$ dimension feature subspace.

$$\boldsymbol{x} \to \boldsymbol{z} = \Theta^T K(:, \boldsymbol{x})$$

where, $K(:, \boldsymbol{x}) \doteq [k(\boldsymbol{x}_1, \boldsymbol{x}), \ldots, k(\boldsymbol{x}_n, \boldsymbol{x})]^T$.

22.3 Experiment Result

In this part, the discussion will be made on the classification performance of EMMC algorithm based on kernel.

22.3.1 Data Set

In the study, CMU PIE, Isolet, and MNIST databases are used, of which CMU PIE face database contains 41,368 face images of 68 classes. All images of five near front postures (C05, C07, C09, C27, and C29) under different lights and expressions are selected, obtaining 170 different images. All face images are arranged in a line and cut into 32 × 32 pixel. For each class, random selection is made on images of L(= 5, 10, 20, and 30) for training and the remaining are used for the test.

Table 22.1 Correct recognition rates (%) obtained by different dimensionality reduction methods on the PIE database

	$L=5$	$L=10$	$L=15$	$L=20$
LDA [1]	69.58 ± 0.86	78.34 ± 0.77	84.53 ± 0.68	92.07 ± 0.18
MMC [5]	68.20 ± 1.04	76.93 ± 0.88	82.61 ± 0.92	91.56 ± 0.55
EMMC	71.72 ± 0.59	85.30 ± 0.92	90.39 ± 0.51	92.82 ± 0.13
KDDA [9]	70.44 ± 1.11	86.48 ± 1.05	93.13 ± 0.36	95.38 ± 0.27
KEMMC	74.16 ± 1.27	88.74 ± 0.71	93.60 ± 0.28	95.73 ± 0.29

22.3.2 Result

The above database is used to compare the performance of different feature extraction methods (as LDA, MMC, and KDDA algorithm), of which the discrimination result is in Table 22.1. The following conclusion can be made:

1. KEMMC is better than KDDA [9], indicating the method in the thesis has higher capacity to extract discrimination information in high-dimension data space. Besides, the consideration is made on between-class local information and regularized information of the data points of the same class. The data classification and the local information between regularized information are of the same class, making it has more capacity to find the discrimination information in the data space.
2. EMMC always shows better performance than MMC, because the regularized technique is used in estimating single-class covariance matrix $\boldsymbol{S}_i$ in the method. It is possible for $\boldsymbol{S}_i$ to be highly deformed in the condition of small size sample.

22.4 Conclusion

In the thesis, a new method is proposed to reduce the dimension, called Maximum Margin Criterion based on kernel (KEMMC), exacting features from high-dimension space. The introduction is made firstly that EMMC can maximize the average margin between different classes and can integrate the local information between classes and regularized information of data points of the same class, guaranteeing the stronger capacity to find the discrimination information. KEMMC realizes nonlinear expansion with kernel, to map the input space to a high-dimension feature space. CMU PIE database is used to compare the method of the thesis with other famous dimension reduction techniques. The results show that the method can have higher identification performance under all experiment conditions, which shows that the capacity to extract discrimination information from high-dimension space is higher than that of other methods.

References

1. Swets DL, Weng J (1996) Using discriminant eigenfeatures for image retrieval. IEEE Trans Pattern Anal Mach Intell 18(8):831–836
2. Belhumeur PN, Hespanha JP, Kriegman DJ (1997) Eigenfaces vs. fisherfaces: recognition using class specific linear projection. IEEE Trans Pattern Anal Mach Intell 19(7):711–720
3. Yu H, Yang J (2001) A direct LDA algorithm for high-dimensional data with application to face recognition. Pattern Recogn 34(10):2067–2070
4. Gashler M, Martinez T (2011) Temporal nonlinear dimension reduction. In: Proceedings of international joint conference on neural networks, San Jose, California, USA, 31 July–5 Aug 2011
5. Li HF, Jiang T, Zhang K (2006) Efficient and robust feature extraction by maximum margin criterion. IEEE Trans Neural Networks 17(1):157–165
6. Lu G-F, Lin Z, Jin Z (2010) Face recognition using discriminant locality preserving projections based on maximum margin criterion. Pattern Recogn 43(10):3572–3579
7. He X, Yan S, Hu Y, Niyogi P, Zhang H (2005) Face recognition using laplacianface. IEEE Trans Pattern Anal Mach Intell 27(3):328–340
8. Liu Q, Tang X, Lu H, Ma S (2006) Face recognition using kernel scatter-difference-based discriminant analysis. IEEE Trans Neural Networks 17(4):1081–1085
9. Lu J, Plataniotis KN, Venetsanopoulos AN (2003) Face recognition using kernel direct discriminant analysis algorithms. IEEE Trans Neural Networks 14(1):117–126

Chapter 23
Research of Big-Data Mining Visualization Application

Chan Zhang

Abstract Big-data value is well recognized. How to extract useful information from the large, high-dimension, massive, diverse forms of big-data in a timely manner, and to help people judge, decision-making, to present new challenges for data-mining technology. Visual analysis of big-data mining has more intuitive, efficient results and can be applied to military, national security, public services, financial, business and other fields. So it is a hot topic in data-mining technology. This paper describes several typical applications of big-data mining visualization and tracks a single example of money laundering by the data-mining visualization and ends with a prospect.

Keywords Big-data · High-dimensional information · Data mining · Data-mining visualization

23.1 Introduction

The data are playing an important role in making judgment and decision. With the coming big-data era, the decision is made increasingly on the basis of the data and analysis but not of experience and intuition in the business, economy, and other fields. In 2012, the term "big-data" is described. The government of Obama announced *Big-Data Research and Development Initiative* [1], which would promote a new competition on the research and development of big-data analysis and processing techniques and talent cultivation in the world. The United States is intended for the research and development of big-data, to establish the new direction and breakthrough of the informatization development to consolidate the statue of the global information power.

C. Zhang (✉)
PLA Institute of Physical Education, 510502 Guangzhou, China
e-mail: zhangjsh@139.com

S. Long and B. S. Dhillon (eds.), *Proceedings of the 13th International Conference on Man-Machine-Environment System Engineering*, Lecture Notes in Electrical Engineering 259, DOI: 10.1007/978-3-642-38968-9_23,

23.2 Features of Big-Data

The globally known consulting firm McKinsey & Company is the first to announce the coming big-data era. It said: "The data have been penetrated into each industry and business functional fields today as an important production factor. The mining and using of massive data indicate a new tide of the production rate growth and consumers' surplus is coming" [2]. It has been a long time for the big-data exists in physics, biology, environmental ecology and military, finance, and communication. However, it becomes the concern with the development of Internet and information industry recently.

23.2.1 Massive Data

In Internet, the big-data means the user network activities data generated and accumulated in the daily operation of companies. Such data are so massive that it is impossible to measure it with G or T but begins with P (1,000 T), E(1 million T) or Z(1 billion T).

23.2.2 Various Types

The big-data are of various types, including network blog, audio, video, images, and geological position information, which makes higher requirement on the data-processing capacity. Compared with traditional structural data, the big-data are mostly high-dimensional nonstructural or semi-structural ones. It will take too much time and cost to unload such data into the relational database for further analysis. Therefore, it is urgent to develop the effective core technique for collection, storage, management, analysis and sharing, meeting the features of non-structural big-data.

23.2.3 Low Value Density

The big-data have a relatively low value density. With the wide development of Internet of Things, there is so massive information with low value density. It is the problem urgent to solve in the big-data era how to purify the data more rapidly with strong computer algorithm.

23.2.4 High Speed and Efficiency

The big-data feature high processing speed and requirement on the time efficiency, which is most remarkable feature different from the traditional data mining.

It is impossible for the existing technical structures and routines to make highly efficient processing of such massive nonstructural data. For some related organizations, it will not be worth it if the collected information cannot be processed and no effective feedback information is made with a huge investment. The big-data era makes a new challenge to the data-processing capacity but provides unprecedented space and potential for the people to obtain more profound and overall observation capacity.

23.3 Visualization of Big-Data Mining

23.3.1 Big-Data Mining

The data mining means the process to make classification and arrangement on massive, incomplete, noise, vague, and random actually used data and to make extraction of the potential, previously unknown but related useful information and knowledge. The original data can be structural, as the data in the relational database, or semi-structural, as texts, figures and images or even nonstructural data in the network. The data mining is actually a deep data analysis method on correlation analysis, clustering analysis, classification, prediction, time series pattern, and variance analysis with the purpose of mining the knowledge from the data to support making decision.

It is a long history for the data analysis. However, the data collection and analysis was mainly intended for the application in a relatively narrow field, as in science research. The samples are structural sampling data, local data, and partial data. With the development of corresponding software and hardware, great improvement of the processing and storage capacity of the computer and the development and popularization of the Internet, it is possible for various industries to obtain and use overall data, complete data, and system data (mostly are nonstructural and high-dimensional data) in quite deep level and quite many fields. The big-data mining is widely used in the state security, business, finance and other common industrial fields, besides the science research.

- In 2009, the engineers of Google made prediction on the time and region of H1N1 flu according to the search records on the Internet, which is in high correlation of 97 % with the official data and more timely than that of the center for disease control [3];
- In 2012, Farecast system made analysis on air ticket booking database and predicted the domestic ticket price of the United States with the accuracy as high

as 75 %. Similar to the search records of Google, such flight records support for "Global aviation spread and Early warning study of H1N1 flue [3]";

- According to consumption habits and products surfing records, Amazon will make the most accurate product recommending information automatically instead of random massive advertisements, improving the purchasing rate effectively;
- Facebook owns active users of 845 million each month. The activity mode of users can indicate the potential of the users changing into the customers more accurately. Facebook can also transform the traditional massive advertisements into an accurate customer search. The report of Marin, a research institution of the United States, shows that the average value of CPC (the unit price paid for each click on the advertisement by the advertiser) of the searching advertisements in the 4th quarter of 2011 was US$0.85. However, that of Facebook was only US$0.23 [4], which indicates that the accurate advertisement has high return of investment than the key words advertisements;
- In 2012, Obama won the president election again, overturning the law of the president election in the history of the United States. The data-mining team of Obama has been making searching, storage, and analysis on quantities of data in the past 2 years. They took different mobilization channels and different interactions to different users, no matter in collecting election fund, decision on purchasing advertisements or voting mobilization. Therefore, it made an election wonder which was impossible in the past.

23.3.2 *Visualization of Big-Data Mining*

More than 70 % of the information is obtained with eyes every day. The sense of sight is most rapid in all sensors, much faster than the sense of touch and sense of smelling. Much information is perceived with eyes from different channels every day. Eyes are extremely sensitive to changing color, shape, and pattern, which are considered as the language of eyes. With integrity of the language of the brain, or words, numbers, and concepts, it can provide fast understanding on the information by work jointly and promote each other.

23.3.2.1 Features of Data Visualization

The data-mining visualization means mining the correlation information from the enormous and complicated data and changing into clear, direct, and effective visualization mode with assistance of graphics. It will make full use of the rapid recognition capacity in the visualization mode and show people the relation between the data and correlation of information with simpler and clearer mode. It has super variety and expressive force, which can be dynamic video or static visualization images.

23.3.2.2 Related Method for the Data Visualization

The data visualization is intended to show highly effective and clear communication information with assistance of graphics. To pass more effective data information and assist the users to understand and response, the visualization, on the basis of present multiple media means, ranges from the plane to 3D with the media from the paper to the network and video, changing continuously in the interaction and time-efficiency. The typical means are as follows [5]:

- Cladogram (phylogeny)
- Color alphabet
- Dendrogram (classification)
- Halo (visualization technique)
- Heatmap
- Hyperbolic Tree
- Multidimensional scaling
- Problem-Solving Environment
- Treemapping.

23.3.3 Application Examples of the Big-Data Visualization

23.3.3.1 Facebook Users Rose Diagram

Figure 23.1 is Facebook users statistics data diagram [6]. The diagram scatters outside with the center as the base point in the visualization, which is also called rose diagram because of the figure in a flower shape. Such diagram can display the high-dimensional data and its relation to the people simultaneously, quite clearly and definitely.

23.3.3.2 We Feel Good: Purchasing and Selling Stocks

Such project is based on statistics results of quantities of Blogs, showing the people emotion [7]. The database containing millions of people emotions with newly added records of 15,000–20,000, provides a series of interesting interfaces for inquiry and data mining. In Fig. 23.2, the particle system is one of visualization modes generated with the interface, extracting 1,500 data, each of which is a particle, indicating an emotion of each people. The colors are corresponding to the emotion types, bright yellow indicating the happiness, deep blue the sadness, red the anger, and light green the calmness. The size of each particle is correlated with the length of the sentence to express the emotion. The figure can show the analysis results clearly.

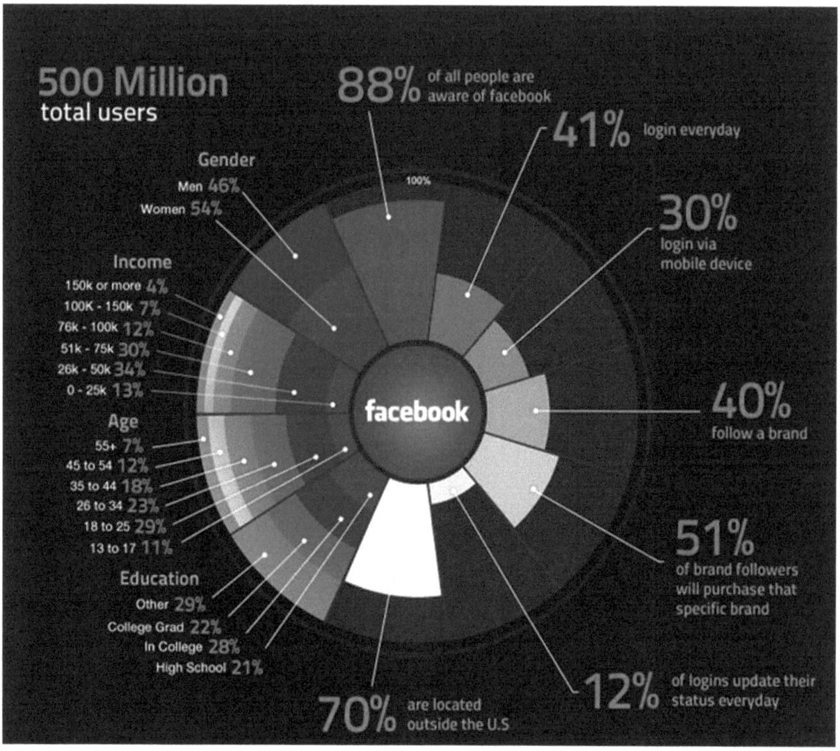

Fig. 23.1 Facebook user data visualization

"He will buy when you are happy while he will sell when you are anxious." Paul Hawtin, CEO of Derwent Capital Market in Wall Street, will make utilization and judgment on such emotions and mark from 1 to 50, one of his daily work contents. Hawtin will decide how to deal with millions of stocks in hand according to the marking results. Hawtin has the simple principle for judgment that he will

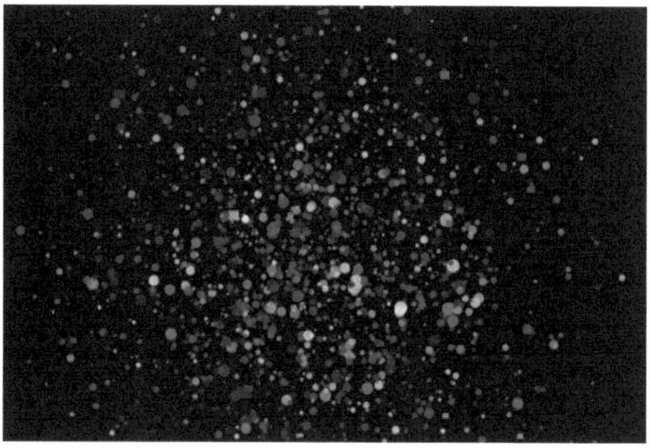

Fig. 23.2 Visualization of human feelings data

buy the stocks when all people seems happy and will sell out when all people seems anxious. It has remarkable achievement. In the first quarter of 2012, the return of investment of Hawtin's company was 7 % [8].

23.3.3.3 Visualink Analysis Sytem (We Can Find You: Bank Money Laundering)

Some day in 2010, Mr. Dai went in a bank as usual, taking a thick envelope of shabby cashes of the face value of RMB ¥ 20–100. He deposited RMB ¥ 185,000 approximately in half week. He was busy every week, living about 12 km away from the bank. He received the service of the bank nearly for one year. He wound not like the bank too near his apartment and he did not like being questioned too much on his transaction.

According to Mr. Dai, he operated the express service and was paid with the cash by most of clients. Each time Mr. Dai deposited the cash of RMB ¥ 180,000–190,000 in the bank. He was informed to fill in a form to explain the nature of his transaction in details when he once attempted to deposit the cash more than RMB ¥ 200,000. The teller told him the currency transaction report (CTR) must be submitted in case of any depositing or drawing accumulated cash more than RMB ¥ 200,000 each day. Then, Mr. Dai deposited the cash less than RMB ¥ 200,000 each time, to avoid compulsory CTR submission requirement.

The bank had examined his self account and could not confirm he had any cause, which Mr. Dai did not know. The bank also checked his credit report and residence history, learning that he had no reliable employment to support such amount of cash deposit. Actually, this transaction, together with all transactions in the bank was recorded with a particular mode, called Suspicious Transaction Report (STR). In STR, all detailed documents on transactions are recorded, including account type, account number, name, address, ID number, transaction date, amount, telephone number, reciprocal account number, name and suspicious transaction feature description.

Visualink analysis tool is a graphic analysis tool unrelated with the platform, which provides a set of solution with powerful functions on data sharing, integrity, analysis, and pattern discovery. It is designed to integrate various different data sources and to mine the pattern, trend, correlation and hidden network of the data sources of any quantity and types. Such tool is used to make analysis on STR data, search for STR records and transfer records of targeted objects. It can be used to make visualization analysis and show the fund flow of the targeted account, providing the evidence for related financial activities or even criminal activities of targeted objects

Correlation analysis on STR reports of Mr. Dai showed that he had more than 10 STR records in 2010, as indicated in the upper part of Fig. 23.3. To avoid attention, a person usually opens different accounts for his financial activities. Mr. Dai often used 2 accounts. The careful reading on suspicious transaction

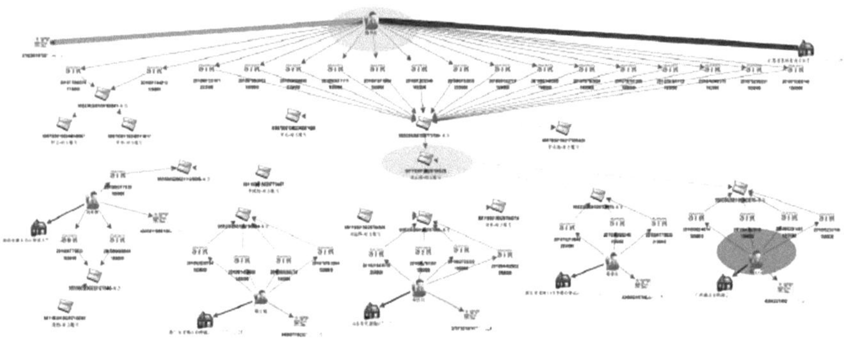

Fig. 23.3 Suspicious transactions associated trajectory of bank

feature description showed that the transaction subject was considered to participate in a certain kind of illegal drug activities.

From here to expand the correlation for several steps (9 in Figure), as indicated in the lower part of Fig. 23.3, showing all involving subjects in the transaction. After careful analysis and recognition, Mr. Deng (indicated with the red circle) was confirmed a drug dealer reported in the recent news. Then the important clue and capital flow were discovered: Mr. Dai → Mr. Liu → Mr. Deng.

23.4 Conclusion

The coming big-data era is changing the thought mode and brining about new opportunity for the social public management and business pattern. In March 2012, the government of Obama announced Big-Data Research and Development Initiative, involving in various sectors of science, healthcare, energy, state defence and geological exploration of the federal government. These sectors promised to make total investment of over US$200 million to promote and improve the collection, organization and analysis tools and techniques related to the big-data, including various projects to create effective and convenient customized visualized human–computer interaction tools. With the increasing mature of the big-data visualization technique, the big-data will not be mysterious and abstruse and will display to the people in the more clear, simple and direct mode.

References

1. http://www.whitehouse.gov/blog/2012/03/29/big-data-big-deal
2. http://wenku.it168.com/d_000739288.shtml

3. Mayer-Schonberger V (2013) Big data: a revolution that will transform how we live, work, and think. Houghton Mifflin Harcourt, Boston
4. http://www.1mfg.com/1mfg/j/StoryAff/show/39019.shtml
5. http://blog.xiqiao.info/tag/data-visualization
6. http://www.yixieshi.com/ucd/10556.ht
7. http://www.wefeelfine.org/movements.html
8. http://bjwb.bjd.com.cn/html/2012-06/15/content_100013.htm

Part III
Research on the Environment Character

Chapter 24
Research on the Background Noise of Ordinary Classrooms

Jie Li, Kang Jiang and Jian-Zhong Guo

Abstract The level of background noise is one of the key factors to influence the classroom acoustic; for students study can be disturbed in noisy environment, 17 classrooms are chosen including 8 old classrooms and 9 newly build classrooms of a college in Beijing. Sound pressure level and noise criteria were calculated. The result of the average background noise in new classroom is 43.1 dB(A) and 45 dB(A) in old classroom. The noise in old and new classrooms exceeds the standards in different levels. The value of the NC of two old classrooms is higher than the limited level and that of others meets the requirements of the standards. Classroom indoor background noise is mainly affected by the indoor air-conditioning system, the surrounding classroom situation, and the outdoor environmental noise. The further discussion is made on the results in the paper.

Keywords Classroom · Classroom noise · Background noise · NC curve

24.1 Introduction

The background noise in the classroom is main part of the acoustic environment in the classroom, including noise from the electric equipment, heating and air-conditioning systems, floor and outside noise (as indicated in Fig. 24.1). High background noise will mask the voice or reduce the clearness of the speech, directly influencing the communication of teachers and students, or even the attention of students in the classroom. The study shows that the acoustic

An erratum to this chapter is available at 10.1007/978-3-642-38968-9_72

J. Li (✉) · K. Jiang · J.-Z. Guo
Capital University of Economics and Business, Beijing 100070, China
e-mail: lijie_jerry@126.com

S. Long and B. S. Dhillon (eds.), *Proceedings of the 13th International Conference on Man-Machine-Environment System Engineering*, Lecture Notes in Electrical Engineering 259, DOI: 10.1007/978-3-642-38968-9_24,

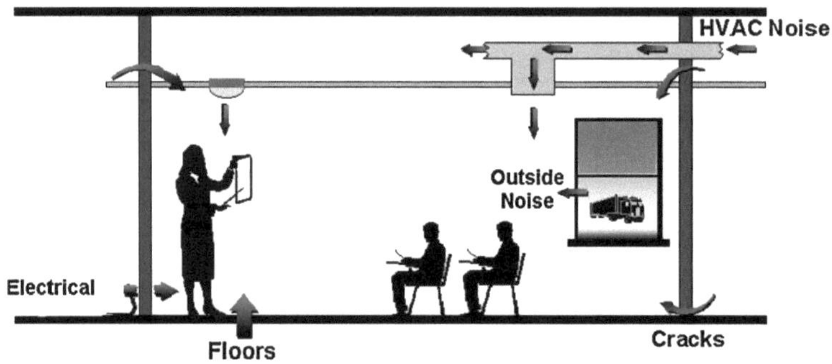

Fig. 24.1 The main source of background noise in the classroom [4]

environment in the classroom will affect the achievements of students [1]. The further study shows that the noise from the air-conditioning or heating systems is the cause affecting test scores or achievements of students [2]. It is easier to get fatigue for students in poor acoustic environment for a long time. The acoustics in the classroom has been a concern of acoustic study in recent years. The study is deepening to improve and increase the acoustic environment quality in the classroom in China. In the study, the college ordinary classroom is selected as the object to measure, analyze, and study the background noise in the classroom. The ordinary classroom means the classroom for general courses teaching, except special classrooms for music, sports, and language. The background noise in the classroom in the study means outside noise and noise from classroom equipment (lighting equipment and air conditioners), excluding noise from teaching activities and teaching equipment [3]. In the paper, the analysis is made on measured results based on the background noise and NC curve.

The related study shows that the noise in the empty classroom is generally 41–51 dB (Manlove Frank 2001; Bess Sinclair 1984; Crandell et al. 1994) [18]. Related stipulations are made on requirements of the acoustic environment in the classroom in different countries at present. Standards related with the acoustics in the classroom is (GB50118-2010) *Code for Design of Sound Insulation of Civil Buildings* Part 5, which requires the allowable noise of the ordinary classroom $\leq$45 dB(A) [3]. The Acoustic Society of the USA started to concern and study the acoustic problem in the classroom in 1994 [5] and made amendment on ANSI S12.60-2002 (R2009) classroom acoustics in 2009 and 2010. It divided the original standards into two parts at present [6, 7]. It will make addition and improvement on the content of Part 3 [8]. The standards stipulate the learning space is no larger than 566 m^3. Otherwise, it will be considered the teaching supporting area. It also requires that the background noise in the classroom no larger than 566 m^3 is no larger than 35 dB(A) and that in the classroom larger than 566 m^3 is no larger than 40 dB(A). Britain Building Bulletin 93 (2003) requires the limit of $L_{Aeq,30min}$ is 35 dB(A) and Japan requires the limit of L_{Aeq} is 40 dB(A). American Speech Language Hearing Association ASHA (1995) [19] stipulates that it is 30–35 dB(A) while British Association of Teachers of the Deaf BATOD (2001) [19] stipulates that it is no larger than 35 dB(A). Besides, other countries or organizations as France, Germany, Brazil, Iran, and WHO made stipulations on the background noise in the classroom [9–11].

NC curve (noise criteria) appearing as early as 1957 is a widely accepted method (Beranek and Ver 1992) [20]. NC is a single-evaluation quantity on the acceptance degree of the background noise in the classroom, involving sense loudness or annoyance and speech interference level. To draw the octave band sound pressure level curve on the reference line, the low curve without any part exceeded by the octave band sound pressure level curve is the required NC value [13]. For classroom NC value, Beranek recommended to be NC35-40, Stumpf and Sharland recommended to be NC30—35, and Parkin and Humphreys recommended to be NC25 approximately [14]. The recommended value of NC in literature [15] is 25–30.

24.2 Materials and Method

Blue-sole modularized integrating sound level meter of 01 dB is used for measurement and calibrated before using in the experiment. The measurement is made at breaks of 11:00–14:00 in accordance with the stipulations of GB50118-2010 when it is sunny. The measurement is made with the windows closed, air conditioner and lighting equipment started and without teaching personnel in the classroom [3].

The measured classrooms have similar features, except the ceiling is cement wall in the old classroom while the ceiling is decorated in the new classroom. In the measurement, two measuring points are selected in the classroom less than 30 m^2, of which one measuring point at the center of the classroom. Three measuring points are selected evenly on the central line in the lengthwise direction of the classroom for the classroom larger than 30 m^2. Typical measuring points are selected, as indicated in Fig. 24.2.

The measuring points are evenly distributed along the central line parallel to the wall with largest window area in case of the room plane being a square. The measuring points are 1.5 m approximately away from the floor and no less than 1 m from the reflecting surface as the wall. The distance between measuring points and the distance between measuring points and the noise sources are larger than 1.5 m. GB50118-2010 stipulates that if the study object is the static noise, the equivalent continuous A sound level for 5 min should be made one time at each measuring point. All measured values at the measuring points should be made for mean energy according to Formula (24.1) [3].

$$\overline{L_{\mathrm{Aeq}}} = 10\log_{10}\sum_{i=1}^{N} 10^{0.1\cdot L_{\mathrm{Aeq}}-i} - 10\log_{10}N \tag{24.1}$$

Fig. 24.2 Layout of background noise measurement points

24.3 Results and Analysis

The measurement is made on the background noise in 17 classrooms, and records are made for measuring results of L_{Amax}, L_{Amin}, $L_{Ceq,5min}$, and $L_{Aeq,5min}$ (as indicated in Table 24.1). *Code for Design of Sound Insulation of Civil Buildings* of China stipulates that the background noise in the classroom shall be no higher than 45 dB(A). With the reference to Fig. 24.3, it shows that the selected objects have exceeded the stipulations at different levels. The values of three classrooms of old and new classrooms are remarkably higher than the standard value. The background noise value in the new classroom are all less than 50 dB(A) with the highest being 48.3 dB(A). The background noise values in three old classrooms are higher than 50 dB(A), with the highest being 54.5 dB(A). Wang Jiqing et al. thought after the study that the background noise higher than 50 dB(A) would impact the teaching sharply [16].

Besides, the US classroom acoustic standards not only stipulates the equivalent continuous sound level A no larger than 35 dB(A), but also stipulates the equivalent continuous sound level C no larger than 55 dB(C). The present standards in China do not stipulate the sound level C. Therefore, it is only considered the reference to the sound level A without any further analysis. However, with reference to Fig. 24.4, we can easily find that the noise in the classroom is mainly of the low frequency, for which the sound level C has much better responding level than the sound level A; therefore, it is suggested that the sound level C has been provided as the reference in the evaluation of the acoustic environment in the classroom in the further upgrading the standards.

Table 24.1 Basic measurements of the noise level of 17 classrooms

Classroom no.		Floor	Volume (m^3)	L_{Aeq} dB(A)	L_{Ceq} dB(C)	$L_{A\text{-}max}$	$L_{A\text{-}min}$	Air conditioner	Remarks
New classroom	5#407	4	448.08	44.4	54.5	55.4	43.4	2-T	B
	5#611	6	185.92	43.0	53.6	60.3	41.7	1- T	B
	5#317	3	185.92	44.9	54.5	55.2	43.1	1-T	B
	5#205	2	448.08	37.7	52.4	55.4	32.4	F	B
	5#733	7	185.92	48.3	56.5	60.0	47.5	1-T	C
	5#725	7	399.15	45.8	55.8	54.8	44.8	2-T	C
	5#723	7	399.15	47.4	57.6	55.5	46.3	2-T	C
	5#728	7	399.15	46.3	56.6	50.2	45.2	2-T	C
	5#307	3	448.08	30.0	48.1	47.2	25.4	F	C
Old classroom	3#213	2	123.86	39.8	49.4	67.5	26.7	F	A
	3#211	2	206.68	44.3	53.8	60.4	38.5	1-T	A
	3#313	3	123.86	54.5	58.0	63.7	46.8	1-T	A
	3#311	3	206.68	51.9	55.6	57.8	41.5	F	A
	3#216	2	206.68	46.0	56.1	52.5	45.0	1-T	B
	3#220	2	298.52	53.6	62.0	54.9	52.9	1-T	B
	3#418	4	298.52	35.9	48.1	53.1	30.4	F	C
	3#214	2	206.68	33.9	48.4	54.8	28.6	1-T	C

Note (1) "2-T" indicates the air conditioner number-normal operation state and *F* indicates the air conditioner cannot operate in the classroom. (2) A indicates that the classroom has windows on one side toward the subway construction site and it has impact from the continuous or interrupted noises during the measurement. *B* indicates the classrooms on two sides or one side are having lessons. *C* indicates no classroom around having lessons

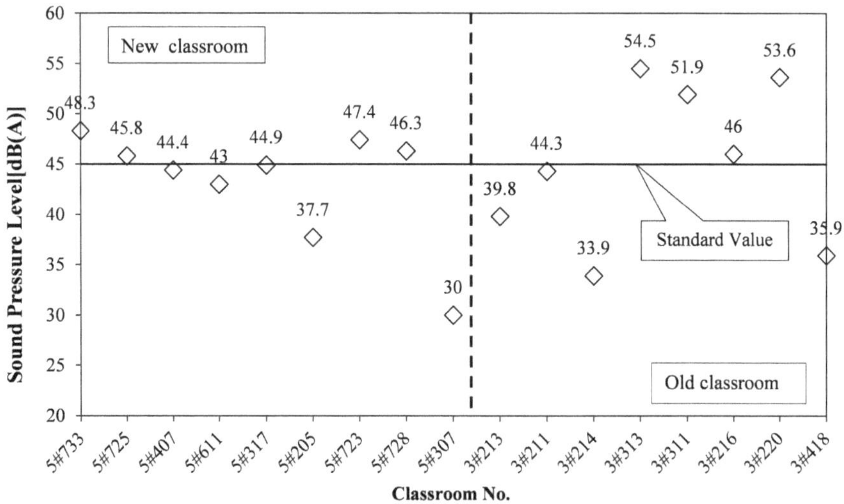

Fig. 24.3 Comparison of the measured values with the standard value

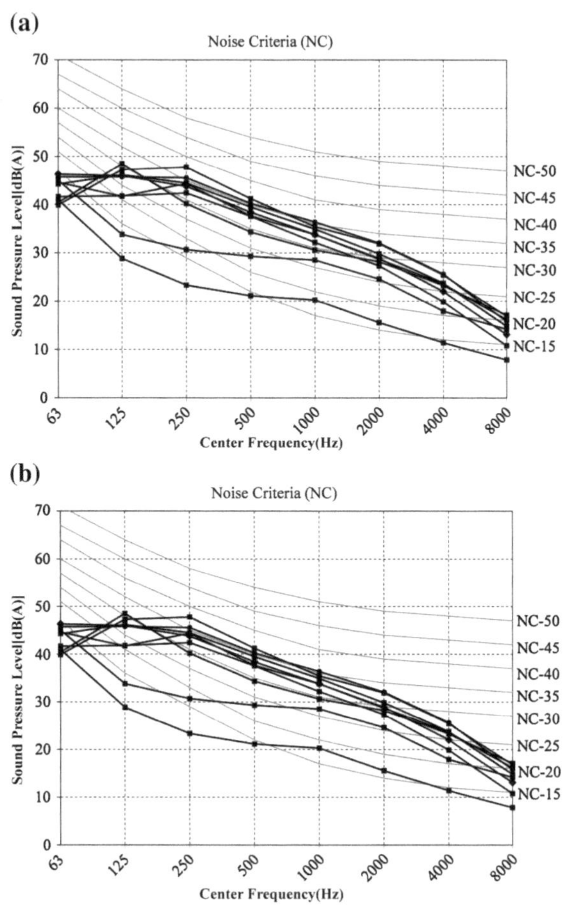

Fig. 24.4 The location of the old and new classrooms octave band sound pressure level in the NC curve **a** The NC value of the old classroom, **b** The NC value of the new classroom

Table 24.2 Octave noise levels of 17 classrooms

Classroom no.		Octave center frequency f/Hz								NC value
		63	125	250	500	1,000	2,000	4,000	8,000	
New classroom	5#205	45.0	33.8	30.7	29.3	28.5	24.6	17.9	14.2	27
	5#307	41.0	28.8	23.3	21.2	20.3	15.5	11.4	7.8	18
	5#317	44.6	41.7	44.4	37.7	33.8	28.7	23.6	16.2	33
	5#407	41.6	41.8	42.4	37.6	32.2	27.3	19.9	10.8	33
	5#611	40.4	48.5	40.2	34.3	30.6	28.2	23.3	17.1	31
	5#723	45.7	45.8	45.6	40.3	36.4	32.0	25.7	15.8	35
	5#725	46.3	46.0	43.8	38.4	33.8	28.9	22.0	13.1	33
	5#728	44.3	46.2	44.5	39.5	34.9	29.8	23.8	14.8	35
	5#733	39.9	47.3	47.8	41.3	35.5	31.9	25.5	16.9	37
Old classroom	3#211	45.8	42.5	39.7	38.1	33.6	29.4	19.5	13.8	33
	3#213	44.0	35.1	33.0	33.0	29.5	26.7	15.6	11.5	28
	3#214	39.1	33.2	32.8	26.8	23.8	17.6	14.0	10.8	22
	3#216	46.8	46.3	41.6	37.1	35.9	30.9	33.0	25.6	35
	3#220	44.3	42.0	37.8	34.0	32.3	27.7	28.3	21.1	31
	3#311	45.7	41.2	41.4	43.5	43.6	39.7	26.6	15.4	43
	3#313	45.2	46.2	44.5	47.3	45.9	42.1	29.3	18.8	45
	3#418	36.4	34.0	29.2	24.8	26.0	23.6	22.6	15.0	26

For further analysis, the octave sound pressure level is measured in each classroom and the further analysis is made on the results with reference to NC curve. The octave sound pressure level and NC value of each classroom are present in Table 24.2. NC curve is calculated according to literature [17]. The study selects the recommended NC value of the classroom of 25–40, according to literature [14, 15]. From Table 24.2 and Fig. 24.5, we can find that the NC values

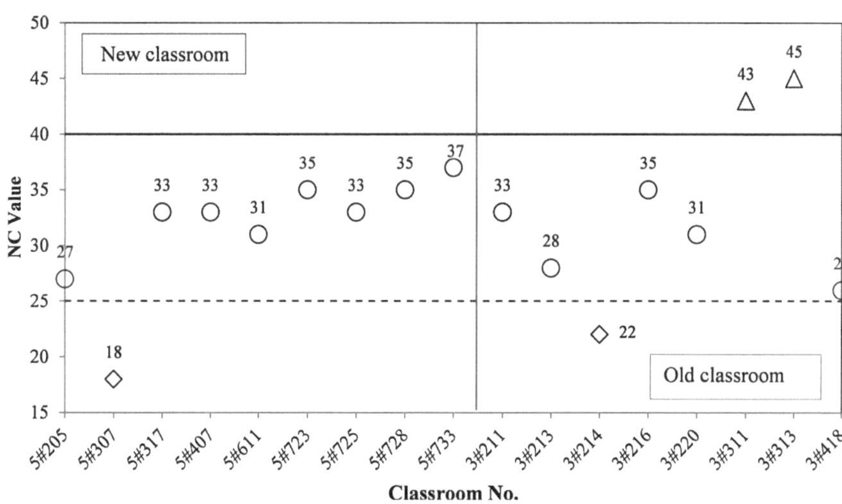

Fig. 24.5 The distribution of the NC value

of new classrooms are within the stipulated range while that of two old classrooms of 3#311 and 3#313 are higher than NC40.

24.4 Discussion

According to requirements of *Code for Design of Sound Insulation of Civil Buildings* of China, there are classrooms, old and new, with the background noise exceeding the stipulations in a different level. The influence is different in a certain level in different classrooms.

The monitoring on the background noise in the ordinary classroom shows that the acoustic environment of the new classroom is better than that of the old classroom by comparison of the equivalent sound level or NC value. The further analysis shows that it is in direct relation with the location of the old classrooms and air-conditioning equipment inside. The subway construction on one side of the old classrooms has direct impact on partial measured data. Comparison of turning on and off the air conditioners in the new and old classrooms shows that the noise from the air conditioner contributes much to the background noise in the new and old classrooms. Besides, the background noise of the classroom 3#331 reaches 51.9 dB(A) due to the subway construction. The background noise of other classrooms with air conditioners off is less than 40 dB(A).

To meet the background noise in the ordinary classroom, the insulation reconstruction on windows facing the street and the maintenance on the heating and ventilation equipment can be made. Furthermore, a certain stipulation can be made to make regular examination on the acoustic environment in the classroom and make targeted solution on the discovered problems.

The study on the acoustic environment in the classroom has been considering as the study in the construction acoustics while there is no adequate systemic study on the acoustic environment in the classroom in China. In the further study, related study blind points require supplement and refining, to provide scientific and reasonable solution for improvement of the acoustic environment in the classroom.

Acknowledgments This study has been supported by the Program for Joint Education Graduate Programs, Capital University of Economics and Business (CUEB).

References

1. Ronsse LM, Wang LM Classroom acoustics affect student achievement. http://www.csemag.com/home/single-article/classroom-acoustics-affect-student-achievement/0e5cd30611.html
2. Study shows background noise affects test scores. PHYSorg.Com.2 Nov 2011. http://medicalxpress.com/news/2011-11-background-noise-affects-scores.html
3. GB 50118-2010 Code for design of sound insulation of civil buildings
4. http://www.fefpa.org/pdf/summer2007/ClassroomAcoustics.pdf

5. Classroom Acoustics Standard TimeLine: http://www.asha.org/uploadedFiles/ASHA/Publications/leader/2010/100921/Classroom-Acoustics-Timeline.pdf
6. ANSI/ASA S12.60-2010/Part1, American national standard acoustical performance criteria, design requirements, and guidelines for schools, part 1: permanent schools
7. ANSI/ASA S12.60-2009/PART2, American national standard acoustical performance criteria, design requirements, and guidelines for schools, part 2: relocatable classroom factors
8. ANSI/ASA S12.60/Part 3, American national standard acoustical performance criteria, design requirements, and guidelines for schools, part 3: information technology equipment in classrooms
9. Yong-min S, Sheng-wo S (2006) Progress in classroom acoustics research. Tech Acoust 01:56–61
10. Shield BM, Dockrell JE (2003) The effects of noise on children at school: a review. Building Acoustics 10(2):97–106
11. Zannin PHT, Marcon CR (2007) Objective and subjective evaluation of the acoustic comfort in classrooms. Appl Ergon 38:675–680
12. Wang LM (2003) Performance review of indoor noise criteria. ASCE 1–4
13. Kang J, Dai G (2011) Urban sound environment theory. Science Press, 28
14. Li C-Y (2009) Classroom acoustics research. Archit Tech Des 10:94–99
15. NC (Noise-Criterion). http://www.engineeringtoolbox.com/nc-noise-criterion-d_725.html
16. Wang J-Q, Gu Q-G (1987) Design of sound of classroom. Inter-nosie, vol 87, pp 1141–1144, Beijing
17. Online Noise Criterion-NC Calculator. http://www.engineeringtoolbox.com/nc-noise-criterion-d_517.html
18. Savage C (2007) Classroom acoustics and intervention strategies to enhance the learning environment.14
19. http://shodhganga.inflibnet.ac.in/bitstream/10603/9572/31/10_chapter%202.pdf
20. Beranek, L. and I. Ver. (1992). Noise and Vibration Control Engineering: Principles and Applications. Wiley-Interscience.

Chapter 25
Cab Suspension Vibration Isolation Analysis Based on Vibration Decoupling Theory

Zhongliang Wei, Minghui Chen, Sijuan Zheng, Xue Shi, Liang Ling, Fang Xie, Ke Bao, Jinlong Zhao and Xi Lu

Abstract A certain engineering vehicle is taken as an example to build up the three-dimensional and full-floating cab 3D model and establish the vibration response model in the Adams platform. The vibration decoupling theory is applied to optimize the vibration stiffness of suspension and arrangement angle. The basic vibration isolation theory is used in the frequency domain for the analysis and calculation of transmission. Optimization results of the suspension system are verified, and the cab suspension system dynamics model of vehicle is created on the RecurDyn platform. The cab suspension system is placed in the vehicle for the comfort simulation to improve the comfort of the driver. NVH performance is improved with satisfactory results.

Keywords Cab suspension · Vibration decoupling · Multiple rigidity system dynamics · Vibration isolation analysis

25.1 Introduction

To guarantee the large load, high-speed, and high-power performance and low self-load, the engineering vehicles are designed with thinner body and lighter components. However, it results in larger vibration and higher noise. According to the limit of the market and related laws and regulations and requirements of customers, the vehicle designers can not ignore how to reduce the vibration and noise in the technical innovation to improve the competence [1]. According to the vibration measurement on vehicles in working at present, the vibration of a certain vehicle reaches as much as 2.5 G, several times higher than that of the present

Z. Wei (✉) · M. Chen · S. Zheng · X. Shi · L. Ling · F. Xie · K. Bao · J. Zhao · X. Lu
China North Vehicle Research Institute, Beijing, China
e-mail: 13811972705@139.com

S. Long and B. S. Dhillon (eds.), *Proceedings of the 13th International Conference on Man-Machine-Environment System Engineering*, Lecture Notes in Electrical Engineering 259, DOI: 10.1007/978-3-642-38968-9_25,

vehicles. It shows the extremely strong vibration in the cab [2]. The driver seat is directly connected with the vehicle base plate with bolts. The vehicle vibration passes to the driver through the seat supporter. It is hard to meet the driving comfort only with the cushion. It requires exploring new vibration reduction method according to the vehicle features to improve the driving comfort [3].

In the thesis, the whole built-in cab is designed within the vehicle according to the thought of modularization design. The vibration reduction optimization design is made on the basis of the vibration reduction optimization of the action system. The decoupling is made in various vibration directions on the vibration decoupling theory, obtaining optimized parameters of the suspension system in the cab.

25.2 Modeling

25.2.1 Structure Model

The 3D structure diagram is made on a certain engineering vehicle after the simplification, as indicated in Fig. 25.1.

In the figure, a capsule is added inside the cab shell to make overall vibration reduction, as indicated by the dotted line. On the basis of the action system, the capsule of composite materials is made with the casting molding technique inside the cab shell to reduce the vibration from the passing access and reducing the components' coupling vibration. The improvement is also made on the design of the operation parts in the cab according to the new can structure of the vehicle. The suspension unit is added between the inner capsule and the base plate, to reduce the vibration and noise and to improve the driving comfort in the whole cab.

Fig. 25.1 The structure model of some engineering vehicle

25.2.2 Dynamic Modeling

The capsule inside the vehicle cab (the capsule) is suspended on the chassis with the suspension system of four groups of spring damping components in the direction of front and back, and left and right. Suppose the capsule only moving vertically, take the balance position of the capsule as the origin and also consider the spring force in the initial deformation stage in balance with the gravity, then [4]:

$$m\ddot{x} = -k(x - y) - c(\dot{x} - \dot{y}) \tag{25.1}$$

If the cabin movement is harmonic vibration in the vertical direction, can be expressed as:

$$y = Y \cos\omega t \tag{25.2}$$

To solve it, then

$$\frac{X}{Y} = \sqrt{\frac{k^2 + c^2\omega^2}{\left(k - m\omega^2\right)^2 + c^2\omega^2}} \tag{25.3}$$

If let the frequency ratio $\gamma = \omega/\sqrt{k/m}$, damping ratio $\xi = c/(2\sqrt{mk})$, then the transfer ratio X/Ywill be

$$\frac{X}{Y} = \sqrt{\frac{1 + (2\xi\gamma)^2}{\left(1 - \gamma^2\right)^2 + (2\xi\gamma)^2}} \tag{25.4}$$

According Formula (25.4), we can obtain the changing curve of the transfer ratio with frequency domain, as indicated in Fig. 25.2. When the frequency ratio is larger, the elastic support will make the vibration isolation. The smaller the damping ratio is, the better the vibration isolation is. However, too small damping ratio will lead to larger resonance peak value.

The capsule structure finite element simulation is made with the inertial parameters and coordinate positions of the equipment within the cab. The suspension system is installed in positions with high strength. The shell finite element simulation shows that the four corners of the base plate of the cab are ideal [5]. The cab vibration analysis model is created on Adams platform. The cab model is directly introduced with 3D modeling software. The vehicle head base plate is replaced with a rectangle vibration platform. The suspension system between the cab and vibration platform is connected with bushing force, with the position, rigidity, and damp of the electric support applied according to design values. The input stimulus is set up as the sinusoidal acceleration on the vibration platform, and the output channel is set up as the vertical vibration acceleration, vertical displacement at the center of mass and vertical vibration acceleration and vertical displacement at the driver's seat.

A multiple rigid body dynamic model of the whole vehicle is set up at RecurDyn platform, including the position of the center of mass and inertial

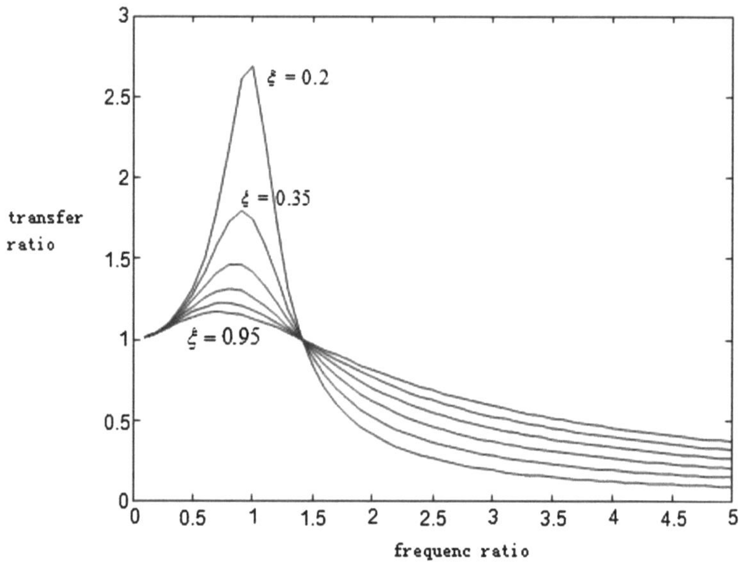

Fig. 25.2 The curve of the transfer ratio with frequency domain

parameters of the front axle, rear axle, chassis, cab, driver's seat, and tank. The comfort analog simulation is made [6].

25.3 Dynamic Analysis and Optimization Calculation

To obtain the optimized results, it requires modeling and optimization on independent cab at first. The sweep frequency mode is used to determine the vibration response curve to decide the point of resonance and effective vibration isolation frequency. For the whole vehicle simulation, generally there is no actual sample vehicle for experiment in the design process. The primary analog simulation can be made with the force curve calculated in dynamics.Optimized design process shown in (Fig. 25.3).

25.3.1 Calculation of Vibration Isolation of Suspension Unit in Cab

According to the simulation model, as shown in Table 25.1, set front suspension and rear suspension stiffness and damping ratio. The acceleration stimulus sweep frequency simulation of 1 g is made within 0.1–100 Hz. The inherent frequency, vibration type, and frequency domain response of the suspension system of the cab can be figured out.

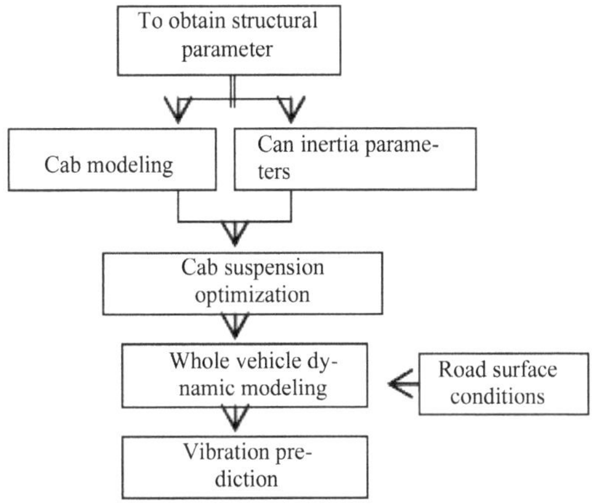

Fig. 25.3 The flow of the optimization design

Table 25.1 Suspension system parameter of cabin

Parameter	Value
Front suspension rigidity	285 N/mm
Rear suspension rigidity	335 N/mm
Suspension damping ratio	0.12

The frequency domain response of the acceleration of the center of mass of the cab is calculated, of which the result shows that the vertical acceleration of the center of mass reaches the peak at 8.75 Hz approximately. The human sensitive frequency 4–8 Hz has larger effect on the driving comfort. The vibration isolation effect of the suspension system displays the frequency range of 18 Hz. In the whole frequency range, the vertical vibration contributes most. The inherent frequency and vibration type can be obtained by the data analysis.

25.3.2 Modality Decoupling Calculation and Optimization

The rigidity matrix [7] can be obtained from the model parameters:

$$[K_{\text{com}}] = \sum_{i=1}^{n} [T_i]^T [B_i]^T [K_i][B_i][T_i] \tag{25.5}$$

where $[K_{\text{com}}]$is rigidity matrix, $[T_i]$ displacement matrix, and $[B_i]$ angle matrix.

Then, the maximum kinetic energy of main vibration in stage i of the system is

$$T_{\max}^{(i)} = \frac{1}{2}\omega_i^2 \{\psi_i\}^T [M] \{\psi_i\} \tag{25.6}$$

where ω_i is the modality frequency of the system and $\{\psi_i\}$ modality matrix vector. The percentage of the kinetic energy distributed to general coordinate k out of the total kinetic energy of the system will be

$$T_p = \frac{T_k^{(i)}}{T_{\max}^{(i)}} = \frac{\frac{1}{2}\omega_i^2 \sum_{l=1}^{6} \{\psi_i\}_l \{\psi_i\}_k m_{kl}}{\frac{1}{2}\omega_i^2 \sum_{l=1}^{6} \sum_{k=1}^{6} \{\psi_i\}_l \{\psi_i\}_k m_{kl}} \tag{25.7}$$

The energy distribution diagram is obtained.

The energy distribution diagram is obtained after determining the vibration decoupling by adjusting suspension rigidity, angle, and position, as indicated in Fig. 25.4:

To make simulation on vibration transfer ratio, we can obtain (Fig. 25.5).

Table 25.2 shows the comparison of the inherent frequency of the vibration isolation system before and after the optimization. The inherent frequency at the highest stage reduces from 13.76 to 12.60 after the optimization, and the inherent frequency of the first stage increases from 3.34 to 3.51. It means that the modality frequency at the highest stage reduces, while the modality frequency at the lowest stage increases, thus reducing the width of the inherent frequency of the system resonance.

To make further investigation on the vibration isolation effect of the suspension unit in the cab after the optimization, the whole vehicle simulation model including the independent cab is created. The straight running simulation is made

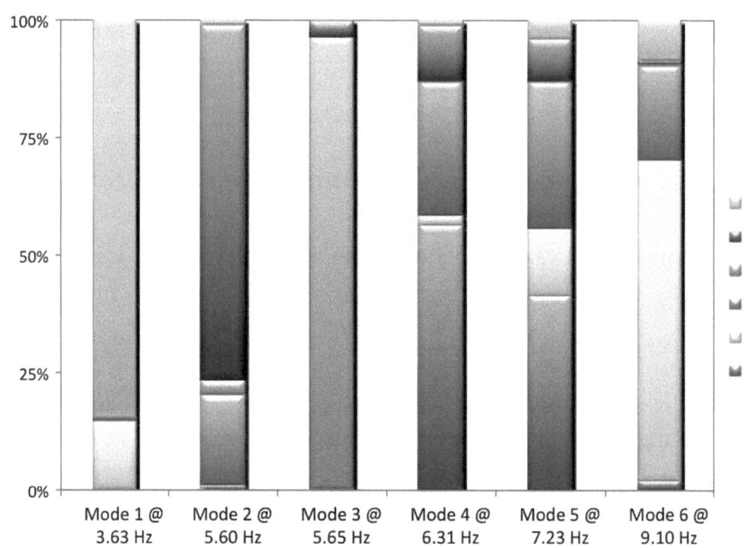

Fig. 25.4 The chart of energy distribution

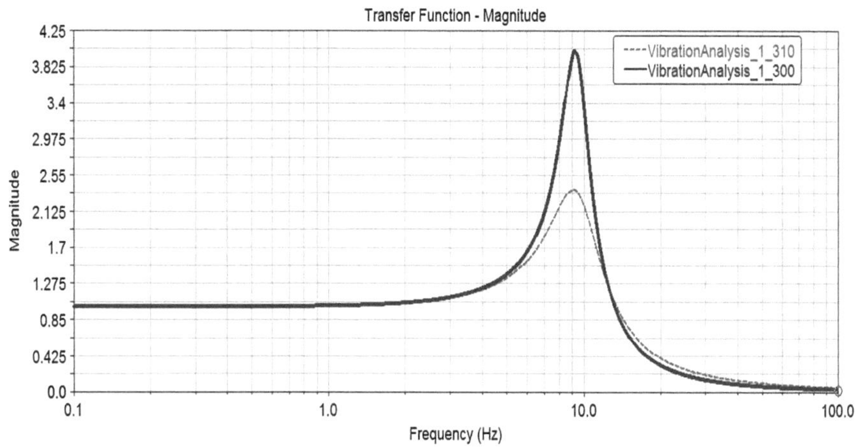

Fig. 25.5 The contrast of transfer ratio curve before and after the optimization

Table 25.2 The contrast of inherence frequency before and after the optimization

Stage	1	2	3	4	5	6
Before optimization	6.2	10.3	10.6	11.0	16.0	17.3
After optimization	6.3	10.5	10.8	11.1	15.7	17.0

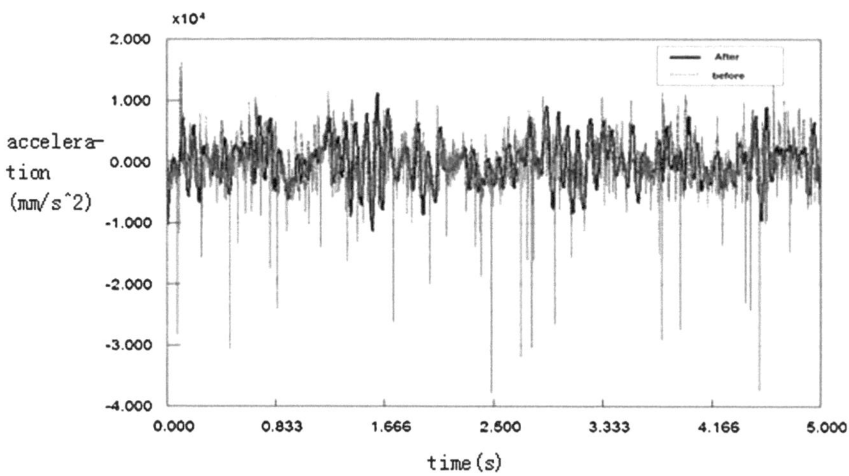

Fig. 25.6 The acceleration curve of upright direction in driver place

in the time domain at the speed of 50 km/h on the road of level E. The simulation calculation results are shown in Fig. 25.6. The acceleration RMS value reduces from 3.8 m/s^2 to 3.2 m/s^2.

25.4 Conclusion

In the thesis, a certain engineering vehicle is taken as an example to create independent cab structure model, cab dynamic model, and whole vehicle dynamic model. The simulation optimization is made on the basis of the vibration decoupling theory. The best state of the suspension rigidity of the cab is decided after the optimization calculation of the suspension rigidity and damping ratio matching. The comparison and simulation experiment show that it improves the comfort of the operator of special vehicles and also improves NVH performance of the whole cab.

References

1. Xu Q (1994) Progress and dynamic vibration isolation technology. Mechanical Strength 1:6–11
2. Li P, Ma L, He T (2005) Research on vibration separate simulation of cabin suspension in commercial vehicle. J Veh Eng 10:740–743
3. Ding Y, Guo Yongde. (1999) Vehicle and driver seat frequency response calculation method. J North China Ins Tech 12:139–146
4. Yu Z (1990) Car theory. M China Machine Press
5. Yan J (1985) Every entry isolation technology. Noise Vibr Control 2:58–64
6. Yang H, Ma L, Zhu Z, Li P (2007) Semi-trailer towing vehicle cab suspension vibration isolation simulation research. Special purpose vehicle 1:37–39
7. Yan J (1985) Mechanical vibration isolation technology. Shanghai Science and Technology Press

Chapter 26
Hardware in the Loop Simulation for Electromagnetic Environment Based on the Software Definable Signal Generator

Qian Zhang and Baohang Shao

Abstract Electromagnetic environment simulation is an important part of the developing and performance testing for communication and communication countermeasure equipments. Hardware in the loop simulation provides true wireless signal. It improves simulation credibility, thus becoming an effective method for the electromagnetic environment simulation. This paper analyzes the hardware and software constitution of the software definable signal generator. To use the signal generator to build hardware in the loop, electromagnetic simulation system was discussed. The system can be used to produce narrowband communication signals. The paper provides a reference for communication electromagnetic environment simulation.

Keywords Hardware in the loop simulation · Software defined radio · Communications · Electromagnetic environment

26.1 Introduction

The electromagnetic environment simulation is an important part of the research and manufacture and the performance test of the communication and communication confrontation equipment. At present, the electromagnetic environment simulation generally consists of mathematic simulation, physical simulation, and hardware in loop simulation. In mathematic simulation, the same set of computer equipment is equipped with different simulation software to make the simulation experiment of different systems, which is based on the mathematic equation similarity and quite general. However, it is quite difficult for modeling in the

Q. Zhang (✉) · B. Shao
PLA Institute of Physical Education, Guangzhou, China
e-mail: zq7106386@sina.com

S. Long and B. S. Dhillon (eds.), *Proceedings of the 13th International Conference on Man-Machine-Environment System Engineering*, Lecture Notes in Electrical Engineering 259, DOI: 10.1007/978-3-642-38968-9_26,

mathematic simulation and it does not produce real electromagnetic signals. Therefore, it is impossible to test the whole performance of the system in the real environment. Physical simulation means using the existing communication equipment or hardware loop combination to simulate the actual electromagnetic environment. However, in the condition with expanded communication band, various communication systems and sharply increasing signal density, such method features large investment, low coverage of signal types and quantities. It requires a lot of hardware to achieve the electromagnetic environment accumulated complexity. Therefore, such method is unfit for the actual application. The hardware in loop simulation integrates the advantages of the former two methods. On the basis of using quantities of software techniques, the hardware is used to replace the components which are hard to create mathematic model. It also provides real radio signals, increasing the reliability of the simulation. With the development of the software radio theory and technique, it is possible to integrate the basic hardware loop and various software modules to make simulation of various communication equipment, creating more complicated communication electromagnetic environment [1]. It is ideal to make A/D and D/A treatment directly on the radio frequency terminal. However, such treatment can only be made in the medium frequency at present due to the limit of the apparatus performance, to reduce the requirements to the components at the radio frequency terminals [2]. Such ideas can be used in designing the software definable signal generator.

In the thesis, the discussion is made on creating hardware in loop simulation system with the software definable signal generator, to provide the reference for the simulation of the communication electromagnetic environment.

26.2 Structure of the Hardware in the Loop Simulation Based on the Software Definable Signal Generator for Electromagnetic Environment

The software definable signal generator can be used to create hardware in loop simulation of the electromagnetic environment in the structure, as indicated in Fig. 26.1.

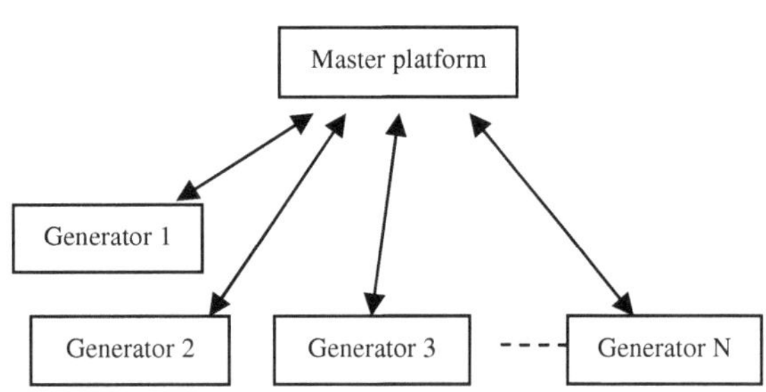

Fig. 26.1 Structure of the Hardware in the Loop Simulation Based on the Software Definable Signal Generator for Electromagnetic Environment

In Fig. 26.1, each signal generator is independent and can be set up at different positions according to different requirements and configuration programs, making different field test environment. Each signal generator receives the setting mode and parameters of the main console and sends out the voices, images, data, and facsimiles of the software signal source database in different parameters and modulation modes, generating different modulated signals. The signal generators make feedback of the working state to the main console, together with sending the signal. The main console is actually a set of network management system, providing a collective monitoring environment. The main console and any signal generator are connected with LAN. In the operation, the main console sends out the information on the signal starting time, duration, modulation type, and parameters of simulation signals to the controller of the signal generator and it will monitor the working state of any signal generator.

26.3 Software Definable Signal Generator System Structure and Working Principle

26.3.1 Software Definable Signal Generator System Structure

The block diagram of the system is indicated in Fig. 26.2.

In the figure, parts in dotted block are generated by the software and parts in continuous block are generated by the hardware. Parts in dotted block are mainly used to create the data of signal sources including voices, images, data, and facsimiles in different modulation type and parameters and store them in the hard disk of the computer in fixed format, making the signal source database. Parts in continuous block are mainly used to send different signal data in the signal source database in fixed format to the hardware modulation circuit and to generate required modulation signals after medium-frequency modulation, medium-frequency amplification, frequency fixing, and playing in high frequency. For application of less real-time requirement, the data can be generated temporarily

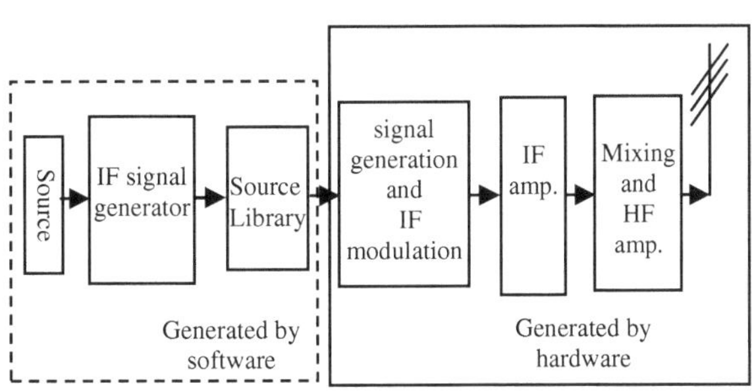

Fig. 26.2 Block diagram of the software definable signal generator system

after selecting modulation type and inputting modulation parameters and added into the signal source database. The modulation type and parameters can be expanded with expanding the modulus of the medium-frequency signal generator software. In this way, the system can be expanded on the basic hardware framework with the upgraded software, meeting the requirements on the generality of the signal. In the system, the frequency mixing and playing in high frequency are mature in techniques. The medium modulation is the key in the design.

26.3.2 Orthogonal Method to Medium-Frequency Modulation [3]

With reference to the design idea of software radio, the ideal method is to make direct D/A transformation on the data with frequency information and to generate modulated signals. However, according to the sampling law, in this case a large amount of data will be produced which is a high challenge to the data processing capacity of the computer and the data transition capacity of the bus interface. It will require more capital and limit the generated signal range.

It is well known that any modulation signal can be expressed with [4]:

$$s(t) = r(t)\mathrm{Cos}[\omega t + \theta(t)] = r(t)\mathrm{Cos}\theta(t)\mathrm{Cos}\omega t - r(t)\mathrm{Sin}\theta(t)\mathrm{Sin}\omega t \tag{26.1}$$

where the range $r(t)$ and phase $\theta(t)$ indicate the features of baseband signal.

Let $I(t) = r(t)\mathrm{Cos}\theta(t)$, $Q(t) = r(t)\mathrm{Sin}\theta(t)$,

Then,

$$s(t) = I(t)\mathrm{Cos}\omega t - Q(t)\mathrm{Sin}\omega t \tag{26.2}$$

Formula (26.2) shows that $I(t)$ and $Q(t)$ contain the range and phase information of the signal source, which do not have high requirements for the data transition of the interface due to less data quantity. After sending out such two parts of information, it is possible to pass the baseband information with lower data transition speed. Then, the orthogonal method is used to generate any modulation signals. The signal generator is designed on such principle, of which the medium-frequency modulation block diagram is indicated in Fig. 26.3.

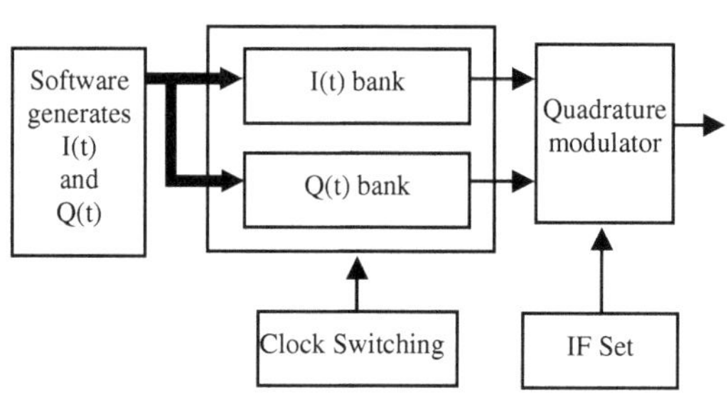

Fig. 26.3 Block diagram of the IF-modulated part of the software definable signal generator

In Fig. 26.3, after defining modulation pattern and parameters, $I(t)$ and $Q(t)$ sampling texts of the baseband signal in various modulation types can be calculated by software. Calculation on the modulation type and parameters of the software definable signal. When the signal is sent out, the data will be sent to the hardware circuit for D/A transformation. The sampling data are recovered into simulation signal and then generated into medium-frequency signal with fixed frequency after the orthogonal modulation.

26.3.3 Software System Structure and Flow

To match the hardware system, the software system makes generation of the signal source database under different modulation types and parameters and the system control and waveform display.

The software system flowchart is as follows (Fig. 26.4):

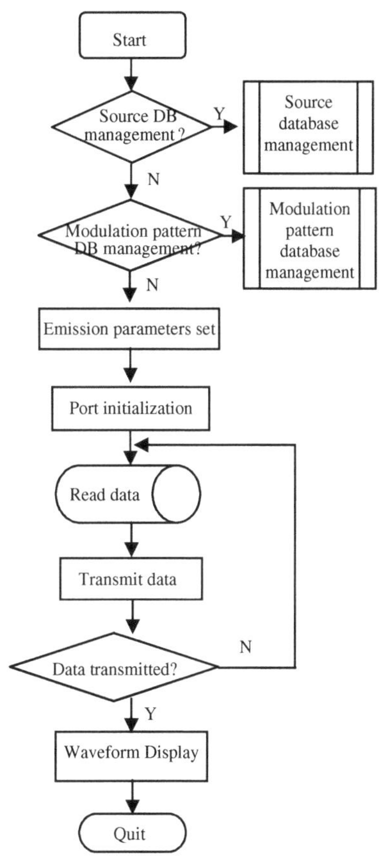

Fig. 26.4 Software system flowchart

In the software design, quantities of signal sources of SystemView of US Elanix can be used to generate baseband signals, including single frequency, sweep frequency, voices, digital baseband, and PN code. The different signal treatment models can be created with different modulus of SystemView [5, 6], thus making corresponding $I(t)$ and $Q(t)$ data.

26.4 Conclusion and Discussion

With the software simulation, the software definable signal generator can generate various common narrowband communication signals of AM, FM, PM, ASK, FSK, PSK, MSK, and GMSK. The signal parameters can be set up freely under the control of the software. The signal types in the signal source database can also be expanded according to the actual requirements, to provide solution to creating electromagnetic environment and the foundation to test the performance of the communication equipment in a different signal environment.

With the continuous development of component manufacture technique, network technique, and programming technique, continuous improvement in the mathematic modeling, especially the development of the software radio theory and technique, and also the application of the modulus design thought, the hardware in loop simulation of the electromagnetic environment will have increasingly expanded frequency range, increasingly strengthened signal density, and more and more flexible function expansion.

The software definable signal generator is quite general and easy to expand in signal types and quantity by expanding modulation type database and signal source database. The key is to find the expression as Formula (26.1) and calculate corresponding components of $I(t)$ and $Q(t)$ for different modulation types. It provides convenience for the upgrading of the system.

To expand frequency and hopping frequency and effect of the weather, topography, and electronic equipment in the signal transmission, the signal hardware in loop simulation modulus can be added at the receiving terminal of the communication equipment, to improve the reliability of the simulation.

References

1. Xiao-niu Y, Cai-yi L, Jian-liang X (2001) Principles and applications of software defined radio. Publishing House of Electronics Industry, Beijing
2. Cheng X-Y (2011) Digital automatic demodulation technology based on software defined radio. J Changchun Univ Sci Technol (Nat Sci Ed) 9:66–69
3. Shao B-H et al (2009) Design of a signal generator based on AD9857. Electron Des Eng 8:120–122
4. Zhang X-Z (1995) Principle of modern communication systems. Publishing House of Electronics Industry, Beijing

5. Zhang J-W, Zhang X-Y (2010) Narrowband FM principle and SystemView simulation. [EB/OL][2010-02-04]. http://www.elecfans.com/article/90/151/2010/20100204161958.html.[EB/OL][2010-02-04].
6. The application of virtual communication experiment based on system view. [EB/OL][2010-02-24]. http://www.elecfans.com/bandaoti/eda/20100224169353.html.

Chapter 27
Research of Detecting Positions Electromagnetic Signal Leakage by Using Sensor Network

Xiaoli Zheng and Digang Jiang

Abstract With the development of Internet of Things and Sensor Technology, sensor network is widely used nowadays; especially in the system of C^4KIR, emphasis is stressed on the system's capacity of perception, integration, and utilization of information. The sensor network becomes an important method for intelligence and reconnaissance. To meet requirements of the battlefield weapons, equipped anti-position electromagnetic signal leakage, the sensor network is used to find the position electromagnetic leakage. By analyzing the networking structure of wireless sensor network, the study is made on the composition and working principle of the system. The simulation test is made on technical parameters of this system. The result of testing proves that the system can protect the security of electromagnetic environment effectively.

Keywords Wireless sensor networks · Electromagnetic leakage · Investigation wiretapping

27.1 Introduction

In the information war, it requires the operation system of clear reconnaissance, quick response, and accurate attack. The sensor network can meet requirements on the real-time, accurate, and overall military information in various conditions. It can assist to realize the effective perception of the battlefield situations, making sure the operation force learning themselves and their enemy well. The sensor

X. Zheng (✉)
PLA Institute of Physical Education Guangzhou, Guangdong 510500, China

D. Jiang
Zhongshan School of Medicine SYSU Guangzhou, Guangdong 510080, China
e-mail: Pig_100@163.com

S. Long and B. S. Dhillon (eds.), *Proceedings of the 13th International Conference on Man-Machine-Environment System Engineering*, Lecture Notes in Electrical Engineering 259, DOI: 10.1007/978-3-642-38968-9_27,

network is an indispensable part of C^4KIRT (command, control, communication, compute, intelligence, surveillance, reconnaissance, and targeting). C^4KIRT system is intended to design an integrated battlefield commanding system of command, control, communication, compute, intelligence, surveillance, reconnaissance, and targeting for the modern war in the future with the assistance of the advanced technology, which is common concern of developed countries in military. The sensor network is made up of intensive and randomly distributed nodes of the low cost [1, 2], of which the self-structure and fault tolerance capacity can guarantee the whole system will not breakdown due to some nodes damaged in the hostile attack [3]. It is such unmatchable advantage from the traditional sensor technique that makes the sensor network quite suitable for the position battlefield, including monitoring our army's force, equipment and materials, the area of conflict, reconnoitering the enemy's topography and defense, positioning attack target, evaluating the loss, reconnoitering and detecting nuclear, biochemical attack field. Besides, the commander requires learning the timely and accurate information on the weapon equipment and military materials supplying. The sensors in the battlefield can collect related information and send the data to the commanding post with the clustering nodes. Then, such data will be sent to headquarters. Finally, the complete battlefield situation map from the integrity of the data from various battlefields is generated. Compared with the independent satellite and ground radar system, the sensor network has the following potential advantages:

1. The distributed nodes can make the information integrity in various angles and dimensions, improving the signal/noise ratio (SNR) effectively. The SNR is a technical problem hard to solve in satellite and radar systems.
2. The sensor network is designed with low cost and high redundancy, which makes sure the strong fault tolerance capacity of the whole system.
3. The sensor nodes are set up near the detection target, greatly reducing the effect of the environmental noise on the system.
4. Various types of sensors are used in integrity in the nodes, improving the detection performance.
5. Various nodes are set up, making a large real-time detection area.
6. Some movable nodes are set up, effectively eliminating the shadow and blind points in the detection area with their adjustment capacity to the network topological structure.

On the basis of such unique performances of the sensor network and requirements on information security of special operations of China, hi-tech weapon equipment positions, the thesis proposes the application of the sensor network in detecting the electromagnetic leakage in the battlefield, with the purpose of improving the capacity to examine the electromagnetic frequency spectrum in key regions and to find the wiretap in time. The wireless sensor network can obtain the information with mini-nodes. The nodes have the capacity for automatic networking and cooperation. In the network, the wireless communication mode is used to collect and process the information within the network [4]. It can make real-time and random monitoring on the near-field. It can also make timely and accurate discovery of the emission sources in near-field and any possible deliberate or accident emission signals in the battlefield and can make automatic identification of normal or abnormal signals. It can find the sensors and wiretaps and monitor normal wireless communication. It can make timely warning in case of discovering abnormally positioned and illegal communication equipment and high electromagnetic leakage source. The sensor network can store all monitoring information and documents. The system can provide theoretical basis for the rapid discovery and positioning of the microsignal source and the development and manufacturing of the anti-wiretap.

27.2 Analysis on Wireless Sensor Network

27.2.1 Hardware

The wireless sensor network is a network monitoring system, made up of various nodes of various sensors, data processing units, and short-distance wireless communication modulus, can complete appointed mission autonomously according to the environment requirements [5].

The hardware of the sensor network depends on the networking mode and application range. The network mainly consists of sensor nodes, terminal nodes, and task manager. The sensor nodes, as main parts of the whole network for their dual functions of network terminal and router, can make collection and simple processing of the local information and can also make storage, management, and integrity of the data from other nodes. The sensor nodes are distributed in the data monitoring region in large quantities. The type is determined by the physical form of the monitoring signal. The terminal nodes can make data communication between two communication networks, the communication protocol transition and management between protocol stacks, and can also send the collected data to the external network. The task manager can make examination and management on the whole network, make data processing, and provide the service for the computer or handheld equipment operating the management software. The wireless sensor network is constructed in such structure, as indicated in Fig. 27.1.

27.2.2 Network Protocol

The network communication protocol stack mainly consists of physical layer, data link layer, network layer, transport layer, and application layer [6]. No generally accepted standards have been made on layers of the sensor network at present. ieee802.15.4 standards are intended for low-speed wireless personal area network, with the design of low energy and cost as the main target, to provide the uniform standards for low-speed networking between different personal or family

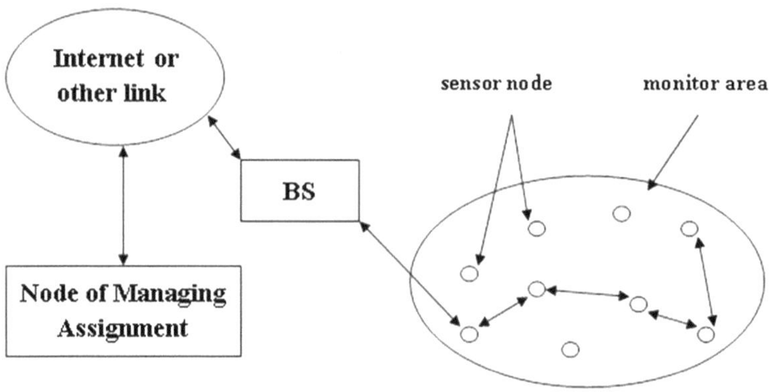

Fig. 27.1 The structure of wireless sensor network

equipment. The network features of ieee802.15.4 standards are similar to those of the wireless sensor network. Therefore, it is considered as the communication of the wireless sensor network by the technical sector at present. ZigBee protocol, based on 802.15.4 wireless standards, is generally accepted as the technical standards for the research and development of the networking, security, and application software. It has the following relation with 802.15.4 standards:

1. ZigNee makes total and full use of physical features with the strong function defined in ieee802.15.4.
2. AigBee is added with logic network and application software.
3. ZigBee is based on ieee802.15.4 RF standards.
4. 802.15.4 team is mainly responsible for formulating standards of physical layer and access control layer, while ZigBee is responsible for the development of the network layer and application layer.

27.2.3 Network Layer Protocol

The protocol is the soul of the wireless network, and it directly determines the structure of the network in case of no other network equipment being required. The network layer protocol consists of the plane protocol and layer protocol.

Plane protocol: All sensor nodes are equal, and it is easy to formulate such protocol. The network is reliable and strong as a peer one without any bottleneck. However, such protocol is poor in expansion, and each node requires knowing the route to the other nodes. It takes quantities of control information to maintain such route information in dynamic changing.

Layer protocol: The network is divided into the head of tree and various tree members. The node of the tree head is responsible for the data transmission between trees. The tree head can be appointed in advance or elected automatically by nodes with algorithm. It has the following advantages: The tree members have simple functions. It requires no maintenance on complicated route information, reducing the quantity of the route control information in the network. Thus, it can be expanded soundly and much strong. However, it has the following disadvantages: The maintenance layer structure requires the nodes carrying out tree head election algorithm. So, the tress head node may be the bottleneck of the network transmission. Therefore, different network layer protocol structures are used in the up and down routes in designing the network route protocol.

27.2.4 Network Layer Route Protocol

1. Topological structure: Such structure is made up of two different configuration methods of fixed topology and self-organized topology. The fixed sensor nodes can be configured manually to set the data transmission route in advance. But the self-organized topology is used in the more flexible sensor network, where the sensor nodes are organized with the communication protocol. The self-organized topology is used in this system.
2. Data transmission mode: The data transmission modes include continuous mode, event driving mode, inquiry driving mode, and fixing mode. For example, the emergency precaution system must send the event actively to the management platform in case of discovering any emergency. The observation node will send the event to the sensor network in case of the system requiring any data. Then, the network system will return the data in the event driving mode.
3. Route selection: The standards include minimum using times, maximum energy, and shortest distance, etc. There are single hopping and multiple hopping routes according to the node distance standards. The RF transmission quantity is in direct property to the square distance. The multiple hopping route consumes less energy than the single hopping one. However, the multiple hopping route costs much in the topology management and link connection. In the distribution of electromagnetic detection system, different detection methods lead to greatly changing detection region and high requirements on the data quality. The multiple hopping route is used is this system.

27.3 System Basic Structure and Technical Parameters

27.3.1 System Basic Structure

The system is a wireless cellular monitoring network covering all fixed region. It can make real-time monitoring with the radio signal, find any possible deliberate and accident emission radio signal in near-field, amend the space confidentiality leakage, and provide the capacity to monitor the electromagnetic spectrum and discover the

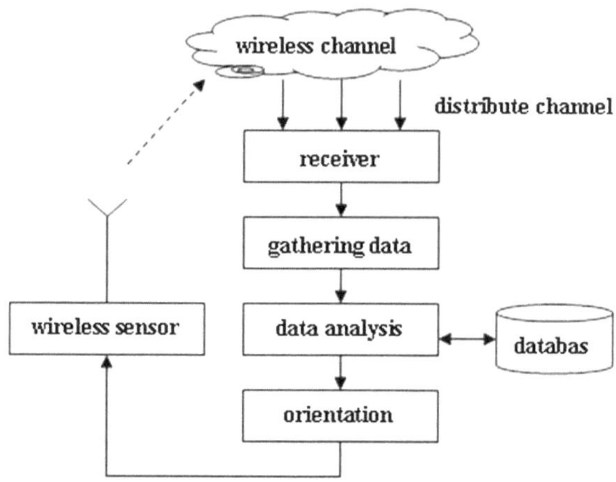

Fig. 27.2 The total structure of system

sensors and wiretaps in key regions. The system consists of the wireless signal monitor in the main system and the wireless signal sensor network of the subsystem. The wireless signal monitor consists of the antenna, receiver, communication equipment, microprocessor, monitoring system software, and database. It can make signal analysis and identification. The wireless signal sensor network is ZigBee wireless sensor network, which can discover the signal in near-field in time, measure the working frequency, and pass to the receiver along the communication channel. The receiver will make the demodulation on the signal and obtain the original signal. The total structure of the system is indicated in Fig. 27.2.

27.3.2 System Technical Parameters

The system is made up of various wireless signal sensor nodes and a wireless signal monitoring station. It is of a cellular network with each cell as a monitoring area to monitor the radio signal from the area. The monitoring station is equipped with signal process system and node communication system, which can receive the real-time radio signals and make demodulation, record, analysis, and display the signal frequency spectrum in real time. The original signal data can be stored in the system database for a long time, providing the reference data to identify suspicious signal.

27.3.2.1 Main System Technical Parameters and Function Parameters

The main system wireless signal monitoring station consists of antenna, receiver unit, networking communication, networking communication equipment, monitoring system equipment, data record, and database.

Main technical parameter index:

Monitoring frequency *fj*: 20 MHz–10 GHz.
Frequency resolution *N*: 1–20 KHz.
Max. bandwidth *B*: 5–20 MHz.
Monitoring sensitivity *V*: 3–8 uV/m.
Frequency spectrum display resolution *N*: 1 Hz.
Monitoring range: set according to situations.
Near-field radio signal identification rate: >90 %.
Communication frequency between nodes: main frequency 2 G and backup frequency 900 M.
Communication speed: 200 KB.

Main function parameters include near-field signal automatic identification, wireless node automatic networking, new signal automatic warning, signal source region positioning, signal modulation and demodulation analysis, sensor changing configuration, original signal record, rapid searching the database, and wireless node networking.

27.3.2.2 Subsystem Technical Parameters

The near-field signal sensor consists of near-field radio signal identification and node network communication equipment, to identify the near-field signal of the identification region and make real-time measurement of the signal frequency.
Main technical parameter index:
Working frequency: 20 MHz–10 GHz.
Monitoring range: 300–600 m.
Communication frequency between nodes: main frequency 2 G and backup frequency 900 M.
Network structure: ZigBee network technique.

27.4 System Basic Workflow

Step 1. The sensor at each node works separately, measuring the signal frequency within the node, informing the monitoring station through the wireless communication channel of the wireless network. The monitoring station makes automatic distribution of receivers, modulates to corresponding communication channel to receive and demodulate the signal, makes real-time record, displays the frequency spectrum, and determines the signal source range.

Step 2. In manual operation, analysis is made on the signal features and ground conditions, deciding whether it is an illegal signal.

Step 3. In automatic operation, the system will make detailed record on communication. After the stipulated working period, the system will examine the signal emission and treat normal and abnormal signals.

27.5 System Simulation and Results Analysis

A movable monitoring system can be established in a certain position. The system consists of a monitoring station and 5–7 wireless sensor nodes in wireless cellular pattern. The wireless signal sensor can find the near-field signal source accurately and eliminate the interference of far-field signals, avoiding the shortcoming of a common monitoring system to process quantities of useless far-field signals. To meet different geological ranges, the wireless sensor nodes can be added or reduced to guarantee the effective monitoring on the radio signal frequency spectrum within the key regions. The whole network can be remotely controlled with real-time orders and data, including node position, order number, single sensor real-time frequency spectrum, original record, analysis data, frequency spectrum parameters, and control orders. The whole system can monitor the range of 1–2 km^2.

27.6 Conclusion

The information technology is promoting a new military reform. The development of the wireless sensor network will surely inspire the development of the espionage. It promises the new electronic eye and ear of the battlefield, which will play an unimaginable part in the battlefield in the future. It is actual requirements to make deepening study on the battlefield, hi-tech weapon position sensors and wiretaps. The position sensor network signal detection system will be set up in hi-tech weapon position region, to make real-time discovery of wireless communication signals and other electromagnetic leakage signals within the position, to make automatic identification of the normal and abnormal signals, to make analysis, discovery, and positioning abnormal illegal communication equipment and high strength electromagnetic leakage source and to make timely warning, guaranteeing the security of the electromagnetic environment in important regions.

References

1. Doherty L, Pister KSJ, EI Ghaoui L (2001) Convex position estimation in wireless sensor networks [CJN]. IEEE 1655–1663
2. Savvides A, Han CC, Strivastava MB (2001) Dynamic fine-grained localization in ad-hoc networks of sensors. ACN Press, New York, pp 166–179

3. Shi Y, Eberhart RC (1998) A modified particle swarm optimizer. IEEE, USA, pp 69–73
4. Eberhart R, Shi YH (2001) Particle swarm optimization: developments, applications and resources. IEEE, USA, pp 81–86
5. Younis O, Fahmy S (2004) HEED: a hybrid, energy-efficient, distributed clustering approach for ad hoc sensor networks. IEEE Trans Mobile Comput 3(4):366–379
6. Heinzelman WB, Chandrakasan AP, Balakrishnan H (2002) An application-specific protocol architecture for wireless micro sensor networks. IEEE Trans Wireless Commun 1(4):660–670

Chapter 28
Examination of the Problems in LAN-Based Instruction Environment: From the Perspective of CLES

Linong Shi

Abstract Multimedia network classroom (MNC) is a LAN-based instruction environment which is widely used in the fields of education and training. However, there are still many problems in the design and the use of the environment, and its performance has not been fully exploited. This paper analyzes the current situation of the use of the environment, from the perspective of constructivist learning environments (CLEs), and discusses the common problems in the design of the environment and MNC-based instruction. Some improvement strategies such as teaching strategy and resource strategy and cooperation strategy are given.

Keywords Multimedia networking classroom · LAN-based instruction · CLEs

28.1 Introduction

Professor Seymour Papert, the inventor of LOGO language, once said: "If everyone can afford a pencil, the learning mode will change accordingly" [1]. Compared with traditional classroom, MNC has great changes in the teaching environment. When each people owns a computer to provide rich teaching resources with strong multiple information processing capacity and real-time interactive function, the teaching is sure to change accordingly.

However, because of the influence of the traditional class, teachers' lack of teaching theory quality and shortage of information technical quality, the technical advantages of LAN-based teaching are far from satisfactory, with some teaching materials, teaching plans, and teaching activities simply copied on MNC screen.

L. Shi (✉)
PLA Academy of Special Operations, Guangzhou, China
e-mail: shilinong@163.com

S. Long and B. S. Dhillon (eds.), *Proceedings of the 13th International Conference on Man-Machine-Environment System Engineering*, Lecture Notes in Electrical Engineering 259, DOI: 10.1007/978-3-642-38968-9_28,

What features does MNC teaching environment has? What influence do such features have on the teaching? What is the actual application condition of MNC teaching environment? What problems are there in the teaching environment, compared with globally advanced teaching environment? What improvement can be made on the basis of technical features of MNC? In the thesis, the discussion is made on such problems one by one.

28.2 MNC Technical Features and Teaching Meaning

Multimedia network classroom (MNC) is actually a LAN and a mini teaching network environment between teachers' and students' computers with the network technique and related equipment. The new teaching environment has two more important factors of network and computers. The communication capacity of the network and intelligence of computers provide rich potential and limitless development space for the teaching in the classroom.

It has the following main technical features:

- Independent node
  The network node is the computer with independent address and perfect functions. The independent computer guarantees that it is possible to make self-control study and individual teaching in the whole teaching environment in the classroom.
- Node communication
  Network nodes are in the same LAN and connected with network equipment, which is convenient for information communication and sharing. The communication provides a new access of the cooperative learning activities.
- Various supporting media
  The computers of the network nodes are complete in functions. The computers can play audio, video, and animation in common formats and can also play media documents of any format with installing corresponding software. Various media provide rich forms to express the knowledge.
- Clear and real-time media demonstration capacity
  MNC with video and audio transmission can pass the images on teachers' computer and teachers' voice to students' computer in real time, making students percept various details clearly and improving the information spreading quality.
  In such sharing learning environment, students and teachers can make interaction and cooperation with the assistance of the computer and software, to improve the learning efficiency and students' experience in the classroom.

28.3 MNC Application Conditions

The above analysis on MNC environment features shows that compared with ordinary multimedia classroom, MNC has the following main advantages: single computer capacity, network interaction capacity, and close multimedia demonstration capacity. MNC teaching environment is essential to various trainings, as information technique courses. According to the survey and years of experience of the author, the widely used functions of MNC are screen broadcasting, experiment platform, and assignment handing in and out.

28.3.1 Real-Time Close Screen Broadcasting

MNC is generally equipped with the software of electronic classroom (e-classroom), of which the core function and technique focus on screen broadcasting. The screen broadcasting generally contains teacher demonstration and students model. For example, in the information technique course, there are quantities of software operations. The teacher can make real-time transmission to students' computers. Each student can see clearly the contents from the teacher closely on his own computer.

However, in common multimedia classroom, there is only a screen for far demonstration. It is impossible for the words and images on the screen to be too small and complicated. Otherwise, it is hard for students to see clearly.

28.3.2 Students' Experiment Platform

Various types of software can be installed on students' computer, to provide real or simulated experiment environments. Students can practice as learning, avoiding the low learning efficiency due to forgetting in the long isolation of the theory course and experiment course.

28.3.3 Assignment Handing In and Out

The e-classroom software generally has such two functions for the teacher to hand out the assignment and students to hand in the assignment.

28.4 MNC Problems Analysis and Improvement Strategy

28.4.1 Constructivist Learning Environments Model

The learning theory develops from the epistemology to behaviorism, cognitivism, and popular constructivism at present [2]. The constructivism stresses emphasis on designing learning environment. D.H.Jonassen, one of the most influential representatives of the constructivism proposed a constructivist learning environments (CLEs) [3], of which the model is indicated in Fig. 28.1.

Such model mainly contains the following six parts:

- Problems (including question, project, assignment): core of the design of CLEs. The target of the learner is to express and solve the problems (or answer question, complete project, and assignment);
- Examples or cases related with problems;
- Information resources: various information resources related to solving problems, including texts, images, audios, videos, and animations;
- Cognition tool: tools generated on the computer to assist and promote the cognition process;
- Dialogue and cooperation tool: tools for learners group to make mutual communication, discussion, and negotiation for joint knowledge construction;
- Social background support: consideration is required to make on the support from social cultural background, objective environment, and material conditions on the present learning in designing CLEs.

28.4.2 MNC Problems Analysis

Compared with common multimedia classroom, MNC has made much improvement. However, there are still some existing problems, in the view point of CLEs.

Generally, two teaching modes are used in MNC, namely traditional mode and task-driving mode, as indicated in Fig. 28.2.

The traditional teaching environment features focusing on knowledge and skills, with the teacher teaching knowledge points and demonstrating skills.

Fig. 28.1 Model of CLEs

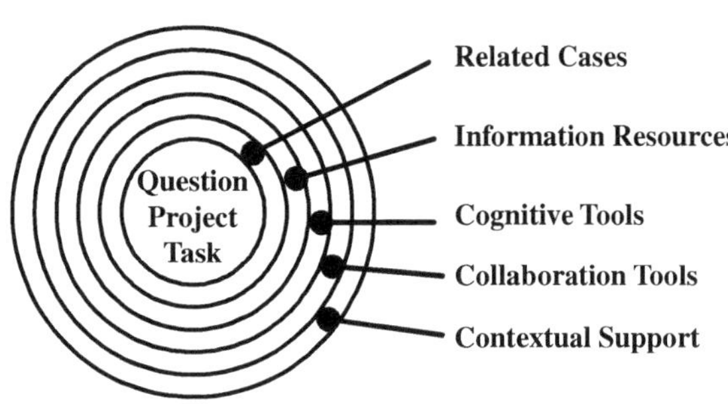

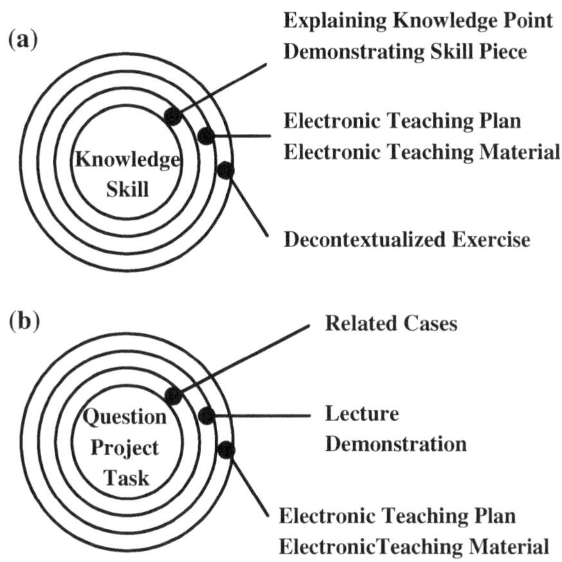

Fig. 28.2 Model of MNC instruction environments

The simplified example is used generally to assist understanding complicated knowledge skills. The constructivism does not accept such traditional teaching mode, thinking in such mode the knowledge and skills are divided into small units, causing the simplification and de-contextualization. It is difficult to put the knowledge and skills into actual application.

The task-driving teaching mode is relatively new, whose idea of task center is consistent with that of CLEs. However, it lacks the information resources to support self-control learning. Two kinds of resources are most important, of which the first is to assist learners to understand problems and the second is to assist learners to solve problems. Besides, it lacks effective support to the cooperative learning.

28.4.3 MNC Improvement Strategy

In CLEs model, the definition is made on general features of CLEs. However, the most important is how to change such features into actual learning environment and how to guarantee the effect on cultivating students in knowledge and skills. MNC is quite different from the traditional classroom in essence. The computer and network provide rich potential and broad development space and may provide sound support to CLEs after the proper design and development. The classroom teaching of the information technology course is taken as an example for discussion on improvement strategy on MNC.

28.4.3.1 Task Center and Visualization Teaching Strategy

It is still quite important for the teacher to make speech in MNC. However, the teacher should not play the courseware for a long time but stress emphasis on the task center and try to express with visualization mode as much as possible.

Quantities of work of cognitive psychology at present show that learners will master better when they participate into problem solving [4]. Real tasks provide real background for the knowledge application, with knowledge points hidden in the task. Learners will find problems in completing the task and such problems will be the actual requirements and motive power to learn new knowledge points. The knowledge the students learn is used for problems solving instantly. The knowledge is verified in completing the task and also the students deepen the understanding of the knowledge. A real task involves various knowledge points, including new and learned knowledge points, which makes sure the knowledge application environment not fragmentary, isolated, and simplified but integrated and non-well formed. It is convenient to establish the relation between knowledge points and assign practical meaning to the knowledge. The task completion provides a natural and direct method to evaluate learning results. Therefore, the task as the situation integrates the knowledge and capacity, emotion, and motivation properly.

The teaching content should be demonstrated according to the principle of demonstration but not explanation. It is easy to confuse learners with quantities of texts in explaining complicated relation. However, it is generally much simple with better results to express with visualization information, including tables, images, and animations. The properly designed illustrations can play the better role than the pure words in showing hidden information, making complicated and vague information easier to understand, clear, and attracting.

28.4.3.2 Information Resources Construction and Utilization Strategy

The information resources here means various information resources related with the teaching in the classroom, including teaching plans, demonstration videos, related teaching materials, examples, or cases related with problems. The information resource environment means MNC environment where learners can make self-control learning and teachers can make individual teaching and hierarchical teaching.

The information resources can be obtained in two ways of the teacher's sending and students positive downloading. The teacher can use the function of document issuing provided by MNC teaching software. The students need a FTP server to download positively. Generally, the materials send by the teacher are for all students while those downloaded by students include two types of the supplementary materials for students with weak learning capacity and the improvement materials for students with strong learning capacity. Students can download the materials according to their own actual learning conditions.

For information technology course, a quite important type of resources is the help files attached to the software, which is often ignored by many teachers. The information technology course is intended for the beginners, who do not know the existence of such held files and do not know how to use them, either. The held files are contextually sensitive and easy to use in cultivating students' independent capacity to solve problems and mastering the computer software effectively.

28.4.3.3 Cooperation and Communication Strategy

Such strategy can realize mutual communication between the teacher and students or between the students themselves, who can construct the knowledge jointly after discussion and negotiation.

The task-centered teaching mode requires students to complete the appointed task in the given period. In completing the task, students will find more problems gradually. The teacher shall make collective explanations to problems involving new knowledge and common problems. Generally, it does not advocate the self-teaching in the classroom due to the limited time in the class. The self-teaching is more suitable after the class. The teacher should advocate the mutual learning between students for non-common problems. To teach is also to learn and cooperation means win–win.

MNC cooperation can not only be made in the actual space but in the virtual space based on the network as well. Students can demonstrate under the guidance of the teacher and also can make positive communication with different means as blog, e-mail, BBS, and chatroom. The multimedia teaching software supports the cooperation quite little. It is quite necessary to establish a platform for the cooperation and communication to MNC without connection with Internet.

The cooperation learning can only improve the capacity to construct the knowledge positively but cultivate the interpersonal communication capacity and team spirit of students as well. In the communication and discussion, students can find partners for learning and study. The teacher can understand and track the learning of students.

The teacher must monitor and participate in the cooperation learning activities of students, to avoid the discussion and cooperation away from the learning subject and to make timely correction of students understanding in the interaction.

28.5 Conclusion

Compared with traditional classroom, fundamental changes have taken place in MNC teaching environment. The new teaching environment has two more important factors of network and computers. The communication capacity of the network and intelligence of computers provide rich potential and limitless development space for the teaching in the classroom. At present, the most used functions of MNC are the screen broadcasting, experiment platform, and assignment handing in and out. MNC still has great potential to display. Compared with CLEs, MNC still has great improvement space in teaching strategy, resource strategy, and cooperation and communication strategy.

References

1. Shengquan Yu (2007) Development and prospect of three generations of mobile learning: knowledge transfer, cognitive construction and situated cognition. Chin Educ Technol 06:7–18
2. He K (2004) On the educational thoughts and philosophical basis of constructivism. Mod Distance Educ Res 71(3):12–16 (Ch)
3. Jonassen DH (1999) Designing constructivist learning environments. In: Reigeluth CM (ed) Instructional design theories and models: a new paradigm of instructional theory [C], vol 2. Lawrence Erlbaum Associates, Mahwah, pp 215–239
4. Merrill MD (2002) The first principles of instruction. Educ Technol: Res Dev 50(3):43–59

Chapter 29
Research of 3D Virtual Sounds in Cockpit: Calculation, Measurement, and Individuation Customization in Aviators' HRTFs

Guowei Shi, Xiaochao Guo, Dewen Cheng, Xianjun Li, Duanqin Xiong and Yu Bai

Abstract Acoustical display in cockpit is an effectual means to reduce potential aviators' visual load in traditionary cockpit. The 3D virtual sounds, including spatial positioning information and displaying by headphone, adapt sufficiently for acoustical display in limited cockpit space. In order to improve cockpit acoustical display level and synthesize cockpit 3D virtual sounds, we exploringly research the key foundation of cockpit acoustical display which is aviators' head-related transfer functions (HRTFs) calculation, measurement, and individuation customization. First, the importance of HRTFs in 3D virtual sounds is analyzed based on the binaural acoustical model and then summarized the calculation, measurement, and individuation customization methods of HRTFs. Second, the primary research of aviators' HRTFs measurement is brought forward according to demand of cockpit acoustical display, designed the free acoustic fields environment and aviators' HRTFs measurement system. Finally, the individuation customization methods of aviators' HRTFs are investigated based on physiological parameter matching in aviators' HRTFs measurement system.

Keywords Virtual auditory · HRTFs · Customization · Acoustical display · Free acoustic fields

G. Shi (✉) · X. Guo · D. Xiong · Y. Bai
Institute of Aviation Medicine, Air Force, Beijing, China
e-mail: bitgwshi@163.com

X. Guo
e-mail: guoxiaochao1965@163.com

D. Cheng
School of Photoelectricity, Beijing Institute of Technology, Beijing, China

X. Li
Department of Computer Science and Engineering, Beihang University, Beijing, China

S. Long and B. S. Dhillon (eds.), *Proceedings of the 13th International Conference on Man-Machine-Environment System Engineering*, Lecture Notes in Electrical Engineering 259, DOI: 10.1007/978-3-642-38968-9_29,

29.1 Introduction

Because of the limited visual perception dimensionality, it is quite easy for the display mode with more and more figures and images within the modern aircraft cockpit to increase the visual load of aviators, resulting in information block in the visual perception channel of aviators. It is an effective solution to the problem to apply the auditory in the cockpit display [1]. Physiologically, the omnidirectional auditory can make up for the limited visual range. It does not require making active search for the target, reducing the visual load. The auditory display will be an important supplement to the visual display in case of the visual display being limited, failing to use or unsuitable to use or continuous multidimensional massive information needing to process.

3D virtual sound changes the traditional auditory display mode, contents, and function completely, enhancing the auditory display from the indirect marking to direct comprehensive display on the direction, position, motion, and signal information [2]. In the limited space of the cockpit, 3D virtual sound in the headphone is the economic and effective auditory display, of which the key is the head-related transfer functions (HRTFs) of aviators.

29.2 Key Techniques of Virtual Auditory Display in Cockpit

29.2.1 3D Binaural Sound Model

In the analysis of physical acoustics, a person is actually in the acoustic transmission system consisting of sound source, transmission access, and receiver when he percepts the sound information in a real environment. According to the system theory, 3D binaural sound model can be structured, as indicated in Fig. 29.1.

Where $x(t)$ is the signal from the sound source at a certain position in the space, $f(t)$ the pulse response in the sound transmission environment, $h_L(t)$ and $h_R(t)$ HRTF at the left and right ear and sound signal perceived at the eardrum. Thus, the auditory nerve will perceive the space position of the sound source. The sound signal $y_L(t)$ and $y_R(t)$ passing $h_L(t)$ and $h_R(t)$ filtering is considered as the

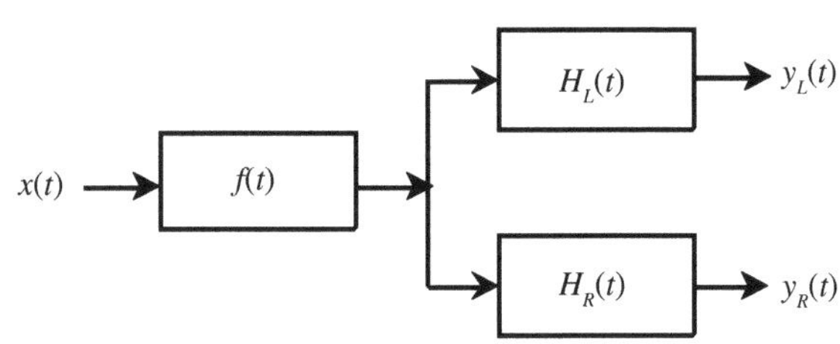

Fig. 29.1 Model of binaural transmission and perception

binaural sound signal. 3D binaural sound is particularly suitable for the outputting with the headphone. It overcomes the cross talk interference of the dual speakers, which is most economic, effective, and scientific method for 3D virtual sound in the cockpit.

29.2.2 HRTFs

The psychoacoustic study shows that the direct sound wave from the sound source reaches ears after the diffraction on the head, resulting in interaural time difference and sound intensity difference [3]. The sound source space positioning is made on the changing frequency spectrum due to the interference at the ear canal between the direct sound wave and the body scattering and diffraction sound waves and such interaural differences. HRTFs indicate the filtering of the external ear, head, and trunk on the sound signals from different direction in the transmission. It contains important position information, including sound frequency spectrum features and interaural difference features, which is related with the direction, distance, and frequency of the sound source and the individual physiological features. HRTFs can be defined to be

$$H_L = H_L(r, \theta, \phi, \omega, s) = \frac{P_L(r, \theta, \phi, \omega, s)}{P_o(r, \omega)}$$

$$H_R = H_R(r, \theta, \phi, \omega, s) = \frac{P_R(r, \theta, \phi, \omega, s)}{P_o(r, \omega)}$$

Each position of the sound source in the space has corresponding HRTFs. Generally, H_r and H_l is the functions of the distance r, azimuth θ, elevation angle φ, frequency ω, and individual difference s. It is the comparison between the sound pressure $P_{l,}$ P_r at the left and right ear and P_o in head center position (in case of no head existing). HRTFs can be obtained after the inverse fast Fourier transform. The equivalent express in the time domain is the head-related pulse response or binaural pulse response, which is a function of r, θ, ϕ, t and s HRTFs or HRIRs can be used to determine the binaural sound pressure of the point sound source at the space position of $(r,\ \theta,\ \varphi)$ [4].

29.2.3 Convolution Treatment of Sound Signals

HRTFs is used to make filtering on the sound signals of the single channel, obtaining the binaural signals in 3D space. The digital processing after the transmission means the linear convolution between HRIFs and sound signals.

Suppose the length of the sound signal $x(n)$ is M, and that of the finite pulse sequence $h(n)$ is L, of which the linear convolution is defined to be

$$y(n) = \sum_{i=0}^{L-1} h(i) \otimes x(n-i) \quad 0 \le n \le L + M - 2.$$

The linear convolution has enormous calculation amount, so the speed of the convolution algorithm is the key of 3D auditory effect. If the pulse sequence is short or the beat number of the filter is small, it is much fast to calculate with the linear convolution. However, generally, HRTFs have large beat numbers, the convolution algorithm requires improvement and optimization according to the beat number of the filter. Generally, there are the circular convolution equivalent calculation linear convolution method of sound signal discrete Fourier transform, window function cutting method, time and frequency extraction Fourier transform method.

29.3 HRTFs Calculation, Measurement, and Individual Customization

HRIFs can be obtained with the calculation method or measurement method. The calculation method means solving the wave equation under a certain boundary conditions. Measurement method means acoustic experiment measurements on HRIFs of different azimuths and elevation angles.

29.3.1 HRTFs Calculation

Generally, the simplified body model is used to solve HRIFs. The common simplified modes are mainly the rigid sphere model and snowman model. The rigid sphere model is one with the head simplified into a rigid sphere to solve the wave equation, of which the result can be expressed into l order Legendre polynomials P_l with wave number of K and l order the second type Hankel function h_l.

$$H_L(\theta, f) = -\frac{1}{(ka)^2} \sum_{l=0}^{\infty} \frac{(2l+1)j^{l+1}(-l)^l P_l(\sin\theta)}{\mathrm{d}h_l(ka)/\mathrm{d}(ka)}$$

$$H_R(\theta, f) = -\frac{1}{(ka)^2} \sum_{l=0}^{\infty} \frac{(2l+1)j^{l+1} P_l(\sin\theta)}{\mathrm{d}h_l(ka)/\mathrm{d}(ka)}$$

In the snowman model, it is simplified into two spheres of different radiuses. Consideration is taken on the effect of the trunk on the basis of the rigid sphere. It can be solved with the spherical harmonics multipolar expansion [5].

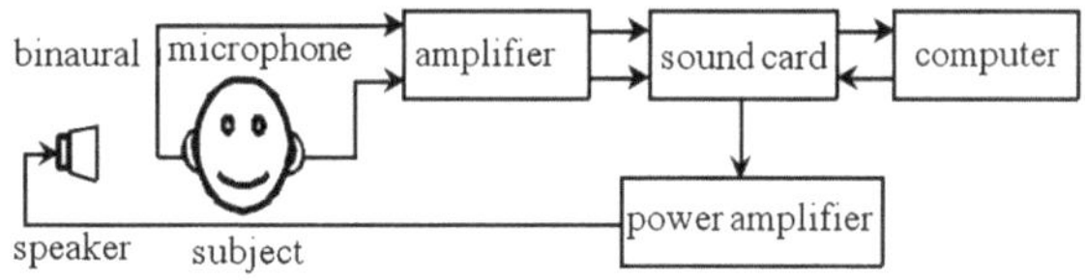

Fig. 29.2 Schematic of HRTFs measurement system

29.3.2 HRTFs Measurement

The error of HRTFs calculation will increase remarkably when the frequency increases until the wave length less than the auricle dimension due to ignorance details as the auricle. The measurement method is more accurate to obtain HRTFs. The pulse excitation signals are sent out from different positions in the space in the special measurement environment. The devices are used to collect HRTFs data [6]. The principle of the measurement system is indicated in Fig. 29.2.

The measured HRTFs are in discrete distribution in the space. The linear interpolation method can be used to obtain HRTFs at any point in the space [7]. Psychoacoustics proves interpolated HRTFs are quite close to the perception positioning effect.

29.3.3 HRTFs Individual Customization

HRTFs are the effect of comprehensive action of the head, trunk, and auricle on the sound wave. They are closely connected with physiological structure and dimensions, thus, showing remarkably individual features. HRTFs individual customization has sound effect, which makes approximate estimation or customization of corresponding individual HRTFs from the application. The individual customization can be made with physiological parameters measurement and non-physiological parameters measurement. The former means selecting HRTFs data which are most close to those of the user from the known HRTFs database by physiological parameters matching. The latter means providing approximate HRTFs with the average value, special person, or subjective evaluation.

29.4 Aviators' HRTFs Measurement and Customization

Aviators' HRTFs play a core role in the virtual auditory and cockpit auditory display. The experiment measurement is an important method to obtain accurate HRTFs. Therefore, we made experiment measurement on HRTFs. The measurement is made in the free acoustic field to eliminate the effect of the sound reverberation. The measurement sketch is indicated in Fig. 29.3.

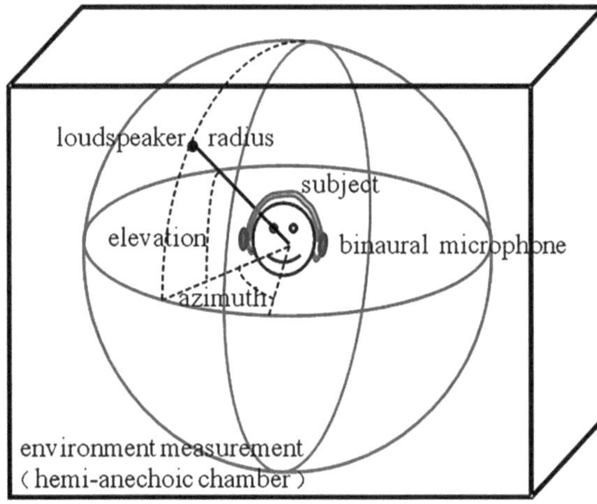

Fig. 29.3 Diagram of HRTF measurement

29.4.1 Aviators' HRTFs Measurement Environment

Aviators' HRTFs measurement is made in the free acoustic environment, which is hemi-anechoic room with no reflection surfaces except the floor [6], as indicated in Fig. 29.4.

29.4.2 Aviators' HRTFs Measurement System

Such system mainly consists of the sound source space positioning system and HRTFs collecting system. The former is used to determine the space position of the speaker in the experiment environment while the latter is used to collect HRTFs data of aviators.

Fig. 29.4 Hemi-anechoic room for HRTFs measurement

29.4.2.1 Sound Source Space Positioning System

Such system is used to position and measure the space position of speakers for HRTFs according to the sound source space resolution. The space resolution of the positioning system is the sound source space resolution, as indicated in Table 29.1.

For the speaker in the system, the horizontal motion is controlled with the horizontal rotation equipment while the vertical motion is controlled with the sliding manual control, of which the mechanism structure is indicated in Fig. 29.5.

29.4.2.2 HRTFs Collecting System

Such system mainly contains head binaural collector, multiple channel data terminals, and multiple channel data analysis system, of which the structure is indicated in Fig. 29.6.

Table 29.1 Spatial resolution of sound source positioning system

Space resolution	Radius Δr	Azimuth $\Delta\theta$	Elevation angle $\Delta\varphi$
	[−0.1, 0.1 m]	[2°, 5°]	[2°, 5°]

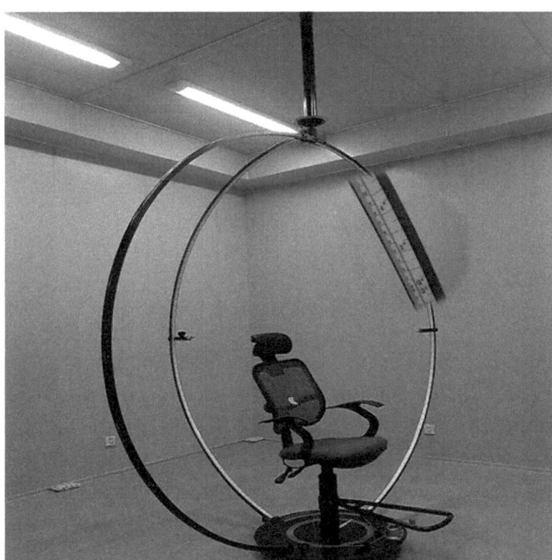

Fig. 29.5 Sound source positioning system of HRTFs measurement

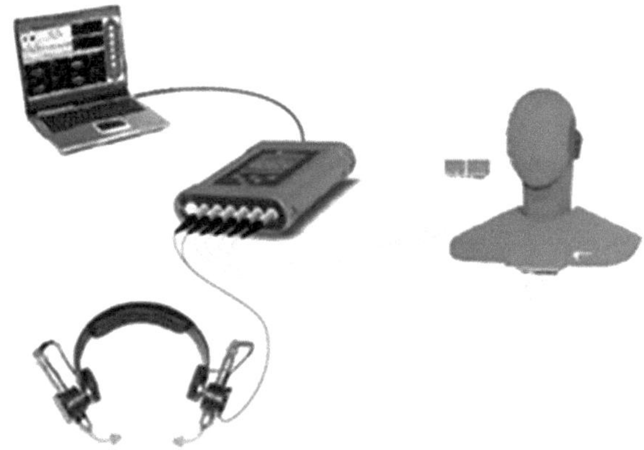

Fig. 29.6 Structure of HRTFs data collection system

29.4.3 Aviators' HRTFs Individual Customization with Physiological Parameter Matching

The physiological parameter matching means selecting HRTFs data which is most close to those of the user from HRTFs database after the experiment measurement. In such method, the first is to screen related physiological parameters with HRTFs in the experiment measurements and to make accurate definition on each selected parameter; the second is to use uniform measurement standards on the physiological parameters. After obtaining approximate individual HRTFs of aviators with the physiological parameter matching, to make further improvement on the display effect of the 3D sound in the cockpit, the virtual auditory drawing and subjective evaluation with headphone replay is made on HRTFs within a certain physiological parameter matching range. It can make further improvement on subjective auditory effect of the virtual 3D sound.

29.5 Conclusion

Aviators' HRTFs are the key and foundation to generate virtual 3D sound and realize virtual auditory display in the cockpit. It is extremely important for the application of the virtual 3D sound in the cockpit auditory display to make exploration and study in advance on aviators' HRTFs measurement, calculation, and customization and to establish aviators' HRTFs database as soon as possible. Despite more equipment and particular measurement and the higher time and economic cost, the measurement method is vitally important for construction of HRTFs database, as an important method to obtain HRTFs data. The auditory display technique based on headphone replay and aviators' HRTFs will greatly reduce the potential information blocks of the visual perception display mode in the modern cockpit. Its application can assist aviators to determine the target direction and search for corresponding visual target and improve the understanding of the language communication by aviators. Meanwhile, the sound space application can reduce the time of aviators to search for visual target and to make decision, thus, reducing the work load. Therefore, it is extremely important to make accurate calculation, measurement, and customization of aviators' HRTFs and to construct aviators' HRTFs database.

Acknowledgments I appreciate deeply for the assistance of Beijing CA Acoustics Technology Co., Ltd. and Landtop Acoustics Technology Co., Ltd. on the measurement environment construction and measurement system design. Subsidized by: Natural Science Foundation of China (61205064).

References

1. Kapralos B (2003) Auditory perception and virtual environments. York University, Canada
2. Guo X (2000) Linear filter constitution means of head-related transfer function. Chin J Ergon 6(3):58–590 (Ch)
3. Blauert P (1997) Spatial hearing. Revised edition. MIT, Cambridge, pp 1–35
4. Xie B (2008) Head-related transfer function and virtual auditory. National Defence Industrial Press, Beijing
5. Kapralos B (2003) Auditory perception and virtual environments. York University, Canada
6. Hao X (2009) Individualization methods of head-related transfer function for virtual three-dimensional sounds. Tsinghua University, Beijing (Ch)
7. Xie B (2009) Head-related transfer function and virtual auditory display. Sci Sinica Pysica: Mech Astron 9(39):1268–1285

Chapter 30
Application of LED in Aircraft Lighting

Duanqin Xiong, Xiaochao Guo, Guowei Shi and Bai Yu

Abstract Aircraft lighting is an important branch of lighting technology in aviation flight domain. Lighting technology in aircraft of every period indicates the feature of lighting technology in certain extent. This paper summarizes the development of LED lighting technology and its application in aircraft cockpit, cabin, and exterior lighting. Thanks to a lot of advantages of LED lighting such as its efficiency, flexibility of optical design, color rendering and lifetime, it is more and more widely used in aircraft lighting, providing some good solutions for the reliability, security, comfort, and economy of aircraft lighting.

Keywords LED lighting · Aircraft · Cockpit · Cabin · Exterior lighting

30.1 Introduction

Lighting technology has been developing rapidly for the past ten decades, including upgrading lighting products from incandescent light, fluorescent light, and metal halide light to LED light. The lighting appliance is also improving in the function and art design. The lighting control is also developing quickly from the simple switch to analog dimmer, digital dimmer to intelligent lighting controller.

With the application of the intelligence lighting and LED lighting technique, great improvement has been made in the lighting mode and functions, basically realizing the lighting environment of any light color, illumination, and color temperature. It is possible to provide more humanized and comfortable lighting. The optimized design of the lighting system for the comfort is becoming a concern, focusing on creating lighting environment meeting the visual requirements as

D. Xiong (✉) · X. Guo · G. Shi · B. Yu
Institute of Aviation Medicine, Air Force, Beijing, China
e-mail: xdqin123@163.com

S. Long and B. S. Dhillon (eds.), *Proceedings of the 13th International Conference on Man-Machine-Environment System Engineering*, Lecture Notes in Electrical Engineering 259, DOI: 10.1007/978-3-642-38968-9_30,

much as possible with the integrity of the visual ergonomics and environment psychology and color psychology [1].

The aircraft lighting is an important branch of lighting technology in aviation flight domain. The lighting techniques in the aircraft in different stages can reflect the features of the lighting technology. The light source is the core of the lighting technique and its reform promotes the development of the lighting technology.

The aircraft lighting develops along the following different stages [2]:

In the first stage, Edison invented the first electric lighting source—the incandescent lamp in 1879. The practically used tungsten lamp in 1908 started the era of the electric lighting. The tungsten lamp with built-in particular reflection structure was used in the aircraft lighting initially. However, because of the evaporation of the tungsten in the high temperature, the lamp service time was only 1,000 h and so it needed frequent replacement. The luminous efficiency was only 10–20 lm/w. After 1960, the halogen lamp appeared and was applied in the aircraft lighting system. The halogen gas inside can make reversible chemical reaction with the tungsten, reducing the evaporation of the tungsten. Thus, the service time of the halogen lamp was improved to 2,000 h with the luminous efficiency of 20–33 lm/w approximately. Compared with the ordinary incandescent lamp, the performance has been improved greatly. The halogen lamp is still widely used as navigate light in the aircraft at present.

In the second stage, the fluorescent lamp appeared in 1940 with the luminous efficiency of 50–80 lm/w approximately and the service time of 5,000 h. However, it is weak in the luminous intensity. So it is mainly used in the interior lighting.

In the third stage, HID lamp appeared in the aircraft lighting system in 1991. HID lamp has the luminous efficiency of 80–100 lm/w, 4 times that of the halogen lamp. It has the service time of 3,000 h approximately, so it is generally used as the exterior lighting of high luminous intensity, including the landing light, taxi light, runway turning light, and signal light. HID lamp is working with high-intensity discharging, making sure its sound anti-vibration performance. However, it is powered in high voltage and there are a certain security potential problems in the cables of the light source and power supply. The strict screening measures are required to make due to the complicated driving power supply and much electromagnetic interference. Besides, it has only a power efficiency of 40 % approximately and most of electricity is changed into the heat and ultraviolet radiation.

In the fourth stage, light-emitting diode (LED) lighting technique is a new type of lighting technique in recent years. It features low voltage driving, without ultraviolet and infrared radiation, high luminous efficiency, favorable color rendering, adjustable chromaticity, energy saving, environment protection, small volume, long service time, strong anti-vibration, and quick response. It is the development direction in the lighting technology and in the necessary trend of the advanced aircraft lighting technological development. LED is used more and more widely in the lighting fields. It is used as the lighting source in different levels in modern civil aviation.

30.2 LED Applications in Cockpit

30.2.1 Cockpit Lighting Profile

The lighting in the cockpit can be used in different parts of the cockpit as top plate, central console, side console, and control system in right front of the pilot, to be more exact, instrument panel lighting, LGP lighting, top light, floodlighting, thunderstorm light, pilot's reading light, fuel supplying light, multi-function lighting, and emergency lighting.

In the cockpit, LED LGP, LED top light/floodlight can provide visual environment of the displaying instruments of favorable ergonomics and low fatigue.

30.2.2 LED Application in Cockpit

30.2.2.1 Instrument Panel Lighting

The instrument panel is mainly distributed in the right front of the pilot, which is considered as one of the most important lighting regions in the cockpit. It includes two lighting sources at least, namely the instrument panel floodlighting source and lighting source inside the instrument.

Compared with traditional light sources, LED has various advantages in the cockpit instrument panel lighting. For example, it is suitable for the energy saving in the aircraft due to its low power of 0.03–1.00 w generally. LED does not interfere with other aircraft equipment due to no infrared radiation. LED is a kind of solid-state light without fragile filament as in the incandescent light, guaranteeing its strong anti-vibration for adverse flight conditions. LED can work stably, which can reduce the failure rate of the lighting system greatly, reducing the maintenance cost of the aircraft.

The survey on new products for the cockpit lighting provided by several globally famous manufacturers shows that Goodrich, WHELEN, and Honeywell all provide LED instrument panel lighting products for the cockpit. Besides, LED light sources are used in instrument panels of Boeing 787 and Airbus A380 [3], indicating that LED has been used for the instrument panel lighting of the aircraft cockpit.

30.2.2.2 LGP Lighting

LGP is one part of the aircraft cockpit, installed on the console in the cockpit, including top control panel, instrument panel, side console, and central console. LGP is a kind of polymethyl methacrylate panel with the coating of diffuse reflection on the surface. The light source is embedded in LGP. The rays passing in the panel illuminate the signs on LGP. The signs on LGP is quite clear whenever in the day or at night with such lighting mode. It can display the state information on the control console to the pilot, increasing the security of the aircraft at night. The LGP signs will be white in case of no power supply, and they will be in required colors with the power supply [4].

The earliest LGP in the cockpit was fluorescent self-illumination. The fluorescent powder was replaced with the lighting at the beginning of the 1950s due to some radioactive elements and some undesirable visual effects.

Traditional incandescent light is used in most present civil passenger aircrafts. LED is used to replace the incandescent light gradually in the modern civil aircrafts thanks to various advantages of LED. Such LGP reduces the failure rate of the lighting system greatly, thus, reducing the maintenance cost of the aircraft. Meanwhile, to get better temperature effect, LED light sources of different color temperatures can be used according to actual requirements.

In the development of LED LGP lighting technology, various dimming modes are used, including voltage mode, PWM pulse mode, voltage-PWM mode, and digital mode.

30.2.2.3 Other Lighting Applications

Besides the above-mentioned instrument panel and LGP, LED is also used in other parts of the cockpit.

For example, Goodrich provides LED products of the top lighting, floodlighting, multiple function lighting, and fuel supplement lighting. Honeywell also provides LED multiple function lighting products.

In the lighting system of the cockpit, LED is used in the top lighting, floodlighting, multiple function lighting, and fuel supplying lighting in Boeing B787, top lighting and pilot reading lighting in Airbus A380. LED is used in all lighting equipment in C919 passenger aircraft independently researched and manufactured by China. LED is also used in the floodlighting, top lighting, reading lighting, thunderstorm lighting, and emergency lighting in some types of aircrafts.

30.3 LED Application in Cabin

30.3.1 Cabin Lighting Profile

Besides the cockpit, there is passenger cabin or crew cabin in the aircraft depending on different missions and functions, including the passenger cabin in the civil passenger aircraft, the cabin for the crew to complete particular military operation in the military transport aircraft, bomber, and AEW. The cabin lighting generally consists of the top lighting, floodlighting, reading lighting, and emergency lighting.

30.3.2 LED Application in Cabin

With the increasing emphasis on the lighting quality and the development of the lighting technology, LED is used gradually in the cabin lighting in recent years. LED light of huge power is used to replace the fluorescent light and reading light in aircrafts of the latest type of Boeing and Airbus. The cabin lighting is made with rich and colorful modes thanks to the extremely flexible color dimming and toning functions of LED. The more accurate and rapid control is made on the light color, color temperature, and contrast of the lighting. The art effect of the lighting in the cabin is improved greatly. In the cabin of Boeing 787, the traditional cabin lighting concept is totally changed. The changeable RGB LED and the diffuser with sound–light mixing performance are used to create magic blue sky effect. Rich and colorful color effect and spatial layering relieve the narrow sense of the cabin successfully. Sky blue lighting in the cabin passageway and ceiling beside creates a cozy atmosphere under the blue sky, relieving the fidget and anxiety of passengers in the trip and eliminating the sense of horror of some passengers on the flight in a certain degree.

The large-scale restructure is made on the cabin of the regional aircraft, giving an impression of large airliners. Two cabins or cabins of three levels with brand-new LED on the interior decoration are made gradually on the regional aircraft and turboprop aircraft [5].

LED is used as the top floodlighting, spare lighting, partial lighting, maintenance lighting, and emergency lighting in the cabin of a transport-type aircraft independently researched and manufactured in China.

30.4 LED Application in Exterior Lighting

30.4.1 External Lighting Profile

The aircraft exterior lighting system means the part of the aircraft lighting system used for the exterior lighting, which is an important part of the aircraft lighting system, guaranteeing the security of the aircraft in taking off, cruising, and landing [2].

In the primary stage, the exterior lighting system is quite simple and crude. To guarantee the normal flight of the aircraft in different weather conditions, the requirements on the exterior lighting system are improved increasingly. The exterior lighting system is completing gradually after the development for a long time, including landing/taxi light, runway turning light, position light, anti-collision light/strobe light, ice detection light, and signal light. Various functions are performed with such lights independently or jointly.

30.4.2 LED Application in Exterior Lighting System

The exterior lighting features high power, strict requirements on light distribution, and high requirement of the adverse environment on the technique. The above analysis shows that the traditional lighting technology can meet the requirements on the exterior lighting. LED has various advantages that the traditional lighting does not have, providing opportunity to improve the exterior lighting system of the aircraft.

LED can improve the efficiency, reliability, and security of the system, providing comfortable and effective visual information for pilots.

The study is being made on LED application by various aircraft integrators and lighting system suppliers in the world in recent years. In the exterior lighting systems of aviation giants in the world of Airbus and Boeing, Airbus typical types of A320, A330, and A340 are equipped with the incandescent light, the halogen light, and high-pressure xenon light of traditional technology. However, LED is used as the position light in the latest type A380-200. The incandescent light, the halogen light, and high-pressure xenon light of traditional technology are used on typical types of Boeing of B727-200, B737-500, and B747-400. However, LED is used in the position light, anti-collision light, strobe light, wing ice detection light, and signal light. High-pressure xenon light is only used in the landing/taxi light and runway turning light. Besides, Honeywell announced LED wing tip navigation lighting solution on March 5th, 2010 [6], which has been the standard configuration of newly produced A320 series of Airbus, including A318, A319, A320, and A321.

30.5 Conclusions

The analysis on LED shows that LED has incomparable advantages of traditional light sources in the efficiency, flexible optical design, color rendering, and service life. It has great application potential in the lighting system of the aircraft. The available LED technology can be used to design aircraft lighting equipment to meet various requirements in the color and optical distribution.

LED is developing rapidly and such technical reform will provide the opportunity to improve the competence of the lighting industry. LED is, thanks to various advantages, more and more widely used in the cockpit, cabin, and exterior lighting system, providing the sound solution on reliable, safe, comfortable, and economic lighting system of the aircraft. With the increasingly improved technology, LED will be more popular in the aviation.

References

1. Chunze W (2010) Optimal strategy research of little space based on vision ergonomics. Master Degree Thesis in Fudan University, Shanghai, pp 4–6
2. Liu H, Wen Y, Jian Z (2010) Application of LED technology in aircraft exterior lighting. Civ Aircr Des Res 4:22–27
3. Qian L (2010) The application research of the LED technology in lighting control system for civil aircraft cockpit. Meas Control Technol 29(11):106–108, 111
4. Huang Y (2011) LED application of instrument panel lighting in commercial aircrafts. Master Degree Thesis in Fudan University, Shanghai, pp 10–20
5. Ni H (2012) Lateral airplanes cabin: a handy tool makes a handy man. Air Transp Bus 3:34–36
6. Han Y (2010) Declaration of LED advanced lighting. Honeywell: LED navigation of wing top making airbus step former. Fortune World 7:54

Chapter 31
Research of Five-Kilometer Armed Off-Road Training Teaching Experiment

Shuai Mu, Zhibing Pang, Honglei Li, Ming Kong, Haitao Zhao and Min Chen

Abstract One of the most important military physical training contents is the 5-km armed off-road training which in conducive to enhance the soldiers body function, improve their ability to resist fatigue and cultivate their willpower. But due to the lack of corresponding theory on the project organization, we always use it as a means of endurance training. What is more, the cadets did not get any drill about this part, it is difficult to be close to the military training. As a result, we organized this teaching experiment research in order to improve the cadets' physical attribute and formulate a more scientific and reasonable training program, then reduce the randomness and blindness of the drill.

Keywords Five-kilometer armed off-road training · Experiment research

31.1 Introduction

The modern war with high technology still makes high requirements on the physical quality of servicemen. The tactic generally requires servicemen enter the combat state after a long maneuvering. The sound anti-fatigue capacity of the servicemen is the premise of the tactic. Otherwise, the good tactic cannot be implemented. Without sound physical quality, the modern equipment and well-trained combat techniques cannot be put into use in the war. At present, the military in different countries is considering 5-km armed off-road training necessary to improve the anti-fatigue capacity of soldiers, making support to the combat tactics and techniques.

S. Mu (✉) · Z. Pang · H. Li · M. Kong · H. Zhao · M. Chen
Air Defense Forces Academy, Zhengzhou, China
e-mail: fkb_mmes@163.com

S. Long and B. S. Dhillon (eds.), *Proceedings of the 13th International Conference on Man-Machine-Environment System Engineering*, Lecture Notes in Electrical Engineering 259, DOI: 10.1007/978-3-642-38968-9_31,

The armed off-road training is intended to train the endurance and field operation capacity of the servicemen with individual arms and equipment on various complicated topologies in the fields. 5-km armed off-road training is one of the major modes in the armed off-road training to build up the combat spirit, strengthen the cardio-pulmonary function, and improve the physical capacity of servicemen, making physical reserve for the operation [1].

At present, such training is quite common in the military of China, which is often organized in the military companies. It is must in the annual physical capacity assessment. However, the training organizers in most militaries consider it only a means to practice the endurance due to incomplete understanding on it resulting from lack of corresponding theoretical knowledge. The cadets from military colleges and universities did not receive such training and they fail to organize such training after coming to the army. Therefore, the study is required to make on the teaching experiment of 5-km armed off-road training to make sure more scientific and reasonable training, which is an important subject for colleges and universities to design the teaching program and the military to make training. It will be of great significance on how to make scientific training program, control the training load and procedure, and improve the training organization capacity, reducing the random and blind training [2].

31.2 Experiment Principle

5-km armed off-road training is one physical capacity training subject of higher strength and exercise quantity, which is vitally important to improve the capacities of running, load carrying, anti-fatigue and to fulfill operations in complicated environment. Therefore, 5-km armed off-road training is an adaptive training, intended to keep up the fighting capacity of the military.

In 2011, 5-km armed off-road training is expressly stipulated to be a must in the physical capacity teaching in the professional education colleges and universities. However, various colleges and universities fail to master the teaching features and laws of such training due to lack of corresponding teaching experience. To optimize the teaching means, improve the training effect and teaching quality, and learn the training laws, it is necessary to make experimental study on the training method, means, and arrangement on exercise load. It will make positive promotion for colleges and universities to organize teaching and the militaries to organize training. The experiment is designed on the teaching and training of such subject with the following five purposes: first, to find the difference between freehand 5,000-m run and 5-km off-road run; Second, to summarize various factors to affect the speed of 5-km off-road run; Third, to explore reasonable exercise quantity and strength, making sure the improvement of physical quality and more scientific and reasonable training program; Fourth, to summarize what training means and methods can be used for different teaching objects (including ordinary students and national defense students) of different foundations and levels to reach excellent level or approach the assessment standards within a certain time. Fifth, to explore the scientific proportion of teaching time in class and after class, maximizing the teaching effect [3].

31.3 Experiment Subjects and Preparations

31.3.1 Experiment Subjects

77 professional training students (48 ordinary students and 29 national defense students), male, with normal blood pressure and heart rate, in sound health.

31.3.2 Experiment Preparations

31.3.2.1 Software

Cattell 16 Personality Factor Questionnaire (16PF) and SPSS11.5 statistic software.

31.3.2.2 Harness and Experiment Apparatus

Weapon fittings: one 95 semi-automatic rifle, one satchel with personal washing ware, one full canteen, one grenade package with four training grenades, one bullet belt with four empty bullet holders, ten stop watches, two sets of clothes with numbers, several achievement statistic forms and 150 questionnaires.

31.3.2.3 Test Devices

Four EW3002 wrist electronic sphygmomanometers to test the basic data as heart rate and blood pressure of the subjects;

One EWOVC-101II electronic pneumatometer to test lung's capacity of the subjects;

One set of POLAR team two heart rate telemeter team training system to make real-time track on changing heart rate of the subjects in the run and to make monitoring and comparison analysis on the heart rate of different types of students in the run;

One CENTER310 temperature and humidity tester to test and measure the temperature and humidity of the external environment of the experiment;

One laptop computer to record and store real-time record results of the heart rate meter.

31.4 Experiment Methodology and Procedures

31.4.1 Basic Methodology

Literature method: to obtain strong argument data on quantities of books and journals related with the armed off-road training.

Questionnaire method: to hand out 150 questionnaires to all test subjects and get back 150, with the rate of 100 %.

Experiment method: to obtain quantities of the first hand experiment data from one freehand and 3 armed 5,000-m tests.

Mathematical statistic method: to make statistic analysis on the test results with SPSS software [4].

31.4.2 Basic Procedure

The experiment is made with the subjects totally without off-road training in three stages.

31.4.2.1 Stage to Collect Basic Data

The data of physiological index (including age, height, weight, heart rate, lung capacity, and step experiment) of the subjects are collected in the laboratory. The test and analysis is made on personal features of each subject. Then, the test of freehand 5,000 m is made in the common environment, obtaining basic physical capacity data for comparison. After a rest of a week, the first 5-km armed off-road test is made, and meanwhile, the real-time questionnaire is made on the subjects. The questionnaire mainly focuses on different feeling in various parts of the body in the freehand 5,000-m run and armed 5,000-m run and factors to affect the speed of 5-km armed off-road run.

31.4.2.2 State to Strengthen Training

The comparison and analysis is made on results of two tests in the first stage and the questionnaire to find out the difference of the freehand run and armed run and to analyze factors affecting running speed and to improve the equipment carriage mode. The different groups are divided according to the different basic capacity level of the cadets. The targeted training programs are made in the normal teaching periods and entertainment periods. The trainings of different exercise quantity and strength are arranged, and proper adjustment is made according to the real-time monitoring effect in the training procedure.

31.4.2.3 Stage to Discover Laws and Summarize Experience

The second test of armed 500-m run is made at the end of the whole teaching cycle to test the training effect of cadets of different levels. The further adjustment is made on the training program according to the test data. The corresponding special quality trainings are made in the entertainment periods. The last test is made at the end of the whole experiment cycle. The systemization and analysis is made on the data of three experiments to summarize the training method of cadets of different levels and the scientific proportion of the teaching time in class and that after class.

31.5 Conclusion

31.5.1 Major Differences Between Freehand Run and Armed Run

The average score of the subjects in the freehand 5,000-m test is 22′54″ and that of the subjects in the first armed test is 26′23″, decreasing by 15.3 %. The test scores show that the running speed of the armed subjects decreases remarkably, which because the freehand run requires only sound cardio-pulmonary function and a certain endurance quality while the armed run requires a certain muscle power operation endurance (especially that of the upper arm and shoulder and back), together with the sound endurance quality. Therefore, there is remarkable difference in running technique and physical capacity distribution.

31.5.2 Major Factors Affecting the Speed of 5-km Armed Off-Road Training

31.5.2.1 Equipment

A whole set of equipment weighs 12.5 kg. The subjects feel much uncomfortable in the first armed run, for which the main cause is improper equipment carriage. Too tight or loose wearing, too long shoulder belt, and infirm fixing (especially satchel, canteen, grenade belt, and rifle belt) are three common conditions in the run, resulting in sway equipment, difficult breathing, nervous action, higher running resistance. Such conditions lead to the consumption of large quantities of physical capacity and early fatigue, which affect the achievement and the running speed. Therefore, it is vitally important for the proper equipment wearing to the off-road run.

31.5.2.2 Physical Capacity

The subjects show different load carriage capacities depending on their various physical qualities. Especially to national defense students who are thick and short, it is difficult for them to run due to their poor basic qualities and added load. Some of them fail to complete the stipulated distance. The observation in the experiment shows that under the same resistance, the much stronger subjects will be relatively easy to make action than those less strong and it is much easier to repeat or continue the training. Therefore, it is necessary to improve the overall basic quality (especially the strength quality) of the subjects before the 5-km armed off-road training, making sure the individual quality reaching a certain level. Otherwise, it will not only affect the training effect but damage the trainees as well.

31.5.2.3 Environment and Weather

The test is made in the temperature of 7 °C. The bearing capacity of the body is quite different in the external environment and particular natural environment (extremely hot or cold and lack of oxygen). The study shows that the nerve conduction velocity decreases by 15 m/s in case of the temperature decreasing by 10 °C. Under the temperature of 8 $\sim$ 10 °C, the working capacity of limbs will reduce rapidly in case of feeling cold. The action will become less flexible and harmonious with the increasing viscosity of the skeletal muscles and less speed of muscle contraction. In such condition, the subjects will show normal exercise capacities [5]. Therefore, the temperature in the natural environment will bring about a certain impact on the subjects. The servicemen will not keep the body function in the best state in case of failing to adapting to the environment in the operation, damaging the fighting capacity seriously. It is must for the servicemen to strengthen the adaptive trainings in various environments.

31.5.3 Scientific Training can Improve the Capacity of the Armed Off-Road Run Significantly

After the test of the first stage, the subjects are divided into three groups according to the capacity on the basis of the test results. The training programs of different intensity and exercise quantities are made according to the actual conditions of each group. The strengthening trainings are made in different stages.

Firstly, adaptive training stage: the modes as varied speed running of low intensity are used to make the body become adaptive to the load. Meanwhile, the proper adjustment is made on the equipment carriage modes, making sure the comfortable and reasonable mode as much as possible according to individual physical conditions and without interference to the running speed and operation preparation.

Secondly, cardio-pulmonary function and muscle strengthen development stage: the training is made with increasing exercise quality and intensity. Emphasis is not only stressed on the endurance training but on related strengthen training as well. The running is mainly the integrated mode of walking and running, focusing on continuous running of medium intensity or interval running of high strength. The strength training is made mainly on muscle groups of the upper limbs, abdomen and shoulder and back, to improve the anti-fatigue capacity. The trainings of running and strength are made alternately.

Thirdly, load training stage: on the basis of the above-mentioned trainings of two stages, the body becomes adaptive to the load and the overall body functions develop. The overall strengthening training can be started. In the training, the reasonable intensity should be kept to reduce the over training and fatigue.

After the training of the above three stages, an armed test is made on the subjects, with the rate reaching the standard of 92.8 %. Only the reasonable training method is used in different stages can guarantee the effect of 5-km off-road training.

31.5.4 Reasonable Teaching and Training Cycle is the Premise of Effect

5-km armed off-road run capacity is improved gradually, requiring adequate time and training quantity. One of the purposes of the experiment is to find the effective method to improve physical quality in the limited time and to obtain the proportion of the teaching in class and training after class. In the whole experiment of a month, 14 teaching periods in class and 40 training periods after class are used in the training. Greater improvements are made in the teaching effect with achievements increased by 28.6 %. However, there are still a few subjects fail to reach the lowest standards. From the angle of human factor, different persons have different capacity levels. Even a same person has different bearing capacity to the load in different periods. Therefore, the armed off-road training should be started from the foundation with the principle of step by step. The real-time adjustment should be made on the exercise load on the basis of consideration of overall physical quantities to improve the training level gradually. Only in such way can ideal teaching and training effect be reached. There is no shortcut in such training.

Therefore, to improve overall capacities of 5-km armed off-road training, it is far from enough for a strengthening training of a month. It requires a gradual improvement procedure for a long time. Besides, it should not consider 5-km armed off-road training only as one method to improve the endurance quality. It shows overall capacities of endurance, strength, speed, anti-fatigue, and willpower. It is vitally important for the servicemen to improve the capacity to fulfill the operation in complicated environment and the fighting capacity.

References

1. PLA army military training and examining outline. PLA Press 2008
2. Wen-sheng W (2009) The systematic teaching procedure for the military 5 km cross-country in battle gear. Guangzhou: J PLA Inst Phys Educ 28(3)
3. Li Zhidao Tradition Education and Modern Education for all-round development of PLA. National Defence University Publishing
4. Zhang S (2005) Air comprehensive experimental researches on the anti-vertigo ability. Defense Forces Man Mach Environ Syst Eng 7(204)
5. Manchao H (2012) Investigation of physical education reform in military academy. Mil Educ Study 11(356)

Part IV
Research on the Man-Machine Relationship

Chapter 32
Research of PowerPoint Interactive Technology

Xiaoyao Wang, Li Li, Shulin Liu, Pinwang Zhao, Jinlong Zhao and Na Jiao

Abstract PowerPoint as a popular tool is active in forums, including stands, kiosks presentations, Internet format, navigation controls in PowerPoint interactive presentation plays a crucial role in the production. Studied the production problems of interactive presentations in PowerPoint multimedia production in this paper, some methods about how to use the navigation controls in creating interactive presentation were introduced, for example, how to contain Web cross-reference in slides, how to set the hyperlink hidden slides and the invisible "back" button, how to custom show. The specific examples were given. Finally, the kiosk mode in interactive show, as well as the major factors to consider, as published as a CD format, Web formats, and so were described in a simple way.

Keywords PowerPoint · Multimedia · Interaction · Hyperlink · Self-defining projection

32.1 Foreword

With fast development of multimedia technology, its application has been extended to industries like advertising, art, education, entertainment, engineering, medicine, commerce, and scientific research. PowerPoint, one of the tools of multimedia production, is always popular with users. The use of PowerPoint can create visual matching materials and help audiences obtain information-no matter what kind of information it is, no matter what form it is. Following the development of technology [1], PowerPoint is able to provide presentations in many forms including kiosk, kiosks presentations, Internet format. However, no matter

X. Wang (✉) · L. Li · S. Liu · P. Zhao · J. Zhao · N. Jiao
China North Vehicle Research Institute, Beijing, China
e-mail: hotjn@126.com

S. Long and B. S. Dhillon (eds.), *Proceedings of the 13th International Conference on Man-Machine-Environment System Engineering*, Lecture Notes in Electrical Engineering 259, DOI: 10.1007/978-3-642-38968-9_32,

what form the presentation is, interactive PowerPoint is necessary to realize an ideal effect, be popular in the audiences, and allow audiences to choose the contents they need, while the navigation control in PowerPoint plays a vital role [2] in the production of PowerPoint. Now we will introduce several methods to create a navigation control during production of interactive PowerPoint.

32.2 Slide Includes Web Cross-reference

Inclusion of Web cross-reference in slide is a basic navigation control function, or text hyperlink. The Web site may be listed in slide to allow user to access and thus get more information about subjects covered by the presentation. When entering a text in any of the following formats in the slide, PowerPoint will convert it into hyperlink automatically.

- Web site: any string starting with http:// or www.
- Email address: any string without blank inside and with a @ in the middle.
- FTP address: any string starting with fttp:///.

No any special effort is required to create such hyperlinks. When entering them and then pushing down Enter or Spacebar, PowerPoint will convert them into hyperlinks, as shown in (Fig. 32.1).

In case PowerPoint cannot create hyperlink automatically, click "AutoCorrect" on "tools" menu and then option "AutoFormat when entering." Choose check box "Replace Internet and network path with hyperlink."

32.3 Hyperlink Hidden Slides and Setting of Invisible "Back" Button

Basic presentation can illustrate basic contents required to be illustrated, while the creation of hidden slides of detailed information and hyperlink with these slides need hyperlink buttons. A "back" button needs to be set in hidden slide to allow user to return back to main presentation easily.

Select a slide thumbnail at the left of the main browser window and then right click it, finally select "hidden slide" in pop-up menu. The color of the slide becomes shallow and the serious number shows a strikethrough, meaning the slide is hidden.

Fig. 32.1 Examples of several internet hyperlink

http://www.microsoft.com/powerpoint

www.baidu.com

example@example.com

ftp://ftp.ppt.com

The action button in PowerPoint allows audiences to switch a slide in the presentation to a hidden slide. The action buttons attached to PowerPoint and their preset hyperlinks are as shown in (Table 32.1).

Course of inserting action buttons is as below [2]:

- Select slide for button setting.
- Click "Shape" in tab "Insert" or "Start." Shape palette appears: action buttons are at the bottom of the palette. Select button like "First slide," "Backward or previous item," "Forward or next item," "Start," "End," or "Previous slide."

Table 32.1 Action button and hyperlink meaning

Button	Name	Hyperlink to
	Custom show	No content by default. May add text or fill button to create self-defining button
	First slide	First slide of the presentation (the first is the start of the slides, a picture of a house)
	Help	No content by default, but it may be set to appoint to help file containing help document or other applications (generally, extension is .hlp, but .chm or .html is okay)
	Information	No content by default, but it may be set to appoint to slide or file containing information
	Backward or previous item	The slide prior to the presentation (it may not the last slide of shown presentation; in comparison with the "previous slide")
	Forward or next item	Next slide of the presentation
	Start	First slide of the presentation
	End	Last slide of the presentation
	Previous slide	The previous slide no matter whether the order is normal. It is useful to show the hidden slide the audience switches to with others (for instance "Help") to help them return back to main presentation at the end
	File	No content by default, but it may be set to open appointed file
	Audio	Broadcast appointed sound. Without selected sound, it will broadcast with the first sound in PowerPoint standard sound list (applause)
	Movie	No content by default, but it may be set to broadcast appointed movie

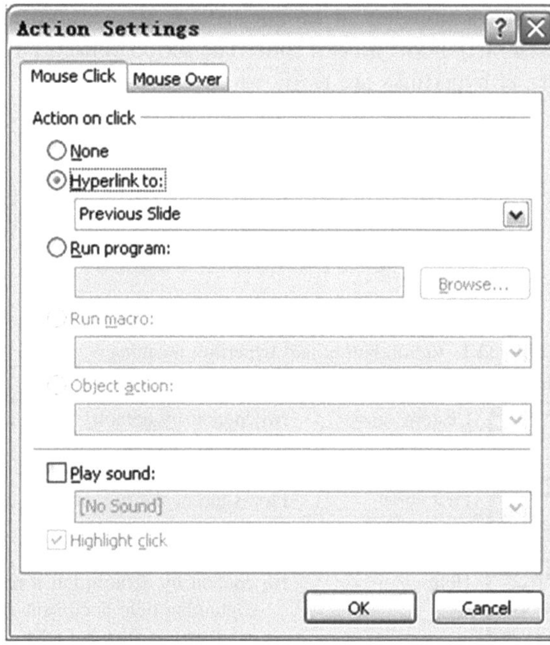

Fig. 32.2 The picture about action setting

- Click the slide and drag to the target place for button.
- Dialog box "Action setting" appears. Ensure tab "Click mouse" on the top as shown in Fig. 32.2.
- Confirm or alter hyperlink set in dialog box "Action setting."

At first sight, it seems there is no reason to use action buttons for simple slide forward or slide backward. However, if we use kiosk mode, we cannot use traditional keyboard or mouse to switch a slide to another. What a mouse can do is to click action button and create hyperlink.

Besides, we can create action buttons composed of random shapes: a shape representing painting, a barbola, a picture, a textbox, and any other shape. For this purpose, we just need to click "Action" in tab "Insert." And then, set "Action of clicking mouse" into "Hyperlink to," "Operation program," etc.

To allow users to know about what they can get when clicking the shape on action button, we may add text to the object directly or make textbox interpreting its function; if we need to mark the item with screen tips, we should use hyperlink instead of action setting.

32.4 Custom Show

You may be required to provide the teams working on two Web sites of the company with a presentation. You may make a presentation suitable for different audiences by creating a custom show. Custom show may be combined slides which can be shown by one by one or a set of slides in presentation pointed by hyperlink.

The slide presentation contains slides 1–5. A "Network 1" custom show may be created for the first set which just contains slides 1, 3, and 5, and then, a "Network 2" custom show may be created for the second set which just contains slides 1, 2, 4, and 5. Of course the slide show may be always operated in original order as shown in (Fig. 32.3).

1 is the slide of Network 1, and 2 is the slide of Network 2.

Also, we may organize contents in the presentation in the form of hyperlink custom show. For instance, create a main custom show about complete organization of this company, then create custom show presenting each department in the organization, and then link these projections with main projection. We may use create a slide contents with hyperlink custom show to navigate to specific slide show. This method may select the part the specific audience requires showing.

The course of creating a custom show is as below:

- Click "Self-defining slide projection" on tab "Slide projection" and then select "Custom show."
- Click "Create" button. Open "Define custom show" dialog box.
- Enter title of for custom show in textbox "Title of slide" to replace default title.
- Click the first slide for custom show in the window "Show slides in the presentation."
- Click "Add" and then "Confirm."

As shown in: (Fig. 32.4).

32.5 Interactive PowerPoint

The interactive PowerPoint may be published after its text is completed. But how to publish depends on the presentation form. Is it published just once or with multiple media in difference occasions? Can it be taken away by audience? Is it published by computer projection or via Web or in kiosk mode?

Fig. 32.3 Custom show

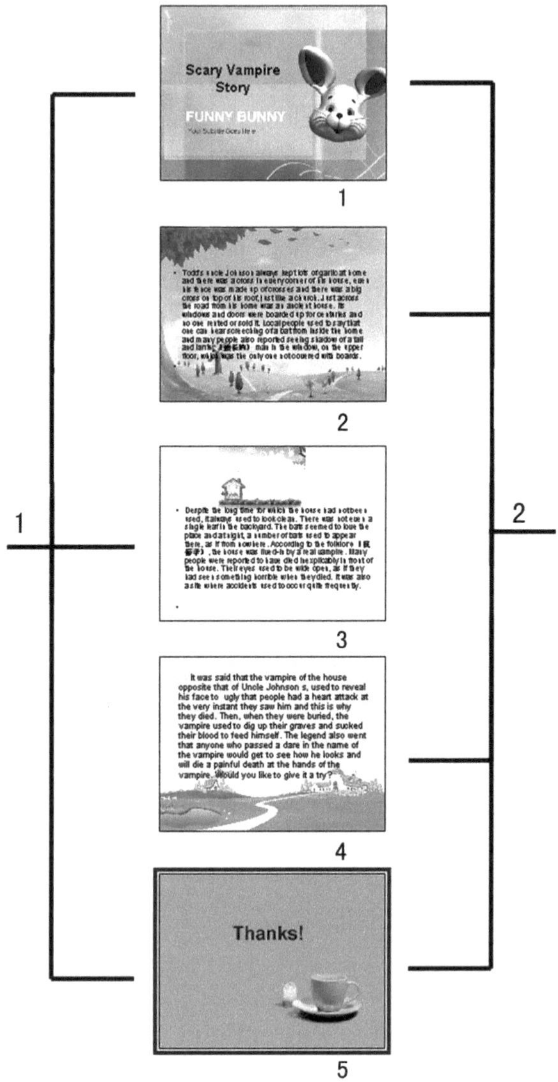

32.5.1 Preparation for Publication

Kiosk is unattended public-operated computer, so set presentation in "Kiosk" mode first.

- Click "Set slide shown" on tab "Slideshow." The dialog box "Set slideshow mode" is open.
- Click "Browse on kiosk (full screen)" in "Show type" zone.
- Click "Confirm."

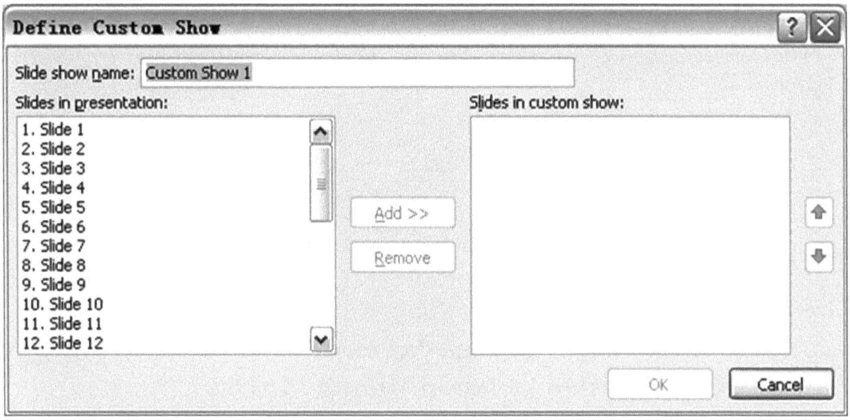

Fig. 32.4 The picture about custom show

Fig. 32.5 The dialog box about package for CD

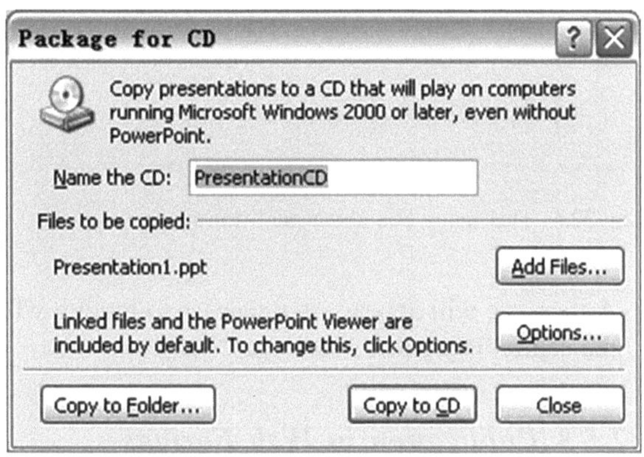

32.5.2 Publication via CD or DVD

A better way to deliver presentation is the use of PowerPoint's "CD Packet" function which allows reading all linked files and associated objects and ensures they are delivered together with main presentation. The operation is as below:

- Insert the empty disk in writable CD/DVD driver.
- Select "Office button," ⇒ "Publication" ⇒ and "CD packet." "Package for CD" dialog box is open, as shown in Fig. 32.5.
- Enter title of CD;
- Click "Copy to CD";
- In case of message inquiry, ask whether to packet the linked files, click "Yes";

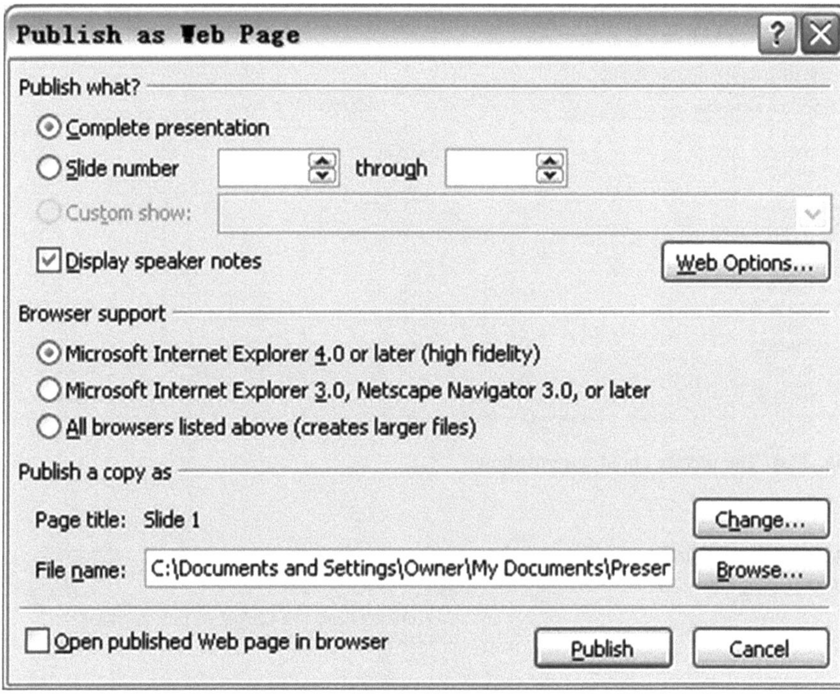

Fig. 32.6 The dialog box about publishing web

- A message will appear after success to inquire whether to copy the same file in the same disk. Click “Yes” or “No.”

32.5.3 Publication in Web Format

Due to convenience of Web, sometimes we need to publish PowerPoint on Web to allow anyone having a Web browser to check the presentation without any special software so that our work is accessible extensively.

Click “Publish” button in “Save as” dialog box to conduct Web page publication with multiple options, as shown in Fig. 32.6.

References

1. Faithe Wempen (ed) (2007) Translated by Tian Y, Hou X. Microsoft PowerPoint, Bible. Posts and Telecom Press, Beijing
2. Microsoft Office PowerPoint Help

Chapter 33
Design and Research of Lunar Rover Human-Computer Interactive Roaming System

Na Jiao, Shulin Liu, Li Li, Qiufang Wang, Jinlong Zhao and Pinwang Zhao

Abstract A more broad area of research for human-computer interaction technology was created with the development of virtual reality technology. The human-computer interaction roaming system is the comprehensive application of virtual reality technology. In this paper for the study of lunar rover, a lunar rover human-computer interactive roaming system was designed based on the Virtools virtual reality software. Study on the system design of 3D modeling technology, collision-detection, human-computer interaction, and other key technologies, the lunar rover model and virtual environment model were established in the virtual roaming system. This paper introduced the hardware and software composing of the total system and the realization. On this basis, a human-computer interactive roaming system of lunar rover was build, which laid the foundations for the next virtual maintenance and virtual assembly of human-computer interaction research.

Keywords Lunar rover · Human-computer interaction · Virtools · Collision detection · Roaming

33.1 Foreword

Human-machine interaction refers to the method by which human and machine affect each other and exchange information, a bridge and link between human and machine. In narrow sense, human-machine interaction mainly refers to information exchange [1] between human and machine. The rise of virtual reality technology creates a new research fields for the development of human-machine interaction. Its main differences from traditional common interactive imitation include

N. Jiao (✉) · S. Liu · L. Li · Q. Wang · J. Zhao · P. Zhao
China North Vehicle Research Institute, Beijing, China
e-mail: hotjn@126.com

S. Long and B. S. Dhillon (eds.), *Proceedings of the 13th International Conference on Man-Machine-Environment System Engineering*, Lecture Notes in Electrical Engineering 259, DOI: 10.1007/978-3-642-38968-9_33,

multidimensionality of information and naturalness of human-machine interaction. The characteristic of such technology is that computer can create an artificial virtual environment which is just as the "same" as or close to the real world so that the user has a feeling of being immersed in virtual environment visually [2].

Human-machine interactive roaming system is the comprehensive application of virtual reality technology with extensive perspectives and well social and economic benefits. Whereas the specialty of lunar rover, based on Virtools software and virtual reality hardware system, a lunar rover human-machine interaction system has been developed in this research so as to create experience of lunar rover roaming in early period of the design.

33.2 Crucial Technologies in System Design

Virtual scene modeling is an important link in human-machine interactive roaming system research. The whole system operation takes virtual scene model as the core and 3D modeling technology is used for virtual scene modeling research.

In virtual reality system due to interaction and movement among objects in scene, they will collide with each other inevitably. Therefore, the system must be able to detect such collision in real time to avoid occurrence of penetration and other phenomena of objects in scene in movement process [3]. The collision gun test is a key technology to solve out penetration after mutual collision among objects.

Human-machine interaction is an interaction between human and virtual environment. When computer acts in response to entry of use, an interaction between user and environment created by computer happens [3]. Human-machine interactive technology is the key to realization of experience of lunar rover roaming in a virtual environment.

33.2.1 3D Modeling Technologies

3D modeling is the basis of human-machine interactive roaming system. Due to human's information perception mainly from visual sense, whether virtual reality can revivify reality effectively has direct influence on the immersion sense of observer in whole virtual system. Therefore, 3D scene modeling plays a very important role in the whole system.

Three-dimensional model can be built up with various modeling methods Pro/E, CATIA, and other CAD design tools, or 3DMax, Rhino, and other modeling tools. In this system, the building of lunar rover model is realized by extracting existing data from CAD system, exchanging in intermediate file format and then entering in virtual environment which is created with 3DMax. Data conversion process is shown in Fig. 33.1.

To realize real-time roaming, virtual scene model must be optimized first. The scene optimization of modeling in virtual environment is mainly for simplify 3D model to enhance display efficiency [4].

Simplification shall follow some principles:

1. Just reserve shape profile of virtual scene according to shelter relation.
2. Delete non-interactive part in the circumstance of no influence on shape completeness.
3. Consolidate parts moving as a whole according to requirement of interaction.

33.2.2 Collision Detection Technology

The collision detection algorithms between two geometric models include space decomposition and hierarchical bounding volumes.

The basic thought of hierarchical bounding volumes is approximately describe complicated geometrical object with bounding volumes with slightly large size and simple geometrical characters. In this way, researcher just needs to make further intersection test on the objects superposed in bounding volume. Approach to geometrical model by building up a tree-structured hierarchy until almost all geometrical characters of objects have been obtained.

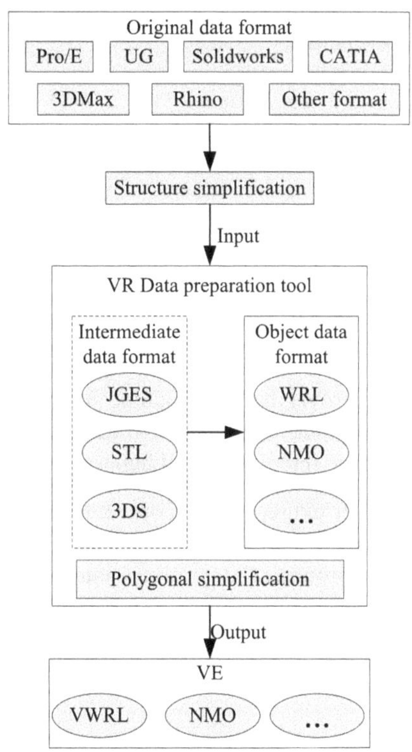

Fig. 33.1 Data conversion process

Space decomposition means dividing the whole virtual space into some unit cases with the same volume and then just conducting intersection test on the geometrical objects in the same unit case or adjacent cases. The typical methods include K-D tree, octree, BSP tree, tetrahedral mesh, regular grid, etc. Space decomposition is realized by hierarchy methods such as octree and BSP tree, which can further improve the speed of algorithm [3].

In lunar rover human-machine interactive roaming system, the collisions mainly include collision between lunar rover and ground, and collision between lunar rover and protuberant objects like rock on the ground. In Virtools, collision test between 3D models in virtual scene may be realized by transfer and connection of corresponding collision test blocks.

Virtools 4.0 realizes collision detection [4] of 3D entity by the following BB blocks

Name of act blocks	Functional description
Collision detection	Collision detection
Multi-collision detection	Multi-collision detection
Object slider	Object slide
Prevent collision	Prevent collision
Sphere slider	Sphere slide
Declare floors	Declare floors
Enhanced character keep on floor	Enhanced virtual character on floor
Floor manager setup	Floor manager setup
Floor slider	Floor slider
Get nearest floors	Get nearest floor
Home on entity	Direction
Box–box intersection	Intersection between boxes
Box–face intersection	Intersection between box and face
Face–face intersection	Intersection between face and face
Frustum–object intersection	Intersection between frustum and object
Sphere–sphere intersection	Intersection between sphere and sphere

33.2.3 Human-Machine Interaction Technology

Lunar rover human-machine interactive roaming system is a real-time dynamic system, when operating virtual scene model by interactive equipment, and human-machine interaction will happen. A complete interaction process is in three stages:

1. User expresses own operation intention to system via output equipment.
2. System interprets user's entry according to entry and current state of virtual environment, determines interaction task, and implements corresponding task.
3. Feed results of task implementation and acts of user back to user via output equipment.

The realization of an interactive process requires interactive character modeling. When system loads geometrical model of virtual scene, the interactive character model is read at the same time. The interactive system may visit interactive character information of object; if conditions of a behavior are satisfied, activate current behavior [5].

Interactive character model in roaming system is composed of the following two interactive characters generally:

1. Object properties. Design attributes of object like geometrical characters, and physical properties, like center of mass.
2. Object behaviors. The response made by description object to every interaction and purpose of the behavior, such as generating act or changing state. The object may have a variety of behavior and implementation of behavior shall be triggered according to conditions of state of object.

Interactive character modeling in Virtools means adopting object-oriented modeling method. The interactive characters of rover system may be used as template with succession and reusability.

33.3 System Design

33.3.1 Composition of Software and Hardware of System

The main hardware devices in this system include DVG graph clustering system, three-channel CADWALL projection system based on Infitec 3D technology, video signal distributor, space ball, and Infitec 3D glasses.

Integration method is shown in Fig. 33.2. DVG graphic clustering outputs three-channel VGA signal, which will be converted into RGBHV signal with VGA/RGBHV signal converter, the signal in each channel will become two channels of same signal through the distributor: one to CRT monitor for view of operator, and another to Barco projector for display on big screen; the 3D signal from a DVG node output is sent to three Barco projectors at the same time, and 3D signal output of each of the projectors is connected to other two. The RS 232 ports of three Barco projectors are interlinked together and connected to DVG graphic clustering control mainframe in parallel.

In this system, a 7.5 m (w) × 2.68 m (h) screen, 7 DVG graphic computers, 3 video signal distributors, and 3 Barco Galaxy 12 HB +12000ANSI projectors compose a three-channel passive 3D projection display system to show picture of lunar rover roaming system on the screen. Besides, there is stereo device to output dimensional sound. The operator wearing a pair of 3D glasses conducts real-time interactive operation in front of lunar rover roaming system of trim size.

The system software mainly includes Pro/E, 3Dmax, Virtools, etc.

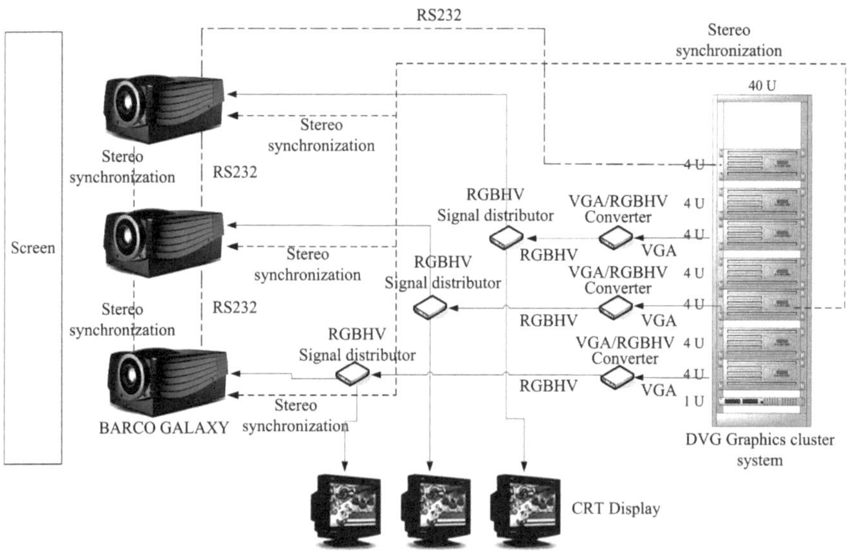

Fig. 33.2 Connection of DVG graphics cluster and CADWALL projection system

Fig. 33.3 Model effects

The lunar rover model is built up by Pro/E software and then imported to Virtools by data conversion; scene model is built up by 3Dmax and then imported to Virtools by data conversion. Virtools software realizes the lighting, materials, collision test, human-machine interaction design, and other functions of the whole roaming system.

33.3.2 System Design

First of all, input scene model to Virtools. To get vivid display effect, the key factors in scene model like lighting, materials, and camera shall be set. The setting effect is shown in Fig. 33.3.

Any of the six wheels of lunar rover can realize movement at six degrees of freedom, to better control movement of vehicle, first of all set several global variables to jointly restrict movement of wheels.

Design of collision test and human-machine interaction is realized by setting and programing BB block in Virtools. In this system, with object-oriented modeling method used in Virtools, the designer built up BB group block which may be transferred repeatedly to alleviate complexity of system and save a lot of work. BB setting and effect of the whole system are shown in Fig. 33.4.

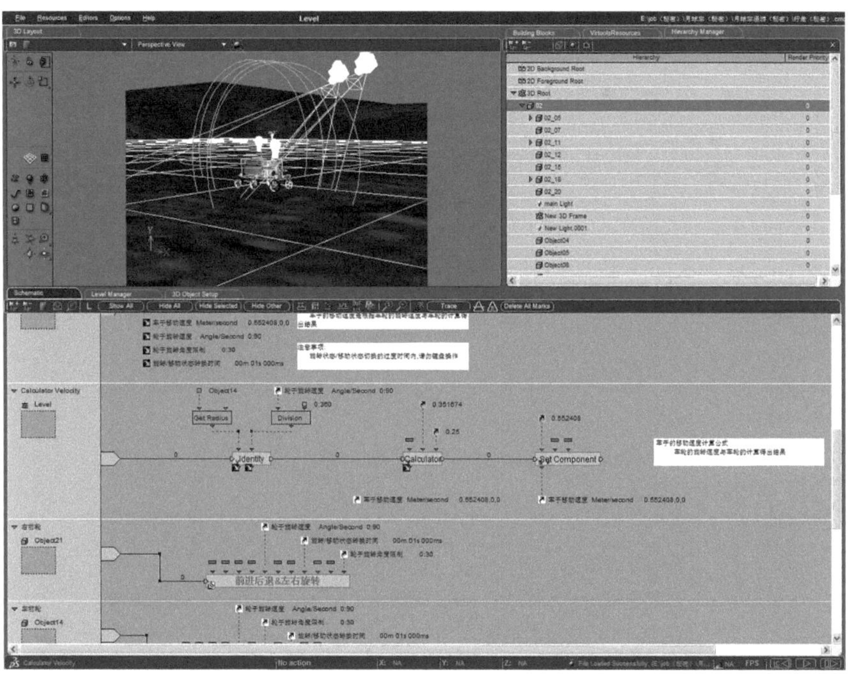

Fig. 33.4 System settings and effects

33.4 Conclusion

This paper makes study on lunar rover human-machine interactive roaming system technology. The author introduces the key technology for building up a human-machine interactive roaming system and the software and hardware environment and building process of human-machine interactive roaming system. Based on conclusion on relevant experience, a 3D visualized human-machine interactive roaming system has been built up to give user reality sense like being on the scene.

The building of this system and the research on relevant technologies lay a foundation for applying virtual reality technology in the research on virtual repair and assembly technologies in the next step.

References

1. Meng X, Li X (2004) Interactive technology—principles and applications. Tsinghua University Press, Beijing
2. Xiao T (2004) Virtual manufacturing. Tsinghua University Press, Beijing
3. Hao J (2008) Virtual maintenance theory and technologies. National Defense Industry Press, Beijing
4. Wang XC, Li B, Li L (2010) Research on key technology of virtual maintenance certain boat diesel engine based on Virtools 4.0. Beijing: J Acad Armored Force Eng
5. Lv C (2012) Design, analysis and verification of maintainability. National Defense Industry Press, Beijing

Chapter 34
The Study of Human-Computer Interaction Virtual Design

Sijuan Zheng, Fang Jie, Zengli Wang, Zhongliang Wei, Xue Shi, Liang Ling and Qiufang Wang

Abstract In the virtual environment, persons are going to design, evaluate, revise, and validate. So, virtual simulation technical was rapid developed. Using the technical method, the combination of the act capture system and simulate software is carried on human-computer interaction interface design, evaluate, revise, validate, and excel the old one for vehicle. In the virtual environment, persons are operating. The designer may intuitional discern matter of human-computer interaction interface. The designer can timely revise. It is create human-computer interaction interface of comfort, safety, shortcut for person. In this paper, it describes design processes that use the virtual simulation technical. By the virtual simulation technique, the time and cost, which makes to design process, has to lower and raise the human-machine environment design level of the whole vehicle.

Keywords Virtual simulation · Human-computer interaction · Act capture system

34.1 Introduction

In today's large manufacturing and research fields, like aviation, ordnance industry, shipbuilding, automobile, and other relevant industries, advanced information technology tools and methods, such as CAD/CAE/CAM/CAPP and the PLM technology, and a lot of simulation technologies are extensively applied in design stage, the beginning of the project. While in the special vehicle human-machine environmental design, simulation method is still staying in vehicle evaluation stage and not really applied in the special vehicle design. As a result, many human-machine environmental issues appear in the use of special vehicles [1].

S. Zheng (✉) · F. Jie · Z. Wang · Z. Wei · X. Shi · L. Ling · Q. Wang
China North Vehicle Research Institute, 100072 Beijing, China
e-mail: clubjully@qq.com

S. Long and B. S. Dhillon (eds.), *Proceedings of the 13th International Conference on Man-Machine-Environment System Engineering*, Lecture Notes in Electrical Engineering 259, DOI: 10.1007/978-3-642-38968-9_34,

Today with continuous upgrade of information technology, special vehicle generates more and more information volume and human-machine system becomes more and more complicated. The physical sample machine and engineering model method causes high cost and long cycle. Besides, once physical sample machine and engineering model setting are determined, any modification and alteration may become difficult. In such cases, the human-machine engineering design, test, and evaluation based on virtual sample machine technology can save a big volume of time and resources while making special vehicle human-machine interface more usable and more personalized. In this way, the stimulation technologies can be really integrated in overall design f special vehicle so as to enhance comfort and long operation of crew and overall special vehicle human-machine environmental level.

34.2 Human Body Model

With continuous development of information technology, people require machines to complete most of works because machines are more accurate, more intellectual, and more operable. While advanced equipment and huge IT will be integrated in the special vehicle to enhance relevant performance of vehicle to a certain extent, make vehicles more intelligentized and more automatized. However, no matter how big or intelligent the information volume in the vehicle is, the dominating role of crew in special vehicle is unchangeable.

To reduce load and fatigue of crew during operation, we must create a comfort operating space for crew. However, certain difficulties in modification and optimization may be caused by time, cost, or other factors once we validate, modify, and optimize the safety, comfort, and accessibility of human-machine interface design in physical sample machine. With gradual mature and application of simulation technology, in vehicle human-machine environmental design, human body 3D digital model shall be used to stimulate operating gestures of crew. However, it requires designer to edit operating gestures of crew manually with low accuracy. Today, connection with simulation system via capture system can reflect the real operating gestures of crew on the simulated 3D human body model in a real time to start simulation process of operating gestures and enhance accuracy of simulation.

34.3 Simulation Equipment

There are many available software and hardware simulation equipment. Through simulation equipment, we can create virtual environment, virtual display, simulation components, etc., with the development of simulation technology, the created virtual environment is more vivid, and researcher can conduct a series of operations in this virtual environment close to reality to optimize human-machine interactive interface. This paper mainly involves act simulation of human during operation and thus uses act capture system and human-machine interface simulation design software.

34.3.1 Act Capture System

Act capture system is used to capture different manipulation postures of crew in different environments, record manipulation postures, and analyze relevant data of each position of active crew.

Vicon optical act capture system is a reflection-based capture system. The system requires sticking an exquisite reflecting ball (or marker) in each specific position of crew. When the red light (or visible light or visible red light) around the optical camera casts to the surface of marker, the marker will reflect red light with the same wavelength to optical camera so as to determine 2D coordinates of each marker. Through processing of analysis software of Vicon capture system, 3D coordinates of each marker will come out. After recording the movement track of each marker, the rigid human body model is generated [2].

34.3.2 Simulation System

DELMIA system software provides designer with a visible virtual digital production environment to allow the primary design, simulation, validation, optimization, monitor, etc., in a virtual environment, designer conducts ergonomic evaluation on the comfort, accessibility, and visibility of the operating gesture of operator so that the designer can conduct validation, modification, and optimization of design in a short period and realize the effective management of virtual model in the whole process from primary conceptual design to detailed design, analysis, evaluation, and validation.

Create a visible virtual digital production environment with DELMIA to conduct plan, design validation, and optimization of human-machine interactive interface of special purposed vehicles so as to integrate human-machine design in whole vehicle design in a real sense and reduce the production cost of physical sample machine and engineering model.

34.3.3 RTID Software Plug-in

RTIDelmia is a bridge connecting act capture system and DELMIA human-machine interface software. RTIDelmia software can help enter the operating postures of crew in DELMIA in real time so as to control virtual human body model in DELMIA software, record posture information of virtual human body model including joint curve degree, and create DELMIA human-machine task with recorded information for follow-up data reappearance and statistics analysis. In addition, the generated human body activity parameters may be entered in DELMIA software for further analysis.

34.4 Simulation Operation

In virtual operation environment, crew makes a series of operation acts. The movement angle, angle acceleration, displacement activity scope, and other relevant data of each position of crew are recorded by capture camera so as to provide concrete data for future cabin human-machine environmental design.

34.4.1 Fundamentals of Simulation System

The whole simulation process is that the researcher captures various operating postures by capture camera within effective capture range of optical capture and creates virtual human body model on DELMIA software platform with Haption software drive, while relevant data of various operating postures are output and analyzed by analysis software so that designer can validate the reasonability of human-machine interface design by data while loan degree of operator can be reflected by activity parameter per min in each position (Fig. 34.1).

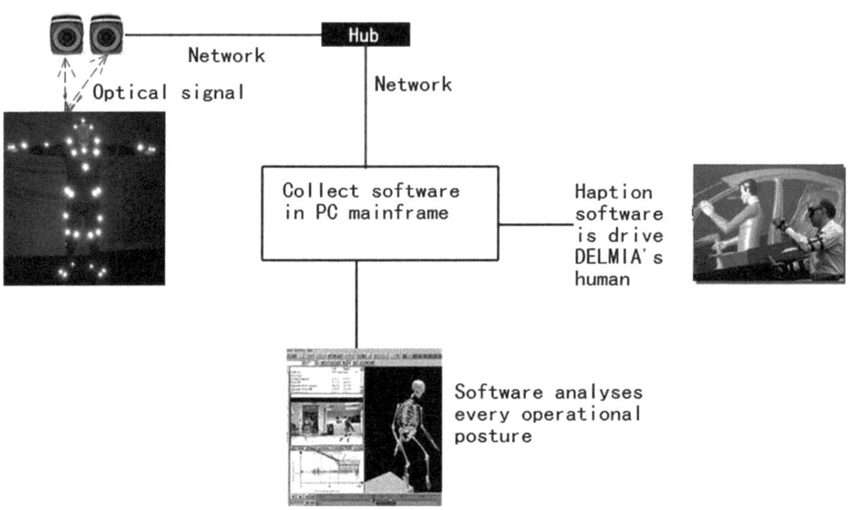

Fig. 34.1 Work principle figure

34.4.2 Build Virtual Operating Environment

To capture the gestures and different angles of crew manipulating the vehicle in an all-round way, 13 optical cameras are deployed in providing scope to capture a series of operation gestures from different angles.

T-type calibration tool shall be used for static calibration during system initiation. Place T-type calibrating level in the right center of the capture scope of optical camera to calibrate the focus of each optical camera while calibrate the 3D coordinates in the area of camera and capture scope and check whether there is any noise and reflection dots in the capture scope; and then, conduct dynamic calibration. A person wears act capture clothes with a certain number of markers on the clothes to dynamically focus on each optical camera. Remove the reflection dots and noises calibrated dynamically statically from the software platform because the coordinates of the marker is effective only if all reflection dots and noise have been removed. Adjust the angle and height of the camera when adjusting focus to ensure all-round capture of act gestures of crew in all positions.

After system correction and database configuration, capture the operation gestures of crew.

34.4.3 Virtual Human Body Modeling

The person reflects the operation acts by struck markers in capture scope. Marker is a kind of global object with reflecting materials on the surface. When the red ray emitted by camera casts to the surface of the reflecting ball, the reflecting ball will reflect the red ray of the same wavelength to camera so that capture camera can determine 2D coordinates of each reflecting ball. Through processing Vicon control software, 3D coordinates of every reflecting ball can be obtained.

In act capture system, Vicon software constitutes a body position by using four markers as a group. That requires operator sticks markers on the body of crew precisely and ensure any two markers cannot be too close to capture the 2D coordinates in bending act because of superposition of two markers (Fig. 34.2).

The simulation design requires putting human body in operation space for design and validation and thus needs driving the human body model built via Vicon act capture system to make it interactive with the human body model on virtual platform in real time. RTIDelmia software is the bridge between act capture system and simulation software. It fully combines real operation acts of human body with virtual operation acts.

34.4.4 Operation Simulation of Crew

During the preparation for simulation, first of all, use virtual human body consistent with the parameters of real human body of crew on DELMIA software

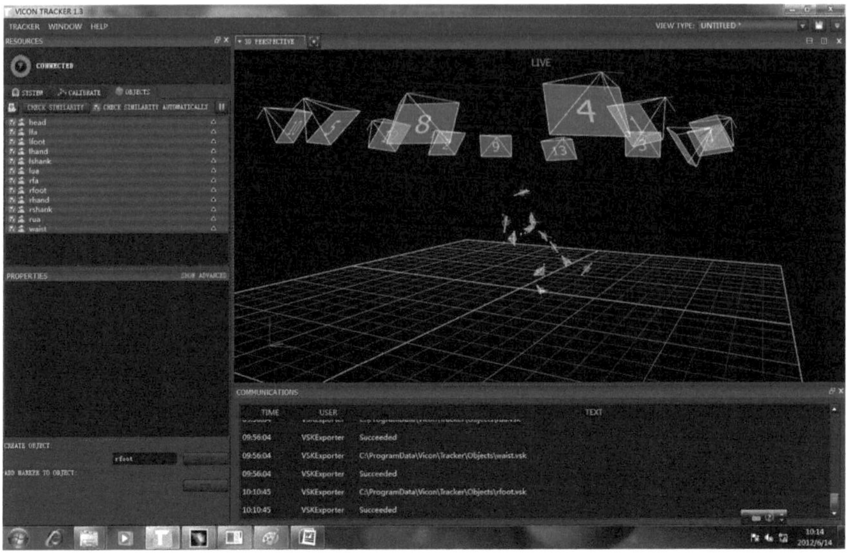

Fig. 34.2 Human body modeling

platform as simulated person; secondly, stick markers on human body precisely and ensure markers will not been superposed with each other when all positions like arms and legs bend; at last, check whether there is no any reflecting dot or noise in capture area and calibrate the focus of capture camera. During the whole simulation process, no reflecting dot other than markers is allowed, so adopted entities shall not contain reflective materials.

In the experiment, conduct operation simulation for crew's operation in the components like steering wheel, shifting handle, accelerator pedal, and switch buttons, record different gestures of crew when he or she operates these components in virtual space with an optical camera, and generate simulation video on DELMIA software platform. The designer will visually check whether there is any intervention in the operation process of crew through simulated operation process.

34.4.5 Data Analysis

Data analysis software is used to process data collected for capturing operating gestures. The mainly course involves data entry, itemized data view, periosteum building, data analysis report template, and generation analysis report. Different from capture simulation software, analysis software needs to re-stick markers on human body because the software has higher requirement on number of markers and capture; it requires less markers than those used in simulation and more precise sticking positions.

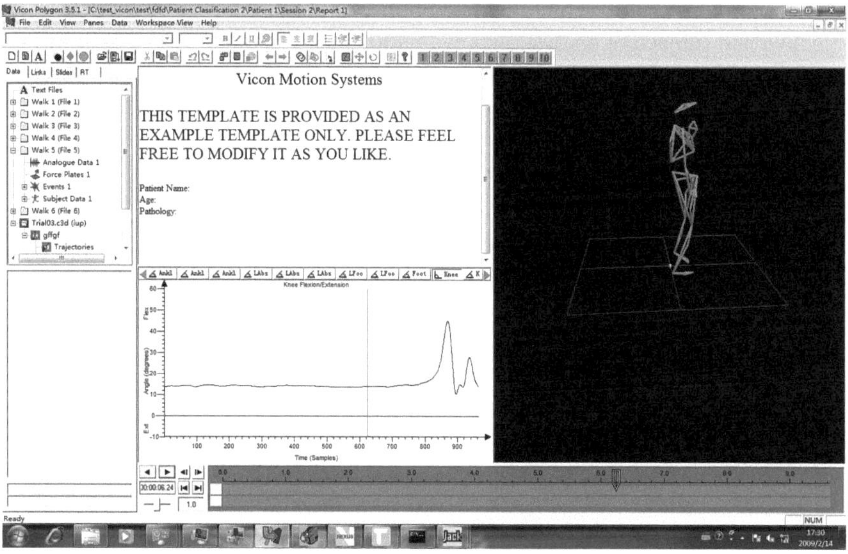

Fig. 34.3 Data analysis

Crew conducts normal operations in erected semi-physical operation environment, and system software makes real time recording for generated data and brings out reasonable suggestions on modification of design plan through analysis on parameters of each position like angle and angular acceleration. These data reflect whether human-machine interface conforms to requirements of ergonomics during operation of crew. The generated report may provide data support for future human-machine interface design (Fig. 34.3).

Through sorting and analysis, the angular acceleration, scope, displacement, and other relevant data generated by different positions of human body in operation which have been recorded through simulation provide data support for analysis, design, optimization, alteration, evaluation, and validation of human-machine interfaces and human-machine interactive interfaces of vehicle and enable human-machine interfaces to further match crew operation and permanent operation with different percentiles.

34.5 Conclusion

Capture of operation gestures of crew by act capture system enables crew to visually observe the situations like whether collision intervention will happen and whether operation is easy, fast, and comfortable during operation of crew and subjectively make analysis and evaluation on the operation gestures of crew while providing specific design basis for human-machine interfaces and human-machine interactive interfaces of vehicle with support of a huge volume of data. Through objective and subjective analysis and evaluation, design, validate, and optimize the human-machine interfaces and human-machine interactive interfaces of vehicle to reduce time and cost for design and enhance the level of human-machine environment design of the vehicle.

References

1. Chen X, Yuan X (2000) Man–machine–environment system engineering total discussion. Beijing University of Aeronautics and Astronautics Press, Beijing
2. Act capture system user manual (2012)

Chapter 35
Design Principles to Improve Man–Machine Friendliness for Interactive PPT Presentation

Linong Shi

Abstract The weak flow control ability of PowerPoint not only led to the inconvenience of the interactive presentation making, but also led to flow confusion and superfluous information irrelevant to the presentation contents due to dedicated control buttons. In this paper, the ideas and principles of structured flow design were proposed, and the whole flow is divided into main flow and branching flows, which were independent of each other. Branching flow was selected (called) by the user in a natural and consistent manner with the structure of presentation contents, and the man-machine friendliness was improved.

Keywords Man-machine friendliness · Interactive presentation · Structured design

35.1 Foreword

Default workflow of PowerPoint presentation is linear, and user does not need any choice during presentation. This is the most common used method of slide.

However, no matter in commercial or in educational application, various connections exist among contents involved. For instance, with regard to the relation between whole and part and the relation between layers, user needs to freely jump between whole and part, parts, as well as layers. In such cases, linear structure cannot realize satisfying presentation. To enhance user friendliness and quality of information delivery, branching selection structure, which can present such logic relation, is required to enable control logics to be consistent with content logics naturally.

L. Shi (✉)
PLA Academy of Special Operations, Guangzhou, China
e-mail: shilinong@163.com

S. Long and B. S. Dhillon (eds.), *Proceedings of the 13th International Conference on Man-Machine-Environment System Engineering*, Lecture Notes in Electrical Engineering 259, DOI: 10.1007/978-3-642-38968-9_35,

The issues about PPT have drawn attentions of a lot of researchers [1–3]. To realize interactive branch structure in PowerPoint, some people brought out some ways to realization [4–6]. However, these methods not only cause inconvenient design and making, but also have unsatisfying presentation effects, such as flow confusion brought by operation error, information irrelevant to presentation contents caused by control or navigation buttons and failure to end branching flow in advance. Therefore, on the basis of analysis on problem of non-structured flow control, this paper puts forward thought and principle of structured design of interactive presentation in order to solve out this issue fundamentally.

35.2 Problems of Non-Structured Flow Control

35.2.1 Realization of Non-Structured Method

Take the branching structure shown in Fig. 35.1 as an example. Slide 2 provides user with two options: A and B present branch A (only slide 6 included) when user chooses A; present branch B (slides 7–9 included) when user chooses B; present slides 3–5 when user directly clicks slide 2 instead of choosing A or B before end.

Following is main realization process of structured method:

Step 1: Change "content object" A and B on slide 2 to "interactive object," respectively, to enable automatically jumping to appointed slide when user chooses corresponding content objective.

Content objects A and B are objects containing presentation contents like texts and pictures generally, shown with solid dots in Fig. 35.1. Through act setting, hyperlink or other means, link content object to appointed slide so that these

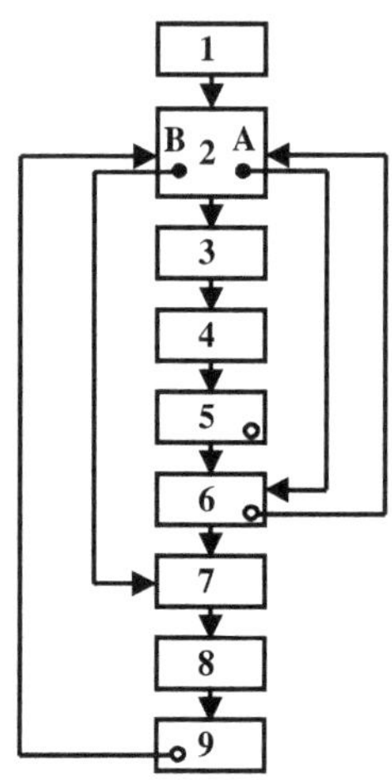

Fig. 35.1 An example of unstructured flow chart

content objects become interactive objects able to interactive selection. In this example, click content object A to jump to slide 6; click content B to jump to slide 7.

Step 2: Set flow "control object" on the last slide of each branch to realize return of branch.

In Fig. 35.1, the last slide of branch A is slide 6 and the last slide of branch B is slide 9, over which, "control object" is presented with a hollow dot. "Control object" is used to control flow, irrelevant to presentation contents, generally under control of an act button, or linked to appointed slide (slide 2 here) by text, picture, or any other object meaning return by means like act setting and hyperlink to realize return of branch.

Step 3: Set flow "control object" on the last slide (slide 5) in the main flow to end projection of all slides.

35.2.2 Problems Caused by Non-Structured Control

Above method may be used to realize branch structure with selection function, but it also bring out following problems.

35.2.2.1 Operation Error may Cause Flow Confusion

Operation error is easy to happen during presentation and will cause flow confusion. Take Fig. 35.1 as an example. Branch A contains slide 6. By choosing "interactive object" A on slide 2, user enters slide 6. After presentation of the slide is over, click "control object" on the slide to return slide 2. If user click place outside "control object" by accident, he or she will enter slide 7 so as to cause flow confusion. The existing solution is cancelling the method of switching slide by clicking the mouse which has been defaulted by slide. That can solve out "lost" problem caused by clicking the mouse by wrong, but there are many ways to switch slides. For instance, user can switch to the next slide by any forward key like N, Enter, Page Down, →, ↓, or spacebar. If user switches to another slide by forward key on slide 6, he or she will enter slide 7 too instead of returning back to slide 2 as wished. In the same way, similar problem is existing for the last slide of the main flow (slide 3 here).

Flow confusion cannot just happen on the last slide in the branch. It may happen on the first slide of the branch too. For instance, on slide 7, by pushing back key like P, Page Up, ←, ↑, or Backspace, user will enter slide 6, while slide 6 is not the content of the branch.

35.2.2.2 Control Information Irrelevant to Presentation Contents may be Required

User has to add control object, set end or back act, or cancel mouse switch method on the last slide of main flow and every branch flow. The steps are complicated, and these special control objects are irrelevant to presentation contents, not only increasing information load, but also affecting appearance of interface.

35.2.2.3 Branch Flow Cannot be Interrupted in Advance

Main flow and branch flows are integrated, so the end of branch flow means the end of main flow.

The essential reason for above problems is that main flow and branch flows are mixed and intertwined together. If each of main flows and branch flows is independent from others, all above problems will be solved out.

35.3 Thought and Realization Technology of Structured Flow Control

35.3.1 Basic Thought

Structured program design limits the use of "goto sentence" for control transfer and advocate structured and modularized program. When main program calls subprogram, the control is transferred to the subprogram and then returns back automatically after operation of subprogram.

Similarly, the use control object shall be restricted in PowerPoint to allow any of main flow and branch flows to be independent from others and branch flows can return back automatically at the end. If we can do that, that means, we have realized structured and modularized flow.

By default, there is just one flow in PowerPoint presentation: main flow. All slides in the presentation belong to this flow. To realize flow modularization of PowerPoint, we have to define relatively independent main flow and branches flows on the basis of defaulted flow and ensure the easy application of defined flows. "Flow definition" means the ability to specify which slides are contained in the flow and what is the order of projection. "Easy application of defined flows" means that main flow can be presented directly and branch flows allow user to interactively select.

35.3.2 Realization Technique

PowerPoint does not provide tools for structuralization and modularization, but such functions may be realized by some other techniques indirectly, like set projection scope, custom show, object linking and embedding (OLE).

35.3.2.1 Set Projection Scope

The technique may be used to define and use main flow on the premise that the numbers of slides contained in the main flow are sequential (this condition can be satisfied in most cases). The specific operation method: enter dialog box "set projection method" via "play slide" → "set projection method," and then specify projection scope in the box-enter the first slide and the last slide in box "From" to box "Until," respectively.

35.3.2.2 Custom Show

Custom show refers to slides in presentation which can be projected, respectively, after combination. Generally, create multiple custom shows for presentation and then project specific part of presentation for specific audiences to make the same one presentation is suitable for audiences of different types.

Main flow and branch flows are also can be defined by creating custom show. If main flow is defined, the custom show shall be defined as main projection. The specific operation method: enter dialog box "set projection method" via "play slide" → "set projection method," select custom show in "play slide," and then select corresponding title of custom show in following dropdown list.

Transfer method if branch is defined: set hyperlink (act setting, hyperlink, etc.) on content object (text, graphics, etc.) requiring interactive selection to link the object to corresponding custom show and input checkbox "display and return."

35.3.2.3 Presentation Object Linking and Embedding (OLE Technique)

OLE is a set of service functions providing a powerful way to create compound documents by information originated from different applications. Objects may cover almost all information types like letter, bitmap, vectorgraph, and even voice annotation, video clip, etc.

The object linked to or embedded in presentation is composed of multiple slides as a whole independent from main flow structurally. Therefore, OLE presentation object may be used to define and use branch flow. If "embedding" is adopted, all information of embedding object will be contained in this document and easy for management.

If "linking," slides contained in branch flow will be stored in another independent PowerPoint document.

OLE presentation object is shown in the form of thumbnail in PowerPoint; therefore, it is especially applicable to the occasions requiring taking thumbnail as hot object.

Click thumbnail in projection, and then, slides in OLE presentation object will automatically show on PowerPoint presentation because system has conducted "act setting": "object's act" is "display" when "click mouse."

35.4 Principles for Structured Flow Control Design

Forgoing paragraphs introduce the definitions and use techniques of main flow and branch flow, respectively. Main flow may be defined by set projection scope or custom show and branch flow may be defined by custom show or embedding of OLE presentation object.

However, how should main flow and branch flow techniques combine together to realize modularization of branch structure? This paper will introduce some application principles. Main flow and branch flow can be independent from each other structurally if only combination mode of main flow and branch conform to any of these principles. In this way, the logics of contents and structure of projection will match naturally without logical confusion appearing in PowerPoint presentation or special control structure set for navigation.

35.4.1 Principle 1

If there is any branch flow is defined with custom show, main flow must be defined by set projection scope or custom show.

The reasons is that the slides contained in custom show are in the main flow of this presentation, so main flow must be redefined by set project scope or custom show to prevent both from being intertwined with each other.

35.4.2 Principle 2

If main flow is defined by set projection scope or custom show and each of branch flows just contains a single slide outside main flow, the branch flows can directly jump to the slide just by interactive object without definition.

PowerPoint presentation will go on in the order defined by main flow, so, when control transfers to a slide outside the main flow, system will directly project this slide and then return back automatically in form feed.

35.4.3 Principle 3

If there is any branch flow defined with OLE text object, the redefinition is unnecessary for main flow.

The reason is that main flow is no longer defaulted by slides contained in OLE text object, so main flow is allowed. Such circumstance is also applicable to branch flow which is hyperlink leading to other presentation.

35.5 Application Cases

The branch structure shown in Fig. 35.1 is realized with structured method brought here, and the result is shown in Fig. 35.2. All work involves definitions of three mutually independent flows: main flow, branch flow A, and branch flow B (presented with a broken-line framework in the figure) and setting of two hyperlinks (presented with solid black dots in the figure) for interaction.

There are many detailed technical plans. This paper will briefly introduce two of them:

Plan 1—The simplest plan

- Define main flow with "set projection scope," the scope here is slides 1–5; not define branch flow A (see "Principle" for the reason); define branch B with custom show.
- Link solid black dot A to slide 6 and link solid black dot b to custom show.

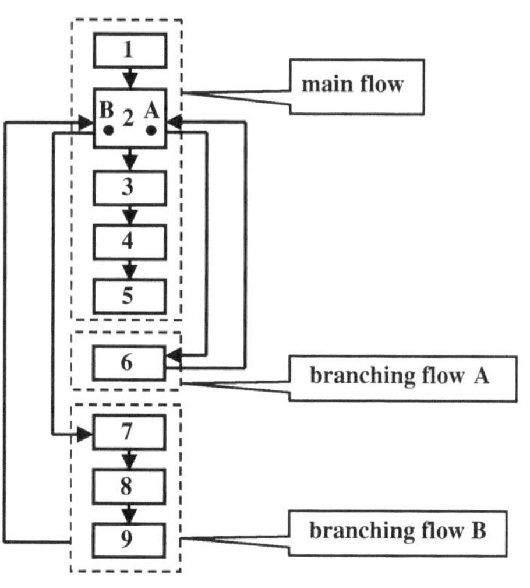

Fig. 35.2 An example of structured flow chart

Plan 2—A plan of taking thumbnail as hot object.

If we wish to show thumbnails of slides 6 and 7 on slide 2 and jump to corresponding branch by selecting thumbnail, we may insert two OLE presentation objects on slide 2 and then remove slides in branch A and branch B to corresponding OLE presentation objects. It is unnecessary to redefine main flow based on principle 3.

During presentation, the operation effects of all plans are the same: when user chooses A on slide 2, slide 6 appears; when choosing B, slide 7 appears. When clicking or pushing any forward key on the last slide of the branch flow (slides 6 and 9 here), the presentation will return back to slide 2; when click or pushing any forward key on the last slide of the main flow (slide 5 here), the display is over automatically.

35.6 Conclusion

Above analysis and application show, the thought and design principles of structured flow control brought here can better improve man-machine friendliness of interactive presentation because they have some advantages: completely avoiding control information irrelevant to content (hollow dot in Fig. 35.1) without requiring setting of special control structure for navigation; thoroughly solving out the "lost" concern caused by operation error (meaning that, flow confusion will not occur when clicking forward on the last slide or back on the first slide in any flow); advance back by clicking ESC in branch flow; simple design, easy use, and extensive practicability.

References

1. Tufte ER (2003) PowerPoint is Evil. Wired 11(9). [DB/OL] http://www.wired.com/wired/archive/11.09/ppt2.html
2. Bartsch RA, Cobern KM (2003) Effectiveness of PowerPoint presentations in lectures. Comput Educ 41:77–86
3. Atkinson C (2011) Beyond bullet points: using Microsoft® PowerPoint® to create presentations that inform, motivate, and inspire. Microsoft Press
4. Yingzi S (2002) Creating navigating system for interactive demonstration manuscript. China Distance Educ (02):56–57 (Ch)
5. Cui Y (2007) Embranchment design of PowerPoint courseware. Comput Know Technol (Acad Exch) (06):1762–1763 (Ch)
6. Yang g, Diao Y (2007) Research on causes and countermeasures of getting lost in using PowerPoint courseware. China Educ Info (16):30–33 (Ch)

Chapter 36
Study on Influencing Factors of Human to the Performance

Zhibing Pang, Honglei Li, Haitao Zhao, Kehua Zou and Cheng Jin

Abstract Aiming at effects of human in the performance, the influences to the performance are analyzed and researched from physiological and psychological two aspects in this paper. In terms of physiological factors, it is researched form five aspects: physical characteristics, auditory function, balance function, visual function, and physiological fatigue; in physiology respect, factors that effect the performance are dissected form three aspects: intelligence factors, will quality, and mental fatigue. Two categories of factors impacting on performance was positioned from the theoretical level, and all of these discussions provide certain reference values to select operators scientifically and quantize specific measures to improve the performance.

Keywords Performance · Physiological factors

36.1 Foreword

The research on operating performance of human is an important part of ergonomic research. There are many factors affecting operating performance, among them, human—the operator and user of weaponry—causes the complexity and randomness of operating effect and huge influence on operating performance due to inherent characteristics and attributes.

Z. Pang (✉) · H. Li · H. Zhao · K. Zou · C. Jin
Air Defence Forces Academy, Zhengzhou, China
e-mail: fkb_mme0s@163.com

S. Long and B. S. Dhillon (eds.), *Proceedings of the 13th International Conference on Man-Machine-Environment System Engineering*, Lecture Notes in Electrical Engineering 259, DOI: 10.1007/978-3-642-38968-9_36,

36.2 Influence of Physiological Factors of Operator on Operating Performance

The influence of physiological factors of operator on operating performance is mainly reflected in five aspects: physical characteristics, auditory function, balance function, visual function, and physiological fatigue.

36.2.1 Physical Characteristics

The physical characteristics of human main refer to physical quality of human, including strength, speed, endurance, agility, and flexibility.

Strength is the ability of human body to overcome resistance when muscle is working. The muscle contraction force is composed of the composition of contraction forces of muscles when completing a movement, coordination ability of muscle contraction, and mechanical efficiency of bone lever. Completion of any movement needs corresponding force, the basis of all activities.

Speed is reflection of comprehensive ability of human body like reaction speed, strength, flexibility, and other qualities, one of basic qualities. Speed quality depends on comprehensive development level of the flexibility and strength of central nerval system agility, speed, and endurance.

Endurance (or hardiness) refers to the ability of human body to resist fatigue after working for a long period. Fatigue is the necessary result of exercise or training. There is no overload recovery and organic function without fatigue. The course of endurance development is the course of keep generating fatigue, overcoming fatigue, and enhancing anti-fatigue ability.

Agility is very important in military battles and combat sports. Agility is the comprehensive reflection of various movement abilities of human body. Well-developed agility can not only help fast, accurately, harmoniously grasp techniques and exercise methods to fully effectively exert physical quality in the practice, but also prevent occurrence of accident.

Flexibility is determined by the many factors like scope of joint movement, stretch, and elasticity of muscles, ligaments, and tendons, and the coordination between intensity and relaxation of muscles. The factors influencing flexibility include age, sex, basic ability of muscle movement, state of bone lever, when exercise is stop every day, fatigue degree, etc.

36.2.2 Auditory Function

Sound can be heard only if it reaches the certain intensity. The min sound intensity is called auditory absolute threshold (auditory threshold); for a specific sound signal, the auditory threshold refers to the min effective sound intensity value when sound can be differentiated in 50 % of total number of times.

The auditory threshold is in reverse proportion to the sentience, and the auditory absolute threshold is different with the change of frequency. People have the highest sentience for the sound within 1,000–4,000 Hz. Once sound is below this range, the sentience will drop fast with the reduction in frequency; once sound is higher than this range, the sentience will fast drop with rise of frequency. Therefore, sounds of different frequencies can give birth to the same loudness feeling only if they have different objective intensities.

Generally speaking, the auditory response duration of a human is 0.115–0.182 s. However, there is big individual difference of auditory absolute sentience. The jobs with high auditory requirements need the human with high auditory absolute sentience.

The hearing of human is influenced by many factors. The people who work in a high noise in a long period will decline in auditory sentience. Age is another important factor influencing auditory sentience. The sentience for alt is every easy to decay with the increase in the age. A child can hear the sound of 40,000 Hz when the drumhead vibration is very small down to 1/10 of diameter of a hydrogen atom. With increase in age, the high-frequency hearing drops at a speed of 100–200 Hz yearly. A young person can hear the sound of 20,000 Hz, but a mid-age person can only hear the sound below 16,000 Hz. Generally speaking, the high-frequency hearing is losing with increase in age remarkably for a man [1].

36.2.3 Balance Function

Another important function of human ear is the balance function of human body. The balance function mainly involves judging direction for human body and maintaining self-balance. The organs responsible for balance function are centralized in inner ear, including semicircular canals, elliptical capsule, and saccule. There are three semicircular canals which are perpendicular to each other, constituting three faces in the space. Canals accept external balance stimulation and send the balance central of the cerebral cortex via vestibular nerve to adjust and manage response to balance.

Well-functioned vestibule of inner ear is one of the basic conditions to produce motion sickness (seasickness, motion sickness, motion sickness). If vestibular function is lost, human body will lose ability to accept various stimulations caused by accelerated/decelerated movements; in other words, the adverse or overlimit stimulations, which can cause the motion sickness, have no effect on human body. For instance, we know some people have car sickness. However, even if they have vertigo, some of them have no carsick symptoms for the loss of vestibular function (ototoxic drugs poisoning, bilateral labyrinthectomy, repeated vertigo attacks, etc.). These people will not feel deceleration, shaking, bump, and other movements involving acceleration in the future, although carsick phenomenon will disappear.

36.2.4 Visual Function

Visual sense of human is caused by light stimulation. It has remarkable adaptability to the change of the intensity of the light. Eyes of human can adapt to the light from dozens of photons to 105 CD/m^2. But the degree of agility of human eyes to light is change with the change of brightness of surrounding environment. Generally, degree of agility is high in dark and low in light. Besides, the degree of agility of human eyes to light is change with the change of the time.

Visual adaption: The first and the foremost visual characteristics of the human is visual adaptability, divided into light adaptation and dark adaptability. Light adaptation refers to the process of visual sense change when human eyes turn from the dark to the light. Light adaptation process lasts for approximately 1 min. Dark adaptation refers to the process that visual sense is enhanced when human eyes turn from the light to the dark. Dark adaptation process is longer than light adaptation process. Studies show that eyes can adapt completely in 30 min or a longer duration from light to dark.

Features of eyeball movement: When seeing an object, people are customized to see from left to right, from top to bottom clockwise; besides, horizontal direction is prior to vertical direction, and the estimation on the dimension and proportion in horizontal direction are more accurate, faster, and more efficient than those in vertical direction. When deviating from visual center, in the same deviation condition, the order of observation is top on the left, top on the right, bottom on the left, and bottom on the right. When observing an object, two eyes are moving at the same time and the visual image will be obtained within 0.07–0.3 s at shortest and 0.17 s averagely; the movement angular speed can be identified only if it is higher than 1–2°. Eyes pay more attention to the profile in comparison with shape (contour effect); and more accept linear profile in comparison with curve profile. Eyes conduct flick in reading and approximately 3–7 viewpoints for a line of letters, approximately 250 ms for a viewpoint when skipping. All these features are very important for arrangement of observation work [2].

36.2.5 Physiological Fatigue

Physiological fatigue is the declining state of work ability caused by large intensity or long duration of movement during working. Biologically fatigue is a natural protective response. People need consumption of stored energy and resources during working and movement, so large intensity and long duration of movement will consume more energy. If energy consumption cannot be supplemented in time and people keep acting, organ will be affected adversely. Therefore, fatigue is a response to prevent physical and psychical overload of organ [3].

When people puts themselves into work after warm-up at the beginning, their activity capacity is brought into full play and task performance reaches the highest level. However, considerable energy and resources will be consumed in this phase; therefore, people will feel tired and cannot insist very long with dropping efficiency, slowed speed, and weakened strength. With continuation of operation, the fatigue is accumulated to interrupt in the work. If serious fatigue is caused frequently, chronic fatigue will form to damage physiologically and psychologically.

36.3 Influence of Psychological Factors Operator on Operating Performance

If we compare physiological factors of human to a complicated "hardware", the psychological factors of human is naturally the "software" of the complicated thing. Unreasonable structure and weak or incomplete function of the software will restrict the exertion of the hardware.

36.3.1 Intellectual Factors

Intellectual factors are one of the basic capacities of human as a whole, covering a lot of factors. In this paper, author just makes a research on the capacities affecting operation: capacities of memory, understanding, reaction, and concentration.

First of all, capacity of memory. According to the rule of generation of military skills, at the beginning of generation of military skills, learner shall make the actions in memory and then kept the memory based on understanding. The good memory is the inborn advantage of learning operation skills.

Secondly, capacity of understanding. Memory is not enough to form operation skills. The learner shall form understanding-based memory through keeping operation actions in mind. On the one hand, this can form deeper memory to automatize skill operation. Even if when the learner is stimulated by an accident, he or she can complete the operation task depending on firm understanding-based memory; on the other hand, the learner can develop more effective operation actions and process in repeated exercise only if he or she keep the operation actions in mind based on the understanding on the fundamentals and interrelation of actions.

Thirdly, capacity of reaction. The capacity of reaction of human is the central reflection of many capacities. For instance, in a digital signal display, the operator shall cover the digital signal information from the display to other information and then take corresponding actions. That requires operator to have better capacity of digital reaction. If a person has prominent capacity of reaction, he can immediately take countermeasures when receiving abnormal sign to ensure the timeliness of military kill operation. By comparison, an operator with weak capacity of reaction has long duration of reaction after receiving abnormal sign and may loss the best opportunity of remedy sometimes.

Fourthly, capacity of concentration. When the operation duration is postponed, the influence on attention will appear. Without specific training, human has limited duration to focus on a single thing and concentration can also greatly impair capacity and energy himself or herself. If a person has been focused on a thing for a certain time, if he or she is forced to focus again, physiological fatigue will be caused. For some operation tasks in military operation, it is necessary to select those operators with long duration of concentration or conduct specific training for selected operators to enhance reliability of the implementation of operation tasks.

36.3.2 Willpower

Will is a kind of very complicated advanced psychological function, the active function of consciousness, the psychological process that a human consciously determines the aim and governs its actions to realize preset the aim. Self-consciousness, perseverance, decisiveness, and self-control are the four basic factors constituting a person's willpower, playing a very important role for operator to maintain efficient task performance in a fixed work environment in a long period.

Self-consciousness refers to that people have clear deep realization on the purpose and motive of action under control of correct belief and view to the world and are able to insist in principle in the activities. It is a character to ensure action can achieve given purpose. The operator with high self-consciousness has high realization and recognition on the work and is able to overcome various difficulties in operation consciously with firm determination to accomplish tasks, playing positive role in pushing for enhancement of task performance.

Decisiveness is a ready-witted character of a person who is good at distinguish right from wrong, able to make and implement decision duly reasonably. It is very important for an operator to make a prompt decision to fast reasonably treat various unexpected events on the basis of all-round consideration of various influencing factors in complicated changing situations [4].

Self-control refers to ability of people to consciously control and adjust the thought, emotion, and behavior. It is a character good at control ego. Such character can not only help operator eliminate interference in operation and firmly implement adopted decisions, but also depress the thought, emotion, and behavior inconsistent with the purpose of operation.

Perseverance refers to a character of people to overcome all difficulties and obstacles diligently in actions to complete established purpose. For an operator, perseverance is the premise to work and complete tough mission and the guarantee to overcome difficulties, remove interference, and resolutely complete task.

36.3.3 Physiological Fatigue

People may feel fatigue and some situations like inefficiency and error may happen in physiological fatigue. If physiological fatigue gets worse, the people will become indifferent and tired about work. Once physiological fatigue occurs, first of all, the work state will be affected and symptoms like distraction, slow thinking and act, bad mood and depression will appear so that people will feel tired about of the operation. Of course, he or she can continue work and maintain task performance by psychological control force in a certain period, but this cannot last long. If the situation cannot be changed, more serious physiological fatigue may appear and needs a longer period to recover. If operator cannot have a rest in such situation, some adverse consequences may occurs, including declining task performance first of all, increased operation errors, dropped quality, slowed work speed, and reduced output. Therefore, many people take change of work performance as an indicator of performance assessment. Sometimes work performance can maintained when fatigue develops to a certain extent mostly because of additional effort of the people. However, additional effort will cause more serious fatigue.

36.4 Conclusion

Human is complicated and uncertain in task performance. Due to physiological and psychological characters, human is easy to be affected by many factors so as to impact task performance [5]. This paper focuses on the qualitative research on the influence of operator on the task performance. The following research will intensify quantitative research by experimental analysis to realize scientization, digitalization, and computerization in the research on the influencing factor of human in task performance.

References

1. Zhu Z (2001) Industrial psychology. Zhejiang Education Press, Zhejiang
2. Pang Z (1999) Air defense forces man-machine-environment system engineering. Zhengzhou Air Defense Forces Academy, Zhengzhou, p 91
3. Benli X (2006) Special operations soldiers mental health education guide. New Times Press
4. Tang C, Pang Z, Zhao H (2011) Study on the influence of stamina to the performance of multiple person operating one machine. Proceedings of the 11th conference on man-machine-environment system engineering, vol 10, pp 332–335
5. Li T, Li H, Pang Z (2012) Experiment and study on stamina of multiple person operating multiple machine operation. Proceedings of the 11th conference on man-machine-environment system engineering, pp 236–239

Chapter 37
Requirement Analyses of City Frequency Management System Based on Man-Machine Interface

Genhua Qi, Chenhui Li, Runfeng Hou, Zhi Bing Pang, Guiqi Liu and Kehua Zou

Abstract City frequency management system is mainly used to analyze deposition of frequency equipment and plan and prepare frequency use in the city. And also by adjusting emergency frequency, this system can understand the situation of frequency use among the city districts in time. At the same time, it can also provide the simulation training platform of frequency equipment monitoring for city frequency management organization. As the body of constituting the Man-Machine interface, it provides the understanding of the main body of the Man-Machine system and service objects, the human can receive running information send by visual, auditory, tactile receptors, and other sense organs' receiver. After the analysis, judgment, and decision of their brain to make response, they will send control signals by moving organs such as hands, feet and also make a direct operational action. In this paper, based on the operating requirements of the system design process, it gives the view and work process and analyzes the requirement of the third users (main body of Man-Machine) in this system.

Keywords Person's machine hands over with each other · The frequency management · Customer's requirement

37.1 Introduction

Human-Machine interface (HMI) is an information medium between human and machine and a platform on which human receives information between machines and takes control actions. Human-Machine interface design shall be scientific, and today's software design also stresses the interactivity. How to solve practical problems with basic theoretic knowledge related to human-machine-environment

G. Qi (✉) · C. Li · R. Hou · Z. B. Pang · G. Liu · K. Zou
Air Defense Forces Command Academy, Zhengzhou, China
e-mail: fkb_mmes@163.com

S. Long and B. S. Dhillon (eds.), *Proceedings of the 13th International Conference on Man-Machine-Environment System Engineering*, Lecture Notes in Electrical Engineering 259, DOI: 10.1007/978-3-642-38968-9_37,

system engineering is the precondition of user-friendly and completely functioned software operation interface [1]. Through friendly human-machine interface and electromagnetic compatibility analysis model, the system provides a platform of operation applications from training to actual management and control in urban spectrum management strengthens scientificity of urban spectrum management, upgrading processing of urban spectrum management and enhance efficiency of spectrum management and control in a complicated electromagnetic environment.

37.2 Overall Requirements

The following respects are included:

1. The agility of electromagnetic spectrum monitoring equipment shall meet practical standard of urban spectrum management.
2. Electromagnetic spectrum monitoring equipment shall have function to monitor and analyze appointed frequency bands and frequency points.
3. System provides user-friendly human-machine interactive ability to realize graphic operation of frequency equipment and device deployment.
4. System provides various operation methods according to environments of applications by different users and provides users with simple easy graphic operation, decreases volume of data entered manually, and reduces manual operation error and abnormality.
5. The frequency plan automatically generated by system and text format must satisfy applicable document standard.
6. The model for analysis on electromagnetic compatibility between frequency equipment and urban areas shall have high degree of simulation, fully in combination with natural, geographic, and land feature environments to realize compatibility analysis among frequency equipment. EMC model requires taking full consideration of the influence of various natural interferences and human-caused interferences on frequency use state in urban area and accordingly generate understandable graphic and file-based result of electromagnetic compatibility analysis.

37.3 System View and Work Flow

37.3.1 System View

The view shows the functional interaction process between system and user, mainly reflecting main work and business flow this system can complete.

Interaction process between system and user is as shown in Fig. 37.1.

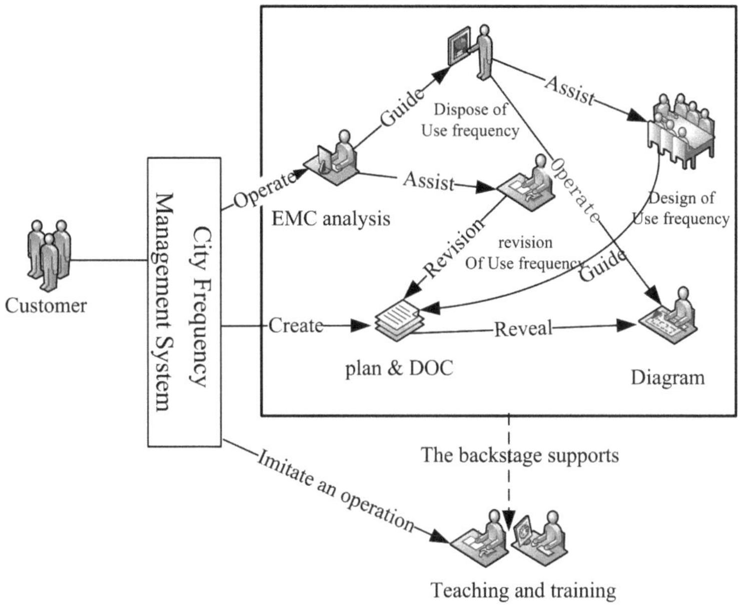

Fig. 37.1 The system–user interactive chart

37.3.2 Work Course

Organization and implementation of urban local electromagnetic spectrum management refers that:

Based on accurate understanding on the urban spectrum planning and comprehensive judgment, electromagnetic spectrum management institutions and professionals bring out electromagnetic spectrum management suggestions in time and promptly makes electromagnetic spectrum management emergency plan, monitors urban electromagnetic environment full-time, global, release electromagnetic state adjustment plan in real time, put priority on the control of electromagnetic environment in main direction, key area and crucial time and use of spectrum resources of frequency equipment, and coordinate and handle problems related to electromagnetic compatibility among various frequency equipment and use of interval frequency to ensure full exertion of actual efficiency of frequency equipment [2]. The basic steps for frequency use are as follows:

1. To set up urban electromagnetic spectrum management commanding center.
2. To collect, analyze, and judge frequency use in key areas. This step is taken based on the analysis and grasp of natural electromagnetic environment, frequency equipment conditions, and electromagnetic spectrum management force.

3. To bring out electromagnetic spectrum management suggestions. This step involves understanding task, analyzing and judging situations, making conclusion through judgment, and bringing out electromagnetic spectrum management suggestions.
4. To make plans related to electromagnetic spectrum management, including electromagnetic spectrum management plan, electromagnetic spectrum management force and spectrum resource mobilization plan and electromagnetic spectrum control construction.
5. To assemble urban electromagnetic spectrum management forces, including establishing urban electromagnetic spectrum management network system, electromagnetic spectrum management commanding system, electromagnetic spectrum monitor network, electromagnetic spectrum detection network, frequency equipment detection system, and electromagnetic spectrum management database system.
6. To organize electromagnetic spectrum management synergy.
7. To organize guarantees for electromagnetic spectrum management, covering guarantees for equipment and technology of electromagnetic spectrum management.
8. To inspect and instruct the preparation for urban electromagnetic spectrum management.

37.4 Users' Demands

37.4.1 Classification of Users

The uses of the system can be classified into three types:

1. Urban spectrum management planners;
2. Urban local frequency managers;
3. Frequency use equipment deployment and planning trainers.

37.4.2 Demands of Planners

The basic tasks the user can accomplish by using system are as shown in Fig. 37.2:

1. To realize collection and monitor the frequency use by frequency monitor devices.
2. To realize dynamic planning of frequency equipment by map-based operation system.

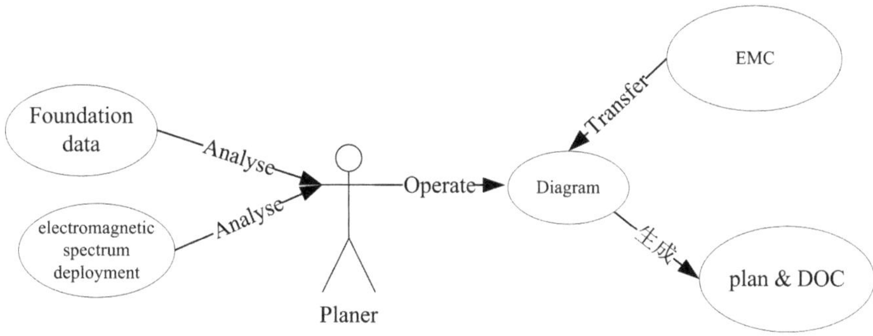

Fig. 37.2 The designer–system interactive chart

3. To control electromagnetic state in urban area by electromagnetic compatibility analysis.
4. To conduct dynamic frequency adjustment with real-time frequency monitor data.
5. To generate urban spectrum management file to provide assisting decisions.

37.4.3 Demands of Managers

The basic tasks the user can accomplish by using system are as shown in Fig. 37.3:

1. To know about electromagnetic state and frequency use conditions in urban area through frequency equipment deployment and result of map-based operation;
2. To realize dynamic frequency use adjustment with map-based operation system according to urban real-time frequency monitor data;

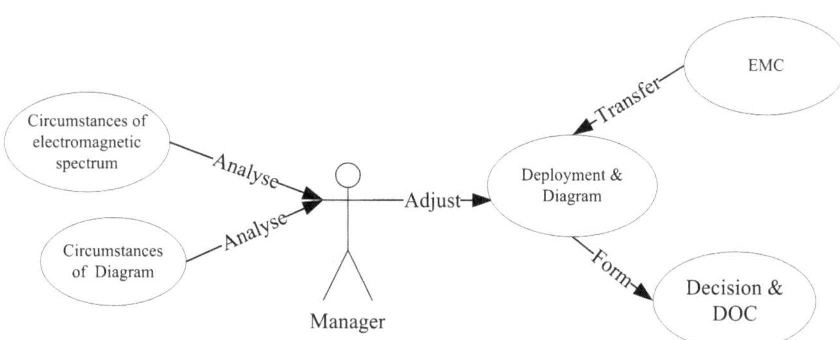

Fig. 37.3 The manager–system interactive chart

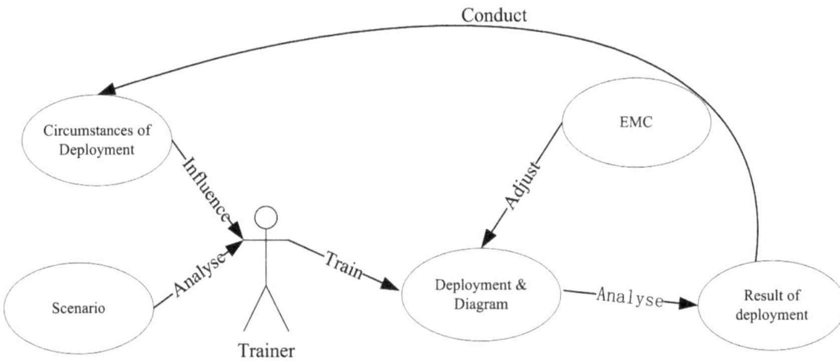

Fig. 37.4 The trainer–system interactive chart

3. To grasp result of dynamic adjustment of frequency equipment with the function of electromagnetic compatibility analysis;
4. Adjust spectrum management documents and assisting decisions in time.

37.4.4 Demands of Trainers

The basic tasks the user can accomplish by using system are as shown in Fig. 37.4:

1. To conduct frequency equipment deployment and map-based operation training according to appointed or simulated urban frequency management emergency plan;
2. To provide environment for real-time frequency use adjustment training by manually or automatic stimulation of dynamic change of urban frequency use;
3. Judge conditions of operation of student with function of electromagnetic compatibility analysis and then conduct circuit training based on follow-up instruction and adjustment and analyze assessment result.

37.5 Functional Requirements

37.5.1 Collection of Urban Local Frequency Use Data

Conduct monitor and collection of various conditions of frequency use in urban area and form database resources is used. The collected data mainly include the installation institution and installation address (detailed address and longitude and latitude) of frequency equipment; the model, work frequency, output power, modulation method, bandwidth, polarization method, antenna type, antenna height (from ground), and antenna gains of each radio transmitter; model, agility, antenna type, antenna gains, and other relevant data of each radio transmitter [3].

Besides, the collection of frequency data of natural electromagnetic environment and civil frequency and non-frequency equipment should also be included.

37.5.2 Real-time Monitor on Local Frequency Use Trend in City

It refers to the measurement of characteristic parameters of spectrum of electromagnetic signals in urban area with monitor devices and technical approaches, scan each frequency point and frequency occupancy rate in each frequency range in the city in real time and form contrastive database resources.

The purpose of this is to provide change in electromagnetic environment for frequency use in the city technologically, maintain order of airwaves, prevent harmful interference, ensure normal operation of different frequency equipment, and make limited spectrum resources exploited in a reasonable, scientific, and effective way.

The main contents under monitor include the following:

1. Quality of radio transmission of frequency equipment, such as use frequency, transmission bandwidth, signal field intensity, frequency deviation, stray emission, modulation method and modulation degree.
2. Conditions of spectrum use, such as statistics and monitor on occupy rate of a certain frequency or frequency range in a long duration, statistics of actual work hours of some radio stations, etc.
3. Unknown signal analysis, such as illegal frequency use, stealing other's frequency, interfering signals.
4. Conditions of change of local electromagnetic environment in city.

37.5.3 Urban Frequency Data System

This data system is to support data support for urban spectrum management, generally composed of database management subsystem, urban frequency basic database, urban frequency—using equipment database, frequency use adjustment and analysis database, geological information database, etc.

37.5.3.1 Database Management Subsystem

This subsystem is a software subsystem of definition, storage, inquiry, update, maintenance, and other management-related operation for battlefield frequency database.

37.5.3.2 Battlefield Basic Database

This database is used for storing urban frequency use data, including civil frequency, frequency distribution data, frequency assignment data, banned frequency data, protected frequency data.

37.5.3.3 Frequency Use Equipment Database

This database is used for storing various urban frequency use equipment data, including name of equipment, producer, owner, type of equipment, work frequency, frequency range, transmission power, type of antenna, method of polarization, antenna gain, agility of receiver.

37.5.3.4 Frequency Use Adjustment and Analysis Database

This database is used for storing data related to variation in frequency used in city, including monitor time, position of monitor point, monitored frequency, frequency occupancy rate.

37.5.3.5 Military Geographic Information Database

This database is used for storing information like landform, land features, relief and other geographic information and natural electromagnetic environment information.

37.5.4 Automatic Generation of Frequency Use Plan and Documents

The main functions include generating various frequency use plans and documents about frequency use in the city after comparative analysis on frequency use deployment and plan and accordingly provide planner with information to assist decision making and spectrum plan.

37.6 Conclusion

Human-machine-environment system engineering brings out overall demand of urban frequency spectrum management system design and demand analysis for interface design to provide a certain human-machine theoretic analysis for creating user-friendly interface in relevant designs.

References

1. Li Z, Wu Q, Zhou R, Zhang L (2007) High-precision method of electromagnetic frequency spectrum and its application in mineral exploration. Journal of University of Science and Technology Beijing, 29(6)
2. Zhang L, Chen M, Yan BH (2012) Research of electromagnetic frequency spectrum management system designing. Ship Electronic Engineering 32(3)
3. Chuanwei L, Wei H, Liu L (2010) Researches on the application of cognitive radio techniques in electromagnetic spectrum management. Electronic Test, (1)
4. Han H, Guo J (2009) Preliminary study on construction of electromagnetic frequency spectrum management on the condition of informationization in joint operation. J Acad. Equip Command Technol 20(5):001

Chapter 38
Study on Man-Machine Interface Multimedia PowerPoint Designing

Runfeng Hou, Genhua Qi, Hongyan Ou, Bingjun Zhang, Chen Jin and Peng Han

Abstract The design of man-machine interface embodied the scientific spirit of man-machine alternate. This text analyzed the principle of multimedia PowerPoint designing, and the text emphasized the mission-oriented principle in the designing of multimedia PowerPoint, and the text emphasized the whole layout sets out from the multimedia PowerPoint interface. Combining the experience of multimedia PowerPoint interface designing several years, the author of this text put forward a series of need of multimedia PowerPoint interface designing. And this text provided a certain design road for the design of multimedia PowerPoint interface designing and optimization the step of the multimedia PowerPoint interface designing.

Keywords Multimedia PowerPoint · Man-machine interface · Designing

38.1 Foreword

The information exchange between software and human is realized by interface, while multimedia PowerPoint production is a human-machine interface by software to conduct assisting class teaching, product introduction, etc. Human-machine interface is the youngest branch subject in computer science, the combination of two major sciences—computer science and cognitive psychology. It involves many hot computer technologies in present like artificial intelligence, natural language processing, and multimedia system while absorbing research outcomes of linguistics, human-machine engineering, and sociology as an interdisciplinary, marginal, and comprehensive subject [2].

R. Hou (✉) · G. Qi · H. Ou · B. Zhang · C. Jin · P. Han
Air Defence Forces Command Academy, Zhengzhou, China
e-mail: fkb_mmes@163.com

S. Long and B. S. Dhillon (eds.), *Proceedings of the 13th International Conference on Man-Machine-Environment System Engineering*, Lecture Notes in Electrical Engineering 259, DOI: 10.1007/978-3-642-38968-9_38,

In this information-based society, multimedia PowerPoint is closely related to our daily life. Multimedia is extensively used in all fields like teaching, research, academy; therefore, it is meaningful theoretically practically to make great effort to enhance researches on the issues in human-machine interface design during multimedia PowerPoint production, positively explore into human-machine design principle during multimedia PowerPoint production course, analyze on demand of human-machine interface design during multimedia PowerPoint production in depth, effectively grasp optimization course of multimedia PowerPoint production, and so on.

38.2 Principles of Human-Machine Interface Design of Multimedia PowerPoint

Human-machine interface is a kind of special software helping realize the appropriate connection, harmony, and consistency among software, hardware, and human. Simply speaking, human-machine interface is a kind of software specially dealing with human-machine interactive activities. It is the cross part among, human, hardware, and software as shown in Fig. 38.1 [1]. The human-machine interface design of multimedia PowerPoint plays an important role in multimedia PowerPoint production course. Therefore, its human-machine interface design must follow specific principles as below.

38.2.1 Give Prominence to User Object

The object of multimedia PowerPoint use determines the demand of interface design. Therefore, multimedia interface designer shall highlight the user object, analyze characteristics of user object, and design interface in combination with the characteristics. The analysis on user object shall be pertinent [3]. Multimedia

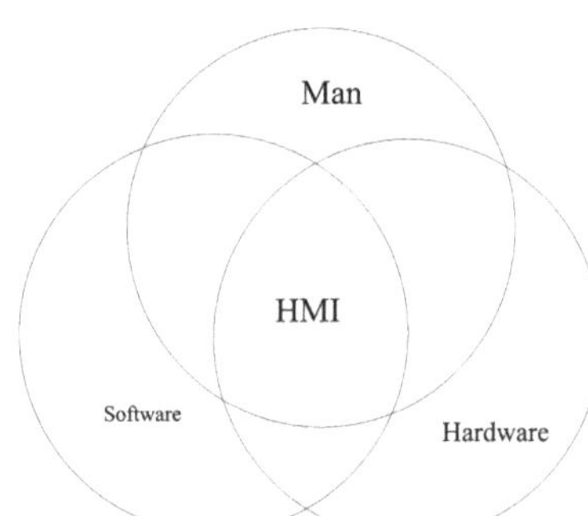

Fig. 38.1 Man-machine interface

PowerPoint faces users; therefore, the personal characteristics, operation ability, and knowledge level of users shall be taken into full consideration during design process. Concretely speaking, multimedia PowerPoint interface design shall be on the premise of users' demands. The designer shall conduct extensive research before interface design to obtain considerable firsthand materials and grasp different demands of different users in depth in order to design human-machine interfaces of different styles. For mainstream human-machine interface design, we should put the emphasis on those computer operators because they are important multimedia user group with knowledge and kills related to multimedia operation and special aesthetic angle of view and requirements on multimedia human-machine interface.

38.2.2 Ensure Consistency

We should pay full attention to the consistency in multimedia PowerPoint interface design [2]. The consistency and harmony shall be ensured for the interface with the same display operation method. Stress on the consistency of the interfaces is to reduce the cognitive burden of user object and help them fully use limited energy in learning of teaching contents instead of wasting time in learning of interface operation.

Consistency is the common principle of human-machine interface field. It shows information of the same type in a consistent, similar way, including show style, layout, position, used color, and simple human-machine operation method. The consistent interactive interface may help learners to promote their updated knowledge and experience to new PowerPoint so as to alleviate burden of learners in re-learning and memory.

Consistency means consistency of whole and locality. A complete multimedia PowerPoint shall be consistent in style, or all chapters of the PowerPoint shall be consistent in style as a whole. Concretely speaking, the style may be measured in the respects like color, composition, and human-machine interactive response. On the major premise of style consistency, appropriate adjustment may be made to change visual effect of users so that learners are clear in mind and not confused. The same interface control elements shall trigger the same behaviors and actions. For instance, the icons with the same function or the buttons with the same picture shall generate the same behaviors and actions so that learners can be released from cognitive burden and avoid confusion or middle of nowhere in mind.

38.2.3 Stress Color Arrangement

The colors of interface shall be closely related the contents, and the color tone shall be clear to give prominence to build up image, created artistic conception, and expressed theme. Different themes and contents shall be expressed with different tones. Some needs brightness and some depression, all depending on artistic conception of teaching. The unreasonable color use may have psychological influence on human like visual fatigue and over rendering will be distractive. Therefore, color use must follow certain principles.

It is necessary to fully consider visual traits of human during multimedia PowerPoint design [4]. Flexible human-machine interface plays an important role in motivate interest of learners and enhance initiatives of learners.

The beauty of color is closely related to teaching contents. The beauty of color is an integral part of the art aesthetics of interface. It may help the learner feel beautiful, increase sentiment in the edification of aesthetics, and get better awareness and understanding. Concretely speaking, the color of the foreground and background of display shall be configured correctly. Foreground shall be greatly different from background in color and both shall form contrast of light and shade. Different colors generate different intensities of visual stimulation. Generally speaking, warm color has intensity bigger than that of cold color, primitive color has intensity bigger than that of supplementary color, supplementary color is bigger than that of achromatic color, and saturated color is bigger than unsaturated color. Red is the biggest in intensity. When designing color of interface, the color shall be consistent with the contents of the learning. The treatment of color must be careful. It is inappropriate to present the content with individual preference. The importance of content, the style, and learning object shall be taken considered to choose appropriate color as main tone. For instance, the bright tones are used to present lively contents generally; pink and purple series are used to deliver mellow light information contents; warm colors and green are used as background of political and cultural contents; blue and gray are used to highlight some technical and specialty contents.

It is necessary to stress that there is a principle of color application in interface design: The tones of background and multimedia shall be consistent basically, and the whole page shall be covered by the same one main tone. Besides, different contents shall form contrast in terms of shadow of color so that media contents look clean, bright under harmonious background, and the whole page is harmonious graceful in color arrangement.

38.2.4 Reflect Interactive Functions

Human-machine interactive function is an important indicator of humanization of multimedia PowerPoint interface. The advantage of multimedia can be taken to the largest extend and the knowledge can be communicated the most efficiently only if a lot of human-machine interactive operations are adopted.

Interface design is one of main contents of multimedia PowerPoint. Interactive interface is the channel for any computer to conduct information exchange. User inquires, manipulates, and controls by entering information in computer via the interactive interface and computer provides users with information via interactive interface for reading, analysis and judgment. Interactive interface generally involves control menu and control, operation and hint, interaction and question, self-examination and exercise, experiment and simulation, etc. The design of different interactive functions can effectively scientifically integrate contents to be presented by the multimedia PowerPoint and powerful give prominence to the meaning of interactive interface design.

38.3 Analysis on the Demand of Human-Machine Interface

38.3.1 Universality of Human-Machine Interface of Multimedia PowerPoint

The applications of computer technologies are keeping updating in the course of development. The new computer technologies play effective assisting roles in the production of multimedia PowerPoint [5].

Of course, the applications of multimedia PowerPoint is not limited to teaching, and the multimedia is relatively common in the new product release and functional presentation [6]. Human-machine interface design is the most intuitionist part of multimedia PowerPoint design. Therefore, human-machine interface design of multimedia PowerPoint is universal to a certain extent.

38.3.2 Importance of Human-Machine Interface of Multimedia PowerPoint

Multimedia PowerPoint is a kind of teaching software integrating characters, graphics, image, animation, video and others together. It can fully mobilize various sense organs of user and help break the limitation of receiving information just by individual organ.

However, anything has two sides. Multimedia PowerPoint is the integration of multiple media; therefore, the interface designer shall not only pay attention to the usage, functional features, and attentions of individual medias, but also stress the coordination and consistency of the whole interface and fully mobilize the attention of users to get relatively satisfying effect. Therefore, human-machine interface design plays a very important role in multimedia PowerPoint production.

38.3.3 *Scientificity of Human-Machine Interface of Multimedia PowerPoint*

Human-machine interface design of Multimedia PowerPoint must be scientific. The content-based design process shall take user as the main body and involve reasonable applications of all relevant technologies.

"Interface" was a common term in ergonomics. A reasonable operation interface enables user to conduct human-machine dialog easily. There are many types of interfaces: main interface, secondary interface, pop-up interface, etc. The key to multimedia PowerPoint interface design is to discuss human-machine interactive operation method. To create a beautiful, understandable, operable guiding human-machine interface, the design shall delight and interest user to enhance efficiency of use. The key to design is a harmonious human-machine environment. The contents of PowerPoint interface design involve title, home page and the last page, menu, panel, icon, mouse, etc. Design interface may be defined as the sum of all information interactions encountered and analyzed in the course of design, reflecting relation between human and object. Design interface reflects the nature of human-machine information communication as the connotation of design art, involving all respects of design.

Currently, there are many technologies applied in multimedia PowerPoint production, like Photoshop popular in graphic design, 3DMAX in 3D field, PowerPoint and Dreamweaver in media production. Of course, human-machine interface designer of multimedia PowerPoint shall reasonably apply relevant technologies according to actual situation to make design more scientific.

38.4 Optimization of Human-Machine Interface Design

The course of multimedia PowerPoint design must follow the principle and give prominence to the practicability. Interface design shall optimize steps to prevent some avoidable error or influence on human operation for blurry design. For instance, graphic design has influence on not only visual effect, but also overall effect of multimedia PowerPoint. Therefore, screen layout is crucial and determined the priority of the elements. Generally, full preparation and commissioning are important in multimedia interface design to form highly operable interactive human-machine interface. In practice, optimization design may be made as shown in Fig. 38.2.

First of all, select appropriate design software. This process is the basis of the multimedia design. Give prominence to user object and analyze characters of user, like education level, overall preference, form a general thought. Actually, it is a process of analysis on "human."

Secondly, designer shall conduct human-based demand analysis under the instruction of principle of human-machine interface design to create a reasonable human-machine interface.

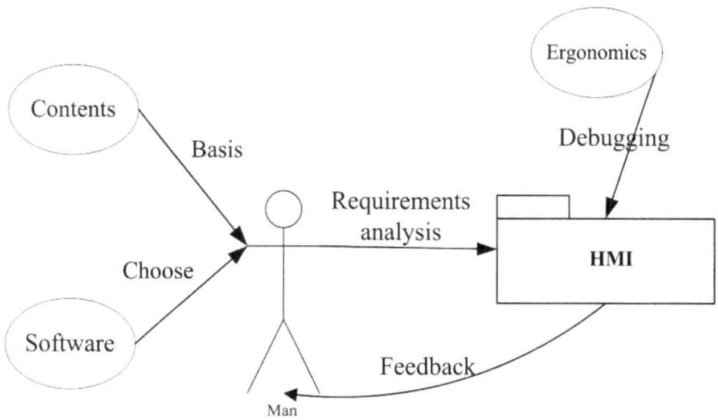

Fig. 38.2 The optimization process of multimedia PowerPoint designing

Lastly, ergonomic analysis was conducted for completed PowerPoint, probed into the issues existing in application, and debugged relevant software is making human-machine interface more scientific.

38.5 Conclusion

To sum up, human-machine interface design of multimedia PowerPoint plays an important role during the progress of the society. Therefore, it is necessary to enhance research on the issues of human interface design during multimedia PowerPoint production. This paper concludes the design principle of human-machine interface during multimedia PowerPoint production, analyzes design demand in this respect, and explores the optimization course of multimedia PowerPoint production. Accordingly, the research outcomes of this research are valuable in theoretic study and practical application to some extent.

Of course, the research here just stays at the layer of qualitative theoretic exploration. In follow-up researches, we should further stress combination of survey questionnaire and practical application and intensify the qualitative persuasion of research to push for in-depth development of human-machine interface design of multimedia PowerPoint.

References

1. Chen Q (2004) The software man-machine interface designs. Higher Education Press
2. Li Z, Hongxia M (2009) The exploration of multi-media courseware based on the man-machine interface. Comput Knowl Technol 5(3):78–80
3. Nie Y (2010) The study of multi-media courseware designing principle and method. Sci Technol Inf 11:153–154

4. Yang J (2011) The man-machine interface in the multi-media lesson piece designs a research. Chinese section teaches innovation to lead to publish. p 188
5. Li Y, Li Q (2005) Military science tutorial. Ocean University of China Press, Qingdao
6. Yu G, Zhao L (2007) Military equipment. Beijing: National Defense University Press

Chapter 39
Research and Design on IETM for a Type of UAV

Aizhi Liu, Hongxia Ji and Baoan Li

Abstract According to the GJB6600 Standard, a set of Interactive Electronic Technical Manual (IETM) system is developed for the Integrated Logistic Support (ILS) practical demand of a type of Long-Range Intermediate Unmanned Aerial Vehicle (UAV); the hardware and software platforms of IETM are constructed, based on general technology; the specific development process of the IETM is elaborated; aiming at the low efficiency problem of the traditional development software, a Data Module (DM) rapid generation software is designed, the efficiency of DM generating is greatly improved, and the development cost is saved; the IETM of the UAV has been in practical application, improving the level of the ILS for this type of UAV, developing a new train of thought for the ILS of the UAV.

Keywords UAV · IETM · DM · ILS

39.1 Foreword

There is a type of Long-Range Intermediate UAV which features complicated system and high requirement of integrated logistic support. It has become a weaponry in army in batch; however, its daily maintenance support is still in the

A. Liu (✉)
College of Mechanical Engineering, University of Science and Technology Beijing, Beijing 100083, China
e-mail: 568524492@qq.com

H. Ji · B. Li
Institute of Unmanned Aerial Vehicle, Beijing University of Aeronautics and Astronautics, Beijing 100191, China
e-mail: jihongxia76@163.com

B. Li
e-mail: superlba@163.com

S. Long and B. S. Dhillon (eds.), *Proceedings of the 13th International Conference on Man-Machine-Environment System Engineering*, Lecture Notes in Electrical Engineering 259, DOI: 10.1007/978-3-642-38968-9_39,

form of traditional paper and electronic file. Such technical material use and management mode are defective in many ways like long editing period, poor timeliness, inconvenient use and keeping, more repeated works, and waste of time, hour, and money so that it cannot satisfy the requirements of modern equipment maintenance support. As an assisting method of advanced equipment maintenance support, IETM can provide user with detailed electronic files by interactive approach, lower down requirement on the experience and level of maintenance person, significantly shorten the duration of maintenance of this type of Long-Range Intermediate UAV equipment, enhance accuracy and support efficiency of maintenance, reduce maintenance cost in whole lifetime, and strengthen ILS level and war preparedness level.

39.2 Basis of IETM Technology

39.2.1 Conception and Classification of IETM

GLB defines IETM in this way: "store in digital form and provide technical publications containing fundamentals, operation and use, maintenance and other contents in forms like text, graph, form, audio and video by means of human–machine interaction"[1].

IETM is in five levels [2]: the practical application of ILS of current equipment is at IV level—the layered database structure by which IETM stores and manages data by database for structured data containing internal logics and connection between information. The data are created in database and then stored in database after being edited; a multi-layer database is created; the database is able to pre-process and provide "user-appointed technical information" according to actual demand. In such architecture, the modularization of information such as text, form, and multimedia allows user to access the technical information by different means. With strong interactivity, the display of system is based on "framework navigation structure".

Web-based IETM is the priority of current IETM research, in four types of architectures [3]. Among the types, S2 Type is "HTTP + application server + database server" architecture, the most extensively used in these years. The architecture of S2 Type is shown in Fig. 39.1.

39.2.2 Standards for IETM Development

Currently, the main standards for IETM development include American military standard MIL series and international standard S1000D. From the perspective of standard-supporting organizations and category of the state and maintenance

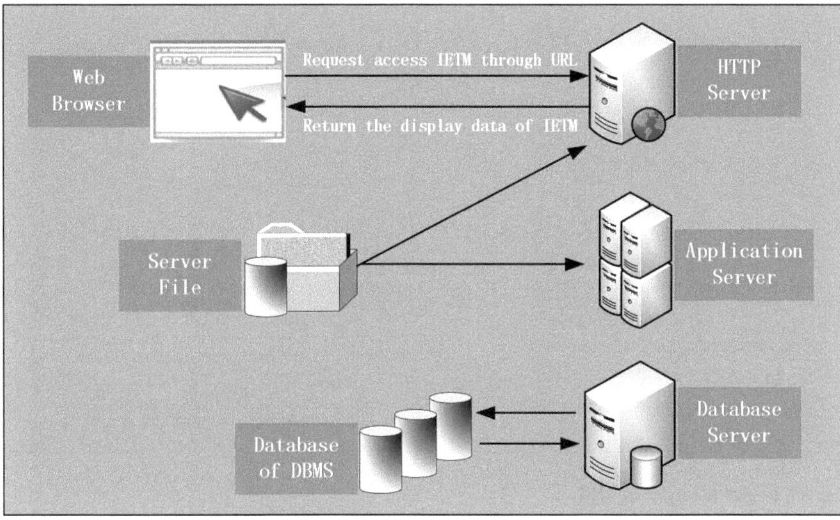

Fig. 39.1 The architecture of S2 type

system, technical basis, and development program of standard, S1000D has obvious advantage [4]. GJB6600 is a military standard of China to guide IETM development, originated from S1000 D standard. In terms of contents, GJB 6600 is in response to V 3.0 of S1000D; in terms of element markings, GLB 6600 is the localized element of S1000D V3.0. Therefore, GJB6600 has been determined as the standard for development of a type of Long-Range Intermediate UAVIETM, while S1000D-related contents are taken as important references.

39.2.3 CSDB-based IETM System

GJB6600 standard adopts modularization design thought, technological materials are split into Data module and information objects, which can be stored in the CSDB.

DM is an important information unit composed of description, program, and operation data of equipment or its component to describe a part of complete information of equipment as the minimal unit for information description, inseparable, atomic [5]. Every DM is composed of two parts: state identification segment and content segment. DM architecture is shown in Fig. 39.2. Information objects refer to DM-attached objects of illustration, multimedia, interactive 3D graph, etc.

Common Source DataBase (CSDB) is a warehouse and management tool of all information objects in a project (equipment) for creating and managing technical publications [6]. The information objects in CSDB are combinable according to

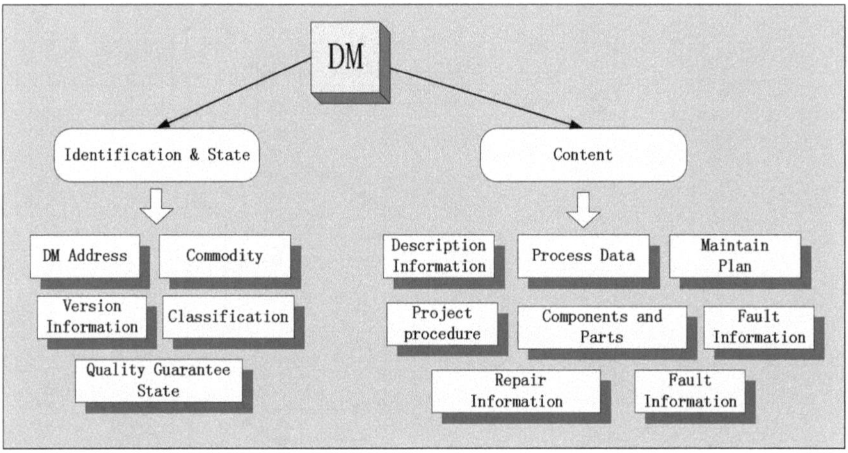

Fig. 39.2 Architecture of DM

demand to satisfy different application requirements of IETM system. In this way, the technical information contents can be created in one time and used repeatedly so as to make repeated use of data contents very easy. The process of editing CSDB-based IETM is shown in Fig. 39.3.

39.3 IETM System Design

39.3.1 Architecture of Development Platform

39.3.1.1 Software Platform

Long-Range Intermediate UAVIETM system software development platform is composed of editing system, management system, release platform, and system validation testing as shown in Fig. 39.4.

The editing system includes independently developed DM fast generation software, IsoDraw vectorgraph drawing software, Flash animation production software, Premier video processing software, and Autodesk Maya 3D modeling software to realize generation of DM's XML file, generation of vectorgraph and heat map, production of multimedia animation, video edit, and production of interactive 3D model.

The management system and release platform adopt trial software of CORE-NA's CORENA S1000D. The software supports business process of technical publication, conforming to S1000D standard in whole lifetime, used for CSDB management and release. The database adopts SQL Server 2005.

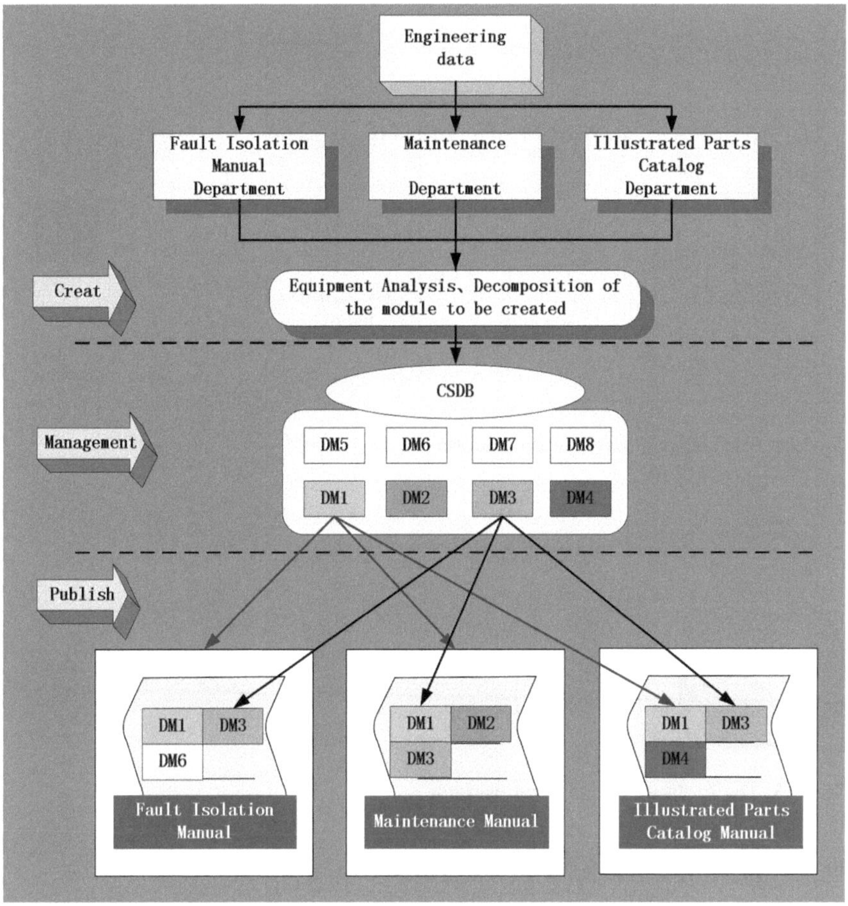

Fig. 39.3 The process of edit for IETM based on CSDB

The reading system adopts universal Web browser, while vectorgraph browsing plug-in IsoView and interactive 3D model browsing control are installed to realize interactive browsing of IETM publications.

39.3.1.2 Hardware Platform

Long-Range Intermediate UAVIETM SYSTEM hardware development platform is composed of server (Lenovo T420 notebook), wireless router (TP-LINK TL-WR740N), and handhold pad terminal (HASEE A10B pad). The hardware system is shown in Fig. 39.5.

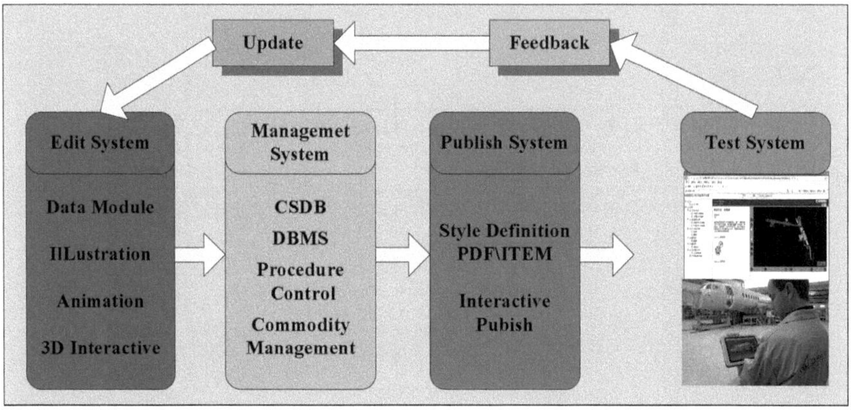

Fig. 39.4 The software development platform

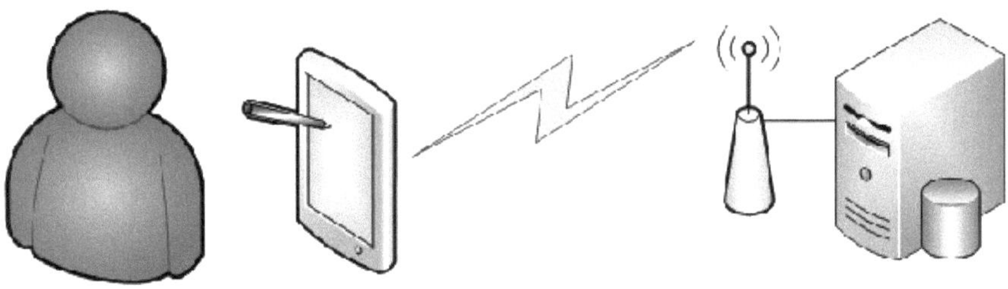

Fig. 39.5 The hardware development platform

39.3.2 Detailed Development Process

39.3.2.1 Analyze Project Requirements

Aiming at the failure information of a type of Long-Range Intermediate UAV in existing technical materials and accumulated in equipment application develop a set of LETM system based on S2 type architecture, meeting requirement of IV level, enhancing the efficiency of trouble shooting, separation, and elimination, and shortening the average maintenance duration of this type of UAV.

39.3.2.2 Make Project Regulations

Development standard is GJB6600, while S1000D is taken as reference, which is the international universal standard (3.0); GJB629-88 and GJB4855 shall be taken as references during DM coding.

39.3.2.3 Determine Information Set and Information Granularity

The technical materials of this type of UAV include the following: technical specifications (1 set), maintenance regulation (1 set), and frequently seen failure manual (1 set); the established information set includes the following: maintenance information, security information, explanation and operation information, maintenance information, diagrammatic part, and component information; involved DM types include the following: description, program, maintenance, failure, and process; granularity of DM is set as 1 local replaceable unit (LRU).

39.3.2.4 Programme DMRL

DMRL is the list of DMs in the whole process, which shall be programed and the basis for program DM to make CSDB management easy.

39.3.2.5 Program DM

With regard to DM coding, based on relevant provisions of GJB6600 about DM coding, the composition and rule of DM coding are shown in Fig. 39.6. The type ID refers to GJB629-88, and system classification code refers to GJB4855. With regard to composition and rule of coding of information objects like drawing, illustration, multimedia, and 3D, except prefix "ICN" and additional 5-digit information sequence number as postfix, the middle part is identical to DM coding.

39.3.2.6 Create Information Objects

- Take photos of steps of maintenance, produce video of equipment maintenance process, build interactive 3D model, and so on.

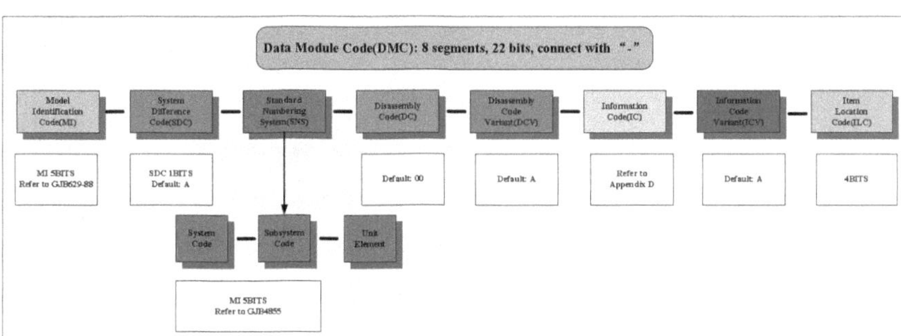

Fig. 39.6 The coding method of DMC

- Conduct Web CGM graphic creation for collected pictures (JPEG\JPG and other formats), including drawing of failure phenomena and heat flow chart of troubleshooting, graphic heat linkage, remarks of part and component diagram, disassembly diagram, etc.
- Assist in video production and editing, including production and editing of video of standard steps of maintenance, addition of off scene, production of caption, generation of special effect, conversion of format, etc.
- Interactive 3D simulation creation, format conversion (mainly into .wrl format); present disassembly process with virtual reality technology.

39.3.2.7 Manage CSDB Contents

CSDB content management includes management of database and DM. The data management includes database type selection, database backup/recovery, user/character management, etc. DM management mainly includes work flow management like programing, testing, quality validation, and release of DM; DM version management; linkage validity management, applicability management; information object management, etc.

39.3.2.8 Release of IETM Publications

Release IETM with trial software of CORENA S1000D. The system release interface is shown in Fig. 39.7.

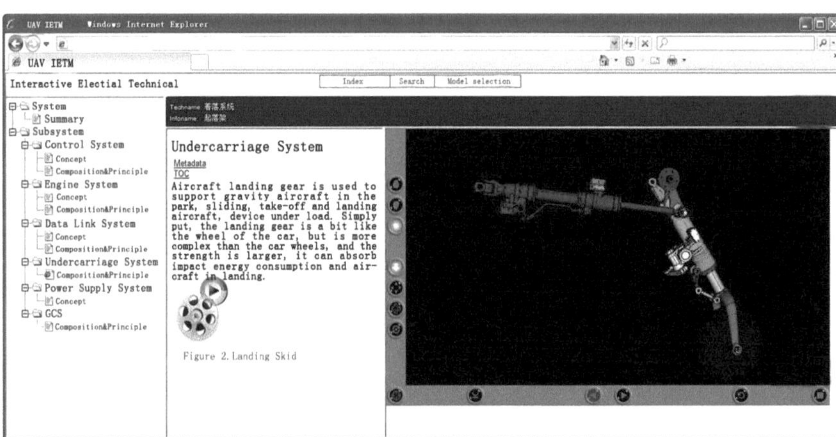

Fig. 39.7 The interface of IETM

39.3.2.9 Update of Feedback Information of User

IETM SYSTEM may be modified and updated according to feedback information of users and the technical materials applicable to this type of UAV to realize ILS in whole lifetime.

39.4 DM Fast Generation Design

There are a lot of DMs in this project. In traditional DM, commercial XML editors (XMLSpy, Arbortext Editor, etc.) are used. Facing bald XML node editing interface, the programmer has low efficiency and high error rate, and DM generation takes most of time for IETM editing. To enhance IETM edit efficiency, DM fast generation software has been developed with vb.net. and the schema to generate DM template with standard framework structure is loaded; finally, DM block node operation (realize addition, deletion, and modification of contents and properties of block node) is realized with application friendly to vb.net development interface. The software flow chart is shown in Fig. 39.8.

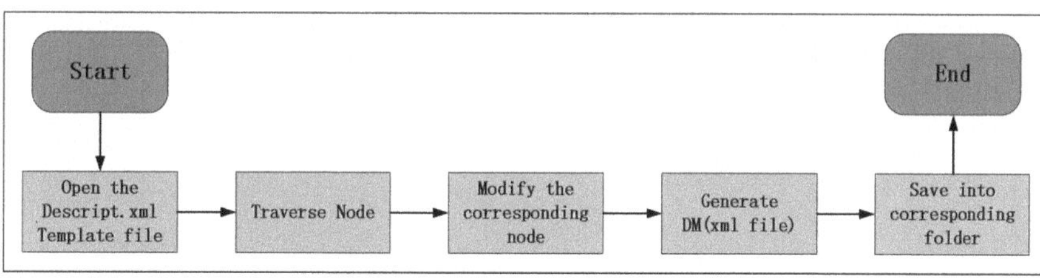

Fig. 39.8 The flow charts of DM rapid generation

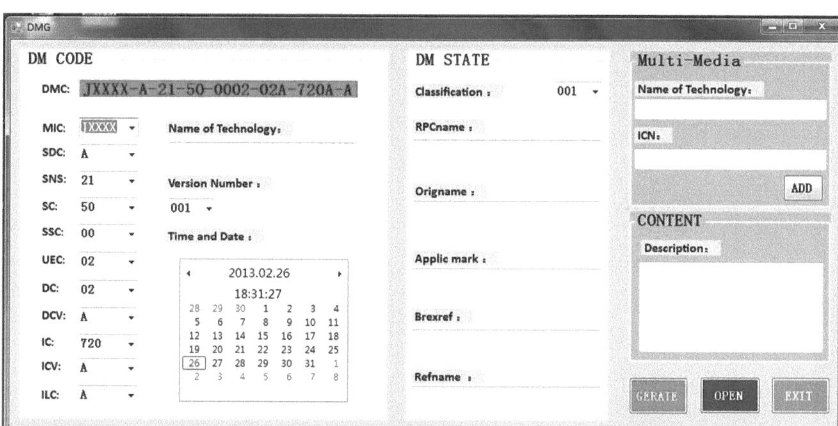

Fig. 39.9 The interface of DM rapid generation

Default value setting is made for DM information with oneness (like confidentiality class, preparer and version no., etc.) in this project to avoid repeated work and enhance DM generation efficiency greatly. In addition, the software interactive interface is user-friendly and operable, avoiding direct operation for XML file and reducing error of edit. The software interactive interface is shown in Fig. 39.9.

References

1. Li H, Jing X, Liu W (2010) Application of IETM in forecasting and health management system of foreign new generation operational aircraft. Ordnance Ind Autom 29(2):9–11, 13 (Ch)
2. HUANG Jian-zhao, LI Feng. Research on IETM for complex weapon system [J]. *Ordnance Industry Automation*, 2008, 27(8):22, 25 (Ch)
3. Gao Wan-chun. Study on A Web-based Cooperative IETM Model And Its Application [D]. *Huazhong University of Science & Technology*, 2007
4. ZHU Ning, DU Xiao-ming, LIANG Bo. Research on development of IETM [J]. *Ordnance Industry Automation*, 2008, 27(11):20–22
5. Research and Application on Produce Flow of IETM Based on S1 000D Standard [J]. *Computer Measurement & Control*, 2011, 19(6):1426–1428, 1465.
6. Chong W (2011) IETM development platform based on S1000D specification. Ordnance Ind Autom 30(12):53–56

Part V
Research on the Man-Environment Relationship

Chapter 40
Study on Safety Analysis of Passengers Hitting Frontally on Rigid Wall

Li Li, Na Jiao, Chao Wang and Xiaoyao Wang

Abstract Starting with the elastic–plastic mechanics theory and the finite element method, the major models of the cockpit, the crew, and others, which have influence on the safety of the crew in the collision, have been established, and then the safety of the crew has been studied. Using the VPG, the free-fall simulation model has been presented, and the test model has been verified in comparison. On this basis, the collision simulation analysis model has been established, and the initial parameters of the speed and position have been set. By analyzing the results, the dynamic movement of the crew under the impact, the dynamic response of the crew under the head-on collision, and the motion curve of the key parts can be obtained, which provide an effective basis for the further study of the crew safety in the head-on collision.

Keywords Collision · Finite element analysis · VPG

With the surging car ownership in China, safety is increasingly critical, particularly the passive safety. The authoritative method to test the vehicle passive safety is real vehicle collision; however, this costs greatly. Thus, computer simulation of vehicle collision becomes the inevitable road for complete vehicle development. Foreign countries have started the simulation research of vehicle passive safety from the 1960s.

Computer collision simulation technology has been greatly facilitated by the rapid development of hardware and software, as well as the fierce competition of vehicle market. Recent twenty years have seen that collision simulation has become an important part in complete vehicle development. With the speedy development of vehicle collision safety technology, especially the dynamic display of finite element method, carrying out in-depth safety study of passengers in head-on collision has practical significance for design of vehicles, safety of passengers,

L. Li (✉) · N. Jiao · C. Wang · X. Wang
China North Vehicle Research Institute, Beijing, China
e-mail: hotjn@126.com

S. Long and B. S. Dhillon (eds.), *Proceedings of the 13th International Conference on Man-Machine-Environment System Engineering*, Lecture Notes in Electrical Engineering 259, DOI: 10.1007/978-3-642-38968-9_40,

and improvement of vehicle safety level. Therefore, this paper is to study the safety of crew in head-on collision by firstly establishing the major models that have influence on the safety of the passengers via VPG, based on the elastic–plastic mechanics theory and the finite element method, and then handling the data obtained [1].

40.1 Calculation and Contact Method Analysis

Vehicle collision is a transient complex physical process, featuring geometric nonlinearity, namely large displacement, large rotation, and large strain; material nonlinearity, namely material elastic–plastic deformation; and boundary nonlinearity, namely contact friction. The conventional implicit algorithm can never meet the requirement of solving large-scale vehicle collision. Collision simulation mostly adopts nonlinear dynamic display finite element algorithm.

40.1.1 Collision Calculation Method

Calculation of large displacement in collision normally expends enormous time. The primary contradiction is how to advance the calculation speed. To avoid complex simultaneous equations and improve the speed, explicit algorithm is commonly used by the collision analysis software [2].

At moment te, the half-discrete version of variant motion equation is

$$\mathrm{M}\ddot{q}^{N} = P^{N} - F^{N}. \tag{40.1}$$

Among them, $\mathrm{M}\ddot{q}^{N}, P^{N}, F^{N}$ are, respectively, mass matrix, momentary acceleration vector, total load vector, and total equivalent nodal force vector of unit stress field.

Unit adopts single-point integration algorithm and sandglass model control generated by quadrangle and hexahedral element is added in the program due to the single-point Gauss integration. Meanwhile, contact force F_C is introduced into the calculation because contact is inevitable in collision. Then, the equation becomes

$$\mathrm{M}\ddot{q}^{N} = P^{N} - F^{N} + H^{N} + F_{C}. \tag{40.2}$$

Among them, H^{N} is structural sandglass viscous resistance vector.

Time integration is solved by display central difference method.

$$\begin{aligned} \ddot{X}^{N} &= M^{-1}(P^{N} - F^{N} + H^{N} + F_{C}^{N}) \\ \dot{X}^{N+1} &= \dot{X}^{N-1} + 2\ddot{X}^{N}\Delta t \\ X^{N+1} &= X^{N-1} + 2\dot{X}^{N}\Delta t. \end{aligned} \tag{40.3}$$

40.1.2 Contact–Collision Algorithm

Collision refers to the large-scale impact contact. Contact refers to that different subjects or different parts of one subject that have common boundary. Contact surfaces of different structures comprise of primary surfaces and secondary surfaces. Program LS-DYNA3D mainly adopts node constraint method, symmetrical penalty function method, and parameters distribution method in dealing with contact–collision surfaces.

Symmetrical penalty function method is mostly used. It carries out circular process for the secondary node and primary node at every time step with the same algorithm. The process is as follows [3–5]:

Calculation steps for any secondary node n_s are as follows:

1. By searching for secondary node n_s, to confirm the closest primary N_m-related main section S_i.
2. Check all main sections related to main node N_m, to confirm the possible main section s_1, s_2 when secondary node n_s goes through the main surface.
3. Confirm the location that secondary node n_s might contact point C on the main section S_i.
4. Check whether the secondary node n_s goes through the main section.
5. If the secondary node n_s goes through the main section S_i, apply a normal direction contact force vector f_s between the secondary node n_s and contact point C. Apply a normal direction contact vector f_s on the secondary node n_s. And as per the action and reaction principle, a reaction force f_s on the contact point C of main section S_i, this force will be equivalently distributed to all primary nodes on the main section S_i according to the shape function.
6. Calculation of friction force. As the normal direction contact force from the node n_s is f_s, the maximum friction is

$$F_y = \mu|f_s|. \tag{40.4}$$

As per the action and reaction principle, calculate the friction forces distributed to all nodes on the main section S_i. When considering friction force in collision analysis, dynamic and static friction coefficients μ_d and μ_s are generally handled smoothly via exponential interpolation function, then

$$\mu = \mu_d + (\mu_s - \mu_d)e^{-c|V|} \tag{40.5}$$

In the formula, V is the relative speed of the contact surfaces and c is attenuation coefficient.

7. Project the contact vector and friction vector onto the overall coordinate axis direction and get the direction components of the nodal force. Collect them into the overall load vector $\{P\}$ to conduct dynamics analysis.

40.2 Collision Simulation Analyses

In order to conduct simulation test of head-on collision and dynamic finite element analysis of passenger safety, a complete vehicle is needed to be set up first of all. As car is very complicated, and this paper only studies the response of passenger and cockpit; to simplify the calculation, finite element models only for the cockpit and dummy are established.

40.2.1 Experimental Verification of Dummy Model

Passengers are directly affected by the seats in the severe impact and vibration. Seat forces include vertical and horizontal ones. Apart from bearing the weight of passenger, other vertical forces apply directly on the body, forming the vertical impact response. Here, the model is verified and compared with the test on the-vertical direction. Passenger sitting on the seat falls freely from the aloft vertical tower to simulate the impact response in this condition.

The simulated models include Hybrid III (50-percentile) dummy model, seats, restraint system, and pedal. Seat is articulated with the pedal. Seat comprises of face and back and is connected by revolute joint to lock all flexibility. The gravity acceleration speed is set at 9.8 m/s^2. The test model constitutes vertical tower, track, hoist and release device, Hybrid III dummy, sensor, and data collector. Seat and dummy are placed on the test table. The dummy posture is adjusted and the test table is raised to a certain height, then the table is released and the response process is recorded.

The data from the lumbar force sensor are compared with the simulation curve as shown in the Fig. 40.1. The two curves have relatively same trend, which means that dummy model is reliable and simulation analysis proceeds.

40.2.2 Set-up of Finite Element Collision Model

The collision model is simplified as the cockpit and dummy to have head-on impact on rigid wall. As the shell unit has accurate structure, all bearing parts can be displaced accurately. It can represent the structural characteristics completely by combining with the beam, damping, multi-point restraint, and elastic unit, etc. The cockpit simulation model comprises of dashboard, steering wheel, base, and seats, 78 parts in total with dozens of materials. Shell unit is the main part, and solid units and beam units are also included.

Passenger model adopts the RIGID 50 % model provided by VPG with seat belt. Firstly, set four nodes around the seat as a nodal group. Connect the nodal group and cockpit in the way of *CONSTRAIN_EXTRA_NODES. Then, set up

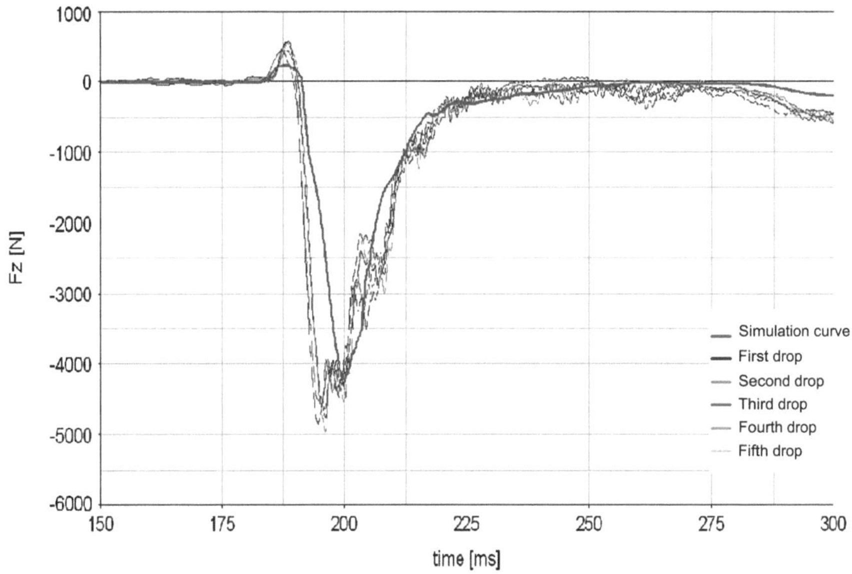

Fig. 40.1 *Curve* test and simulation

the passenger model and adjust the location and posture of the passenger to keep in driving stature. Click the four nodes by sequence following the notices, then the passenger model is set up.

Collision adopts the 208 RIGID WALL 0/30DEG provided by VPG. Set the initial speed, location, and other parameters to establish the collision analysis model.

Define related features and parameters for contact in LS_DYNA and confirm no going through of the contact surfaces. Consider the friction force during the relative motion of contact surfaces. Contact types are mainly single-surface contact, point-surface contact, and surface-surface contact. The contact of passenger and seat is automatically the single-surface contact [4, 5]. Because the actual collision duration is around 80 or 120 ms, this simulation only calculates the collision response at 90 ms, output corresponding animation, node acceleration and passenger data, etc.

40.2.3 Calculation Results and Analysis

The calculation results give out the dynamic response of passenger in head-on collision and the deformation condition of cockpit.

Figure 40.2 shows the dynamic status of passenger in the collision. At time 0.01 s, there is no big motion of various parts, and seat belts are not strained.

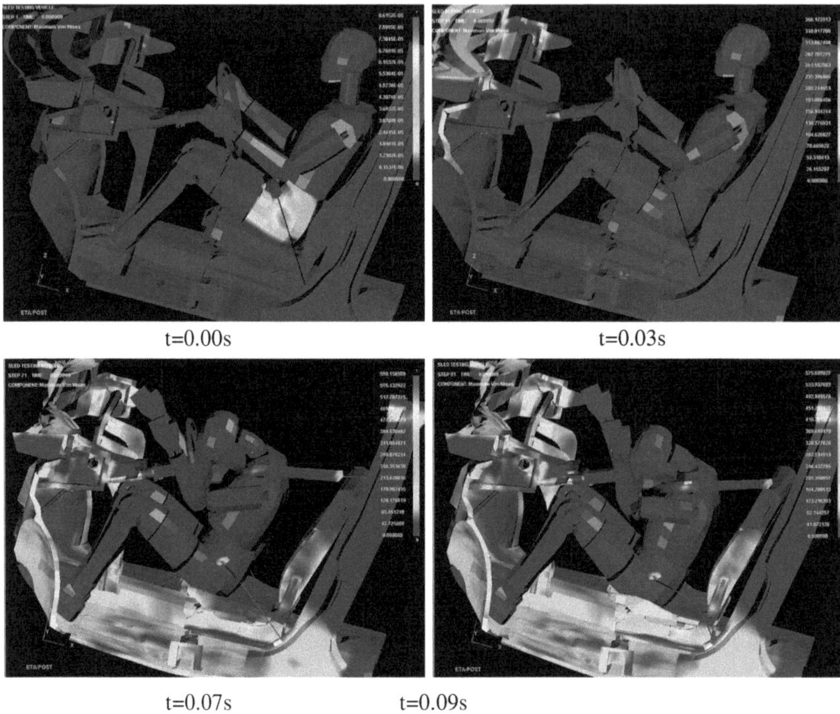

Fig. 40.2 The state of motion of the occupant and the cockpit of the force at each time

However, from 0.03 to 0.09 s, big changes occur that seat belt is in tension, head rushing forward onto the steering wheel roughly. There is no large difference of the resultant velocity of the occupant's head and chest as shown in Fig. 40.3, which states small relative movement.

For reference of standard FMVSS208, a series of curves are extracted from the VPG (Figs. 40.4, 40.5, 40.6, 40.7). Head Injury Criteria (HIC) value is obtained from the head acceleration curve. After converting units, HIC is 689, less than the limit value 1,000 of head impact. Chest acceleration is over 60 g, and chest

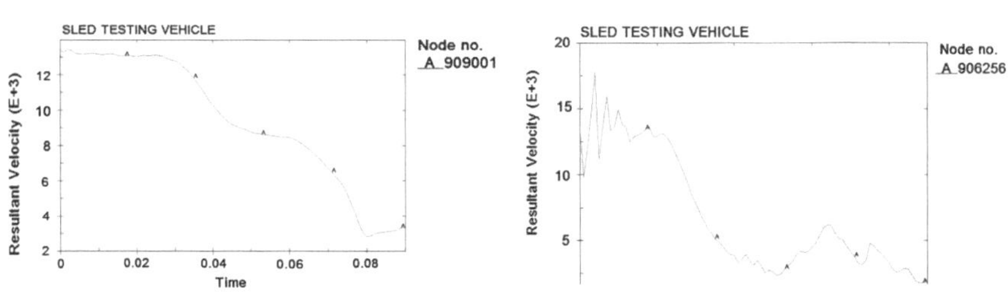

Fig. 40.3 *Curve* the resultant velocity of occupant's head and chest

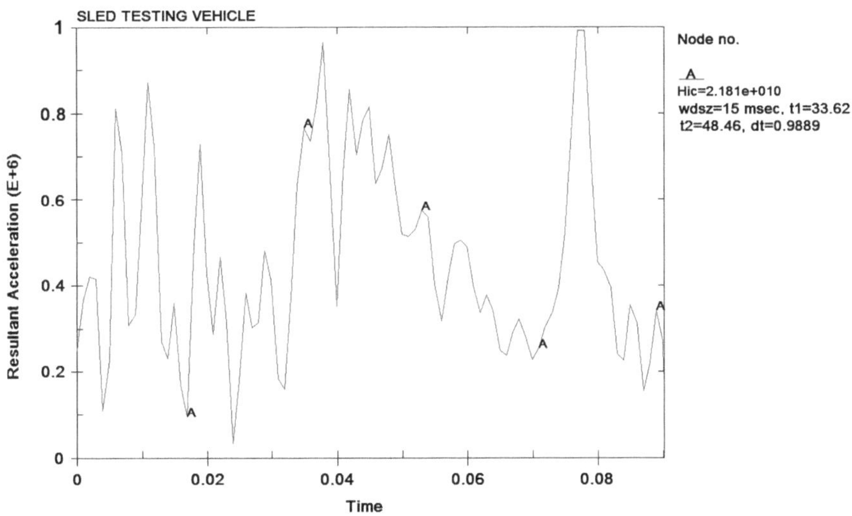

Fig. 40.4 *Curve head acceleration*

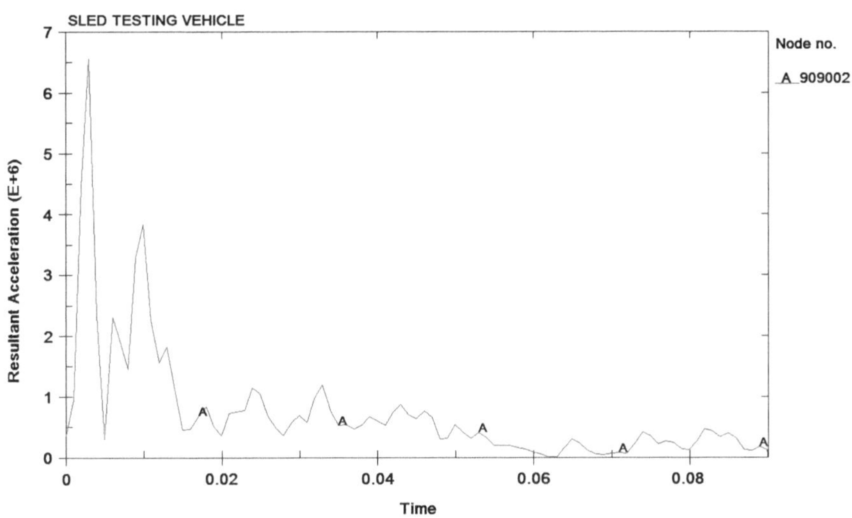

Fig. 40.5 Thorax acceleration >60 g

deflection is 38 mm, less than the 76 mm. The force on the thigh is nearly 20 KN. Among the above curves, HIC and chest deflection are within the limit, but the others have exceeded the limit value specified in FMVSS208. Therefore, it is obvious that head-on collision on the rigid wall will impact the passenger heavily. Although the actual collision will have buffer action by the front vehicle head and

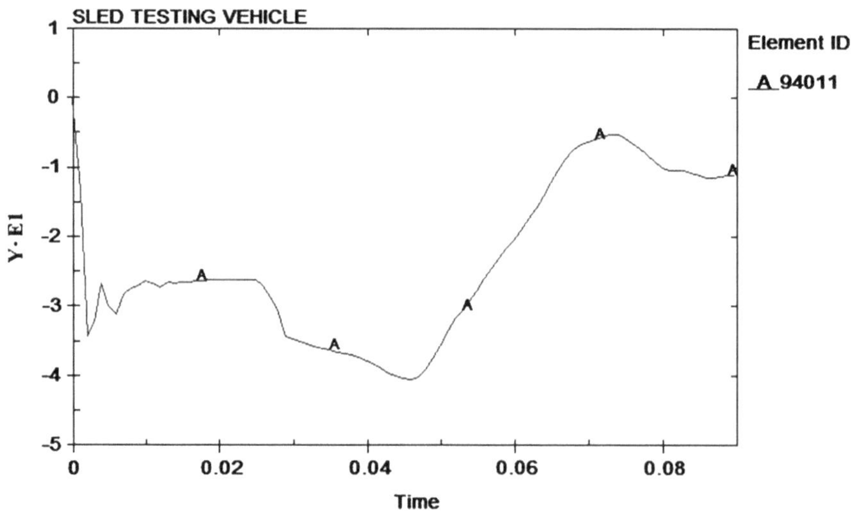

Fig. 40.6 Chest deflection <76 mm

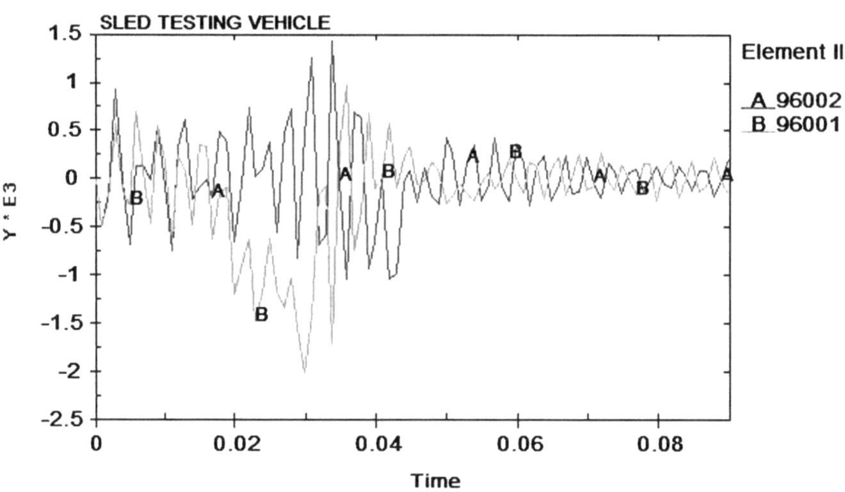

Fig. 40.7 Maximum force >10 kN

the values will decrease certainly, however, the impact is still large and necessary protections should be taken for the head and legs, etc.

40.3 Conclusions

This paper conducts simulation analysis of head-on collision to study the passenger safety by the elastic–plastic mechanics theory and finite element method. The finite element simulation model for the cockpit and passenger is set up in VPG. By comparison of lumbar force of dummy in free-fall model and test model, transient impact response of passenger is analyzed and deformation condition of cockpit is obtained. In addition, dynamic movement of passenger and motion curves of critical parts can be referenced in the later study of passenger safety.

References

1. Hongjian Li (2005) A study on the side crashworthiness of passenger car body. Master degree thesis of Jilin University, JiLin
2. Xiaojiang Shang, Jianning Su (2005) LS-DYNA Dynamic analysis method and engineering practice. China198 Water Power Press, Beijing
3. Fanzhong Kong, Shilin Huang, Jinhuan Zhang, Fengwu,Liu (2002) Advance in the numerical simulation methods of automotive impact. Proceedings of the 7th automotive safety technology conference of society of automotive engineers of China
4. Hongbo Jia, Jinling Huang, Antao Gu, Zhongxiao Wang (1998) The structural crash simulation of the typical thin-walled beam and its parameter selection. Transactions of the Chinese Society for Agricultural Machinery
5. Wei Guo (2008) The research of commercial vehicle body impact performance based on numerical simulation. Master degree thesis of HeFei University of Technology, Anhui

Chapter 41
Research on the Relationship Between Threat and Crew Error in Civil Aviation

Yuchuan Luo, Jun Chen, Jiazhong Yang and Xuemei Deng

Abstract To study the relationship between threat and crew error, the brief description of 84 civil aviation accidents and incidents whose primary causes were flight crew were obtained and coded. Chi-square test was performed to explore the strength of association between the threat and crew error. The results show that adverse weather and other procedural error, aircraft fault and manual flying error, dispatch/paperwork fault and documentation error, ground maintenance fault and flight control error, and manuals/charts fault and ground navigation error had strong association; those threats significantly increased the prediction of occurrence of crew error; the accuracy of prediction was 5.3–20 %. It is suggested to reduce the crew error by enhancing crew task management training and strengthening the management of other occupation's error in civil aviation.

Keywords Civil aviation transport flight · Threat · Error · Association · Flight safety

41.1 Introduction

Investigation of aviation accidents shows that around 60–80 % of the accidents and incidents in civil aviation have direct relationship with crew error or failure. In order to effectively manage the crew error, threat and error management theory

---

Fund Info: NSFC (30700237/c010805); Soft Science Project Fund, Civil Aviation Administration of China (MHRKX200560).

Y. Luo (✉) · J. Chen · J. Yang · X. Deng
Aviation Human Factor and Ergonomics Lab, Civil Aviation Flight University of China, Guanghan, China
e-mail: reven-lyc@163.com

S. Long and B. S. Dhillon (eds.), *Proceedings of the 13th International Conference on Man-Machine-Environment System Engineering*, Lecture Notes in Electrical Engineering 259, DOI: 10.1007/978-3-642-38968-9_41,

comes into being [1]. This theory distinguishes the threat, error, and undesired aircraft state and emphasizes the prevention of threat and error in civil aviation as well as aircraft status facing the undesired aircraft state [2, 3]. It plays a significant role in the control of risks in civil aviation and guarantee of flight safety. Civil aviation of China has applied this management theory of threat and error in a wide range of trainings, including crew resource management training [4], line operation safety audit and air traffic control, and dispatch and maintenance personnel training. However, current study of this theory in China mainly focuses on qualitative research; even if there are some quantitative researches, air traffic control is mainly studied [5, 6]. Thus, flight of civil aviation, as an important part, is rarely involved. It will be very valuable for formulation of policies and regulations, design of training methods, and improvement of flight safety if the relevance between threat and crew error in flight can be specified. This study aims to explore the correlation between threat and crew error in actual flight by quantitative and qualitative analysis, so as to provide a basis for the design of China crew resource management training scheme and management of crew error.

41.2 Research Techniques

41.2.1 Data Source

2006–2010 *Flight Accidents, Ground Accidents and Flight Incidents Statistics* issued by Aviation Safety Office of Civil Aviation Administration of China are adopted [7]. The data shows that 719 flight accidents and incidents took place from 2006 to 2010, in which 640 are freight (passenger and cargo) flight accidents and incidents. Eighty-four were involved with crew error. This study will base on the 84 flight accidents and incidents.

41.2.2 Analysis Tool

The original data are coded by adopting the research framework of threat and error management theory [8]. Threat is divided into environmental treat and airline treat. Environmental threat includes adverse weather, airport condition, air traffic control, and environmental operation pressure. Airline threat comprises aircraft failure, airline operational pressure, passenger cabin condition, dispatch/paperwork, ground/ramp, ground maintenance, and manuals/charts.

Crew error comprises aircraft handing error, procedural error, and communication error. Aircraft handing includes automation, flight control, ground navigation, manual flying, and system/instrument/radio. Procedural error includes briefings, callouts, checklist and documentation error, PF/MF duty, SOP cross-verification, and other procedural errors. Communication error includes internal and external communication errors.

41.2.3 Content Analysis and Coding

The study adopts the combination of content analysis and coding methods. The brief description of the 84 accidents and incidents are analyzed and coded one by one, so that qualitative data is converted to quantitative data. Three teachers of aviation human cause will analyze and code the original data. Firstly, the three teachers will select randomly 10 pieces of data for discussion and analysis, so as to ensure similar coding standard. Any threat or error that emerges in the data will be coded as one; otherwise, it will be zero. Three teachers will code the 84 accidents separately. After coding all the data, they will discuss together all of the 84 accidents to determine the final coding results.

41.3 Results

This study uses SPSS18.0 to conduct statistic analysis.

41.3.1 Descriptive Statistics

Table 41.1 shows the frequency and percentage of threat and error of 84 flight accidents and incidents. There are 82 threats and 202 crew errors. For each flight accident, 0.976 threats and 2.404 crew errors exist. This shows that threat and error are not the simple one-to-one correspondence.

Among these accidents and incidents, adverse weather is mostly faced by the crew, 44 % of them involved in this cause. Passenger cabin condition is the least threat. Among the crew errors, manual flying takes up 61.9 % and SOP cross-verification the least.

41.3.2 Chi-Square Test Statistics and Results

From the perspective threat and error management, errors may occur naturally or by threat [8]. If the correlation between threat and error in civil aviation can be elaborated and specific management is carried out, China civil aviation will benefit from this enormously.

Table 41.1 Frequency of threat and error of 84 flight accidents and incidents

	Type	Frequency	Percent	Type	Frequency	Percent
Threat	Adverse weather	37	44	Passenger cabin condition	0	0
	Airport condition	13	15.5	Dispatch/ Paperwork	3	3.6
	Air traffic control	7	8.3	Ground/Ramp	2	2.4
	Environmental operation pressure	5	6.0	Ground maintenance	5	6.0
	Aircraft	6	7.1	Manuals/Charts	2	2.4
	Company operation pressure	2	2.4			
Error	Automation	5	6.0	Briefing	8	9.5
	Flight control	19	22.6	Callouts	9	10.7
	Ground navigation	10	11.9	Checklist	5	6.0
	Manual flying	52	61.9	Documentation	9	10.7
	System/instrument/radio	6	7.1	PF/MF duty	24	28.6
	Internal communication	10	11.9	SOP cross-verification	4	4.8
	External communication	9	10.7	Other procedural	32	38.1

Chi-square test takes threat as independent variable and crew error as dependent variable. As shown in Table 41.2, five sets' relationship is significant, between 0.225 and 0.425. λ refers to the predictability of the former variant to the latter one, between 5.3 and 20 %. This indicates the evident correlation between some threat and error in China civil aviation.

41.4 Discussions

Based on the research data, there are 0.976 threats and 2.404 errors for each accident or incident among the 84 accidents or incidents. This result is similar to Merritt's [8] study. Threat does not have one-to-one corresponding relationship with the error, i.e., error might be caused by people or threat. The possible reason why one threat corresponds to several errors is that researchers tend to regard the crew as the last part in defense line and attribute the accident to the crew [9].

Table 41.2 Chi-square and correlation analysis of threats and crew error

Combination	X^2 value	*P*	φ	λ
Adverse weather * other procedural errors	7.142	0.008	0.292	0.094
Aircraft failure * manual flying	10.500	0.001	0.354	0.188
Dispatch/paperwork * documentation error	10.181	0.001	0.348	0.111
Ground maintenance * flight control	4.244	0.039	0.225	0.053
Manuals/charts * ground navigation	15.161	0.000	0.425	0.200

Note insignificant combinations unlisted

There is evident relationship between the adverse weather and other procedural errors in Table 41.2. That means the crew is liable to make mistakes and forget or omit the procedures, regulations, and SOP during adverse weather. This relates to the human capability of information processing [10]. Pilots will deal with greatly growing information and operation during adverse weather. If the pilot cannot assign some tasks to other crew members or the automation system timely, he/she will bear tasks beyond the limit. Therefore, apart from the training of profession ability, enhancing the task management and distribution capability can prevent the omission or errors of procedures and regulations in high workload conditions, thus improving the flight safety.

Situation awareness refers to the accurate cognition and prediction of aircraft, crew, environment, and task nature during flight. It will directly affect the decision and behavior of the pilots [9]. The reason why there is evident correlation between aircraft failure and manual flying, between dispatch/paperwork and documentation error, between ground maintenance and flight control error, and between manuals/charts and ground navigation error is that these errors will cause the pilots fail to understand and predict the situations. Therefore, in order to prevent crew error, potential external factors that might induce crew error should be reduced.

PF/MF duty error refers to the duty distribution issue between the operational pilot and the monitoring pilot. Twenty-four cases emerge in the 84 accidents. This study has not found any obvious relationship between the error and any threat. That is possibly caused by the internal crew, like pre-flight briefing, crew organization, member personality, technical condition, etc. Any external threat might cause this error.

Based on this study of division of threat and error, there are 11 threats and 14 errors, namely 154 combinations. However, there are only five combinations that reach the significant level. If having a longer period of data about threat and error, more correlations might be revealed. Those correlations will offer more help for the actual management of crew error in flight.

41.5 Conclusions

1. Through qualitative analysis, this study discovers that there are no one-to-one corresponding relations between threat and error, although there is relationship between them. To manage the threats is the first step for safe flight.
2. Through the chi-square analysis of coding the data, such five combinations as adverse weather and other procedural errors, aircraft error and manual flying, dispatch/paperwork and documentation error, ground maintenance and flight control error, and manuals/charts and ground navigation have evident correlations. These threats will obviously increase the error percentage.

3. Apart from improving the technical capability, paying attention to task distribution and other non-technical abilities, as well as the management of maintenance, dispatch, and air control industry, will help to decrease the possibility of errors.
4. More data might be helpful in discovering more correlations between threat and error.

References

1. Luo X (2002) Cockpit resource management. Press of Southwest Jiaotong University, Chengdu, pp 3–4
2. ICAO (2002) Line operations safety audit (DOC 9803-AN/761)
3. Huo Z, Lv R, Shi Y (2007) Threat and error management in civil aviation operation. J Saf Sci China 17(12):60–65
4. Flight Standard Department of Civil Aviation Administration of China (2011) Crew resource management advisory circular (AC-121-FS-2011-41)
5. Lv R, Zhou Y, Zhou M (2009) Analysis of team error in air traffic control. J Saf Sci China 19(1):64–70
6. Cao H, Li T (2011) Study on threat, error and undesired states in air traffic control based on Bayesian method. J Saf Sci China 21(8):85–89
7. Civil Aviation Administration of China (2006–2010) Flight accidents, ground accidents and flight incidents statistics.
8. (2006) Defensive flying for pilots: an introduction to threat and error management. The University of Texas Human Factors Research Project
9. Chen J, Yang J, Li Q et al (2010) Research on the relationship between crew resource management and unsafe acts of flight crew. J Saf Sci China 20(3):92–96
10. Wang D (2009) Influence of time pressure, and number of attributes on information processing in decision making. J Shandong Normal Univ 54(6):126–130

Chapter 42
Test and Analysis on Insulation of Sleeping Bag

Chenming Li and Yuhong Shen

Abstract This paper describes three methods for the testing and evaluation of insulation of sleeping bag, theoretical calculation method, thermal manikin method, and small group testing method. The thermal manikin method is employed to measure the heat retention of sleeping bags, and the result is 5.7 clo. The sleeping bag is tested by small group testing method at −28 °C in the wild, and the result shows that this kind of sleeping bag is fully able to meet the requirement on insulation. The insulation of sleeping bags tested with thermal manikin method and small group testing method is authentic, and the tests were well repeatable.

Keywords Sleeping bag · Insulation · Thermal manikin

42.1 Introduction

As a military product, sleeping bag plays an important role in field exercises and training, border patrol, and rescue and relief work. And for the outdoor enthusiasts, even ordinary people, sleeping bag has become the common equipments in traveling and recreation. Insulation is one of the most important performances for a sleeping bag. Any sleeping bag has an appropriate temperature range for usage, i.e., different sleeping bags have their own temperature marks [1, 2]. There are normally two temperature marks on the sleeping bag; one is comfort temperature that refers to the ambient temperature when the user feels comfortable; and the other is the extreme temperature (freezing to death). Once it happens, a sleeping

C. Li (✉) · Y. Shen
The Quartermaster Research Institute of the General Logistics Department of CPLA, Beijing 100010, China
e-mail: lichenming82@163.com

S. Long and B. S. Dhillon (eds.), *Proceedings of the 13th International Conference on Man-Machine-Environment System Engineering*, Lecture Notes in Electrical Engineering 259, DOI: 10.1007/978-3-642-38968-9_42,

bag with extreme temperature of −20 °C is intolerable at −5 °C. Because of lacking standards, there are few researches on the insulation performance of sleeping bag currently.

42.2 Theoretical Calculation of Heat Retention of Sleeping Bag

Factors that affect the comfort when using a sleeping bag include environment, heat retention of sleeping bag, and continuous using time. Common method of predicting the comfort temperature is using the human body heat consumption model. Human body must absorb more heat than that dissipated to keep temperature constant . The body exchanges heat with the external environment via convection, transmission, radiation, and evaporation to retain the heat balance [3, 4].

Transmission: heat transmitted to the ground is the main loss. The mattress between the sleeping bag and ground is especially crucial. Mattress with high insulation coefficient should be selected.

Convection: it is the main factor of heat loss. High-quality insulation materials can be used to keep more static air around the human body, meanwhile windproof and air-permeable fabric should be used to prevent forced convection by the wind.

Radiation: it accounts for a small part of the heat loss.

Breathing: breathing will lose much heat in cold weather.

Thermal balance equation of human body

$$M - W = K + C + R + E + C_{res} + E_{res} + S \tag{42.1}$$

In the formula,

M Human metabolism, W/m^2;
W Human body working to the outside, W/m^2;
K Heat dissipation by skin transmission, W/m^2;
C Heat dissipation by skin convection, W/m^2;
R Heat dissipation by skin radiation, W/m^2;
E Heat dissipation by skin evaporation, W/m^2;
C_{res} Heat dissipation by breathing convection, W/m^2;
E_{res} Heat dissipation by breathing evaporation, W/m^2;
S Heat accumulation of human body, W/m^2

GB/T 24254 applies the above thermal balance Eq. (42.1) to calculate the demanded thermal resistance of clothes. Some literature adopts the iteration method based on the human physiological data to regress the Eq. (42.1) into linear Eq. (42.2), so as to predict the demanded heat retention of sleeping bag in different temperatures.

$$T_a = 32.79 - 4.82 \times I_t \tag{42.2}$$

Table 42.1 Comfortable temperature contrast between EN13537:2002 and theoretical prediction value

Heat retention (clo)	Comfort temperature stated in EN13537:2002 (°C)	Predicted comfort temperature by Eq. (42.2) (°C)
4.26	12.0	14.6
5.03	8.1	11.3
5.81	4.3	7.9
6.58	0.6	4.6
7.35	−3.1	1.3
8.13	−6.8	−2.0
8.90	−10.4	−5.3
9.68	−14.0	−8.6

In the formula,

T_a Comfort temperature of sleeping bag (°C);

I_t Total heat retention value of sleeping bag and surrounding air (clo).

Please note that normal sleeping bags are not used directly in contact with the ground. Mattress is used to coordinate with the sleeping bag. Meantime, insulation clothes have direct influence on the effect of using sleeping bag. Comparing the Eq. (42.2) and the comfort temperature Table 42.1 of using sleeping bag stated in EN13537:2002, it is found that there exists a large difference. Therefore, theoretical calculation can only be for reference. In actual research, production and usage of sleeping bags, thermal manikin test method and small group test method can be adopted to determine the heat retention of sleeping bag and appropriate usage temperature.

42.3 Heat Retention Test of Sleeping Bag for Thermal Manikin

42.3.1 Test Method of Heat Retention

There are two standards related to the heat retention of sleeping bag, namely *GB/T 18398-2001 Test Method of Clothes Thermal Insulation* and *ASTM F1720-2004, Standard Test Method for Measuring Thermal Insulation of Sleeping Bags using a Heated Manikin*. Both standards use the thermal manikin test method. But, the test conditions and the manikins used are slightly different [5]. During the test of heat retention using a thermal manikin, the manikin is placed in an artificial climate laboratory and heated with certain power with an average surface temperature of 33 °C. The temperature difference between the manikin and air is over 20 °C. Wind speed is set at 0.3 m/s. Relative humidity is controlled within 30–70 %. The sleeping bag is shaken for 1 min before test to get a puffy bag. The thermal manikin is put into the bag on its back and zipped up. The control system is started. Surface temperature of thermal manikin is T_s. Air temperature T_a is adjusted to approach the set value step by step. The whole process is controlled by the computer. Collect the data every 20 s. When the average skin temperature of the thermal manikin, air temperature, and heating power reach the balanced status, keep running for 100 min. Take the data at last 30 min to calculate the heat retention of sleeping bag.

$$I_t = \frac{KA(T_s - T_a)}{H} \tag{42.3}$$

In the formula,

I_t total thermal resistance of sleeping bag (including the air thermal resistance on the sleeping bag surface), clo;
K Constant, 6.45 clo · W/m^2 · °C;
T_s Skin temperature of thermal manikin, °C;
T_a Ambient temperature, °C;
A Surface area of manikin, m^2;
H Rate of heat flow, W/m^2.

Adopt this method to the measure sleeping bags. Measure three samples for each kind of sleeping bag. If the variation is less than 3 %, take the average value as the heat retention of sleeping bag.

42.3.2 Test Result of Heat Retention for Thermal Manikin

During the test of heat retention using thermal manikin, the ambient temperature is set at −10 °C; wind speed 0.1 m/s; relative humidity 30 %; and the manikin wears underwear, winter sockets, and cold-proof mask. Three tests are carried out. Average heat retention is 5.7 clo. Variation is less than 3 %.

42.4 Small Group Test of Heat Retention of Sleeping Bag

42.4.1 Test Subjects and Methods

1. Subjects: select 6 males as test subjects, 22 ± 4 years old, 71 ± 10 kg by weight, and 169 ± 3 cm by height. They are from Hunan, Henan, Hebei, and Shandong provinces. In the test, the subjects wear underwear, woolen vest, woolen clothes, cotton-padded clothes for cold region, and training clothes in winter, caps, inner gloves, outer gloves, and winter sockets. Insulation sleeping bag is adopted. The heat retention is 5.7 clo.

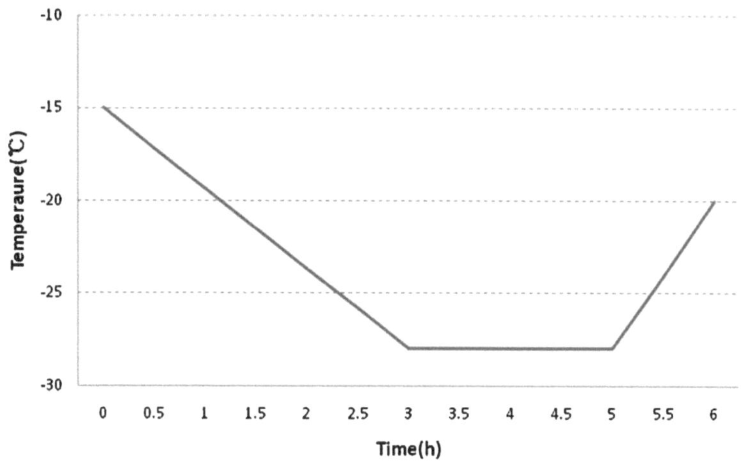

Fig. 42.1 Ambient temperature curve

2. The ambient temperature is controlled by computer within the range of −15 to −28 °C as shown in Fig. 42.1. The test period is from 23 to 5.
3. Auxiliary device: mattress.
4. Test parameters: use the human body cold and warm stress real-time monitoring system to collect the skin temperature at the chest, back, thigh, leg, and finger and ambient temperature. After the test, the subjects fill in the subjective feeling evaluation table, rated as seven degrees of "very cold", "cold", "slight cold", "comfortable", "slight hot", "hot", and "very hot".
5. Preparation before sleep: train the subjects on how to use mattress and sleeping bag correctly (including sufficiently puffy and position direction of sleeping bag) and how to locate the exterior clothes (fill in the head and feet area). Before entering the sleeping bags, the subjects will wear the real-time monitoring system and wait until the signal is stable.
6. Test termination condition: to prevent accidents, the test will be terminated in the following cases:

(a) Average skin temperature of the subject is lower than 32 °C;
(b) Feet toe temperature of the subject is lower than 15 °C;
(c) The subject feels very cold and cannot persist.

42.4.2 Test Results

The average temperature curve of the subject's skin and toe are as the Figs. 42.2 and 42.3.

After test, the subjective feelings of the subjects are summarized in Table 42.2.

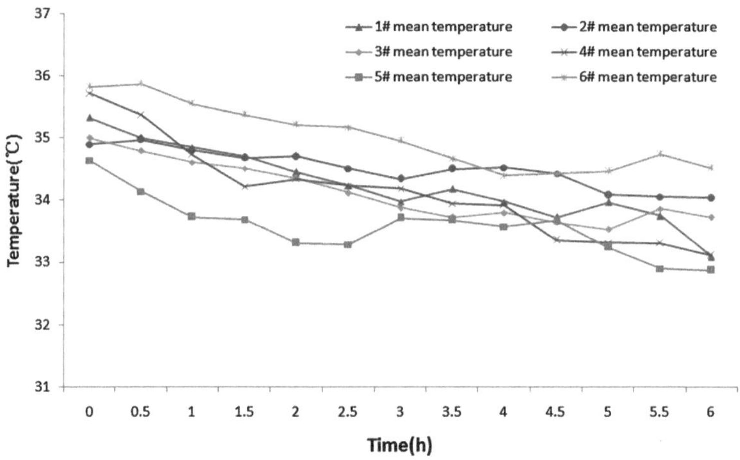

Fig. 42.2 Mean skin temperature curve of six subjects

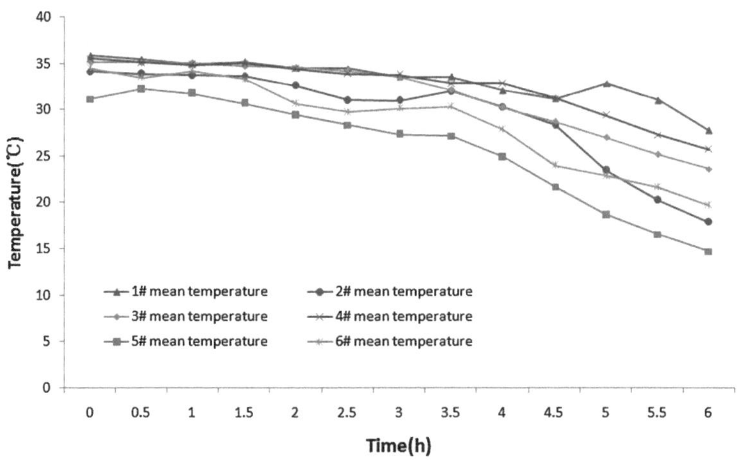

Fig. 42.3 Toe temperature curve of six subjects

Table 42.2 Subjective feeling of subjects

Subjects	1#	2#	3#	4#	5#	6#
Feeling	Comfortable	Comfortable	Comfortable	Comfortable	Slight cold	Comfortable

From the figures, it is easy to see that the average skin temperature is not lower than 33 °C and feet temperature is not lower than 15 °C, which corresponds to the subjective feelings of the subjects. The sleeping bag used in the test with 5.7 clo heat retention can totally meet the insulation requirements.

42.5 Conclusions

In general, whether people feel comfortable in using the sleeping bag or not is affected by the heat retention of the sleeping bag and the gender, age, freeze resistance capability, and adaptive capacity to climate. Therefore, the calculation results from the human body thermal balance equation can only be for reference as theoretical values. Thermal manikin test method is similar to the clothes insulation test method with high repeatability and accuracy. Small group test method can provide the most authentic and reliable data with high representativeness.

References

1. Huang J (2006) Study on the insulation and comfort temperature ratings of sleeping bags. Chin J Ergonomics 12(2):19–21
2. Nie X (2009) A study and filed research of CAPF sleeping bag for the Plateau and Paramos Area. China Pers Protective Equip 25(4):24–27
3. Zhang H, Zhou Y (2009) Clothing ergonomics. China Textile & Apperal Press, Beijing, p 44
4. Havenith G (2002) Moisture accumulation in sleeping bags at subzero temperatures-effects of semipermeable and impermeable covers. Textile Res J 72(4):281–284
5. ASTM F1720-2004, Standard test method for measuring thermal insulation of sleeping bags using a heated manikin

Chapter 43
Dynamics Response of Head and Neck Under -Gx

Minglei Wu, Xiaoyan Sun, Guang Zhao, Heqing Liu, Jinhuan Zhang and Jinwen Zhang

Abstract The aim of this study is to investigate the dynamics response of head and neck and helmet stability under –Gx. The test method to swing a dummy head to make an impact was established. The acceleration, force, and torque within head and neck were measured by transducers. The stability of helmet was measured by high-speed camera. As the impact acceleration peak amplitude is about 6.8G and the duration is about 42 ms, the dynamics response of head and neck is in safety limits. But the displacement of helmet is apparent. It is of important meaning to the fighter plane landing or taking off from a ship.

Keywords Biodynamics · Acceleration · Human response · Head and neck · Helmet

43.1 Introduction

When the plane is taking off and landing, especially the catapult-assisted take-off or hindered landing of the shipboard aircrafts, the speed will accelerate to 300 km/h in 3 s or will slow down from 300 km/h to stop in 3 s. The acceleration peak value is 2.8G, 4.5G, and 5.7G, respectively, by the calculation of constant acceleration, half sinusoid, and triangular wave straight line acceleration. The impact acceleration features relatively low peak value and high frequency. Flight protective helmet will not only increase the burden on head, but also affect the safety operation if it is not

M. Wu (✉) · X. Sun · G. Zhao · H. Liu
Institute of Aviation Medicine, Air Force, PLA, No. 28 Fucheng Road Haidian District, 100142 Beijing, China
e-mail: wuminglei@sina.com

J. Zhang · J. Zhang
Tsinghua University, Beijing, China

S. Long and B. S. Dhillon (eds.), *Proceedings of the 13th International Conference on Man-Machine-Environment System Engineering*, Lecture Notes in Electrical Engineering 259, DOI: 10.1007/978-3-642-38968-9_43,

stable [1]. It is of great significance for the human safety and head protective device usage to study the biodynamics response under –Gx impact acceleration and the helmet stability.

43.2 Test Measurement Devices and Methods

43.2.1 Test Measurement Devices

Test devices: pendulum impact test apparatus, including test bench body, head pendulum comprised of HYBRID III dummy head, neck and rod, pulling device, waveform generator, and releasing device.

Electric measurement device comprised of speed, acceleration (7264C-500), force, and torque transducer (IF-210) and impact measurement recording system. The recording system is made up of amplifier, data acquisition unit (NI USB-6251), computer, and acquisition, record, and analysis software. There are 10 lines of electric measurement signals: one line for speed before collision, three lines for impact acceleration (one line for the horizontal direction of the impact area (area connected to neck, correspond to T1 area of human chest), and two lines, respectively, for the horizontal and vertical directions of the gravity center of dummy head), three lines for impact force, and three lines for torques (X, Y, Z direction of neck).

Optical measurement device: high-speed camera (MotionXtra HG-LE) and affiliated record analysis software. Shooting speed is 1,000 frames per second. Photo resolution is 752×544.

All test devices are in good condition and inspected by authorities. Indication error meets the national measurement requirement. Among them, error for speed is 2 %, for impact acceleration 3 %, for impact force 2 % (static state), and for torque 2 % (static state).

43.2.2 Test Methods

The head pendulum is upward at one side and hanging in the air and is fixed by bearings at the other side. Use tension spring to connect at the center of pendulum and restrict at two flanks, so as to ensure the head pendulum is swinging in the plane. Put the helmet on the head. Pull the head pendulum 15° backward and release suddenly. The pendulum swings 30° forward and the impact area collides with the waveform generator. As the helmet is not rigidly fixed on the head, there will be a relative displacement between them. Measure the relative displacement, acceleration of the gravity center of head, neck force, and torque, and then analyze the stability of helmet and the biodynamics response of head and neck.

43.2.3 Dummy Head

The adopted HYBRID III dummy head is 4.54 kg by head and 1.54 kg by neck. Head circumference is 585 mm, length 198 mm, and width 156 mm. After putting on the wig, head circumference is 595 mm, length 200 mm, and width 160 mm. Select the helmet test sample according to the HYBRID III dummy head type.

43.3 Test Results

43.3.1 Determination of Impact Acceleration

The main task is to determine the wave form of impact acceleration. This study is about the acceleration pattern of triangular wave accelerated movement.

Impact acceleration firstly relates to tension spring (number of tension springs, elastic coefficient, etc.), and secondly to the material of waveform generator (rubber of different hardness or foam of different density).

After research, three pieces of tension spring (240 mm long and 16 mm of diameter) are proper. Steel wire diameter of the tension is 1.5 mm.

Similarly, after repeated tests, it is good to select the combination of rubber (127 × 127 × 28 mm) and rubber foam (160 × 75 × 50 mm) for the waveform generator. The impact acceleration wave is similar to triangular wave, showing peak at 6.8G and width at 42 ms. Main acceleration peak of head response is around 5.0G and width is around 170 ms.

43.3.2 Basic Characteristics of Electric Measurement Data

Pendulum impact test bench restricts the swing movement in the plane as an arc. The measurement results of acceleration, force, and torque accord with the movement with the following characteristics:

Impact acceleration (A_{T1}) is along the tangential direction of the movement track, corresponding to the acceleration at the chest location.

Acceleration at the gravity center of head: both components at tangential direction (main acceleration A_x) and radial direction (A_z), and the resultant acceleration is A_{x+z}. Horizontal acceleration (A_y) is perpendicular to the plane, so this study ignores it.

Impact force in neck: tangential force (main force) is F_x and radial direction force is Fz, and the resultant force is F_{x+z}. The horizontal force (F_y) is ignored.

Torque in neck: horizontal torque (main torque, M_y). The tangential (M_x) and radial (M_z) direction are ignored.

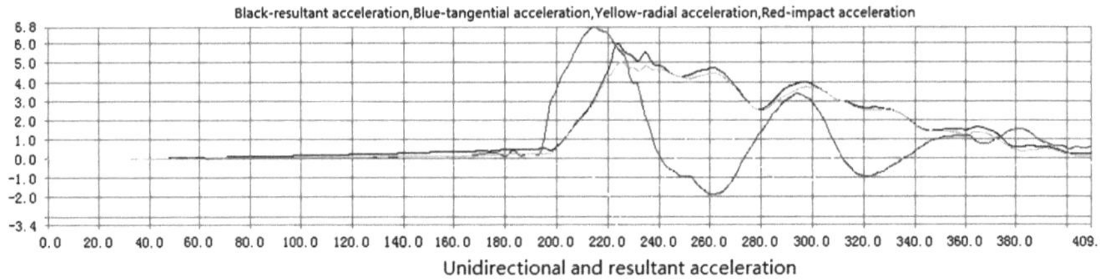

Fig. 43.1 *Curve* accelerations in the head

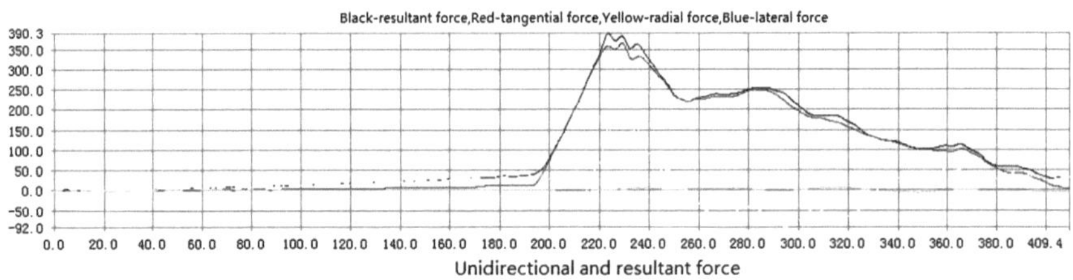

Fig. 43.2 *Curve* forces in the lower neck

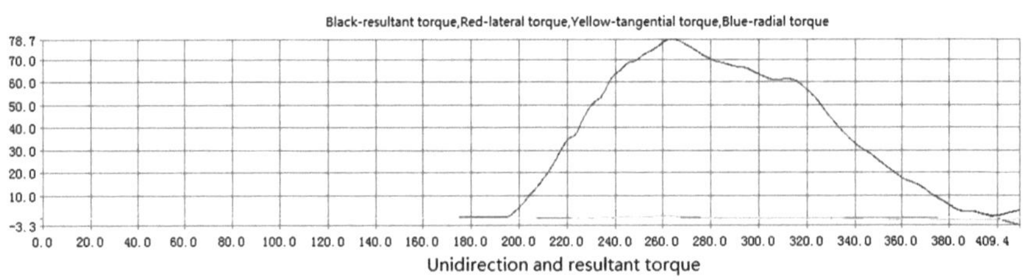

Fig. 43.3 *Curve* torques in the lower neck

In the test, impact acceleration is the input and the acceleration of head, force, and torque of neck are the response of the impact acceleration. And the response time prolongs.

The measurement of impact speed is used only for monitoring the test status, which is only for references (Figs. 43.1, 43.2, 43.3).

43.3.3 Electric Measurement Data Processing Method: Filtering

The measurement data is processed and analyzed by the direct-pass (no filtering), 1,000 Hz filtering and 200 Hz filtering. The results show that apart from several main impact forces with high frequency, other data vary little. Comparing the

1,000 Hz filtering and direct-pass filtering, the variation of peak and width is less than 1 %. And comparing the 200 Hz filtering and direct-pass filtering, the variation of peak is less than 3 % and width is less than 5 %. In accordance with the human low-frequency response characteristics [2] and GJB963 *Measure System Specialty Requirements for Ejection Acceleration* [3], the 200 Hz filtering method is adopted after comparison of the three methods.

43.3.4 Service Restrictions of Waveform Generator

There are various waveform generators, like solid type, hydraulic type, and barometric type. Simple and daily-used waveform generator is generally the high-polymer solid material (like rubber, PS) and special metals (like honeycomb aluminum). As these materials age or their performance changes by impacts, they have service restrictions. In this study, rubber foam is used. After testing and exploring repeatedly, its change rule is grasped. With the increasing impacts, the peak value of main acceleration will increase a little and the width decrease a little from 185 to 120 ms. Take the wave width as the control index; the width of the main acceleration is restricted between 185 and 165 ms (or between 40 and 45 ms for the impact acceleration wave width). The number of impact that meets the requirement should be controlled within 20 times.

43.3.5 High-Speed Photography and Helmet Displacement Calculation

Locate and adjust the camera and light according to the location of head and helmet in the impact and the movement trend. The shooting range must cover from the initial impact moment to the largest displacement of helmet. Before the test, at the initial impact location, record respectively a static mark panel (coordinate system) and a static dummy head with helmet. Mark clear points on the head (nose point and face) and helmet (optical sights soleplate and flank). See Fig. 43.4.

Calculate the displacement of helmet with the help of supporting software of the high-speed camera. The main indexes are the angular displacement and linear displacement.

Angular displacement refers to the angle difference between the initial angle of the mark lines on the head and helmet before collision and the last angle of the two mark lines after collision.

Linear displacement refers to the combined difference between the initial mark point of helmet (X1, Y1) relative to the point on the head before collision and the last mark point (X2, Y2) after collision at the maximum displacement moment.

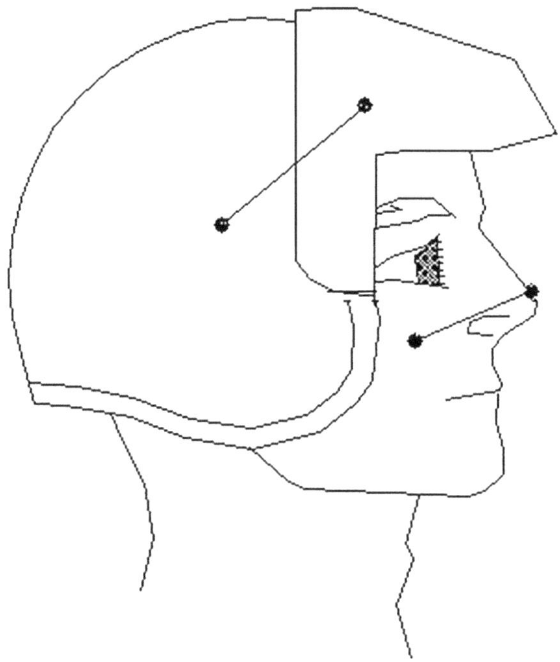

Fig. 43.4 Marks in the face and helmet

43.3.6 Impact Acceleration, Force, and Torque in Head and Neck

The impact acceleration, force, and torque in head and neck accord with the pretest results. As the helmet weight varies a little, there is no evident influence to the impact acceleration, force, and torque. The peak value and width are summarized in Table 43.1.

Table 43.1 Acceleration, force, and torque (ave. ± stand. deviation, $n = 20$)

Index	Ave. ± stand. deviation
A_{T1} peak (G)	6.84 ± 0.19
A_{T1} width (ms)	42.46 ± 0.80
A_x peak (G)	4.96 ± 0.09
A_z peak (G)	3.24 ± 0.15
A_{x+z} peak (G)	5.69 ± 0.14
A_{x+z} width (ms)	177.65 ± 4.75
F_x peak (N)	369.38 ± 8.4
F_z peak (N)	159.53 ± 14.07
F_{x+z} peak (N)	393.47 ± 8.81
F_{x+z} width (ms)	187.60 ± 7.10
M_y peak (Nm)	78.86 ± 2.19
M_y width (ms)	168.23 ± 5.34

43.3.7 Calculation Results of Helmet Displacement by High-Speed Photography

The relative displacement of the helmet is obtained by the high-speed photography. Six groups are divided according to the helmet type in Table 43.2.

43.4 Discussions and Conclusions

43.4.1 Test Result Analysis

Human safety analysis: according to statistics, human neck can sustain pressure of 1,862 N (190 kg) [4] or 2,097 N (214 kg) [5] and shear force of 902 N (92 kg) [6]; safety index of torque is 90 Nm [7], and in cases of ligament or bone injury, it is 190 Nm. This study shows that human safety is guaranteed in this condition. But, the torque is close to the safety index.

Helmet stability analysis: in general, the linear displacement and angular displacement are consistent. Larger linear displacement is along with larger angular displacement. Regarding the same type helmet, group C and D have good repeatability, while group E and F have large data discretization. The helmet is taken off and wore again each time after collision. So the different status of each wearing is one of the causes. Besides, although the impact condition is controlled to be the same every time, the actual condition cannot be totally the same. Lastly, the fitness of helmet and head is also an important factor of stability. Though the test helmets are selected with the same circumference, length, and width as the dummy head, it is hard to ensure the identical curve-shaped head.

Comparing A and B, C and D, and E and F, only E and F has large difference. However, the linear displacement ($t = 0.88$) and angular displacement ($t = 1.31$) of E and F show that P is greater than 0.05, no significant difference. In fact, before the helmet is equipped with optical sights (A, C, E), its stability is bad; after equipping the optical sights (B, D, F), there is no evident variation.

Table 43.2 Displacements of helmet (ave. ± stand. deviation)

Group	Num.	Angular (°)	Linear (mm)
A	3	8.54 ± 2.24	28.47 ± 5.43
B	3	6.87 ± 4.70	27.53 ± 6.25
C	3	16.15 ± 1.45	25.51 ± 2.59
D	3	16.94 ± 1.35	24.46 ± 1.82
E	4	10.91 ± 5.03	18.51 ± 13.84
F	4	15.19 ± 2.57	26.53 ± 8.49

43.4.2 Conclusions

The impact test shows that there is a evident displacement of helmet when the impact acceleration peak is 6.84 ± 0.19G and wave width is 42.46 ± 0.80 ms. Angular displacement is mostly over 10°, and linear displacement is over 20 mm. The acceleration, force, and torque of the head and neck are within the safety limits of human body.

References

1. Wu M-l, Li Zhao G, Wu F et al (2011) Influences of displacement of the pilot helmet with mounted display to information viewing. Chin J Aerospace Med 22(3):184–188
2. Wu M-l, Li B-h (2000) Analyse of frequency band in acceleration measurements during ejection and parachuting. Space Med Med Eng 13(2):140–142
3. Chinese Military Standard, Measure system specialty requirements for ejection acceleration. GJB 963-90
4. Weiss MS, Matson DL, Mawn SV (1989) Guidelines for safe human exposure to impact acceleration. AD/A215287, pp 1–12
5. Yang Q-w, Yu M-s (1983) The tolerance of human spine to impact resulted from opening of parachute. PLA Med J 8(6):405–407
6. Liu B-g, Yang Q-w, Cao B-p et al (1991) Study of the transverse tolerance strength of atlas and axis. Chin J Aerospace Med 2(1):13–15
7. Perry CE, Buhrman JR (1996) Effect of helmet inertial properties on the biodynamics of the head and neck during +Gz impact acceleration. Safe J 26(2):34–41

Chapter 44
Effect of Deep-Breath Biofeedback on Heart Rate Variability and Blood Pressure at High Altitude

Qingfeng Liu, Huamiao Song, Yi Du, Zhengtao Cao, Yubin Zhou, Fei Peng, Liu Yang, Lei Yang and Yongchang Luo

Abstract The objective of this study is to study the effect of deep respiration on heart rate variability and blood pressure at high altitude. Methods: Experiments were conducted in 74 male military operators who have been deployed to 3,780-m-high altitude for 60 days. Heart rate variability and blood pressure were monitored at rest state and deep-breath biofeedback state. Result: Heart rate of deep-breath biofeedback was significantly lower than that of the rest ($t = 2.01$, $P = 0.043$). SDNN and LF were significantly higher ($t = 3.70, 5.40$, $P < 0.001$). There was no difference in HF ($P > 0.05$).Both systolic pressure and diastolic pressure of biofeedback state were significantly lower ($t = 4.06, 7.63$, $P < 0.001$). Conclusion: Deep-breath biofeedback can increase heart rate variability and reduce heart rate and blood pressure in high altitude. It is an important assistant method used to acclimatize high altitude which can bring positive psychophysiological change to military operators.

Keywords Hypoxia · Deep-breath · Biofeedback · Heart rate variability · Blood pressure · Military

44.1 Introduction

Hypoxia environment at high altitude will affect the capability and health of the military operators. The morale might go down owing to the negative neuropsychological and emotional influences. Necessary measures must be taken to assist

Q. Liu · H. Song · Z. Cao · Y. Zhou · F. Peng · L. Yang · L. Yang · Y. Luo (✉)
Institute of Aviation Medicine AF, Beijing 100042, China

Y. Du
Troops 95455 PLA, Chongqing 563127, China
e-mail: Littleponds@163.com

S. Long and B. S. Dhillon (eds.), *Proceedings of the 13th International Conference on Man-Machine-Environment System Engineering*, Lecture Notes in Electrical Engineering 259, DOI: 10.1007/978-3-642-38968-9_44,

the soldiers to acclimatize to the high altitude. The most common measures include hypoxia pre-acclimatization, additional oxygen supply to cope with the oxygen deficit [1], medicines [2], and nutritional intervention [3] to enhance the adaptive capacity of high altitude. Besides, mental intervention is also one of the effective measures, which ameliorates the feared state of mind of high altitude and the altitude reaction symptoms through cognitive adjustment and relaxation training. However, there are few research reports about the application of biofeedback training at high altitude. This study aims to conduct deep-breath biofeedback training among the military operators at high altitude, so as to provide technical support to the military operations.

44.2 Subjects and Method

44.2.1 Subjects

There are 71 male military operators at high altitude taking part in this test. Age ranges between 19 and 42. They have lived at altitude of 3,780 m for over 60 days. No mental or anti-hypoxic medicines were taken a week before the test. And they did not take any oxygen inhalation therapies. The test was finished at August 2012.

44.2.2 Method

44.2.2.1 Test Tool

Biofeedback Equipment

SPCS system developed by Beijing Haofeng Digital Technology Co., Ltd. was used to complete the experiment, which was operated on Microsoft Windows XP SP3, Lenovo V370 notebook. The infrared sensor can record the heart rate and HRV index (including M-HRT (bpm), SDNN (ms2), energy total LF (ms2) between frequencies 0.04 and 0.15 Hz, and energy total HF (ms2) between frequencies 0.15 and 0.4 Hz, LF/HF). Meanwhile, this system can calculate the mental regulation training score, including coordination index, stability index, and assessment score, according to the physical and mental parameters.

Blood Pressure Monitor

Panasonic electronic blood pressure monitor was used for recording the blood pressure during the test, type EW-3006, measuring error $\leq$3 mmHg.

44.2.2.2 Test Process

Adopt the 2 × 2 within-subjects design.

The experimenter firstly explained to the military operators the unfavorable influences that the high altitude environment has on the heart rate and blood pressure and how the deep-breath biofeedback method can work against the influences. The test will not begin until the operators have grasped the method. First of all, the operators take 5 min rest, and the basic physiopsychological parameters are recorded by the system automatically during the next five minutes, including heart rate and HRV signal. Meantime, blood pressure is measured two times, and the average value is regarded as the basic value. Secondly, the operators take the deep-breath biofeedback training for 10 min. The operators are requested to breathe according to the system signals. The frequency is around 0.1 Hz. The effects are shown on the display. HR and HRV are recorded automatically by the SPCS system and blood pressure is measured two times during training and taking the average value as the training value. The physical and mental indexes are compared between the rest state and biofeedback state.

44.2.3 Data Analysis

The data are analyzed by the software SPSS 12.0 for Windows. Various test indexes are in the form of "average value ± standard deviation ($\bar{x} \pm s$)". The paired t test statistical method is adopted to compare the differences in physical and mental indexes between the rest state and deep-breath biofeedback state. Significance test level α is equal to 0.05 and 0.01.

44.3 Results

44.3.1 Comparison of Heart Rate and HRV at Rest State and Deep-Breath Biofeedback State at High Altitude

The heart rate at deep-breath biofeedback state is lower than that of rest state ($t = 2.33$, $P < 0.01$), while SDNN and LF at the former state is higher than that of rest state ($t = 3.70$, 5.40, $P < 0.001$). The results have statistical significance. There is no difference in HF between two states ($t = 0.12$, $P > 0.05$). The mental score of deep-breath biofeedback state, including coordination index, stability index, and evaluation score, is higher than that of rest state. The results have statistical significance. For details, please see Table 44.1. The Figs. 44.1 and 44.2 show the heart rate curve and respiration frequency spectrogram for the baseline and biofeedback state.

Table 44.1 Comparison of evaluation score and HRV between baseline and biofeedback ($\bar{x} \pm s$, $N = 71$)

Item	Rest	Biofeedback	*T* value	*P* value
Coordination index	36.76 ± 24.59	93.32 ± 13.08	16.63	0.000**
Stability index	38.51 ± 24.47	94.56 ± 12.64	16.53	0.000**
Evaluation score	55.34 ± 16.79	91.68 ± 8.43	16.34	0.000**
M-HRT	84.48 ± 10.62	82.36 ± 11.39	−2.01	0.043*
SDNN	54.38 ± 28.24	75.95 ± 36.69	3.70	0.000**
LF	200.01 ± 320.34	712.44 ± 765.91	5.40	0.000**
HF	72.33 ± 124.79	69.48 ± 181.63	−0.12	0.906
LF/HF	6.13 ± 6.01	25.71 ± 24.35	6.49	0.000**

* $P < 0.05$, ** $P < 0.01$

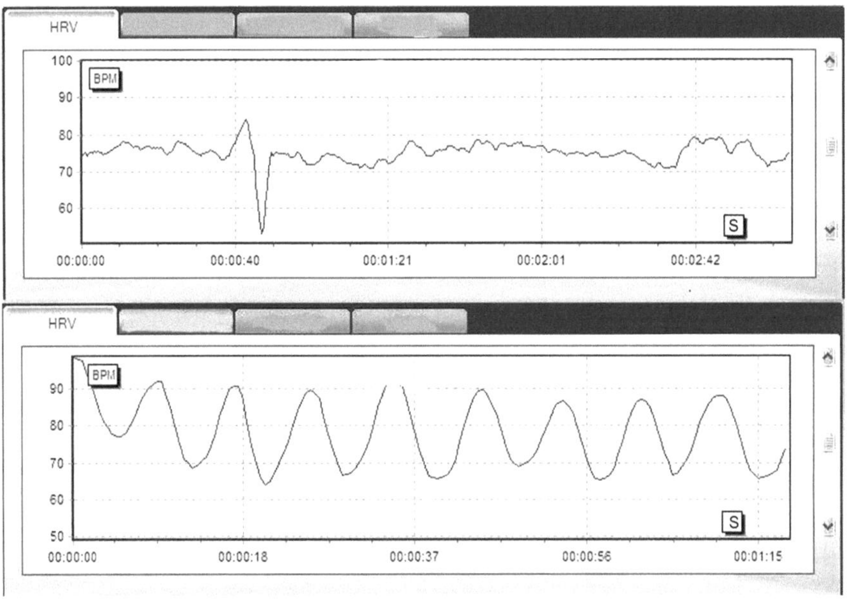

Fig. 44.1 Example of heart rate curves for baseline and biofeedback

44.3.2 Comparison of Blood Pressure at Rest State and Deep-Breath Biofeedback State at High Altitude

The systolic pressure at deep-breath biofeedback state is lower than that of rest state ($t = 4.06$, $P < 0.01$), so as the diastolic pressure ($t = 7.63$, $P < 0.01$). The results have statistical significance. It shows that deep-breath biofeedback can effectively lower the blood pressure at high altitude. For details, please see Table 44.2.

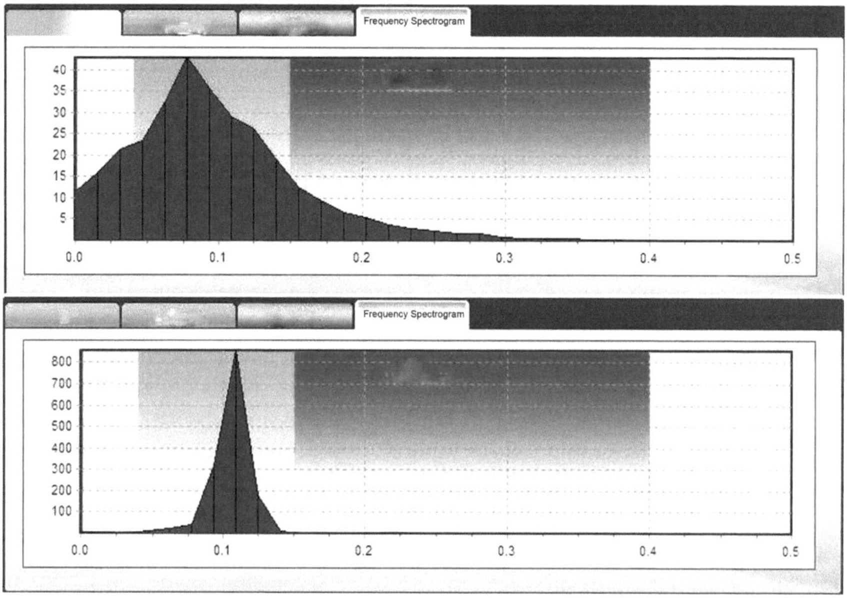

Fig. 44.2 Example of respiration frequency spectrogram for baseline and biofeedback

Table 44.2 Comparison of blood pressure between baseline and biofeedback ($\bar{x} \pm s$, $N = 71$)

Item	Rest	Biofeedback	T value	P value
Systolic pressure	126.59 ± 10.32	122.42 ± 10.78	4.06	0.000**
Diastolic pressure	85.79 ± 10.20	80.78 ± 10.21	7.63	0.000**

* $P < 0.05$, ** $P < 0.01$

44.4 Discussions

The military operators will have adaptive responses by the body compensatory mechanism to reduce the unfavorable effect caused by the hypoxia when they first enter the high-altitude environment. The blood oxygen concentration will be improved by increasing the breathing frequency, heart rate, blood pressure, and other compensatory responses of the respiration system [4]. These physical compensatory responses will cause mental reactions like fear and anxiety. According to the emotional recognition evaluation theory [5], unhealthy emotions and physical response will form a vicious circle. Due to these physical responses, the unhealthy emotion cannot recover by itself easily. The deep-breath biofeedback method is a common way to relax. Study shows that deep-breath can improve the cardiovascular system, increase the HRV, and decrease the physiological activation level and blood pressure. Therefore, deep-breath biofeedback training at high altitude can not only ameliorate the physiological response, but also relieve the mental emotions.

HRV is an important physical and mental index, indicating the activity and balance of autonomic nerves system. Clinically, HRV is the prediction factor for MI and CF. Patient with low HRV has bad prognosis [6]. HRV is closely related to mental pressure. Anxiety and fatigue can easily cause HRV to decrease. High-altitude environment can affect HRV too. Study shows that SDNN will decrease in hypoxia environment [7]. Therefore, HRV can comprehensively reflect the mental stress and adaptation to hypoxia environment of the military operators.

Deep-breath is the most effective and common way to relax. It can effectively restrain the accelerated breathing and heart rate. Respiration is closely related to HRV. Respiration mainly affects the HRV through autonomic nerves system. While inhaling, vagus nerve is restrained and heart rate speeds up; while exhaling, vagus nerve is unrestrained and heart rate slows down, namely RSA. Because the systems involved in breathing are complicated and respiration has various stimulation responses, the influences that respiration has on HRV is very complex [8]. Starting from the normal respiration frequency (0.2 Hz), HRV decreases when respiration frequency rises, and when frequency decreases to 0.1 Hz, HRV gets to the peak value. The biofeedback system set this frequency as the most relaxing state. During the deep-breath biofeedback training, the mental regulation index, including coordination index, stability index, and evaluation score, is higher than that of rest state, and HRV rises, which indicate that this training is effective at high altitude.

Hypoxia will cause the rise in blood pressure, which is a normal compensatory response. However, elevation of blood pressure is not only a physical response; it can severely affect the mental state of the military operators. The confidence to adapt to the high altitude will be brought down, and anxiety for healthy problems will increase. Study shows that deep-breath biofeedback method can lower the blood pressure and be used in clinical cases [9]. Our test shows that it can be adopted to control the blood pressure, especially obvious to the diastolic pressure.

The drawback of this experiment is that it was conducted only once. So, the continuous result and long-term results cannot be obtained.

Respiratory system and circulatory system will have direct responses to the high-altitude environment during the adaptation process. Long-term living in high altitude will definitely affect the mental state and morale of the military operators if without effective safeguarding measures. This study adopts the deep-breath biofeedback method to cope with the high-altitude environment by incorporating the psychological education. It is simple and effective and appropriate for the military operators. Therefore, this method should be studied more thoroughly.

References

1. Muza SR, Beidleman BA, Fulco CS (2010) Altitude preexposure recommendations for inducing acclimatization. High Altitude Med Biol 11(2):87–92
2. Hackett PH (2012) Caffeine at high altitude: java at base camp. High Altitude Med Biol 11(1):1–5
3. Department of the Army.U.S. Army Technical Bulletin Medicine 505, Altitude Acclimatization and Illness Management. Washington, D.C. 2010
4. Rock PB, Iwanyk EJ (2002) Military medical operations in mountain environments. In: Lounsbury DE, Bellamy RF, Zajtchuk R (eds.) Textbooks of military medicine: medical aspects of harsh environments, vol 2. Office of The Surgeon General Borden Institute, Washington, pp 854–869
5. Lieberman P, Protopapas A, Reed E et al (1994) Cognitive defects at altitude [letter]. Nature 372(6504):325
6. Task Force of European Society of Cardiology (1996) The North American Society of Pacing, and Eleetrophysiology, Heart rate variability[J].European Heart J, 7(3):354–381
7. Cheng CF, Lin HM, Tsai HC et al (2005) Analysis of heart rate variability during acute exposure to moderate altitude and rowing exercise. J Exerc Sci Fit 3(1):25–32
8. Al-Ani M, Forkins AS, Townend JN et al (1996) Respiratory sinus arrhythmia and central respiratory drive in humans. Clin Sci (London) 90(3):235–241
9. Pinheiro CH, Medeiros RA, Pinheiro DG et al (2007) Spontaneous respiratory modulation improves cardiovascular control in essential hypertension. Arq Bras Cardiol 6:651–659

Chapter 45
Effects of Microwave Radiation on the Blood Routine of Different Subjects

Ying Qi, Ji Feng, Hua Guo and Baisheng Jing

Abstract The objective of this study is to investigate blood routine changes of aerotransport crew and mice under the microwave radiation (MW) exposure for health promotion strategy. According to their duty, airfield staff were divided into two groups: radar operator group ($N = 25$); and common worker group ($n = 48$), and their blood was collected for blood routine analysis. Thirty Kunming male mice were randomly divided into three groups ($n = 10$): the sham MW group, MW one-time group, and MW three-time group. Microwave radiation conditions are as follows: average surface density of MW is 10 mW/cm^2 and the frequency of MW exposure is once every two days and 15 min every time. After the final MW exposure, the blood of mice was collected for blood routine analysis. Results: compared with common worker group, the case of low WBC quantity in radar operator group increased significantly ($P < 0.05$); and compared with the sham MW group, RBC quantity of MW three-time group decreased obviously ($P < 0.05$). Conclusion: MW radiation can injure WBC of radar operators or RBC of mice, and it may threat the health of airfield staff.

Keywords Airfield staff · Microwave radiation · WBC · RBC

45.1 Introduction

Microwave technology is widely applied in radar, satellite communications, radio navigation, and microwave ovens, including the radiators and the transmission lines between the equipments and radiators. As various microwave equipments

Y. Qi (✉) · H. Guo · B. Jing
Institute of Aviation Medicine AF, 100042 Beijing, China
e-mail: littleponds@163.com

J. Feng
Clinical Laboratory, General Hospital of AirForce, 100142 Beijing, China

S. Long and B. S. Dhillon (eds.), *Proceedings of the 13th International Conference on Man-Machine-Environment System Engineering*, Lecture Notes in Electrical Engineering 259, DOI: 10.1007/978-3-642-38968-9_45,

operate with other electronic devices simultaneously, ever-changing electromagnetic pollution is produced and it significantly affects the health and efficiency of the workers. Study shows that microwave radiation has the following influences: (1) high-intensity or long-time low-intensity microwave radiation might cause functional change of the nervous system, including functional disorder of the central nervous system and autonomic nerves system; (2) microwave radiation may have direct or indirect influences on the cardiovascular system, causing angiectasis, hyperemia, blood flow increase of related organs, as well as heart rate and blood pressure decrease; (3) the influences of microwave radiation on the hemopoietic system and immune system are bi-directional, i.e., small dose can enhance the systems in short term, while large dose can restrain the systems in long term; (4) microwave can cause hyposecretion of the digestive glands and disorder of gastrointestinal function; (5) microwave can cause eye lens opacity; and (6) microwave can directly increase the temperature of the testicles and damages of the sperm ducts. This paper aims to explore the influences of microwave radiation on the blood routine through two groups of experimental results.

45.2 Subjects and Methods

45.2.1 Subjects

There are 73 workers on the aerotransport, including 25 radar operators and 48 normal workers. So, 73 blood samples are collected. The operators are 35.9 ± 9.0 years old and have worked 805.1 ± 631.2 h. The normal workers are 31.5 ± 4.9 years old and have worked 720.1 ± 542.3 h.

Thirty Kunming male mice were randomly divided into three groups ($n = 10$): the sham MW group, MW one-time group, and MW three-time group. Microwave radiation condition: average surface density of MW is 10 mW/cm^2 and the frequency of MW exposure is once every two days and 15 min every time. Duration is 1 week.

45.2.2 Methods

45.2.2.1 Equipments

High-power pulse microwave equipment: *S*-wave band, average surface density of MW is 10 mW/cm^2.

XYMEX-XI2100 type (made in Japan) full-automatic complete blood cell analyzer.

45.2.2.2 Test Indexes

Blood routine indexes: HGB, RBC, HCT, MCV, MCH, MCHC, and WBC.

45.2.3 Data Analysis

Fisher's accurate test and variance analysis and one-way analysis of variance are conducted by SSPS18.0. Significance test level is $\alpha = 0.05$ and 0.01.

45.3 Results

45.3.1 Comparison of Blood Routine Between Two Groups Working on the Plane

In the blood routine test of the workers, there are evident differences. The WBC number of the radar operators is lower than that of the normal workers ($P < 0.037$) (Table 45.1).

45.3.2 Comparison of Blood Routine Between Three Groups of Mice

In the animal test, RBC decreases evidently after radiation ($P < 0.05$) (Table 45.2).

Table 45.1 Comparison of abnormal blood frequencies between two human groups

Indexes	Radar operators			Normal workers			χ^2	P
	Low	High	Normal	Low	High	Normal		
HGB	0	9	16	0	7	41	4.406	0.071
RBC	0	3	22	0	2	46	1.581	0.331
HCT	1	1	23	0	0	48	3.674	0.114
MCV	0	1	24	0	0	48	1.947	0.342
MCH	0	1	24	0	0	48	1.947	0.342
MCHC	0	2	23	0	2	46	0.466	0.603
WBC	3	0	22	0	0	48	6.007	0.037

$P < 0.05$, $P < 0.01$

Table 45.2 Comparison of blood routine index among three mice groups ($\bar{x} \pm s$, $n = 10$)

Index	Sham MW	One-time MW	Three-time MW
HGB	128.91 ± 19.83	123.00 ± 8.21	113.51 ± 16.06
RBC	7.88 ± 0.52	7.60 ± 0.89	6.91 ± 0.68*#
HCT	0.38 ± 0.03	0.38 ± 0.04	0.35 ± 0.04
MCV	48.54 ± 2.13	49.50 ± 2.45	49.93 ± 1.27
MCH	17.32 ± 3.99	16.28 ± 1.36	16.47 ± 2.18
MCHC	358.55 ± 92.40	329.30 ± 23.22	329.80 ± 44.17
WBC	3.50 ± 0.46	3.85 ± 1.78	3.02 ± 1.42

* $P < 0.05$, to sham MW group # $P < 0.05$ to MW one-time group

45.4 Discussions

Among the occupational adverse factors, environmental factor is the most crucial one. And frequently, various adverse factors affect the workers simultaneously, aggravating the results. Types and intensity of occupational adverse factors vary with the development of technology, social economy, and production processes. With the introduction, production, and application of high-tech materials and products, adverse factors during the working process have become the focus of occupational health management in China [1].

Researches on the hazard of microwave radiation are mostly done on the animals under specific radiation doze in the laboratory. In these animal experiments, some factors are controlled or changed by human actions, and complicated physiological process and pathological process are simplified for the convenience of experiment. The experiments are predesigned, condition-controlled, and factors applied so as to initiatively cause, reproduce, or transform the natural process of diseases. Results obtained from these experiments might present opposite cases because of various differences in the lab equipments, experiments design, species of subjects, temperature, and humidity.

As the most common and meaningful test, blood routine examination can promptly reflect the metabolic activity in normal and pathological states and gives great clinical significance to the understanding of human life and prevention of many diseases. This study shows that only the RBC amount changes and other indexes have no evident changes in the blood routine examination of the animals under microwave radiation, while in the experiments for the two human groups, WBC amount changes. The hazard degree of microwave is related to the field intensity and application duration, namely the total amount received by the living body. The larger the amount is, the higher the morbidity is. Schulz et al. [2] think that some human organs are very sensitive to the radiation. And the direct damages caused by the evenly distributed radiation in long time are lighter than that of the concentrated radiation in short time. This is because living body can reproduce by itself. However, there might be some later period damages of the cell tissue, which can only appear after several months or years, like leukemia and cancer. These are not relevant to the radiation pattern, but to the total absorbed dose. This research result is crucial. It makes no sense to simply compare the animal and human experiments because the two are different after all. Results from the animal experiments do not always correspond to the human cases and should be eventually verified clinically. Although it is certainly not easy to study the occupational hazardous factors, the animal experiments cannot be simply applied to human cases. In this study, the blood routine examination is conducted in the normal working conditions of the airfield workers without any design or selections of specific prerequisites. So, the data obtained is reliable and representative, as well as have valuable referential significances.

It is obviously found in the blood routine examinations that the influences of radiation vary according to the changes in the distances exposed to the radiation. White blood cell is the initial defense cell in the immune system. Polak et al. [3] think that electromagnetic field can reduce the number of WBC, cause disorder of cellular metabolism and affect its proliferation, lead to chromosome abnormality and deficiency, as well as synthesis disorder of DNA. Besides, cell dynamics and NKC activity are affected. Decrease in hemoglobin content is one of the important objectives of this study. The relation between RBC number and hemoglobin is that the hemoglobin amount increases when the absoluteness or relativity of RBC rises. In the microwave radiation, the blood routine indexes change, like decrease in WBC or increase in hemoglobin, which agrees with the results reported by Gong et al. [4] and Wang et al. [5]

This study shows that the microwave radiation has caused certain negative effects to the workers and further investigation should be conducted regularly. Following measures are advised: (1) strengthen personal protection, including wearing protective clothes, protective glasses, protective hats, etc. (2) improve the working conditions, including necessary screening; (3) enhance dietary nutrition, including adding appropriate nutritious supplementary or improve the meal standard; (4) guarantee sufficient sleeping and rest and offer regular rehabilitation or holidays; and (5) publicize the damages of microwave radiation and upgrade the protection awareness.

In conclusion, scientific research methods are required for activities exploring the unknown. And the methods for the medical researches demand higher thanks to the complexity of the subjects. Repeated experiments under the strictly controlled conditions can obtain the wanted information and verify the facts or hypothesis by changing or combining various terms systematically. Many uncertain mechanisms or phenomena exist in the effects caused by microwave radiation. Further researches and explorations are necessary.

References

1. Yu X (2009) Hygiene of military labour. Military medicine and science publishing house, beijing. 5:37–38
2. Schulz H, Vogt HG et al (1984) The utility of radiation protection principle. Science Press, Beijing, pp 46–47

3. Polak A, Frlanek A, Taradaj J et al (2006) Estimation of magnetic radiation effects on leucocytes. Pol Merkuriusz Lek 20(117):350–354
4. Gong J et al (1995) The influence of pulse microwave radiation on the health of operating workers. Disease Surveillance 10(4):107–108
5. Wang P, Wu L, Xiang Y et al (2012) Effect of high intensity electromagnetic radiation on the health of exposed population. J Prev Med Clin PLA 6(30):409–411

Chapter 46
Protection and Ergonomics Analysis About Two Types of Partial Pressure Suits

Zhifeng Qin, Liyong Shi, Li Ding and Huajun Xiao

Abstract The aim of the study was to observe the difference of protection and ergonomics between the capstan partial pressure suit and the bladder pressure suit. The physiological index and body surface pressure of the subjects wearing different partial pressure suit were recorded under different pressure. At the same time, the articulation motion range was recorded by Vicon three-dimensional motion capture. The mean arterial pressure (MAP) of the subjects wearing the capstan partial pressure suit DC-4 was higher than that of the subjects wearing the Bladder pressure suit DC-7. The CO (cardiac output) of the subjects wearing the capstan partial pressure suit DC-4 decreased slower comparing to that of the subjects wearing the Bladder pressure suit DC-7. The ergonomics performance of DC-7 was better than DC-4 with little effect to the head, elbow joint, and knee joint and the mobility of the DC-4 was not as good as that of DC-7.

Keywords Partial pressure suit · Positive pressure breathing · Protection performance · Motion range

46.1 Introduction

When the cabin loses gas tightness or the pilot pops out the plane over the height of 12 km, pressurized oxygen supply is necessary for the pilot and the partial pressure suit needs to be pressurized to keep the internal and external pressure

Z. Qin (✉)
School of Aeronautic Science and Engineering, Beihang University, Beijing, China
e-mail: qinzhifeng@hotmail.com; 11776687@qq.com

Z. Qin · L. Shi · H. Xiao
Institute of Aviation Medicine Air Force, Beijing, China

L. Ding
School of Biological Science and Medical Engineering, Beihang University, Beijing, China

S. Long and B. S. Dhillon (eds.), *Proceedings of the 13th International Conference on Man-Machine-Environment System Engineering*, Lecture Notes in Electrical Engineering 259, DOI: 10.1007/978-3-642-38968-9_46,

balance, ensuring the normal breathing. Capstan partial pressure suit takes the primary place in China; however, the bladder partial pressure suit is taking its shape in recent years. Both suits have their own advantages and disadvantages. They are employed in different types of planes at different periods. There are various researches around the world to compare the two suits in many aspects, like protection performance and thermal load [1]. In order to provide further theoretical bases and new referential proposals, this study aims to compare the two typical types of partial pressure suits from the perspective of physiology and ergonomics.

46.2 Subjects and Methods

46.2.1 Subjects

According to the personal protective devices table, ten male youngsters are selected, 24.6 ± 4.9 age, 171.6 ± 2.9 cm high, 66.0 ± 4.0 kg by weight. The subjects are voluntarily to participate in the test and are healthy with normal cardio-pulmonary function, without psychiatric history.

46.2.2 Equipments

46.2.2.1 Non-Invasive Continuous Hop-By-Hop Blood Pressure Detector Portapres 2.1

Portapres can record data for 23 h and 55 min. Analyzed by the computer, HR, MAP, CO, TRP can be obtained.

46.2.2.2 Body Surface Pressure Acquisition System

This system collects the body surface compensatory pressure by the pressure sensor CGYL-202. The range of the sensor is 0–1 bar with accuracy of 0.25 %FS.

46.2.2.3 Three-dimensional Motion Capture System VICON460

Vicon460 system is comprised of infrared camera and software. The camera and other devices provide the real-time optical data, and the software conducts the automatic 3D data reconstruction and automatic identification by the tracking device [5].

46.2.2.4 Pressurized Breathing and Protection System

Capstan system comprises DC-4 capstan partial pressure suit, TK-4 air-locked helmet and YHX-1 respiratory training device. Bladder system comprises TK-10 protective helmet, DC-7 bladder partial pressure suit, YM-6 oxygen supply mask, and KHX-1 pressurized respiratory training device.

46.2.3 Methods

Cardiovascular indexes are monitored and recorded by Portapres 2.1 under normal conditions (wearing normal clothes) and the joint motion trajectory is captured by Vicon460. This is used for comparison. Then, in different partial pressure suits, pressure goes from 0, 200, 400, 600, and 800 to 1,070 mmH_2O. Pressure time at 600 mmH_2O is 3 min, and over this pressure is 2 min. Three minutes interval should be guaranteed between different pressures.

46.3 Results

46.3.1 Comparison of Physiological Index Under Different Pressure Level

Table 46.1 shows the physiological indexes of the two partial pressure clothes at different pressure level. With the increase in pressure, the two partial pressure clothes begin to have evident influences on various physiological indexes ($P < 0.01$). At pressure of 1,070 mmH_2O, there is no obvious differences between HR, CO, TPR of the two kinds of clothes, while for MAP, DC-7 affects smaller than DC-4. In contrast, at low pressure of 200 mmH_2O and 400 mmH_2O, DC-7 affects larger than DC-4.

46.3.2 Comparison of Compensatory Pressure Under Different Pressure

Body surface compensatory pressure is the direct index showing the protection performance of the partial pressure clothes. From Table 46.2, we can see that with the increasing pressure, body surface compensatory pressure rises. For DC-4, the pressure level ranks from the largest at calf, to thigh, to abdomen, and last to chest, which reflects good protection performance. However, for DC-7, fluctuation emerges at the pressure of thigh and calf positions, and evidently lower than that of DC-4 ($P < 0.01$).

Table 46.1 Physiological index at different pressure level

Control			Pressure/mmH$_2$O				
			200	400	600	800	1,070
HR							
DC-4	82.4 ± 9.9	87.6 ± 7.4	88.8 ± 10.6	91.0 ± 9.5*	92.8 ± 9.6*	100.4 ± 9.8**	
b/min							
DC-7		85.6 ± 8.6	89.4 ± 9.5	95.2 ± 12.9*	101.9 ± 1.0**	103.2 ± 6.8**	
MAP							
DC-4	85.0 ± 13.9	127.9 ± 21.6**	142.4 ± 21.7**	152.7 ± 28.5**	185.4 ± 16.4**	181.2 ± 18.6**	
mmHg		DC-7	120.3 ± 22.5**	132.6 ± 16.8**	149.4 ± 23.0**	164.3 ± 31.5**	164.1 ± 19.8**
CO							
DC-4	6.7 ± 0.9	6.1 ± 0.8	5.7 ± 1.4	4.8 ± 1.0**	3.5 ± 1.2**	3.6 ± 1.1**	
L/min		DC-7	5.2 ± 1.6*	4.2 ± 1.1**	3.8 ± 1.0**	3.4 ± 1.0**	3.5 ± 1.0**
TPR							
DC-4	0.8 ± 0.1	1.4 ± 0.4	1.7 ± 0.5*	2.2 ± 0.9**	3.6 ± 1.2**	3.5 ± 1.1**	
MU		DC-7	1.6 ± 0.7	2.2 ± 0.8*	2.7 ± 1.0**	3.6 ± 2.0**	3.4 ± 1.4**

Compared to control, $*p < 0.05$, $**p < 0.01$; 1MU = 1 mmHg.s/ml

Table 46.2 Compensatory pressures of the subjects

Excessive pressure/mmH_2O	DC-4/kPa				DC-7/kPa			
	Chest	Abdomen	Thigh	Calf	Chest	Abdomen	Thigh	Calf
200	4.8 ± 1.9	7.1 ± 3.6	9.7 ± 4.2	11.2 ± 6.6	3.2 ± 1.4	3.2 ± 1.4	2.3 ± 0.6**	4.6 ± 2.5**
400	5.7 ± 1.8	8.7 ± 3.5	10.7 ± 3.7	12.5 ± 6.6	4.7 ± 2.2	4.2 ± 1.7	4.4 ± 1.2**	5.9 ± 1.8**
600	6.1 ± 2.7	10.0 ± 3.5	11.0 ± 3.8	13.2 ± 7.4	5.2 ± 2.8	5.6 ± 2.0	6.1 ± 1.2 **	5.7 ± 2.3**
800	7.1 ± 2.0	9.3 ± 5.6	9.4 ± 4.0	14.5 ± 14.5	6.1 ± 3.9	7.2 ± 1.8	8.2 ± 1.8	6.8 ± 2.6**
1,070	7.9 ± 2.9	10.2 ± 5.8	10.3 ± 4.9	14.0 ± 9.8	9.5 ± 5.5	9.1 ± 2.4	10.3 ± 2.4	6.0 ± 2.9**

Compared to control, $*p < 0.05$, $**p < 0.01$

46.3.3 Comparison of Ergonomics Under Different Pressure

This research group has established a series of evaluation methods for the clothes ergonomics in previous researches and obtained good results. This paper selects several motions for each joint under different pressures, as shown in Table 46.3. The analytical parameters of the shoulder joint are the vertical shift of the elbow joint, while other analytical parameters are the motion angles in the plane. The base group is also in the normal clothes. The results show that DC-4 restricts greatly on various joints motion ($P < 0.01$) and DC-7 restricts greatly on the shoulder and hip joints ($P < 0.01$), while restricts little on the elbow and knee joints ($P > 0.05$).

46.4 Discussions

In the condition of high-altitude anoxia or pop-out for escape, pressurized breathing is necessary for life safety, which causes the internal chest pressure increasing and then venous blood refluxing affected greatly. Next, when the venous blood accumulates in the limbs, the filtration pressure of the blood capillary will increase and liquid in the blood will infiltrates into the tissue fluid, namely hypovolemia. These factors lead to the decrease in CO and increase in HR, MAP, and TPR. Ackles et al. [2] has done researches about the pressurized breathing by the bladder partial pressure suit and discovered that systolic pressure and diastolic pressure of the artery and average artery pressure increase with the ascending excessive pressure, while SV and CO decrease evidently. This paper has got the same results as the above research. Han et al. [3] has also conducted researches on the pressurized breathing by the anti-load trousers with different bladder areas. He discovered that the stroke volume decreases in the case of less than 45 % coverage while in the case of 65 and 90 % coverage, stroke volumes increase with the excessive pressure. As this paper only tested one set of bladder partial pressure suit, we cannot compare the research of Han Xueping.

Goodman et al. [4] has evaluated the US tactical life-support system by determination of the physiological indexes. It shows that cardiac function is affected by the excessive pressure. And the larger the pressure and MAP is, the better the protection performance is. From the results of this paper, at each excessive pressure, the MAP of DC-4 is higher than that of DC-7, which indicates that the protection performance of DC-4 is better than DC-7 from this perspective.

Gradient change in body surface compensatory pressure can better help the backflow of venous blood. DC-4 can achieve relatively high compensatory pressure under low pressures and the gradient change meets the protection requirements. In the same conditions, DC-7 increases slowly due to the air inflation of the bladder, which leads to the less decrease in CO of DC-4 than DC-7. However, with the increasing pressure, the body surface compensatory pressure of DC-7 basically keeps the same level with DC-4 except at the calf position.

Table 46.3 Motion range

	Control	DC-4/mmH_2O			DC-7/mmH_2O		
		0	400	800	0	400	800
NR/°	71.9 ± 23.3	37.6 ± 12.8**	24.9 ± 10.4**	17.5 ± 6.6**	64.8 ± 8.9	56.2 ± 8.6	52.6 ± 9.5
SA/°	129.9 ± 11.9	115.8 ± 13.0**	84.0 ± 14.3**	71.9 ± 6.2**	119.6 ± 4.8*	113.8 ± 12.2**	118.2 ± 7.2*
SF/mm	638.8 ± 34.8	588.0 ± 32.8**	478.6 ± 31.9**	410.1 ± 40.3**	605.2 ± 29.9*	592.1 ± 22.1**	592.8 ± 28.1**
EF/°	119.0 ± 8.0	110.0 ± 8.2**	90.4 ± 7.9**	85.3 ± 10.8**	114.7 ± 4.5	114.4 ± 5.8	115.3 ± 6.6
HF/°	48.8 ± 10.4	38.3 ± 10.2	20.4 ± 5.1**	17.7 ± 5.5**	33.9 ± 12.1**	32.1 ± 5.2**	23.2 ± 7.1**
KF/°	149.2 ± 24.7	131.4 ± 18.4*	104.7 ± 11.4**	81.4 ± 21.2**	137.7 ± 17.8	133.6 ± 15.6	130.1 ± 25.8*

Compared to control, $^{*}p < 0.05$, $^{**}p < 0.01$

NR head turning, *SA* shoulder abduction, SF shoulder front stretching, *EF* elbow bending, *HF* hip stretching, *KF* knee bending

Finally, the CO and TPR index of the two suits are close. In short, DC-4 is better than DC-7 in terms of responsive speed.

Regarding ergonomics of clothes, scholars around the world have conducted researches for special-purpose protection suits. Aitor Coca [6] has carried out comparative study on the flexibility of firefighter uniform and normal suit and thinks that there is no evident difference of flexibility between the two kinds of suits, but the firefighter uniform is not as comfortable as the normal suit. Zhang Chunguang [7] has done researches on the ergonomics of the capstan partial pressure suit and thinks that the high-altitude pressurized suit will affect obviously the movement and operation ergonomics of the pilots. This paper has compared the ergonomics of DC-4 and DC-7 and shows that DC-4 restrains some joint motion more severe than DC-7. For the head, as DC-4 is equipped with air-locked helmet, its tension device severely restrains the turning of head. And the issue of heating load is also very serious. DC-7 affects the head turning much smaller and less heating load; however, the respiratory mask cannot meet individual requirements, which should be paid attention to. Both suits restrain the shoulder joint very seriously. Without pressure, DC-4 has caused evident restraints. Although DC-7 alleviates a little, the restraints still exist. Shoulder joint is the most complicated joint in the human body. As DC-7 does not have no sleeves, it has not much restraints on the upper limb muscles. But pressurization on the chest, shoulder, and bones will lead to decline of movement. It is DC-7 without sleeves that the elbow joints are not restricted like DC-4. The reason is that there are two vertical rubber capstans at both sides of the capstan partial pressure suit, which pressurizes the body surface by expanding the capstan. Bladder design lightens this influence and makes the heating load severe for DC-7. And the weight of bladder will also have certain influences on movement, mainly on the hip joint stretching (uplifting of thigh). Even in the condition without pressure, the restraint of DC-4 is already obvious. After pressurization, due to the capstan of DC-4 and bladder weight of DC-7, the movement of hip joint in both cases is almost the same. For the knee joint, DC-4 affects it greatly thanks to the capstan design. In general, DC-7 affects movement less than DC-4.

This paper has conducted preliminary researches on the protection performances and ergonomics for DC-4 and DC-7. In terms of protection performance, both suits function well in compensatory pressure and DC-4 is better than DC-7. In terms of ergonomics, both suits certainly restrict the movement. For some joints, DC-7 is better than DC-4. This paper has certain application values for the evaluation methods and standard formulation of the partial pressure suit.

Acknowledgments Acknowledgements to researcher Xiao Huajun and assistant researcher Shi Liyong from Institute of Aviation Medicine, Air Force; and teacher Ding Li and Master Zhang Chunguang from Beihang University, all of whom have imparted us lots of valuable knowledge and test guidance, as well as a great amount of data and test equipments.

References

1. Burn JW, Balldia UI (1988) Gz protection with assisted positive pressure breathing (PPB). Aviat Space Environ Med 59(3):225
2. Ackles KN, Porlier JAG, Holness DE et al (1978) Protection against the physiological effect of positive pressure breathing. Aviat Space Environ Med 49(6):753–758
3. Han XP, Wang Q et al (2003) Cardiovascular effect of positive pressure breathing using G-suits with different bladder coverage. Space Med Med Eng 16(1):24–27
4. Goodman LS, Fraser WD, Ackles KN et al (1993) Effect of extending G2suit coverage on cardio- vascular responses to positive pressure breathing. Aviat Space Environ Med 64:1101
5. Mavrikios D, Karabatsou V (2006) An approach to human motion analysis and modeling. Int J Ind Ergon 36:979–989
6. Coca A, Roberge R, Shepherd A et al (2008) Ergonomic comparison of a chem/bio prototype firefighter ensemble and a standard ensemble. Eur J Appl Physiol 104:351–359
7. Li ZCD, Huajun QZX et al (2011) Mechanical ergonomics analysis of capstan partial pressure suit under pressure. J Beijing Univ Aeronaut Astronaut 37(8):953–957

Chapter 47
Change Rules of Performance of Multi-Person and Multi-Machine Operation Under Low-Temperature Conditions

Qiang Gao and Zhibing Pang

Abstract The factors affecting the performance of multi-person and multi-machine operation under low-temperature conditions are very perplexing, and various factors are not independent, but interrelated and interactive, also the influence degree of each factor is not identical. Combining the experiment, the model of multi-person and multi-machine operation is introduced. SPSS 19.0 data processing software and gray relational analysis method are used to carry on scientific statistic analysis to the experiment data, analyze the influences of low temperature on the operation performance, quantify the operation process, and put forward the measures which could exalt the performance. It provides a basis for making training standards, realizing the military training management timely, standardized and scientific.

Keywords Low-temperature conditions · Multi-person and multi-machine · Operation performance · Change rules research

47.1 Introduction

Antiaircraft weapons, due to its special requirements, sometimes need to be operated in the wild- or cold-temperature environment. Therefore, to understand and grasp, the factors affecting the operation performance under low temperature and to offer countermeasures are beneficial to sustain the combat effectiveness of operators and machines (normally, 21 ± 3 °C is the comfortable temperature range.

Q. Gao (✉) · Z. Pang
Air Defense Forces Command Academy, Zhengzhou, China
e-mail: fkb_mmes@163.com

Q. Gao
Unit71271 of PLA, Qingdao, China

S. Long and B. S. Dhillon (eds.), *Proceedings of the 13th International Conference on Man-Machine-Environment System Engineering*, Lecture Notes in Electrical Engineering 259, DOI: 10.1007/978-3-642-38968-9_47,

So, when the temperature is lower than 18 °C, it can be deemed as low temperature. But, the temperature that will influence the operation efficiency is normally below 10 °C). This paper focuses on the relation between person and environment and combines the system engineering theory of person, machine, and environment [1], taking the example of multi-person and multi-machine operation experiment, so as to analyze the large amount of data obtained from the operators under low-temperature environment, and put forward some factors affecting the operation performance, as well as some feasible countermeasures.

Multi-person and multi-machine system refers to two or more operators simultaneously operate two or more weapons under specific environment. Its characteristic is that the coordination between machine and machine, operator and operator, as well as operator and machine must all be taken into consideration. Any failure by any person might make the whole system out of control.

Operation performance reflects the process and results. It is closely related to the operational motion number, speed, and accuracy, etc. By introducing the performance model of multi-person and multi-machine operation, on the one hand, various factors affecting the performance can be analyzed and weak links during the operation process can be found. By strengthening the targeted training, the operation skills can be improved. On the other hand, by studying the change rules of operation performance, the operation process is quantified, so as to formulate the training standards and realize timely, standardized, and scientific military training management.

47.2 Experiment of Performance of Multi-person and Multi-machine Operation

47.2.1 Experiment Design

Three operators are needed for each group in this experiment, one of them operates one machine (single-person and single-machine system), the other two operate one machine (multi-person and single-machine system). Thus, the two systems constitute a simple multi-person and multi-machine system. Before the experiment, the operators have not been trained in the coordination of the two machines. Temperature, operation accuracy, finish time, etc. are recorded in the experiment. Based on the performance model of multi-person and multi-machine operation, scientific mathematical method is employed to study the formation rule of performance of the multi-person and multi-machine operation under low-temperature condition as well as the factors affecting the operation performance.

47.2.2 Experiment Subjects

Subjects are 18 male students of a university. They are divided into 6 groups, each with 3 subjects. They are 22 years old, 172 cm high, and 68.9 kg in weight in average. Physiological condition like blood pressure and heart rate is in good condition. Meantime, the subjects take the 16PF test, which shows that all operators are good psychologically.

47.2.3 Experiment Preparation

Four pieces of machine A and B which can be operated collaboratively are selected for this experiment. The PF16 software and SPSS 19.0 software, model [2] of multi-person and multi-machine performance calculation, are prepared. Besides, a CENTER310 thermometer and five stopwatches are needed. The experiment is conducted on the smooth, open, and illuminated field.

47.2.4 Experiment Methods

There are three phases of this experiment. The first phase is preparation, i.e., mobilization of the subjects and understanding of the experiment process and significance. Meanwhile, the physical and mental indexes are necessary to be collected for knowing their health conditions. The second phase is training and testing period, i.e., collaborative training of machine A and B and carrying out the experiment after the subjects are familiar with the rules. The third phase is data processing and analysis, i.e., analysis of the data from the two phases by using the SPSS 19.0 and gray relational analysis method [3], and obtaining results.

47.3 Performance Model of Multi-person and Multi-machine Operation

47.3.1 Structure of Multi-person and Multi-machine System

This experiment is comprised of three persons and two machines, as the following Fig. 47.1.

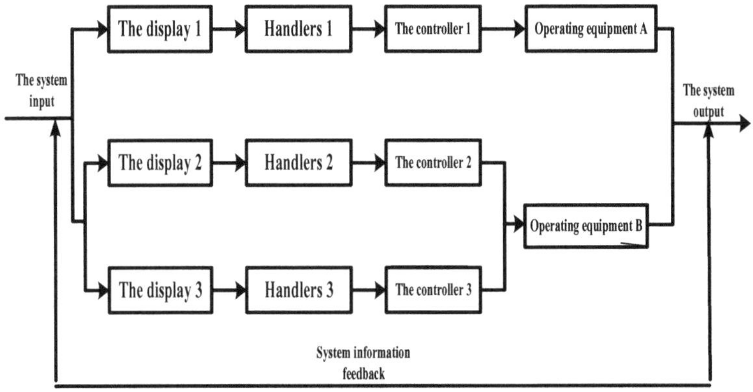

Fig. 47.1 Structural system of multi-person and multi-machine

47.3.2 Assumptions in Building Model

The assumptions in building the multi-person and multi-machine model are as follows: (1) unchanging process and certain number of operation motions; (2) non-relational between operation motions; (3) completeness of the machines; (4) no influence by environment on the performance; (5) when one operator has deviated operation, others cannot remedy.

47.3.3 Model Building

Based on the above assumptions and related data, the performance model of multi-person and multi-machine operation is introduced [3]:

$$E_i = P^i_{zm} \times P^i_{rj} \times U^i_{rj} \times \frac{R\left\{1 - \sum_{j=1}^{h}\left[w_j p_{j(ybcw)}\right]\right\}}{h \times t^i} \tag{47.1}$$

In the formula, E_i is the performance value of the multi-person and multi-machine operation at the i time. The larger the value is, the better the performance and operational skills are. P^i_{zm} is whether there is fatal failure during the operation at i time. P^i_{rj} is the failure motion accumulation degree of the No. operator in j motion at $i$time. U^i_{rj} is the accuracy deviation at i time. R is the weight ratio between operation accuracy and speed. w_j is the weight of the j motion in the whole process. $p_{j(ybcw)}$ is whether there is normal mistake at the j motion. h is the total number of operation of the system. t^i is the duration at i time operation.

47.4 Experiment Data Process and Analysis

Adopt the above performance model to calculate the experiment data and get the relation between the temperature and operation performance of each group as well as the comprehensive performance [4], as showed in Table 47.1.

47.4.1 Analysis of Temperature and Performance for Each Group

Adopt the data processing software to analyze the data in Table 47.1 and get the relation between temperature and performance as shown in Fig. 47.2.

From the Fig. 47.2, we can see that (1) the performance value generally rises with the temperature; (2) the variation amplitude of the fourth group is large and frequency is high; (3) the maximum performance value of all groups is less than $4.5(1/\mathrm{s} \times 10^{-4})$; and (4) the performance decreases slowly between 0 and 6 °C and rises after 6 °C. It rises quickly at 10 °C.

From the curve of the third group, we can see that (1) at first, the performance at 3.6 °C is higher than 5.2 and 5.9 °C, and then rises a little bit; (2) from 9.1 to 10.3 °C, the value is stable; (3) from 10.4 to 13.2 °C, the value rises evidently; and (4) from 13.2 °C, the value rises slowly.

Reasons::

Firstly, the operators take trainings at the low temperature of 0–6 °C and are not skilled in operation, the performance value is inclined to decrease.

Table 47.1 Comprehensive performance of multi-person and multi-machine operation $(1/\mathrm{s} \times 10^{-4})$

No.	Temp.	G1	G2	G3	G4	G5	G6	Comprehensive performance
1	3.6	1.7993	1.6038	1.7366	2.6923	1.9852	1.8026	1.9366
2	5.2	1.6619	1.5988	1.5182	2.5026	1.9437	1.8626	1.8479
3	5.9	1.7024	1.8153	1.8369	2.5103	1.8024	2.3744	2.0069
4	6.8	2.3684	1.9972	1.8384	2.6205	2.2261	1.8139	2.1440
5	7.1	2.4627	2.3594	2.6437	2.9410	2.6083	2.9718	2.6644
6	7.2	2.3673	2.3510	2.5196	2.7046	2.5316	2.6436	2.5196
7	7.9	2.5181	2.3286	2.3040	2.9385	2.5897	2.9408	2.6032
8	9.1	2.4469	2.3349	2.9035	3.0231	2.7360	2.8641	2.7180
9	9.1	2.5826	2.8134	3.2469	3.3038	2.8829	3.1538	2.9972
10	10.3	2.5924	2.8359	3.3814	3.2872	2.7938	3.2718	3.0270
11	10.4	3.2059	2.7961	3.0286	3.2938	3.4564	3.7821	3.2604
12	11.7	3.1685	3.0951	4.0834	3.7846	3.5846	3.8897	3.6009
13	13.2	3.7465	3.0872	3.4877	4.1205	3.4872	3.9359	3.6441
14	13.6	3.4973	3.3681	4.0637	3.9641	3.7462	3.9641	3.7672
15	17.3	3.6642	3.6197	3.7721	4.0897	3.9051	4.1205	3.8618

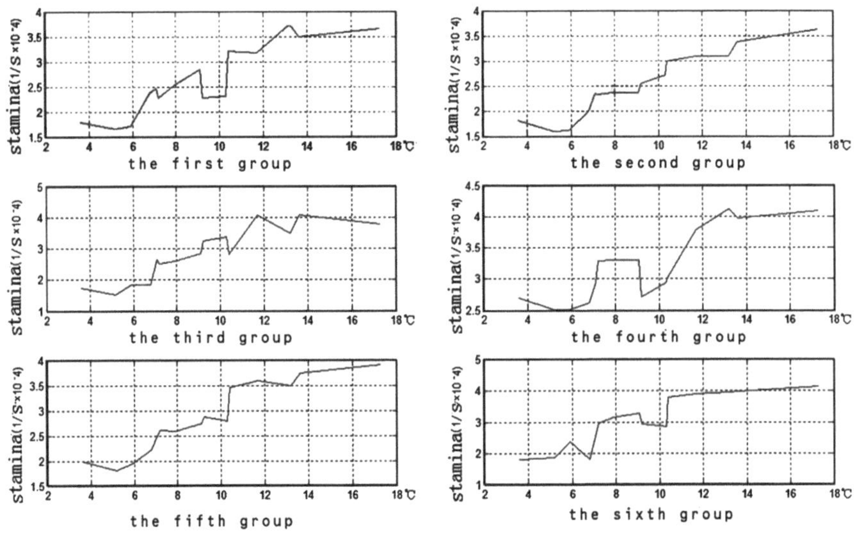

Fig. 47.2 Relation curves of temperature and comprehensive operation performance for each group

Secondly, the group 4 does not pay attention to coordination between operators, sometimes with fast and accurate operations, sometimes with fast and inaccurate operation, and other situations. The data also show the result and indicate the importance of coordination in the multi-person and multi-machine operation.

Thirdly, with the repeated exercises, the skills and perception are stable, so the performance value tends to be stable.

47.4.2 Analysis of the Average Comprehensive Operation Performance

For the convenience of examination and analysis, take the average comprehensive operation performance from the six groups. Adopt the gray relational analysis method to draw the average comprehensive operation performance curve as Fig. 47.3 shows.

Analyses are as follows based on the Fig. 47.3.

Firstly, at the beginning of the experiment, although the operators have strong thirst for knowledge, healthy mental, and physical conditions, they cannot quickly get the hang of the operation at the temperature lower than 6 °C and make more mistakes, even fatal ones with performance at zero. After training for some time, the operators can gradually grasp the skills and the performance value rises evidently.

Secondly, the performance value rises with the increasing temperature and the operators begin to become self-satisfied and fatigued, so as to make the performance level stable temporarily.

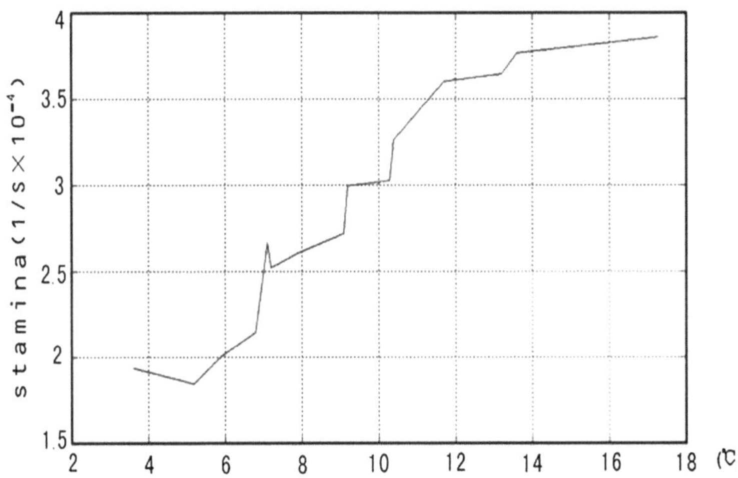

Fig. 47.3 Subjects' comprehensive average operation performance curve

Thirdly, with the decreasing performance, the differences between groups are becoming larger and the groups start to compete with each other. In order to improve the performance, each group pay close attention to the problems in the training and earnestly summarize the breaking points for improving skills with the guidance of the teachers, so that an unprecedented high efficiency appears.

47.4.3 Relation Analysis Between Temperature and Time

The analysis of relation between temperature and time is as follows.

Firstly, operators will have negative feelings about the training due to long-time repeated exercises and stable performances. Meanwhile, the performance goes down because of the limited knowledge, physical and mental conditions of the operators, and training management.

Secondly, in high temperature, the performance is normally higher than that of the beginning at low temperature and stays stable in long time.

47.5 Discussions

47.5.1 Factors

Based on the person and environment relation, and combining the experiment data and analysis, factors that affect the performance of multi-person and multi-machine operation in low temperature are as follows:.

Firstly, the ambient temperature. Under the low temperature (0–18 °C), the higher the temperature is, the higher the performance is.

Secondly, body parts affected by the low temperature and speed. In low temperature, the first part that feels uncomfortable is hand, while the operation is done by hands. So, the lower the temperature of hand is, the lower the performance is.

Thirdly, the exposure time of operators under low temperature. From the above results, performance decreases with the extension of exposure to low temperature. But, to some certain extent, the performance stays stable which is related to the adaptation to the environment and improvement of skills.

Fourthly, human body ability and condition. With the gradually skilled and automatic operation, the influences of low temperature can be compensated to a certain extent. Moreover, with the extension of exposure to low temperature, the distraction caused by nervousness at the beginning starts to disappear thanks to psychological adaptation. Concerning the body response to the low temperature, it not only depends on the environmental stress level and exposure time, but also relates to the adaptability and endurance capacity of human body to low temperature.

47.5.2 Measures

In conclusion, in order to retain the performance under low temperature for the multi-person and multi-machine operation, the following measures should be taken.

Firstly, keeping warm. Measures to keep warm include wearing warm clothes and gloves and socks with battery to protect the hands and feet.

Secondly, guidance by teachers. The teachers firstly explain the operations and relations, then regulate the motions by effective training methods, lastly, strengthen the coordination training so as to gradually improve the capability of operation in low temperature.

Thirdly, reasonable work and rest program. People will get cold and exhausted after long-time training in low temperature. The training should be targeted at improving efficiency and afternoon nap and breaks are necessary.

Fourthly, human factor to be included in the design of machines. From the perspective of combining people and machine, in the design process of weapons, human body features and operation comfort should be fully considered to improve the coordination of operator and machine, and the maximum capability of the weapons, so as to enhance the combat effectiveness.

References

1. Pang Z (1999) Air defense forces man-machine-environment system engineering. Air Defense Forces Academy, Zhengzhou (Ch)
2. Liu S, Dang Y, Fang Z (2008) The theory and application of the grey systems, vol 01. Science Publish, Beijing, pp 50–83
3. Zhao H (2008) The model and software of engineering performance about the combination of human and machine. Air Defense Forces Command Acadamy, pp 22–25
4. Li T, Li H, Pang Z (2012) Experiment and study on performance of multiple person operating multiple machine operation. In: Proceedings of the 11th conference on man-machine-environment system engineering, pp 236–239

Chapter 48
Comparative Analysis on the Performance of Different Students in Five Kilometers Armed Off-road Running

Min Chen, Shuai Mu, Zhibing Pang, Haitao Zhao, Hongyan Ou and Honglei Li

Abstract In order to get detailed information about student officers (including army-oriented students and pre-officers) in 5 km armed running course, we organized 5 km armed running test to analyze their different performance according to the results of the test. By comparing and analyzing the results, we find out the causes which lead to the difference on 5 km armed running of the two kinds of student officers, army-oriented students and pre-officers. Based on the results and data of this test, we come to the conclusion that different training methods and training intensity for the army-oriented students and pre-officers should be applied during the daily trainings. Meanwhile, the theoretical support reflected by the results and data of the test raises the possibility of making training plans more scientific.

Keywords Army-oriented students · Pre-officers · Armed 5 km running · Performance comparison

48.1 Introduction

Five kilometers off-road running is an important sports program in the army and military schools, normally in two types, i.e., unarmed and armed running [1]. Off-road running plays a significant role in improving the trainers' endurance and fatigue resistance, as well as the cardiovascular system, respiratory system, and aerobic metabolism.

M. Chen (✉) · S. Mu · Z. Pang · H. Zhao · H. Ou · H. Li
Air Defense Forces Academy, Zhengzhou, China
e-mail: fkb_mmes@163.com

S. Long and B. S. Dhillon (eds.), *Proceedings of the 13th International Conference on Man-Machine-Environment System Engineering*, Lecture Notes in Electrical Engineering 259, DOI: 10.1007/978-3-642-38968-9_48,

In order to know the difference of the comprehensive physical ability between the army-oriented students and pre-officers, we selected some students to conduct the 5 km test in the unarmed and armed way. Causes leading to differences are to be explored by comparing and analyzing the test results.

48.2 Test Subjects and Methods

48.2.1 Test Subjects

Seventy-five male students of 21–25 years old are selected as test subjects, including 28 army-oriented students and 47 pre-officers. They are in good condition and qualified in the basic physical fitness test.

48.2.2 Test Methods

Test subjects are divided into four groups to take part the unarmed 5 km running and then the armed one. The four groups are as follows: one group of army-oriented students, one group of pre-officers, and two groups of mixed students.

48.3 Test Data Process and Analysis

48.3.1 Comparative Analysis of Individual Performance

48.3.1.1 Comparative Analysis of Individual Performance in the Unarmed 5 km Running

After data process of the scores for the army-oriented student and pre-officers in the unarmed and armed 5 km running, the score (Figs. 48.1 and 48.2) are obtained for the unarmed running and the scores (Figs. 48.3 and 48.4) for the armed running are generated. From Fig. 48.1 for the army-oriented students, we can see that the best score is 20 min and 49 s and worst score 27 min 13 s. Most army-oriented students spend around 22 min or 26 min. Similarly, from Fig. 48.2 for the pre-officers, we can see that the best score is 20 min 14 s and the worst score 24 min 48 s. And the concentrated zone is between 21 and 23 min. Based on the two figures, we can see that pre-officers spend less time than the army-oriented students in general in this unarmed 5 km running.

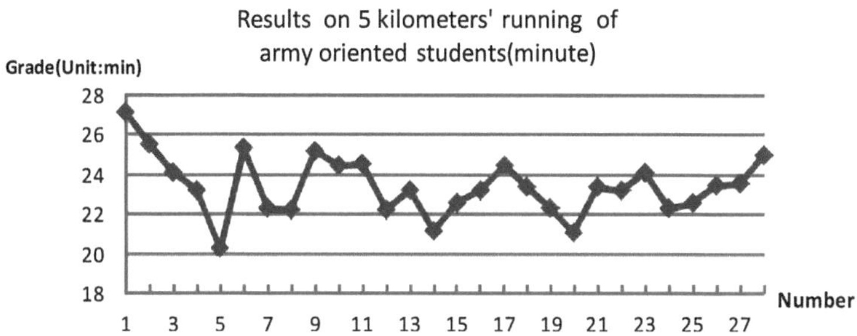

Fig. 48.1 Individual performance in unarmed 5 km running of army-oriented students

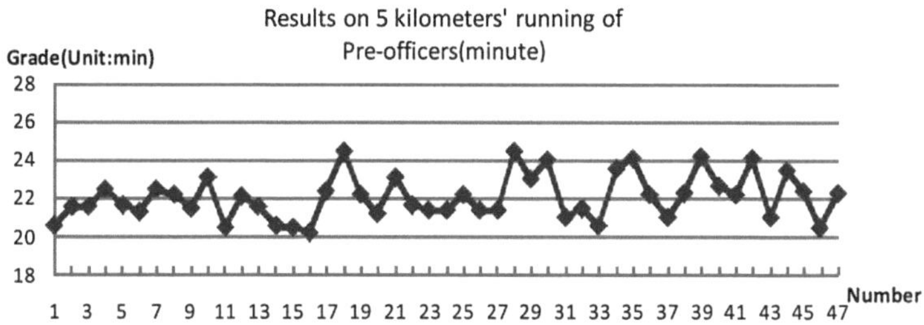

Fig. 48.2 Individual performance in unarmed 5 km running of pre-officers

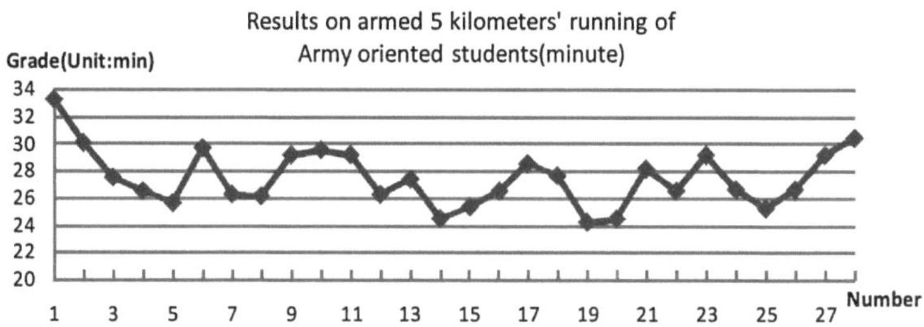

Fig. 48.3 Individual performance in armed 5 km running of army-oriented students

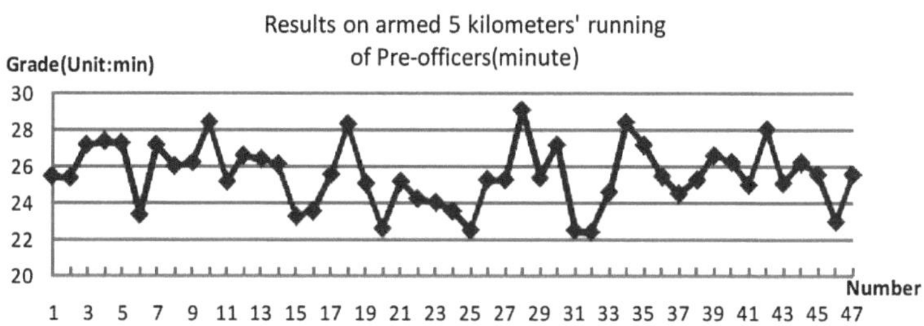

Fig. 48.4 Individual performance in armed 5 km running of pre-officers

48.3.1.2 Comparative Analysis of Individual Performance in the Armed 5 km Running

From Fig. 48.3 for the army-oriented students, we can see that the best score is 24 min 17 s and worst score 33 min 19 s. Most army-oriented students spend around 26 min or 30 min. Similarly, from Fig. 48.4 for the pre-officers, we can see that the best score is 22 min 46 s and the worst score 29 min 15 s. And the concentrated zone is between 23 and 28 min. Based on the two figures, we can see that pre-officers spend less time than the army-oriented students in general in this unarmed 5 km running.

48.3.2 Comparative Analysis of Performance Distribution in the Unarmed and Armed 5 km Running

According to the score distribution for the unarmed and armed 5 km running in the CPLA Physical Standard, the score distribution for the army-oriented students and pre-officers in the unarmed 5 km running is compiled into Table 48.1. From the table, we can see that the qualification rate of pre-officers is 76.6 %, while the rate of army-oriented students is only 35.71 %, which indicates that the performance of the army-oriented students in the unarmed 5 km running is unsatisfactory. As to the pre-officers, the largest number falls into the qualification group, taking up 59.57 % and excellence rate is zero. And for the army-oriented students, the unqualified students take up the largest percent and the qualified students only one good score and zero excellence. Therefore, we can infer that the performance of the pre-officers is better than that of the army-oriented students; however, both have few good or excellent scores.

Similarly, the score distribution for the army-oriented students and pre-officers in the armed 5 km running is compiled into Table 48.2. From the table, we can see that the qualification rate of pre-officers is 61.7 %, while the rate of army-oriented students is only 18 %. As to the pre-officers, the largest number falls into the qualification group, taking up 53.19 % and excellence rate is zero. And for the army-oriented students, the unqualified students take up the largest percent (23 students) and the qualified students zero good or excellence. Therefore, we can

Table 48.1 Classification in unarmed 5 km running

	Pre-officers	Army-oriented students	Percent for pre-officers	Percent for army-oriented students
$T \leq 19$ (excellent)	0	0	0	0
$19 < T \leq 21$ (good)	8	1	17.03	3.57
$21 < T \leq 23$ (qualified)	28	9	59.57	32.14
$23 < T$ (unqualified)	11	18	23.4	64.29
N (samples)	47	28		

infer that the qualification rate for both groups decreases in the armed running comparing with the unarmed running. And the performance for the army-oriented students in both types of running is not satisfactory with low qualification rate.

48.3.3 Comparative Analysis of Average Performance

48.3.3.1 Comparative Analysis of Average Performance in Unarmed 5 km Running

After data process of the results of the army-oriented students and pre-officers in the unarmed 5 km running, we get the Fig. 48.5. We can see that the average score for the army-oriented students is 23 min 41 s, 41 s over the qualification line, and most students are not qualified. And the pre-officers average score is 22 min 3 s, 57 s less than the qualification line, and most students are qualified. In general, the average performance of the pre-officers is 1 min 38 s less than the army-oriented students.

48.3.3.2 Comparative Analysis of Average Performance in Armed 5 km Running

After data process of the results of the army-oriented students and pre-officers in the armed 5 km running, we get Fig. 48.6. We can see that the average score for the army-oriented students is 27 min 49 s, one minute 49 s over the qualification line, and most students are not qualified. And the pre-officers average score is 25 min 48 s, 12 s less than the qualification line, and most students are qualified. In general, the average performance of the pre-officers is 2 min less than the army-oriented students.

48.3.3.3 Comparative Analysis of the Average Additional Time Between the Unarmed and Armed 5 km Running

In order to further compare the score between the pre-officers and army-oriented students, we get Fig. 48.7. We can see that the average additional time for the pre-

Table 48.2 Classification in armed 5 km running

	Pre-officers	Army-oriented students	Percent for pre-officers	Percent for army-oriented students
$T \leq 20$ (excellent)	0	0	0	0
$20 < T \leq 23$ (good)	4	0	8.51	0
$23 < T \leq 26$ (qualified)	25	5	53.19	17.86
$26 < T$ (unqualified)	18	23	38.3	82.14
N (samples)	47	28		

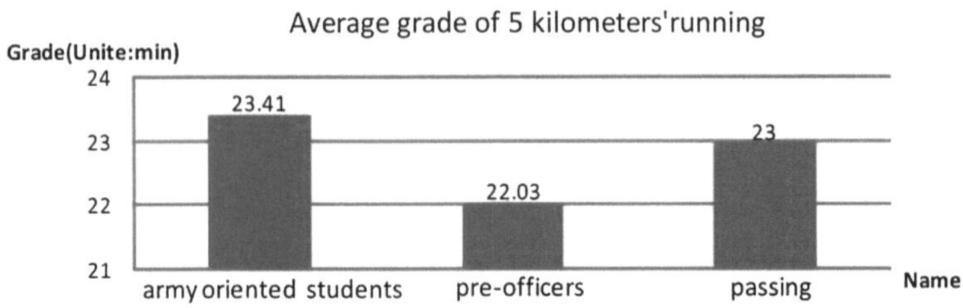

Fig. 48.5 Average performance of unarmed 5 km running

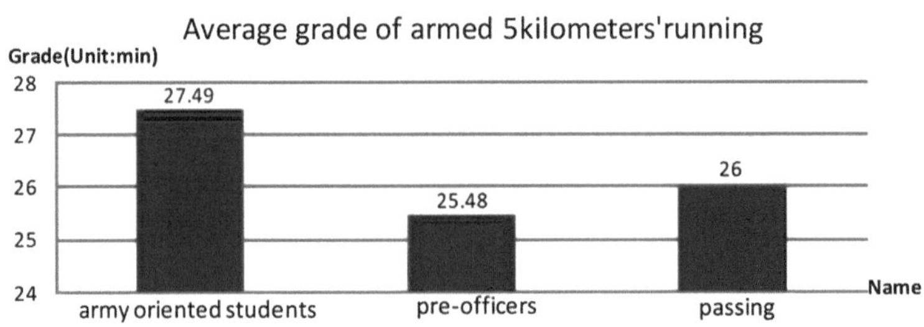

Fig. 48.6 Average performance in armed 5 km running

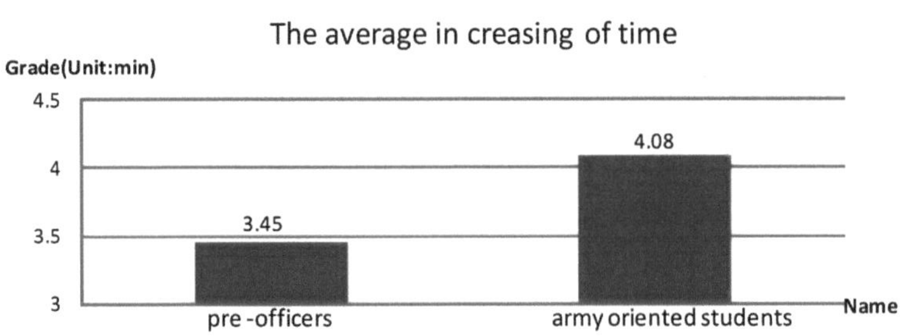

Fig. 48.7 Average additional time

officers is 3 min 45 s, while for the army-oriented students, it is 4 min 8 s. The reason behind this difference is that pre-officers have been trained systematically in military university with better physical ability, while the army-oriented students are trained by normal institutions following the traditional physical training methods [2]. Therefore, we think that the pre-officers have obtained the better scores thanks to their adaptive training in the military schools, and the local normal institutions should formulate targeted training for the army-oriented students so as to enhance the physical ability and comprehensive abilities [3].

48.4 Conclusions and Suggestions

48.4.1 Conclusions

In conclusion, we found that whatever unarmed or armed 5 km running, army-oriented students are not as good as the pre-officers in terms of qualification rate and the difference is evident. The following reasons are as follows.

Firstly, the pre-officers have been trained for almost four years of systematic military exercises, while the army-oriented student are neglected of trainings in regular institutions of higher learning [4].

Secondly, armed running means 12.5 kg equipments, which causes the army-oriented students seriously unadapted to the new condition due to their weak physical ability. For some army-oriented students and pre-officers, inappropriate wearing of equipments also leads to slower speed.

Thirdly, we found that the army-oriented students in the mixed groups perform better than that of all army-oriented students group, which indicates that other fellow's performance can influence the army-oriented students.

Moreover, weather condition and mental condition can also have certain influences.

48.4.2 Suggestions

Firstly, army-oriented students in the local universities should strengthen physical training to meet the requirements of Military Standard.

Secondly, the trainers should take full consideration of the differences of physical ability between the army-oriented students and pre-officers and arrange appropriate training amount and intensity [5].

Thirdly, mixed groups with different physical ability can improve the total performance.

Fourth, the trainers should encourage the students mentally in the training so as to tap the potential to the largest extent.

References

1. Sun X (2005) Adopting the view of scientific development to guide the training in the army. J PLA Inst Phys Educ, Guangzhou
2. Yu X (2003) Strengthening the military physical training to improve the physical quality of defence cadets. Value Eng, Shijiazhuang
3. Zhang Z, Lian J, Wang X (2003) The national defense students' military physical quality. J PLA Inst Phys Educ, Guangzhou
4. Long B, Li D (2013) Scientific interpretation of functional training. J Wuhan Inst Phys Educ 47(2), Wuhan
5. Ren H, Xing W (2007) A theoretical study on structure-dimensional of human physical adaptation. J Beijing Sport Univ 30(11), Beijing

Chapter 49
Analysis of Influence on Operation Reliability by Interior Environment of a Certain Self-propelled Anti-air System

Xiang Gao, Haiwen Zheng, Zhimin He, Ming Kong, Maowen Wang and Min Chen

Abstract Started with analysis of factors affecting the operation reliability, the paper aims to research the noise, vibration, harmful gases, and temperature of the interior environment of a certain self-propelled anti-air gun system; 10 drivers accepted the tests of memory capacity, reaction speed, and operation accuracy on the indicator instrument panel. With the noise and vibration getting worse, the memory capacity, reaction speed, and operation accuracy get worse, while temperature is not the main factor. Conclusions from the analysis of tests results will promote the training and management; limitations of research method are also pointed out at the last of the paper.

Keywords Armed vehicle driver · Operation reliability · Analysis of factors

49.1 Introduction

The self-propelled anti-air weapon system tested in this study is the primary equipment in the air defense force of China army. It plays an important role in the future military battle. This weapon system adopts the armored chassis, featuring fire power, flexibility, defense, etc.

As a special artificial environment, the interior cabin of the system is designed for fulfillment of the mission. This cabin is a place for the soldiers to live, work, and battle, and its condition will directly affect the performance of the system. It has great significance for enhancing the army air defense effectiveness by

X. Gao (✉) · H. Zheng · Z. He · M. Kong · M. Wang · M. Chen
Air Defence Forces Command Academy, Zhengzhou, China
e-mail: fkb_mmes@163.com

S. Long and B. S. Dhillon (eds.), *Proceedings of the 13th International Conference on Man-Machine-Environment System Engineering*, Lecture Notes in Electrical Engineering 259, DOI: 10.1007/978-3-642-38968-9_49,

improving the safety and comfort of this system based on the test results of the interior environment.

Operator reliability is one of the basic conditions for sufficient fulfillment of the mission of armed vehicles. And scientific evaluation method is the foundation of human reliability researches. As the concept of human reliability is initially confined to the engineering viewpoint, its stereotype description of human in the model often leads to large deviation, or even opposite results, although this system can offer certain instructive results from the narrow sense of reliability engineering by partial assumption and certain simplification. Therefore, more and more scholars start to explore the human reliability from the broad sense.

Human is a relatively perfect self-adaptive and self-learning feedback system, featuring nonlinearity, time-dependent nature, discreteness, continuity, dependency, randomness, ambiguity, and hysteresis quality, etc. There is certain similarity between human and machine, but the reliability of both is substantially different thanks to the human individuality, self-adaptation, and analysis capability. Therefore, the individual characteristics must be considered in the research of human reliability, instead of totally imitation of the research method for mechanical system [1].

49.2 Factors Affecting the Operation Reliability [2]

49.2.1 Influence of Noise

In order to acquire fine air defense capability, the self-propelled anti-air weapon system must be highly flexible, which demands higher for the engine. However, high-power engine will generate noises accompanying the dynamic performance. Meanwhile, the transmission system is close to the cabin and also produces noises. The interior noise level has far surpassed the allowed value regulated by GJB-50. Noise is a crucial factor for the operation efficiency. People cannot easily receive effective information in the noisy environment, and the normal operation is affected accordingly. In addition, this kind of environment will aggravate the fatigue degree or even cause temporary or eternal hear damage.

49.2.2 Influence of Vibration

Vibration mainly comes from the friction between running parts of the engine, impact between the caterpillar and road, bumping, and air turbulence in firing, etc. Its damages are evident that low-frequency vibration is easy to resonate with body organs, making people dizzy, while high-frequency vibration will cause the operation accuracy decline, finally affecting the combat effectiveness.

49.2.3 Influence of Harmful Gases

Many types of equipment are installed in the small cabin of the armed vehicle. Ventilation is bad when the windows are closed. The air tightness between the cabin and engine is not good, and the continuous driving makes the temperature higher, especially the diesel engine causes the temperature increasing rapidly. Large amount of exhaust gas penetrates into the cabin, including CO, NO_2, SO_2, NH_3, and TSP, and accumulates in the cabin due to bad ventilation.

The harmful gases become denser and denser with the increasing driving time and successive ejaculation of shells. Although mechanical ventilation can decrease the density of harmful gases to some extent, in some operation situations, the harmful gases can retain at certain level. Deterioration of air environment will lead to imbalance of vegetative nerves in serious conditions and slowness of judgment and control ability, even dizziness.

49.2.4 Analysis of Interior Temperature

It is mainly the summer heat that causes the interior temperature very high, while in winter the engine will ensure the cabin temperature staying at a temperature not very low. High temperature in summer will severely affect the efficiency of human and equipment.

49.3 Test of Factors Affecting the Operation Reliability of the Driver in the Cabin

Mobility of armed vehicles is an important combat performance. It is determined by three factors, firstly, the inherent design of the armed vehicle, including maximum off-road speed, climbing gradient and acceleration; secondly, the comfortableness of the man-machine interface, including whether the allocation of steering and instruments is within the range, and the operation force is within the driver's capability; and lastly, the operation reliability of the driver.

This paper focuses on the analysis of the interior environment affecting the operation reliability. The driver's fatigue degree, memory ability, response ability, vision, and hearing might be affected by the degrading interior condition. This study has conducted tests on the driver's memory reliability of meters, response speed, and operation accuracy taking the noise, vibration, harmful gases, and temperature into consideration. Ten subjects take part in this test, all of which are drivers of self-propelled anti-air weapon system. Measurement for the factors in different working conditions is carried out.

During the 120-min driving test in the training field, noise and vibration are measured by every minute, harmful gases, temperature, memory capacity, response speed, and operation accuracy are measured by every 10 min. MK224 noise sensor is used to test the noise. YD51 acceleration sensor is used to test vibration. PT100 temperature sensor is used to measure the temperature. XK-MC4 tester is used for the harmful gases.

The subjects are shown a series of meters, including odometer, engine tachometer, water thermometer, oil temperature gauge, and oil pressure gauge. They are required to read out the displays and record in 3 s. Ten groups of data are measured for each subject. Take the average value as memory reliability. The calculation formula is as follows [3]:

Memory reliability $= \left(\sum_{(n=1)}^{10} (N/5)\right)/10$. N is the number of correct memory in each test.

The response speed is measured by the time gap between the driver pushing the pedal and the vehicle starting to breaking when there are obstacles emerging suddenly before the vehicle. All subjects will drive the same vehicle to guarantee the accuracy. Test 10 times and take the average.

Operation accuracy is tested by actual driving, including restricted road between piles in the straight line, down slope, and double rectangles, track bridge, uphill designated parking and starting. They are scored by the scoring system of our school. Each score for each subject is the average by four evaluators.

Ten drivers are selected to perform the test. Data are listed in Table 49.1, and curve is shown in Fig. 49.1; 77 male students (48 pre-officers, 29 army-oriented students) are healthy with regular blood pressure and heart rate (Fig. 49.2).

The noise and vibration values in this table are the average value of the ten measurements by every 10 min. And the memory reliability and response time and operation accuracy are the average value of the ten drivers performances.

Table 49.1 Test data

Time (min)	10	20	30	40	50	60	70	80	90	100	110	120
Noise (dB)	38	42	35	39	41	37	30	45	34	37	35	32
Vibration (m/s^2)	5.2	4.1	3.8	4.5	3.9	4.6	5	4.5	4.1	3.7	3.9	4.7
Harmful gas	0.72	0.83	0.91	0.9	1.12	1.11	1.21	1.25	1.32	1.31	1.45	1.44
Temperature (°C)	30.2	30.3	30.3	30.8	31	31.4	31.8	31.8	32	32.1	32.1	32.2
Average response time (s)	0.85	0.86	0.86	0.87	0.89	0.92	0.92	0.98	1.02	1.05	1.05	1.08
Memory capacity (reliability)	0.97	0.99	0.97	0.97	0.96	0.94	0.94	0.93	0.93	0.92	0.88	0.89
Operation accuracy	0.95	0.99	0.97	0.94	0.93	0.91	0.9	0.9	0.88	0.89	0.86	0.85

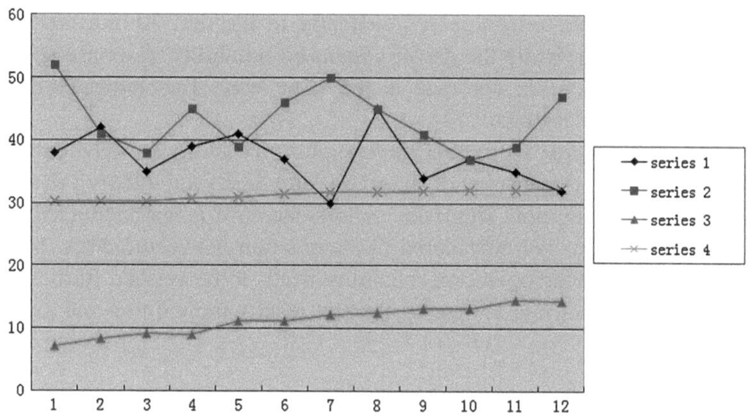

Fig. 49.1 Curve of noise, vibration, temperature, and harmful gas in 120 min. *Series 1* noise level, *Series 2* vibration acceleration (ten times the actual value), *Series 3* density of harmful gas (ten times the actual value), and *Series 4* cabin temperature

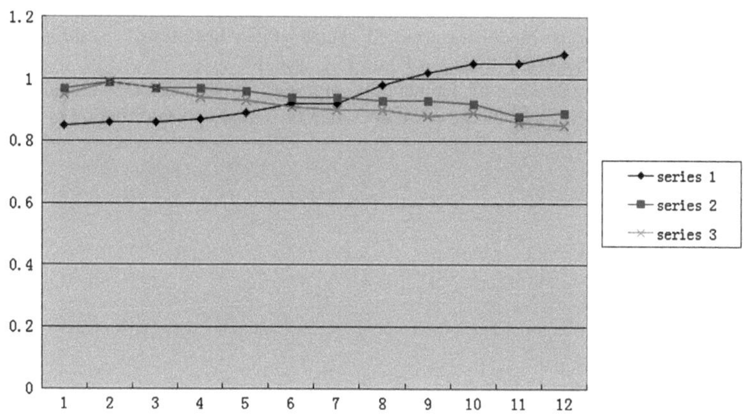

Fig. 49.2 Curve of response speed, memory reliability, and operation accuracy. *Series 1* average response time, *Series 2* memory reliability, and *Series 3* operation reliability

49.4 Conclusions

From the above data and curves, we can get the following conclusions.

Firstly, the driver's memory reliability, response speed, and operation reliability will decrease with long time staying in the large noises and vibration environment.

Secondly, the driver's memory reliability, response speed, and operation reliability will decrease while staying in the cabin with accumulated harmful gases for a long time.

Thirdly, the temperature increases evidently in the first 30 min and changes little after the 30 min, while the driver's memory reliability, response speed, and operation reliability do not decrease in the same way. This indicates that temperature is not a primary factor [4].

Fourthly, driving time will affect the operation reliability to some extent.

There are some drawbacks of this test and analysis. Firstly, due to the restrictions of test condition, single-factor influence test is not conducted. So, the actual test results have not considered the interaction between factors. Secondly, scoring by people for the operation reliability needs to be verified further and the operation items are not much, making the test results limited to some extent.

References

1. Liu W (2000) Study on drive reliability of armored vehicle driver. Proceedings of conference on man-machine-environment system, 10
2. Wang H (2010) The combat effectiveness comprehensive evaluation of the crew in the armored vehicle micro-environment. Fire Control and Command Control, 06
3. Cao W (2011) Evaluation model for armored vehicle driver operational reliability based on analysis of influence factors. J Armored Force Eng Inst, 10
4. Qu R (2002) Data collection of temperature humidity in armed vehicle man-machine–environment test. Proceedings of the 6th conference on signal and information process, 10

Chapter 50
Research on How to Improve Combat Efficiency of Air Defense Forces Under the Complex Electromagnetic Environment

Kun Wang, Peng Han, Zeqiang Qi and Zaochen Liu

Abstract With the unprecedented improvement of informatization in modern air defense operations, more complex electromagnetic environment has become the foremost objective environmental conditions that influence and limit the capability of ground air defense combat forces and realization of air defense combat effectiveness. Started with electromagnetic environment that the future air defense forces will face, the paper puts forward the new understanding and creative ways on training innovation and improving combat efficiency of air defense forces under the complex electromagnetic environment. And it is the premise and foundation of correct guidelines for the air defenses that how to correctly understand the complex electromagnetic environment, how to identify the basic elements, how to know the mechanism of action well, and how to recognize the influence toward ground air defense combat.

Keywords Complex electromagnetic environment · Operations of air defense forces · Theoretical innovation

Informationalized air defense operation features various corps, diversified technical equipments, and complex electromagnetic environment. It is of great significance for the air defense operations to formulate reasonable guidelines, to make scientific preparations, to utilize various forces, and to take effective measures against electromagnetic influences in such a complex environment. Therefore, we must carry out researches to enhance combat efficiency in complex electromagnetic environment.

K. Wang (✉) · P. Han · Z. Qi · Z. Liu
Air Defense Forces Academy, Zhengzhou, China
e-mail: hahaguai777@163.com

S. Long and B. S. Dhillon (eds.), *Proceedings of the 13th International Conference on Man-Machine-Environment System Engineering*, Lecture Notes in Electrical Engineering 259, DOI: 10.1007/978-3-642-38968-9_50,

50.1 Formulation of Scientific Combat Guidelines Based on Clear Understanding of Importance of Electromagnetism Control

Air defense forces are the vital force for air attack and strike. In mechanized wars, air defense forces mainly depend on the fire power to destroy or damage the aerial targets. Fire power is the key factor of combat success or failure. However, in contemporary information age, advantages in electromagnetism have become the crucial part in military edges. Air defense firstly starts with electromagnetic confrontation [1]. The initiative relies on the control of electromagnetism. Therefore, to take the control of electromagnetism is the priority task. And formulation of scientific combat guidelines becomes the most significant issue. Firstly, be ahead of the enemy. Make time and response quickly to seize the control of electromagnetism with initiative advantages. For this purpose, researches on the electromagnetic combat capability of the powerful enemy should be focused on at normal trainings and its development trend be known well. Electromagnetic intelligence scout should be conducted actively regarding the critical war zones and directions so as to formulate thorough electromagnetic combat proposals. In the air defense combat, electromagnetic situations should be understood well at all times. Only with timely intelligence scout, prompt decision-making and speedy interference and suppression can the initiative of electromagnetism in air defense combats be obtained. Secondly, the integrated operations, i.e., employment of electromagnetic forces by all units and comprehensive utilization of various methods so as to form the integrated combat operations. All units of air defense force must be united together to make integrated plans of various aspects including soldiers, weapons, patterns, methods, frequencies, and power. Integrated operations should combine the electromagnetic attack and defense, the soft and hard damage, the professional and non-professional electromagnetic countermeasures, so as to lay the foundation for control of electromagnetism. Thirdly, flexibility, namely flexible strategies, should be adopted in the combat. The commander of air defense force must consider the actual situations of electromagnetism and get hold of the key points of electromagnetic countermeasure. Tactics must be flexibly employed and alternated with the changing situations.

50.2 Guarantee of Air Defense Capability by Analysis of Influences of Electromagnetic Environment

First of all, the problem that the reconnaissance and pre-warning system cannot scout explicitly and correctly stands out in the complex electromagnetic environment. Electromagnetic interference results in dramatic performance decline of the pre-warning system. The detection range of radar reduces greatly and detection capacity decreases or even fails. Even limited interference can drastically bring down the target detection rate, which brings about heavy mental pressure on the operators and thus causes mistakes and negligence. Secondly, the command and control system cannot operate stably in air attacks. As the "nerve center" of the air defense combat, the system is often attacked in both "soft and hard" ways by the enemy relying on its electronic information technology advantages. Meantime, self-interference and mutual interference of our own electronic equipment, as well as bad weather and special landforms will affect the operation of the command and control system to some extent. Thirdly, problem of inaccurate attacks becomes distinct. Existing various air defense weapons, except the antiaircraft guns with optical sights and missile with infrared guidance, are guided by fire control radar, which are liable to transmit wrong data, delay, and block the communications under the strong electromagnetic interferences. These weapons will have shorter detection range and cannot detect the targets accurately. In the case of failure of radar systems, the antiaircraft guns will also be affected with lower accuracy. Fourthly, maneuvers of forces are exposed to the advanced reconnaissance technology of the enemy. The enemy can have high-precision, all-day, all-around surveillances of the battlefield. The various air defense equipments are liable to be found by the enemy. In addition, the heavy air attack, high-accuracy assault, multi-failure mechanism, and strong damage efficiency make it more difficult to take cover. Once found by the enemy all will be destroyed at once.

50.3 Rational Operational Preparations Based on Thorough Comprehension of Battlefield Electromagnetic Environment

In the battle against Taiwan Army, they were not only capable of all-range and multi-dimensional electromagnetic reconnaissance, but also able to generate strong power and three-dimensional electromagnetic interferences. Particularly in recent years, they have paid much attention to improve the capability of new electromagnetic combats. They have had the anti-radiation attack ability and developed the information network assault ability. With the assistance of US Army, they might also have such strong electromagnetic attack ability as EMP bomb and new-type graphite bomb. For this reason, we are forced to establish scientific, rational, and systematic air defense operational preparations to face the new changes. Firstly, comprehensive task allocation. Normally, the primary task of air defense soldiers is to protect and cover the weapons and forces, known as the "fixed tasks." However, in complex electromagnetic environment, their tasks need to extend to electromagnetic reconnaissance, interference, and protection work, known as "flexible tasks." Both types of tasks should be combined closely and alternate according to the actual combat progresses. Secondly, united forces. Control of electromagnetism will become the "commanding point" in the complex electromagnetic combats [2]. Except for the necessary forces and weapons for air defense, such electromagnetic countermeasures as electronic, network, and radiation will definitely break the boundaries of army forces. The team will be selected with the best soldiers and weapons. In short, united force is a must in the air defense wars with complex electromagnetic conditions. Thirdly, reasonable allocation of forces and weapons according to the tasks, enemy situation, and landform. The forces and weapons should be integrated with the electromagnetic counterforces to combine the hard and soft forces, the attack and the defense. In terms of allocation emphasis, the primary and secondary directions, as well as the forces attack and EPM, should be handled well. In terms of allocation pattern, the relation between dynamic and static operations should be managed well depending on the actual confrontation situations. In terms of allocation effect, attack and defense should be integrated thanks to the various electromagnetic countermeasures.

50.4 Judgment of Enemy Electromagnetic Attempts by Adopting Various Methods Comprehensively

Recent local wars show that electromagnetism has been used in the way of focused all-time attack based on the war attempts in modern wars instead of the traditional successive attacks along with the battle progress. At the beginning of the Gulf War, multinational forces applied long-time strong electromagnetic interference on the communications and radar systems of Iraq Army and achieved the unexpected air raid. Only by accurate judgment of the enemy attempt and comprehensive countermeasures can the air defense soldiers confront the hi-tech air attack. Firstly, peacetime preparation and wartime electromagnetic information acquisition [3]. To obtain the real time, accurate electromagnetic information is to acquire the advantages and initiative in the combat of control of electromagnetism. Therefore, the military and local forces, the professional and unprofessional forces should be fully coordinated to collect the enemy intelligence and get the accurate enemy position and features, functions, frequencies, and technical parameters. Before and during the war, the real-time information about the enemy deployment, tasks, weapon types and amount, as well as other tactical and technical information should be obtained and shared. Secondly, soft and hard methods to damage the enemy electromagnetic interference system. Systematic and non-systematic instruments, modern and traditional methods should be combined to damage the enemy electromagnetic interference system. In terms of soft damages, active and passive interferences, "virus" attack, and exhaust can be used to interfere, attack, and break the enemy electromagnetic system. In terms of hard attacks, elite forces and weapons should be concentrated to attack the enemy electromagnetic interference system based on the accurate understanding of enemy attempts and deployment. Thirdly, confuse the enemy in various ways to weaken the enemy electromagnetic interference effects. Confronted with the enemy with electromagnetic advantages, we should adopt the interference, feigned activity to create opportunities for our air defense forces. Combination of feigned electromagnetic and forces activity can conceal our purposes, deployments, and confuse the enemy.

50.5 Guarantee of Smooth Control Via Various Communications Networks

First of all, technological convergence to achieve ceaseless command and control. The ultra-shortwave radio is directly connected to the channel of shortwave radio via its pilot frequency port. Then, they are connected to the wired telecom channel to realize the communications network. The command relations are concealed, and radio communications are guaranteed. Based on the local communications network, hierarchical circuitous compound communications network is formed by combining the radio network, satellite network, duplex network, wired network, and data transfer network, as well as the corresponding automatic information process system, nodal switching system, and radio inlet network switching sys. Secondly, comprehensive networks to keep the control system stable. Wireless command network is set up on various radio stations. Air intelligence notification and command network are set up on the local communication networks. Intelligence command network is centered on the intelligence command system. Communication tools like Beidou handset and high-power radio station should be particularly distributed to troops with crucial tasks. Messages are transmitted during the March by quick positioning way [4]. Thirdly, reasonable deployment to prevent interference and suppression. Various communication tools should be deployed by cleverly using the terrain. In the battlefield, electronic equipments should be allocated dispersedly. The way of combing fixed and flexible communications, as well as high-power and small power radio stations, should be used together to resist interference. Messages are sent by one button at different frequencies, some frequencies for blind transmission, some for real information, so as to confuse the enemy.

50.6 Organize Rigorous Electromagnetic Protection on the Most-attacked Targets

Although the enemy will implement electromagnetic attack throughout the air raid, they normally will select key targets depending on their attempts so as to paralyze the whole system by only focusing on one point. Therefore, to find out the key targets of the enemy and organize effective electromagnetic protection is the premise and foundation for the air defense forces to sufficiently fight against the enemy [5]. Specific measures include: firstly, control the service time and range of radar, guide, and radios to the minimum limit, so as to prevent electromagnetic interception or anti-radiation destroy by decreasing the exposure time of radio wave in the air. Secondly, to evade the frequency band that the enemy often attacks after knowing the enemy interference features. By using the time and spatial gap of enemy electronic interference, change the frequency in a flexible way and rationally plan the range to minimize the interference effects. Connect the radar, guide, and radios together by using the existing networks, nodal switching technology, and radio inlet equipment, so as to realize the whole system functioning well all the time even though one of them is destroyed. Thirdly, to conceal the real target by using camouflage net, painting, metal corner reflectors, and setting up covers for anti-radar, anti-infrared, anti-photoelectric, and anti-laser reconnaissance. Use the ground objects to set false targets against the radar detection. Widely use the systematic or simply instruments to fake the radio stations and radars. Set up simply radiation sources to send false signals on the artificial battlefield. In addition, smoke screen and mist can be taken advantage of so as to affect the reconnaissance of photoelectric instruments. Fourthly, the instruments like radar, radios, and guides should be dispersedly allocated based on the landform and the air defense requirements, so as to prevent concentrated radiation zone and minimize the damages.

50.7 Flexible Air Defense Deployments to Improve Attack Effects

First of all, ambush flexibly and look for chances to attack. After knowing the activity routines of enemy raid, conceal the forces at some location and make preparations to wait or seduce the enemy. The air defense force is mixed of a company of antiaircraft guns and a platoon of missiles. They enter the ambush location in advance, close to the covered target and the lag line of the enemy guided missile, or close to the airspace of the enemy electronic attack planes, or close to the area liable to air raid. Secondly, defend the key targets like command center and artillery, and deploy nearby forces to reinforce the air raided area. When the enemy air raid is regular over a nearby area, partial forces will be deployed to the area rapidly during the gap to destroy the enemy weapons. Thirdly, many small groups will be deployed around the covered targets without affecting the general air defense arrangement. They will cover more area and take the initiative to confuse the enemy, seduce the enemy and disperse the enemy. Fourthly, the air defense forces and electronic defense will work together to resist the attacks. There are three patterns, namely combination pattern, trick, and cover. Combine the active electromagnetic interference and fire power to prevent discovered by the enemy and increase interception success rate. Utilize the abundant frequency resources and strong power to simulate the frequencies of the primary weapons. This trick will deceive the enemy electronic reconnaissance. Employ the electronic equipment and smoke discharge equipment to cover the air defense forces and prevent discovered by the enemy.

References

1. Liu F (2007) Air defense forces tactics. Military Science Press
2. Wang F, Li X (2008) Modern air defense. Aviation Industry Press
3. Li Z, Song H (2010) An analysis of air defense forces reconnaissance warning system. Military, 6
4. Wu W (2009) Research on the U.S. air raid and transformation of air defense forces. Haichao Press
5. Su W, Yao M (2010) Reflection on how to improve air defense forces reconnaissance warning capability under the complex electromagnetic environment. J Command, 1

Chapter 51
Discrimination Characteristics of Physical Determinants for Tactile Roughness Sensation of Fabric Surface Textures

Qun Zhao, Jiyong Hu, Xudong Yang and Xin Ding

Abstract The lack of understanding of the threshold on the basic physical factors forming rough sense of basic hinders the study of tactile roughness sensation. In this paper, differential threshold and Weber fraction are investigated for clearer understanding of how physical factors affect the roughness sensation by the constant stimulus method and the paired comparison method. The results show that the differential threshold for the mean deviation of surface profile is 0.86 μm and that the differential threshold of texture spatial period is 2.48 mm. As there is an interaction between the roughness indexes, any of the indexes alone cannot represent roughness sensation.

Keywords Fabric · Tactile roughness · Spatial period · Differential threshold · Weber fraction

51.1 Introduction

With the development of Internet and information technology, online shopping becomes the new trend and breaks the structure of retail industry. Nevertheless, online shopping confronts an immense challenge, i.e., how to make the customers feel the products genuinely which might affect their decisions, like tactile sense [1], particularly for the fabrics. Roughness sensation is one of the basic features constituting the tactile texture attributes space. Researchers are widely concerned

---

Q. Zhao · J. Hu (✉) · X. Yang · X. Ding
Ministry of Education, Key Laboratory of Textile Science & Technology (Donghua University), Shanghai, People's Republic of China
e-mail: hujy@dhu.edu.cn

J. Hu · X. Ding
College of Textile, Donghua University, Shanghai 201620, People's Republic of China

S. Long and B. S. Dhillon (eds.), *Proceedings of the 13th International Conference on Man-Machine-Environment System Engineering*, Lecture Notes in Electrical Engineering 259, DOI: 10.1007/978-3-642-38968-9_51,

on the roughness sensation [2]. This study focuses on the discrimination characteristics of inherent physical determinants for tactile roughness sensation of fabric surface, so as to guide the design of fabric tactile texture simulator.

Currently, some scholars have conducted researches on the discrimination threshold and Weber fraction of physical stimulus affecting material surface roughness sensation. Among the researches by directly touching the material surface with fingers, Louw found out the discrimination threshold of authentic Gaussian distribution surface (σ: 150–240 μm) is 1–8 μm, whatever raised or sunk surfaces. Estimated threshold of sine fence surface (spatial period 2.5–10 mm) is 0.64–4.99 μm [3, 4]. Nefs [5] has proved that the discrimination threshold of the sine grating surface amplitude by active dynamic touch can reach 2 μm (amplitude of reference surface is 12.8 μm, and spatial period is 2.5 mm). The amplitude resolution of surface with spatial period 2.5–10 mm increases proportionally to the spatial period.

The above researches are all about the hard materials surface, few about fabrics. In order to simulate the tactile sense of fabric roughness, the inherent physical stimulus attributes and psychological physical attributes for the roughness sensation should be firstly studied. We have discovered that the fabric roughness is related to the height and period of texture element in the previous researches. Moreover, considering the definition of stimulus pattern in the simulated tactile field, the surface topography will be firstly Fourier transformed into the spectrogram, and the peak and corresponding wave length will be selected to represent the fabric surface topography. Therefore, the four indexes are selected as the physical stimulus in this study. The discrimination threshold and Weber fraction of fabric roughness sensation are explored by the paired comparison method.

51.2 Experiment

51.2.1 Experimental Materials

Thirteen pieces of normal fabrics are selected for this experiment. Mean deviation of surface (SMD), maximum harmonic amplitude (MHA), harmonic wavelength (HW) of MHA, and texture spatial period (TSP) are taken as the physical stimulus indexes, as shown in Table 51.1.

51.2.2 Subjects

21 students, 8 males and 12 females, took part in the experiment, ages between 20 and 30.

Table 51.1 Test values of each of four physical roughness indexes

No	SMD (μm)	MHA (μm)	HW (mm)	TSP (mm)
1	3.69	5.21	1.4	3.03
2	3.35	3.61	1.51	1.84
3	2.08	1.98	1	0.39
4	4.14	8.36	2.2	2.35
5	2.81	3.93	0.85	0.89
6	2.39	1.8	0.83	0.39
7	1.68	1.7	0.63	0.56
8	7.22	8.86	1.97	4.38
9	3.58	2.94	0.78	0.78
10	1.79	1.32	0.69	0.37
11	2.73	3.5	0.82	1.57
12	2.6	2.02	0.71	0.59
13	2.16	1.59	0.57	0.39

51.2.3 Experimental Methods

First of all, the guide will rank the above 13 pieces of fabrics according to the stimulus SMD, MHA, HW, and TSP in the ascending order. Then, take one of the fabrics as the reference fabric and compare it with the others about the roughness sensation. The larger the ranking of the stimulus is, the strong the roughness is. Next, change another piece of fabric as the reference, in order to finally make sure any two of the fabrics have been compared for once. If the subject judges the same as the ranking regarding the degree of roughness, it is considered correct judgment and recorded as “1”, otherwise, as “0”. Draw the psychological physical measurement curve with the differences of each roughness stimulus as the x-axis and the correction rate of judgment as the y-axis. The discrimination threshold of roughness sensation can be calculated by the paired comparison method. And according to the difference of reference fabrics, Weber fraction is obtained.

51.2.4 Experimental Steps

Put the 13 pieces of fabrics into the constant temperature and humidity laboratory for over 48 h. Firstly, the guide ranks the 13 pieces by the stimulus SMD in the ascending order. Take the No. 7 fabric as the reference and compare it with the others one by one. The guide will record the judgment results. Make sure all the fabrics have been compared with the No. 7 reference fabric.

Then similarly, take the Nos. 4–10 fabrics as the reference one by one. Make sure all the 13 fabrics have been compared to each other and calculate the judgment correction rate.

51.2.5 Statistical Analysis

The differential SMD value is the difference of the compared fabric and the reference one. As SMD value is only tested in the above experiments, MHA, HW, and TSP will also be used to rank the fabrics. Because any two fabrics in the above experiments have been compared, we can judge if the comparison results obtained after the change of stimulus are correct or not, based on the results of SMD experiments. Besides, draw the psychological physical measurement curve for each roughness stimulus in the same way as the SMD. The discrimination threshold of roughness sensation can be calculated by the paired comparison method. And according to the difference of reference fabrics, Weber fraction is obtained. Most psychophysical researches think that the fitted curve of threshold is in the shape of S [6]. The following Boltzmann fitting formula is adopted,

$$y = \frac{A_1 - A_2}{1 + e^{(x-x_0)/dx}} + A_2 \tag{51.1}$$

in which A1 is the low y limit, A2 is the high y limit, x0 is the inflexion (half amplitude) point and dx is the width.

There are two kinds of psychological physical measurement curve. The first kind is that the influence of reference fabric is not considered. Take the physical stimulus differences as the x-axis and the corresponding judgment correction rate as the y-axis, so as to draw the curve. The second kind is that data points are divided according to the different reference fabrics, and for the same reference fabric, data are recorded in one diagram.

51.3 Results and Discussions

51.3.1 Psychophysical Measurement Chart of Roughness Without the Impact of Reference Fabric

Firstly, the influences of reference fabrics are not considered. The curves of stimulus and correction rate are shown in the following Fig. 51.1. The horizontal coordinates are the stimulus difference in the paired comparison tests. And the vertical coordinates are the judgment correction rate.

The diagrams in Fig. 51.1 show that the curves are all in the shape of S. Adopt the scattered points in the Boltzmann fitted curve to calculate the 75 % discrimination threshold of each stimulus. Heller [7] took the sand paper as the experimental object to study the roughness sensation and found that the subject can discriminate the surface of 2–3 μm sand diameter. Nefs found that the discrimination threshold of amplitude can reach 2 μm. In this study, the 75 % discrimination threshold of SMD stimulus is 0.86 μm, which is smaller than that of hard

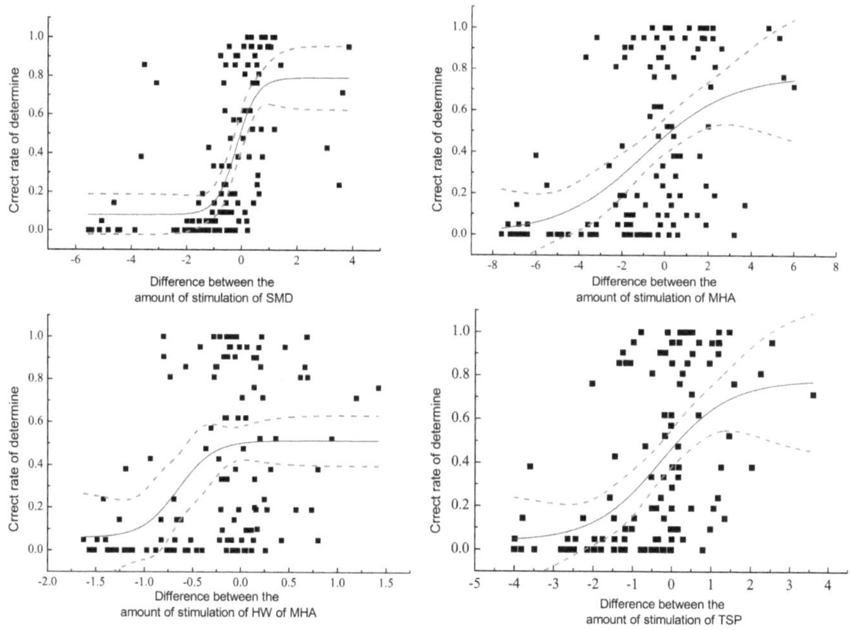

Fig. 51.1 Psychophysical measurement chart of roughness without the impact of reference fabric

material. This is because the scholars who studied the surface roughness of hard material have selected materials with larger amplitude than that of the materials used in this study.

Morley [8] found that regarding fence with 770–1,002 μm slot width, Weber fraction is 5 % and discrimination threshold is 38.5–50.1 μm. Nefs has found that regarding surface with 2.5-mm spatial period, Weber fraction is 6.4 % and discrimination threshold is 0.16 mm. In this paper, when taking TSP as the stimulus, its 75 % discrimination threshold is 2.48 mm, larger than that of the hard materials. As the stimulus selected in this study is not single factor variance, there are interactions between the stimuli, which lead to the above results. In addition to this, because fabrics used in this study have many other surface characteristics including thickness of thread, twist, and feather, it is not as sensitive as the hard material surface. Thus, the roughness threshold of fabric is larger than that of the hard materials.

As to the other two stimuli of single harmonic component, including the peak and corresponding wave length, the judgment correction rate is less than 0.75. And with the increasing stimulus difference, the correction rate tends to stay constant. This means that the 75 % discrimination threshold of the harmonic wave peak and corresponding wave length cannot be obtained.

51.3.2 Psychophysical Measurement Charts of Different Reference Fabrics

Take Nos. 2, 5, 6, 9, 11, 12, and 13 as the reference fabrics, respectively, and compare them with the others about the roughness sensation. Calculate the judgment correction rate and draw the psychological physical measurement curves; 75 % discrimination threshold is calculated by the fitted curve, as shown in Fig. 51.2.

From Fig. 51.2, we can see that 75 % MHA discrimination threshold is zero in the case of using No. 2 as reference fabric; and 75 % HW is also zero in the case of using Nos. 6 and 9 as reference fabrics. This shows that, to some extent, the two stimuli cannot describe the fabric roughness sensation correctly based on the macro-dimension of fabric.

Weber's law is correct in most cases in psychophysics. Weber fraction can be calculated by the discrimination threshold and stimulus value. In Fig. 51.3, with the increasing SMD, MHA, HW, TSP, the Weber fraction of each stimulus tends

Fig. 51.2 75 % differential threshold of physical stimulation

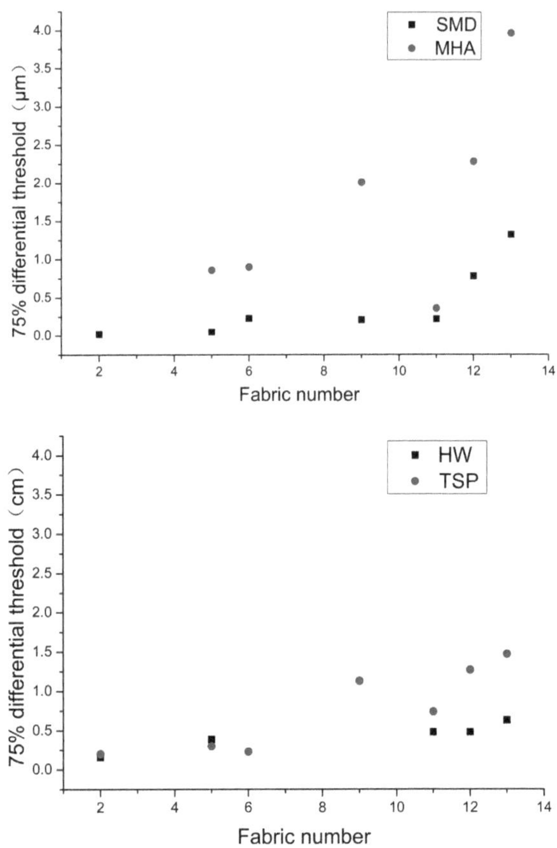

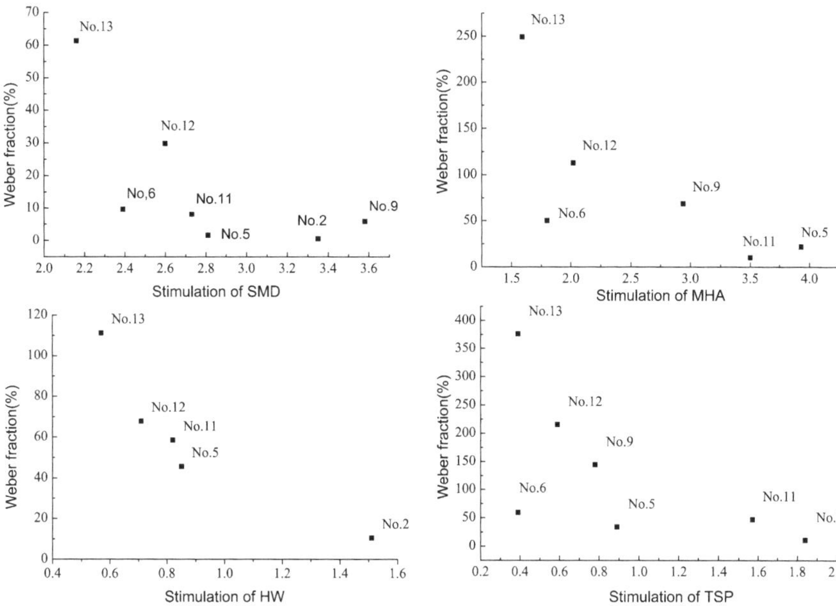

Fig. 51.3 Weber fraction of different fabrics' stimulus

to decrease. Researches about hard surfaces also show that Weber fraction changes little while the texture height increases. With the increasing texture spatial period, Weber fraction decreases from 11.8 to 6.4 % [5]. However, in this study of fabrics, Weber fraction scatters widely, some even reaching 3.70. When taking texture spatial period as the stimulus, the Weber fraction of No. 13 fabric is largely different with that of the hard surface. There is one common point of the four diagrams, i.e., Weber fraction is very high when using Nos. 12, 13, and 9 as the reference fabrics. Reasons might be:

1. When taking one piece of fabric as the reference, data amount is small and it is difficult to fit the psychological physical curve accurately, resulting in the abnormal discrimination threshold and Weber fraction.
2. Because the stimulus of Nos. 12 and 13 fabrics is small and 75 % discrimination threshold is large, Weber fraction is relatively large. Although the two fabrics have low ranking of medium height, they feel granular strongly and make the subjects think they are rough. For this kind of fabric, according to the researches of physiology, when the texture spatial period of hard material surface is less than 200 μm, the roughness sensation is determined by the vibration sensed by the PC receptor. High frequency can stimulate PC receptor to release action potential within certain range and sense the roughness [9–11].

For Nos. 9, 12, 13 fabrics, stimulus generating the roughness sensation might not be the four adopted in this study.

51.4 Conclusions

This paper has explored the physical stimulus related to the fabric roughness sensation and Weber fraction by the paired comparison experiments. The differential threshold of fabric surface roughness is 0.86 μm. And the differential threshold of texture spatial period is 2.48 mm. These are different with the hard metal surface, which is caused by the various features of fabric. Any single roughness index cannot represent the fabric roughness sensation, because these indexes interact with each other.

Acknowledgments This study was funded by NSFC (No. 51175076), Shanghai Natural Science Fund (No. 12ZR1400500), and the Fundamental Research Funds for the Central Universities.

References

1. Hollins M, Bensmaiea SJ (2007) The coding of roughness. Can J Exp Psychol 61(3):184–195
2. Picard D, Dacremont C, Valentin D et al (2003) Perceptual dimensions of tactile textures. Acta Psychol 114(2):165–184
3. Kantowitz BH, Roediger III HL, Elmes DG (2008) Experimental psychology. Wadsworth Publishing Company, 167–198
4. Louw S, Kappers AML, Koenderink JJ (2000) Haptic detection thresholds of Gaussian profiles over the whole range of spatial scales. Exp Brain Res 132(3):369–374
5. Nefs HT, Kappers AML, Koenderink JJ (2001) Amplitude and spatial-period discrimination in sinusoidal gratings by dynamic touch. Perception 30(10):1263–1274
6. Leek MR (2001) Adaptive procedures in psychophysical research. Atten, Percep, Psychophys 63(8):1279–1292
7. Heller MA (1982) Visual and tactual texture perception: Intersensory cooperation. Atten, Percep Psychophys 31(4):339–344
8. Morley J, Goodwin A, Darian-smith I (1983) Tactile discrimination of gratings. Exp Brain Res 49(2):291–299
9. Bueno MA, Lamy B, Renner M et al (1996) Tribological investigation of textile fabrics. Wear 195(1–2):192–200
10. Bolanowski SJ, Gescheider GA, Verrillo RT et al. (1998) Four channels mediate the mechanical aspects of touch. J Acous Soc Am, 84:1680–1694
11. Hollins M, Risner SR (2000) Evidence for the duplex theory of tactile texture perception. Atten, Percep Psychophys 62(4):695–705

Chapter 52
Design and Development of Novel Ventilated Clothing

Xinmei Guo, Xingjuan Zhang and Xiugan Yuan

Abstract In the modern industry, operators have to face more and more extreme high-temperature environment. There is an urgent need to study how to reduce heat stress effectively. Air cooling system is an important way to reduce heat stress, but it is seldom used in general industry with high-temperature environment. In this study, a novel tube-style ventilated clothing with front opening and unsealed ventilation was designed and developed. The novel ventilated clothing has the characteristics of wearing easily, portability, small resistance, and better ergonomics, which can be integrated into various operation modes to solve the heat stress for workers. This helps air cooling system to be more widely applied in high-temperature environment.

Keywords Ventilated clothing · Heat stress · Clothing structure · Design · Development

52.1 Introduction

Whatever in the summer heat or workshop industrial high temperature, operators cannot work normally and their health and ergonomics will be severely affected under the heat stress [1, 2]. Methods for alleviation of heat stress include artificial control of working conditions, adjustment of labor intensity, and thermal acclimation to high temperature. These methods can alleviate the heat stress to some

X. M. Guo
School of Fashion Art and Engineering, Beijing Institute of Fashion Technology, Beijing 100029, China

X. J. Zhang (✉) · X. G. Yuan
Institute of Man-Machine-Environment, Beihang University, Beijing 100191, China
e-mail: xingjuanzhang@163.com

S. Long and B. S. Dhillon (eds.), *Proceedings of the 13th International Conference on Man-Machine-Environment System Engineering*, Lecture Notes in Electrical Engineering 259, DOI: 10.1007/978-3-642-38968-9_52,

extent; however, the working conditions cannot be controlled by human in many cases, like outdoor high temperature or metal smelting and casting factories. The most effective way to solve this issue is to adopt personal cooling equipment.

Personal cooling method is to cool the microenvironment between the skin and exterior clothing and bring out the heat load, thus cooling the operator. There are three kinds of personal cooling system, namely air cooling, liquid cooling, and phase changing, according to the use of coolants. Due to the technical difficulties and high costs, research of personal cooling system started late and the devices are applied in a small range. Currently, they are mostly used in the national defense, like aviation and aerospace, and seldom used in other industries [3]. This study has designed and developed a ventilated clothing and conducted evaluation on its ventilation effects. This study aims to extend the application of personal air cooling system in the high-temperature industries and solve the heat issue for workers.

52.2 Objectives

Personal air cooling system is comprised of ventilated clothing and ventilation device. Ventilation device brings in natural wind or air cooling wind. Ventilated clothing distributes the air between the underwear and exterior clothing throughout the whole human body surface. Air convection brings out the heat load and prevent the body temperature rising.

The structure of ventilated clothing directly influences the air distribution performance as well as the heat transfer efficiency. Ventilated clothing is usually divided into sealed and unsealed types. The sealed ventilated clothing has its branch tubes going throughout to the distal end of the body, including the limbs and head. Then, the air from the distal end of the body flows back to the waist or chest, and outflows through the manifold [5]. This type of ventilated clothing is more widely applied in engineering, for example the space suit [6]. However, sealed ventilation clothing is costly and difficult to match with other working uniforms, thus uneasy to be applied in the normal high-temperature environment. The unsealed ventilation clothing distributes the air to the body surface, and the air outflows from the openings of clothing naturally. This study is to design and develop an easy-to-wear and highly adaptable unsealed ventilated clothing, for the solution of high-temperature issue in the normal industries.

52.3 Research Methods

52.3.1 Material Selection for Ventilated Clothing

Novel ventilated clothing comprises of three layers, namely exterior, ventilation pipe network, and lining. The exterior fabric should protect the ventilation clothing and human body as a protective layer, which is selected based on the different working conditions, such as in the case of radiation, heat radiation protection exterior should be selected. Air and moisture permeable materials with high heat resistance are good for decreasing heat load and increasing the cooling efficiency if without any special requirements. In this study, without specific to some kind of working condition, the comfortable and good-looking polyester–cotton camouflage fabric is selected. The lining is warp-knitting Dacron fabric with 2-mm eyelets. It is soft, elastic, well-ventilated, and comfortable, easy for the sweat going out and without blocking the wind from the ventilation pipe network. For the ventilation pipe network in the middle, flexible and soft PVC pipe should be adopted. Pipe diameter of the distributing manifold at the waist is 10–18 mm, and the branch pipes are 6–8 mm [4, 7]. In addition, ventilated clothing is not good to wear directly over the human skin. A piece of underwear should be worn, like cotton T-shirt or vest. This is to prevent discomfort if ventilation air directly blows on the skin on the one hand, and on the other hand, the cotton underwear can absorb the sweat and beneficial to transport of the evaporation latent heat.

52.3.2 Style Design of Ventilated Clothing

Style design of ventilated clothing lies with the pattern, namely whole body, half-body or vest, or ventilation hat, which style is selected depending on the working conditions and body parts necessary for cooling. Various factors including body surface area, physiological properties, heat transfer characteristics, clothing structure, and working conditions should all be considered during the design.

Convection is the main form of heat exchange between the ventilation air and the body. The larger the area air goes through, the better the cooling effect is. From this perspective, whole-body style is the best. However, this type of clothing is complicated in design and has severe impact on operator ergonomics. Moreover, it is difficult to be compatible with other operating uniforms so that cannot be widely used. Therefore, it is not advised of using this whole-body type unless in the special conditions. The novel ventilated clothing in this study was designed based on thoughts of the half-body style, easy to wear, and to be easily popularized in the normal industries.

The area ratios of each body part for the Chinese adults are as follows: head 9 %, upper limbs 18 %, lower limbs 36 %, and trunk 37 % [8]. Head and upper limbs take up the small percent and they move often when working. Hence, considering from the cooling area, the ventilated clothing covering the lower limbs and trunk can cool the most parts of the body and meantime reduce the impact on the working efficiency to the minimum. Besides, the heat transfer property of human body relates with the basic metabolism, heat capacity, blood flow rate, thermal conductivity, and distribution characteristics of sweat glands. Each body part has different cooling efficiency, and thighs have the lowest cooling efficiency [9]. Besides, in many working conditions, thighs have to move regularly. Based on the above analysis, the novel ventilated clothing will be in the pattern of vest, both with high cooling efficiency and with working efficiency. For the convenience of wearing and taking off, it is designed with front zipper. The collar, armhole, and hem are loose style for the exhausted cooling air to cool the neck, upper limbs, and thighs again at a certain extent [10].

52.3.3 Design of Ventilation Pipe

The sketch of ventilation pipe distribution is drawn based on the vest pattern structure. Pipes structure is comprised of the distributing manifold and the branch pipes. For this easy-to-wear ventilated vest, the distributing manifold is set at the hem and air supply main pipe is at the middle of back hem; 12 pieces of the branch pipes go out of the distributing manifold from the front to the back (see Fig. 52.1). *f*1–*f*6 are the six the branch pipes at the front and *b*1–*b*6 at the back. Each the branch pipe is sealed at the end. Ventilation holes of 1–2 mm diameter are set at the branch pipes at the side close to the skin, every other 3–4 mm. Cooling air

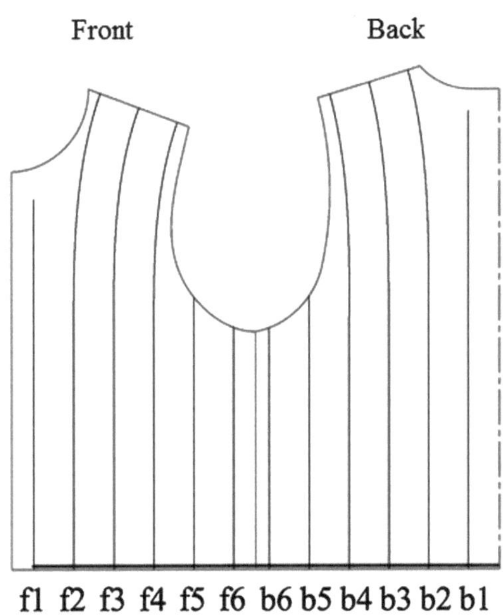

Fig. 52.1 Ventilation pipe distribution

blows out from the holes onto the underwear and then flows out from the fabric gaps and the openings of the ventilated clothing, such as collar, armholes, and hem.

However, we found that pipes f5-f6, and b6-b5 rubbed with the armpits and upper arms when body exercised with ventilated clothing. In order to improve this issue, the pipes f5 and b5, f6 and b6 were connected together. In addition, the pipes f2 ~ f4 and b2 ~ b4 interfered with each other so that it was not conductive to sewing at the shoulder and keeping clothing in good shape. Thus, f2 and b2, f3 and b3, f4 and b4 were also connected together separately (see Fig. 52.2). Two pipes at front center and back center respectively were independent with sealed ends.

Bracing bars are used on the inner side of pipes at the shoulder to prevent pressure from exterior working uniforms or the weight of ventilated clothing itself, which might block the ventilation holes on the branch pipes. There is no need to add bracing bars for the other parts. The overall design pattern is shown in Fig. 52.3.

Fig. 52.2 Ventilation pipe network wearing

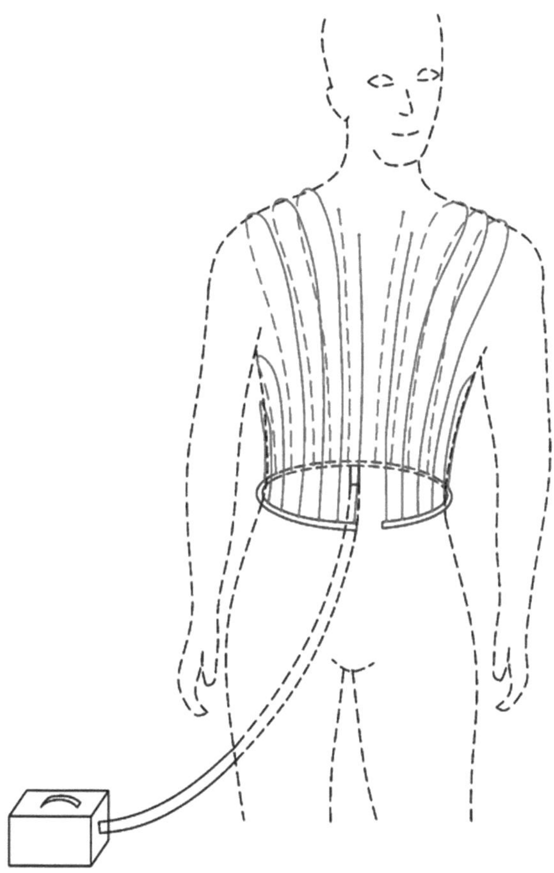

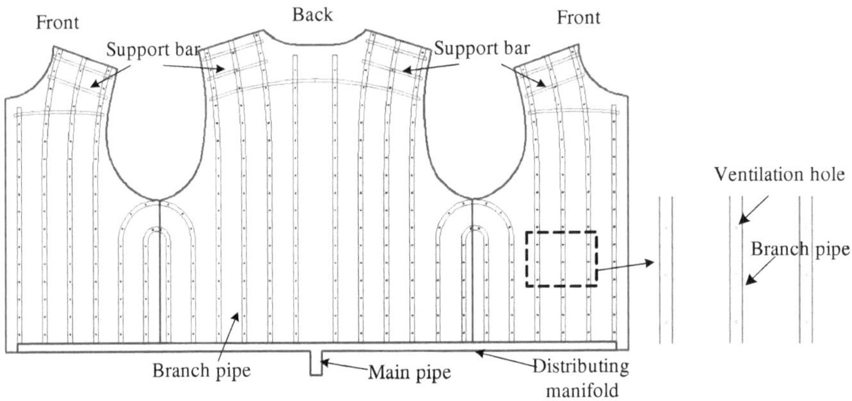

Fig. 52.3 The overall design of the ventilated clothing

52.3.4 Development of Ventilated Clothing

First of all, cut the exterior and lining fabric according to the design pattern. Sew the side seams under oxter and shoulder seams. Stitch the exterior and lining fabric together at the front middle position. Leave the arm hole and hem unstitched, for the convenience of putting into the ventilation pipe network.

Secondly, prepare the distributing manifold and the branch pipes according to the design sketch. And assemble them together by tri-branch tube or plastic welding.

Before combining the ventilation pipe network and clothing fabric, adjust the pipe network for the comfort and suitability, because pipes are not soft as fabric. After putting on the pipe network, adjust them to feel comfortable and match with the body shape (see Fig. 52.4). During adjustment, fix the branch pipes with coarse cotton lines. Meanwhile, check and adjust the length whether the pipes affect the neck or arms activity. Take off the pipe network gently when finish adjustment, then drill ventilation holes on the inner side of the branch pipes and fix the bracing bars on the shoulder as designed.

Finally, assemble the ventilated clothing of exterior fabric, pipe network, and lining together. Put the pipe network between the exterior and lining fabric. Fix them at some key positions, like shoulder, front edge, back center, and oxter. Firstly, stitch the exterior fabric and lining at the arm hole. Then let subjects to try on the ventilated clothing. If no other problems, stitch the exterior fabric and lining at the hem. Final product is shown in Fig. 52.5.

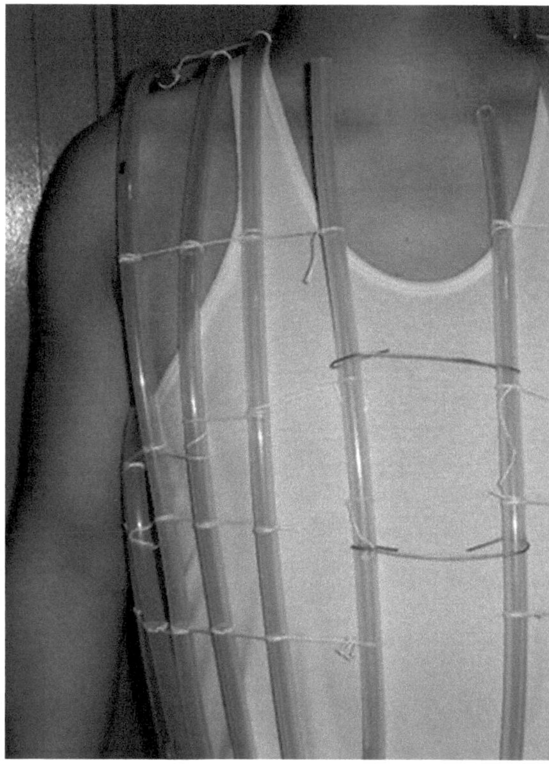

Fig. 52.4 Ventilated pipe network adjusted

52.3.5 Evaluation of Ventilation Resistance

Ventilation resistance is a key index for assessing the performance of a ventilated clothing. The smaller the ventilation resistance is, the better the performance is. Large resistance will increase power consumption for the air supply device and also enlarge the volume and weight of the air supply device. The literature [11] regulates that the ventilation resistance should not exceed 1.47 kPa when the ventilation volume rate is 250 L/min. The ventilation resistance of this novel ventilated clothing was tested, and results are shown in Fig. 52.6. It can be seen that the resistance of this novel ventilated clothing is less than the range of national standard.

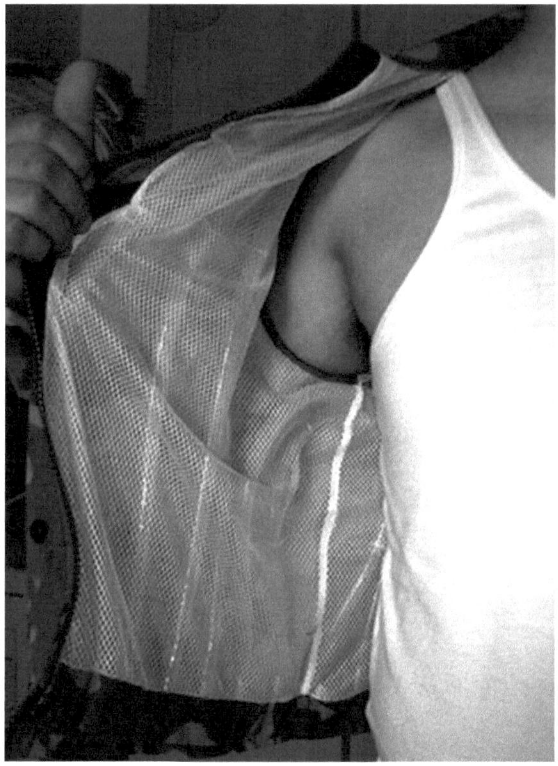

Fig. 52.5 The impression of ventilated clothing

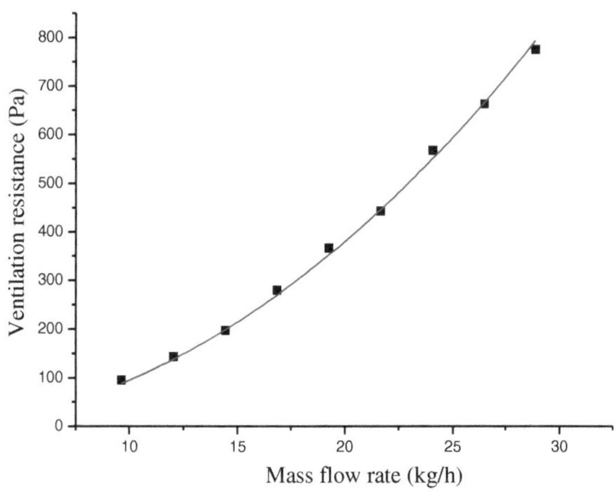

Fig. 52.6 Ventilated clothing resistance testing results

52.4 Conclusions

Personal cooling equipment is an effective way to solve the heat stress in the high-temperature working environment. The study has designed and developed a novel unsealed tube-style ventilated clothing and analyzed its structure and pipe network design. Finally, the sample clothing is tested on the ventilation resistance. This novel ventilated clothing has the following features:

1. Easy-to-wear pipe network structure. Considering the body shape and clothing structure, the branch pipes are connected from the front to the back, making it easy-to-wear clothing.
2. Convenient and compatible. The ventilated clothing is designed as unsealed vest style with front zipper, which is portable, lightweight, easy to wear and take off. It is very easy to match with other uniforms in various working conditions.
3. Small ventilation resistance. Ventilated clothing with small resistance can be combined with small, portable, and low-energy consumption air supply device, so that it can be widely applied.
4. Better working efficiency. This ventilated clothing pays attention to body parts with high cooling area and efficiency. Limbs and neck are not restrained and working efficiency guaranteed.

References

1. Parsons KC (2003) Human thermal environments. Taylor & Francis, London
2. Nunneley SA, Stribley RF (1979) Fighter index of thermal stress: guidance for hot weather aircraft operations. Aviat Space Environ Med 50:639–642
3. Chen X, Yuan X (2000) Physiological basis on system engineering of man-machine-environment. BeiHang University Press, Beijing (Ch)
4. Ou Y (1997) Development of aviation and aerospace ventilated clothing. Int Aviat (4):59–60(Ch)
5. Frim J, Bramalea (1993) Micro-climate conditioning clothing. US Patent No. 5243706
6. Guo X, Yang C, Yuan X (2010) Analysis on some key techniques of air cooling personal heat protective system. Space Med Med Eng 23(4):262–266
7. Liu C, Fang R (1994) Ventilated clothing. China Pers Prot Equip 3:10–14
8. Siguang J (1989) Hypoxia and oxygen of aeronautics and astronautics, physiology and protective equipment. People's Military Medical Press, Beijing (Ch)
9. Daixiu C, Xianzhang W (1994) Physiological basis and medical requirements for the design of liquid-cooled garment equipment 7(3):223–228(Ch)
10. Mekjavic IB, Banister EW, Morrison JB (1988) Environmental ergonomics. Taylor & Francis Ltd, London
11. GJB 2642-96 (1996) General specification of ventilation suit (Ch)

Chapter 53
Experimental Study of Inattentional Blindness Under Altitude Environment

Yajuan Bai, Yuping Luo, Heping Wang, Guansheng Huang, Yaofeng He and Haiyan Niu

Abstract With the purpose to study the occurrence probability of the inattentional blindness at high altitude, inattentional blindness experiments at high altitude are carried out in this paper, and the physiological and psychological changes of subjects are monitored at the same time, such as heart rate, degree of blood oxygen saturation and so on. The results show, in the plateau of low-oxygen environment, inattentional blindness more likely happen, meanwhile, subjects physiological and mental state changed a great deal. The reason may be that narrow scope of people's attention or severe damage to people's short-term memory. In the vehicle visual information interface design, people's cognitive characteristic changes in the plateau of low-oxygen environment must be considered.

Keywords Inattentional blindness · Plateau · Man-Machine interface

53.1 Introduction

Inattentional blindness refers to that people cannot see some evident and distinct objects when they focus on certain object or event [1].

This term comes from the book *Inattentional Blindness* published in 1998 written by Mack and Rock. They got the result that "without attention, without consciousness" [2]. Be noted that "consciousness" refers to explicit conscious awareness, instead of subliminal and unconscious or implicit cognition.

This phenomenon happens not only in the laboratory, but in the real world. For example, we did not recognize our friend when we were walking by the friend, or

Y. Bai (✉) · Y. Luo · H. Wang · G. Huang ·
Y. He · H. Niu
Beijing Special Vehicle Institute, Beijing 100072, China
e-mail: zhjbjs@163.com

S. Long and B. S. Dhillon (eds.), *Proceedings of the 13th International Conference on Man-Machine-Environment System Engineering*, Lecture Notes in Electrical Engineering 259, DOI: 10.1007/978-3-642-38968-9_53,

we cannot hear someone else calling us when watching TV. Actually, inattentional blindness might cause serious damages, like car accidents and air crash [3, 4].

Green thinks that there are four factors, namely conspicuity, mental workload, stereotype, and expectation, as well as process capacity [5], also known as the four factors model by Gu et al. [6].

Current researches of inattentional blindness are conducted according to the four factors model. In terms of conspicuity, stimulus location, color, movement, and stimulus meaning are studied for the possibility of inattentional blindness. From the perspective of mental workload, researches are carried out by considering conscious load and work memory. While considering the stereotype factor, researches are on the stereotypes and expectations. Soberness and age are studied in terms of the process capacity.

However, all the above researches are conducted at the normal conditions, how about the special conditions? This paper aims to explore the possibility of inattentional blindness in the classic experiments at the plateau environment for the unacclimatized subjects and find the primary factors.

53.2 Experimental Methods

53.2.1 Subjects

20 males around 20–35 are selected randomly from the group for plateau vehicle performance tests. They all take the medicine of anti-altitude reaction. The subjects have no evident altitude reaction at the plateau and can take part in the experiments. Their vision or corrected vision is normal without color blindness or weakness. They are the first time taking part in the inattentional blindness experiments.

53.2.2 Experimental Materials

The classic experiment of inattentional blindness designed by Simons is adopted. It is a 75-s video. In the video, two groups each with three people are playing basketball, one group in white and another in black. They are passing the basketballs in the way that player 1 to player 2, player 2 to player 3, and player 3 to player 1. After 44 s, a man with chimpanzee clothing walks from the left to the right and thumps his chest. See Fig. 53.1.

Fig. 53.1 Flow chart of vision attention model

53.2.3 Experimental Methods

The subjects will be explained about the experiment before watching the video. The subjects will be told that two groups are playing basketball, each with three people, and in white and black. The subjects are required to count the number of passing the ball by the white group. After watching the video, the subjects need to write down the number immediately. Then, the following questions need to be answered:

1. Have you noticed any special things in this video?
2. Except the people, have you seen any others?

3. Have you seen anybody else?
4. Have you seen the chimpanzee? If yes, specify the details.

Finally, watch the video again and tell the subjects the answers.

53.2.4 Experimental Results

1. The 20 subjects have counted the ball-passing number between 12 and 15 and the correct count takes up 15 % (Table 53.1).
2. All the subjects answered that they have not noticed any special people or event in the first three questions.
3. Only one subject answered that he has seen a black shadow, but cannot specify the details.

53.3 Physiological and Psychological Influences on the Subjects at the Plateau with Low Oxygen

53.3.1 Test of Cognitive Ability

Adopt the Montreal Cognitive Assessment table [7] to investigate the cognitive ability of the subjects, both at the plateau and plain. Use SPSS 13.0 software to analyze the data and check the standard normal distribution. Compare the difference of cognitive ability at both conditions. Adopt the independent sample to inspect.

Results analysis:

The MoCA score for both conditions are, respectively, 26.0 ± 1.5 and 25.8 ± 1.4, without evident difference ($P = 0.07 > 0.05$). Besides, the sub-investigation items, including visual space, executive function, name, language, abstract, delayed memory, and directional function, have no differences ($P > 0.05$). But the attention at both conditions are evidently different ($P = 0.04 < 0.05$).

Table 53.1 The number of passes recorded by subjects

	12 times	13 times	14 times	15 times
Num.	2	7	8	3

53.3.2 Test of Sleep Quality

Adopt the Pittsburgh Sleep Quality Assessment Table to compare the sleep quality at the plateau and pain.

Use the Pittsburgh Sleep Quality Index (PSQI) to assess the sleep quality of the subjects. PSQI comprises of 19 self-assessment and 5 peer assessment. The 19th self-assessment item and 5 peer assessment items do not score. The 18 self-assessment items will be introduced here. The 18 items are in 7 groups, each group with 3 grades. Accumulate all the groups to score the PSQI (0–21). The higher the score is, the worse the sleep quality is.

The total score of PSQI ($x \pm s$) is regarded as the standard of sleep quality. In plateau, 15 % of the subjects have good sleep (PSQI $\leq$ 4) and 85 % bad sleep (PSQI $\geq$ 1.1). In plain, 20 % of the subject have bad sleep (PSQI $\geq$ 1.1), while have 80 % good sleep (PSQI $\leq$ 4). The high-altitude low-oxygen environment has evident influences on the sleep quality.

At the high altitude, the subjects score the most at the item of sleep latency, daytime dysfunction, and subjective sleep quality.

53.3.3 Test of Degree of Blood Oxygen Saturation and Heart Rate

Adopt the US oxyhemoglobin saturation tester (Onyx-9500) to test the oxygen saturation of the subjects (SaO_2). At the plain, it is (98.15 $\pm$ 2.87) %, and at the plateau, it is (82.6 $\pm$ 3.28) %.

Adopt the US dynamic electrocardiogram recorder (Elite-3A) to test the heart rate of the subjects. At the plain, it is (79 $\pm$ 12) times/min. and at the plateau, it is (98 $\pm$ 21) times/min.

53.4 Discussions

1. Physiological and psychological conditions change greatly of the subjects at high altitude.

The degree of blood oxygen saturation and heart rate change at the high-altitude environment, which indicates the subjects make the stress response via the body compensatory function. There is no evident altitude reaction.

2. Inattentional blindness tends to happen at high altitude.

Foreign researches show that the possibility of inattentional blindness at normal conditions is 50 %.

This method is also used in this study and the passes recorded are within the effective range. Although the samples are less, the validity can be guaranteed. There is nobody noticing the chimpanzee. So, inattentional blindness tends to happen at high altitude.

3. Analysis of inattentional blindness

There are two explanations about inattentional blindness. One is failure of consciousness; the other is failure of memory. The first view thinks that if the observer does not give attention to the unexpected stimulus, the stimulus can never be noticed. The second view thinks that the subject might have noticed the object and be aware of it, but forget it when asked.

Foreign researches on the influence of high altitude having on the human cognitive ability show that low-oxygen environment have big influences. The main reason is that lack of oxygen will affect the normal metabolism and function of the brain although the body compensatory function can alleviate to some extent. Thus symptoms like mental disorder, distraction, forgetfulness, inertia of thinking and movement, low efficiency, and easy to fatigue emerge.

In the experiment of inattentional blindness, the subjects are required to record the passes for the people in white. The reason of inattentional blindness is that they are overly focused and neglect the evident changes. However, the attention of the subjects is distracted at the plateau. If the reason of inattentional blindness is overly concentrated on the objects, then the possibility of inattentional blindness tends to decline in the high-altitude environment.

The experimental results show that the possibility of inattentional blindness increases obviously. The explanation is that the attention range on the high altitude is narrowed or the short-term memory capability falls greatly.

4. The design of human-machine interface of the vehicles should consider the inattentional blindness at high altitude.

At high altitude, as people have narrowed attention range and bad short-time memory, they tend to have inattentional blindness. The design of vehicle human-machine interface, especially the visual display, should take the changes of cognitive ability into consideration. The visual information amount should be controlled, so as to lower the possibility of inattentional blindness.

5. Further researches are necessary regarding the inattentional blindness at high altitude.

This study incorporates the special condition of high altitude into the traditional researches of inattentional blindness, thus making it closely connected to the engineering applications of vehicle visual displays design. Due to the small amount of samples tested and only the unacclimatized subjects considered, large amount samples researches should be conducted in the future and the acclimatized people should be considered, so as to genuinely reflect the actual influences of high-altitude environment having on the possibility of inattentional blindness.

References

1. Simons DJ (2000) Attentional capture and inattentional blindness. Trends Cognitive Sci 14(4):147–155
2. Mack A, Rock I (1998) Inattentional blindness. MIT Press, Cambridge
3. Spence C, Read L (2003) Speech shadowing while driving on the difficulty of splitting attention between eye and ear. Psychol Sci 14:251–256
4. Haines RF (1991) A breakdown in simultaneous information processing. Plenum Press, New York, pp 171–175
5. Green G (2004) Inattentional blindness and conspicuity. Retrieved 10 Nov 2004. http://www.visualexpert.com/Resources/inattentionalblindness.html
6. Gu E, Stocker C, Badler NI (2005) Do you see what eyes see? Implementing inattentional blindness. Lect Notes Comput Sci 3661:178–190
7. Liu X (1999) Manual of mental health assessment scale. Chinese Mental Health magazine, Beijing, pp 375–378

Part VI
Research on the Machine-Environment Relationship

Chapter 54
Development of a Centralized Lubrication System for Vehicle Chassis

Qing Gao

Abstract Traditional auto-chassis lubrication is done manually for individual parts. The amount of lubricant to use and the cycle of service are difficult to determine. Other shortcomings include the high work load, long service and maintenance time, and the difficulty to check the lubrication condition. The automatic, centralized auto-chassis lubrication systems, on the other hand, can effectively extend the lifetime of the auto-chassis, reduce the chassis maintenance cost, and enhance the overall performance and quality level of the vehicles. This paper, by describing and reviewing the principle and implementation methods of the automatic centralized lubrication systems for auto-chassis, demonstrates that the new technology can effectively solve many problems in the manual lubrication, improve the technical conditions of the chassis, and have great economic values.

Keywords Vehicle chassis · Centralized lubrication system · Chassis lubrication · Automatic lubrication

54.1 Introduction

The vehicle-concentrated lubrication technology came originally from Germany. Since the late twentieth century, China has imported many high-grade commercial vehicles from Europe that are equipped with centralized lubrication system. After more than 10 years of development, the centralized lubrication technology has become widely used in China; and the domestic-vehicle-concentrated lubrication technology has already become the mainstream of the Chinese market [1].

Q. Gao (✉)
Shandong Jiaotong University, Jinan, China
e-mail: qing917@126.com

S. Long and B. S. Dhillon (eds.), *Proceedings of the 13th International Conference on Man-Machine-Environment System Engineering*, Lecture Notes in Electrical Engineering 259, DOI: 10.1007/978-3-642-38968-9_54,

Considering the popularization and application of centralized lubrication system for vehicles in China in recent years, it is necessary to discuss and review the status, problems, and innovative developments of the technology.

54.2 Centralized Lubrication System for Vehicle Chassis

Chassis lubrication is a process that requires accurate and reliable supply of clean lubricant to the vehicle parts that need lubrication at particulate time and under specific driving conditions [2]. It aims at reducing friction resistance and mechanical wearing, cleaning, and cooling the part surface, and providing better sealing and antirust at the same time.

Automobile is composed by many parts, the quality of each moving parts of lubrication condition decides the car performance, running condition, and its parts service lifetime [3]. The parts without timely and well-controlled lubrication will wear quickly and cause all kinds of problems of the vehicle; On the other hand, excessive lubrication will generate unnecessary heat and waste and cause more pollution. Therefore, only well-controlled lubrication of vehicle chassis will maintain the good lubrication conditions for the vehicle without negative environmental impacts.

To ensure normal conditions of all vehicle parts, a lubricating system should meet the following requirements:

(1) Supply required amount of lubricant at given pressure and ensure uniform application for each lubricated part. The supply must be sufficient and can be adjusted based on demand.
(2) Use high-efficiency sealing and filtering device to maintain the cleanness of lubricant, prevent the dust and moisture from entering the system, and to prevent environmental pollution caused by leakage.
(3) The system should be simple, low cost, and standardized, which will make it easier for inspection, maintenance, lubricant refill, and quick adjustment.
(4) When lubrication system needs to assure proper working temperature, preheating, or heat exchange device can be easily added.
(5) Adequate display with system condition monitoring and alarm protection that can identify potential problems in time.

For the lubrication of entire vehicle, the chassis lubrication plays a pivotal role. However, for a long time, the lubrication of vehicle chassis has been overlooked. It has been even rare to investigate the problem scientifically and rationally. In the domestic automobile chassis, it has been found that many moving parts showed early severe wear, high rate, and shortened lifetime, which are all related to inadequate chassis lubrication.

54.3 The Advantage of Centralized Lubrication System for Vehicle Chassis

The centralized lubrication system for vehicle chassis is a closed system that can achieve timing, quantitative to filling grease for each chassis friction pair and make the friction pair minimized wear so that can be use in relatively long time, and ensuring the friction pair keeping good work condition, so as to extend the service life and improve the performance of the vehicle.

Compared with manual lubrication, adopting centralized lubrication system will allow to reduce the amount of oil at one-time injection, reduce the oil injection period, and achieve the timing and quantitative injection oil lubrication [4].

The centralized lubrication system for vehicle chassis by oil pump or lubrication device sends lubricant to all lubricating points through the style of mandatory, through each node to lubricate at timely right amount, continuous “get”, keep the oil film is pure and fresh and the best lubrication state of friction pair. Lubrication device is supply system, and the lubricating points to the amount of oil are rare, so the oil pump working time is short, generally less than 10 s; and oil pump rest time is longer, generally greater than 30 min. If the amount of oil to friction pair is larger, improve with the quantitative flow of oil, not by improving the oil supply frequency. This will prevent lubrication oil supply too much and caused oil waste, pollution of the environment.

The centralized lubrication system for vehicle chassis has the following advantages:

(1) Timing and quantitative automatic, according to the need to adjust the fatliquoring time interval and. punctually, automatically, and quantitative to add fat.
(2) Accurately to add fat. According to the different friction pair, each lubricating points of automatic lubrication system of concentrated fatliquoring choose different amount of oil distributor to set, avoid the fatliquoring too much or too little cause the adverse conditions.
(3) The lubrication is reliable. Since each lubricating points with the hose and system to together and ensures that all lubricating points have a certain amount of grease, avoid the possible situation of leaking or difficult to access to parts by manual operation.
(4) Through the closed and automatic precise, lubricating can improve vehicle performance grades, solve a series of problems of manual periodically lubricated, improve the rate and prolong the service life of the chassis, improve the chassis technology situation, and improve the vehicle performance.
(5) Reducing parking time of maintenance and improving the labor productivity. Because the system of fatliquoring is automatic operation in the automobile run, so reduce the parking time of maintenance, so as to increase the productivity of labor.

(6) Reducing the workload and operating more security. Due to the process of centralized lubrication system is automatic operation in the automobile running, greatly to reduce labor workload and avoid the unsafely and unhealthy work under the car.
(7) Lubrication condition is easily to check. At the same time, there is advanced microprocessor control device monitoring the entire system and store data, such as the system appear any breakdown, controller will be issued a warning signal, easy to find out lubricating points fault.
(8) Reducing parts wear and saving grease. In auto-centralized grease lubrication system, the friction pair can get reliable lubrication and reduce parts wear. And according to the need of the friction pair, compared with artificial fatliquoring, it saves grease.
(9) Compact structure, easy installation, simplify the vehicle chassis maintenance procedures, reduce maintenance costs, and ensure the larger economic benefits [5].

Automobile chassis automatic centralized lubrication system as the vehicle important device at saving time, saving work, prolonging the service lifetime, the technology has been widely accepted and adopted.

54.4 The Working Principle and Control Method of Centralized Lubrication System for Vehicle Chassis

54.4.1 The Working Principle of Centralized Grease Lubrication System for Chassis

The main working principle of centralized lubrication system is by car batteries to power sources, through the break of ignition switch to calculate engine work running time. When the engine running time reaches the electronic monitors of injection interval time, motor received command from the monitors instructions, drive the oil pump starting rotation, produce vacuum and suck fat; grease was sent to export by pump pressure and quickly get through the unloading valve with the oil, oil distributor change into the l storage conditions, the lord of the high pressure compress the spring of the oil distributor, make its oil and energy store; when the inside pressure of main oil rose to pressure of switch closing pressure, electronic monitors sent a halt instruction to oil pump motor, and the oil pump stops; unloading valve to be automatic unloading, lord of the hydraulic system tubing falling rapidly, oil distributor changes into supply conditions, spring will send storage oil of oil distributor to all lubricating points [6].

According to the friction pair load and the size of the relative velocity of lubrication points, with the oil in the plunger displacement of oil supply determined quantitatively. The hydraulic principle diagram of lubrication system is as shown in Fig. 54.1 [7].

54.4.2 The Control Method of Centralized Lubrication System for Vehicle Chassis

The control method is currently using single chip microcomputer and hydraulic control of combining the quantitative fatliquoring timing. The system through set pressure sensor to control the main lines pressure and automatic adjustment of the pump operating time, through oil distributors achieved quantitative storage and transport grease to all lubricating points, through the electronic monitors set fatliquoring interval time, achieve automatic fatliquoring and fault alarm. Centralized lubricating system of the main control methods is classified into pressure control, fatliquoring quantity control, and time control.

1. Pressure control

① The necessary conditions of meeting the normal work pressure control of the system are as follows:

$$P_f > P_m > P_s > P_d$$

P_f	Relief valve pressure control
P_m	The maximum output pressure of oil pump
P_s	Pressure switch set pressure
P_d	Work pressure of oil distributor
$P_s > P_d$	Purpose is to give the oil distributor with certain pressure holding time.

② Between the oil pump output pressure and pressure switch setting pressure should meet the following conditions:

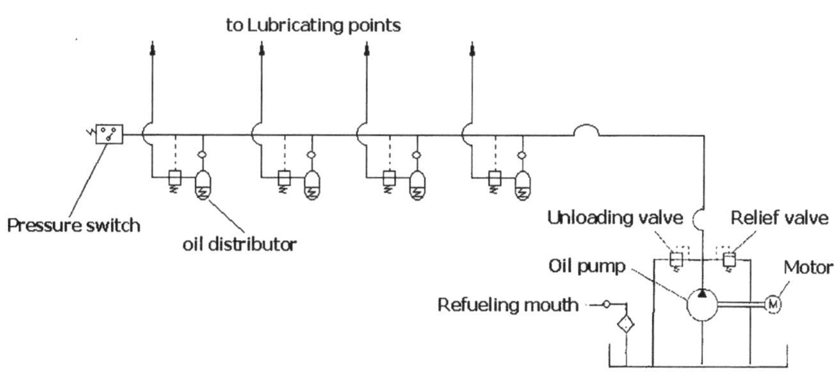

Fig. 54.1 Centralized lubricating system hydraulic schemes

$$P_m \geq P_s + P_{wl}$$

P_{wl} The lord loss pressure pipe and tube connector pressure loss of electric pump export to pressure switch.

③ Work pressure of oil distributor shall meet the following conditions:

$$P_a \geq P_p + P_{w2}$$

P_p Sending fat pressure of lubricating points

P_{w2} Accept pipe and pipe joint pressure loss of oil distributor export to lubricating points.

Because the distance between oil distributor and lubrication points can allocate, in other words, accepting tube length is defined and not affected by changes in the chassis species, so the oil distributor work pressure can get setting value, accordingly, the pressure switch closure pressure can also take value.

Figure 54.2 shows the system pressure control curve, when the pump starts-up, the lubrication system starts pressure (t_1 stage), makes it stores oil and energy. The oil of the piston stroke are defined, and the design is of relevant that their trip size and the node flow. In order to ensure the oil distributor reliable to store oil, given the oil distributor of certain with pressure holding time (t_2 stage). When the pump continues to work, the pressure rises to the pressure (P_0) that switch has set. The pressure switch is closed. There is a transition period, P_0 (t_5 stage) between the oil pump stop working to unloading valve oil and the discharge mouth automatically open again. Lord began discharging pressure, until P_0 (t_6 stage). Discharge to match the oil distributor pressure, the oil distributor sends oil to all lubricating points (P_d, t_3 stage). Because the lord has a certain length of pipe and all kinds of connectors, elbow, it makes the lord exist some pressure P_0. When the pump stops working, monitors indicate the pump into the intermittent stage, until the second circulation starts.

2. Fatliquoring quantity control

Oil distributor accurate the quantity of lubrication grease in the lubrication of system and transports the stored oil to all lubrication node. The storage of lubricant grease and the amount of transported grease are determined by the reciprocating displacement of oil distributor storage chamber piston. The distribution of lubricating grease is based on the lubrication points and the relative speed of friction pair. When the pump sends oil to oil distributor, in hydraulic function, self-control balance valve (check valve) will shut down the oil hole of oil distributor, and filler opening connected with storage chamber. As oil input cavity, the storage piston is

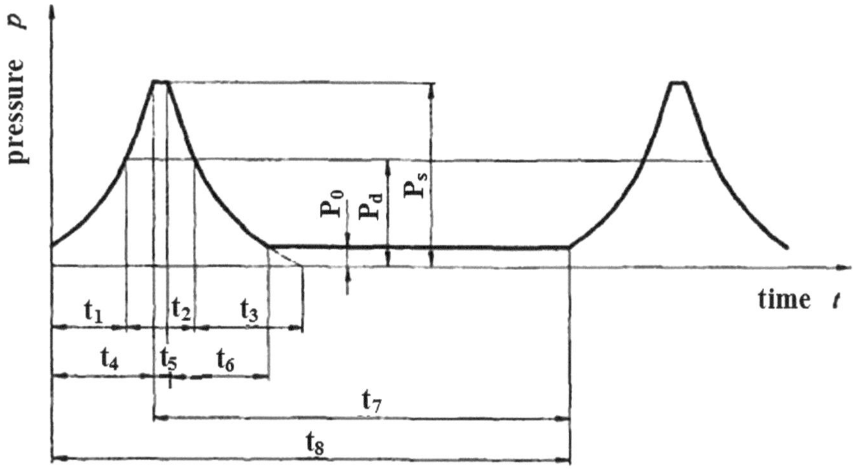

Fig. 54.2 Automatic centralized grease lubrication system pressure characteristic curve. t_1: Oil distributor storage time; t_2: Oil distributor stable time; t_3: Oil distributor supply time; t_4: Pump working time depends on the performance or specifications of the pump and the number of oil distributor; t_5: Transition time; t_6: System unloading time; t_7: Intermittent time; t_8: A lubrication cycle time; P_0: Maximum residual pressure 0.1 MPa; P_d: Oil distributor action pressure interval; P_s: Pressure switch set pressure

passed, at the same time promote the spring compression, and store energy. When the pump stops working and the main oil way unloading, in function of different pressure main lines with storage chamber, the check valve is closed filler opening and opened the oil export, and makes connect with storage chamber and oil export, cavity in spring pressure oil from the oil export to all accept lubricating node through receiving tube, realizing the purpose of automatic fatliquoring.

3. Time control

The existing technology is through the electronic monitors to set the time interval and time monitor, and achieve the function of timing fatliquoring and fault detection of the monitoring system. If fatliquoring fault is occurred , the pressure switch in monitors, accumulative total time, cannot close monitoring, the monitors sent out a warning signal. The setting monitoring time of monitors must be greater than the oil pump normal working hours.

54.5 The Application and Development Prospect of Centralized Lubrication System for Vehicle Chassis

In developed countries, since the 1980s, automatic centralized grease lubrication device has been used in variety of highway transport vehicles. Automatic centralized grease lubrication system of vehicle and engineering machinery chassis is rapidly developed and widely applied in foreign countries. Many heavy vehicles such as the Mercedes Benz, Volvo are already focusing on the lubrication system as a car must be installed.

In China, the centralized auto-chassis lubrication technology has just begun to be adopted by many automakers. Several companies had designed and developed their own systems by referring the similar products from abroad. However, because of the technological and economic reasons, the centralized auto-chassis lubrication systems are still not widely used in China [9].

The centralized auto-chassis lubrication technology represents the advances in automobile design and manufacturing. It can automatically and precisely supply lubrication grease to the lubricating points according to the driving conditions of the vehicle and also allow automatic monitoring of lubrication conditions. Therefore, the research and development of automatic, centralized auto-chassis lubrication systems play important role in the automobile industry and have significant application prospects. The advance of the technology has become an increasingly important trend in improving auto-chassis integrity and enhancing the performance and grade level of the vehicle.

References

1. Zhao Z (2009) The present situation, problem and technical innovation of centralized lubricating of China's vehicle. Limited-Liability Company of Zheng Zhou Otlet Vehicle Technology 12:58
2. http://baike.baidu.com/view/434400.htm
3. Liu Z (2008) Research on bus chassis centralized lubrication system. Highways Transport Inner Mongolia 4:27–29
4. Sheng K Song H, Qian X The development of centralized lubrication system for vehicle chassis device, 4:22
5. Zhang G (1997) The centralized lubrication system for vehicle chassis. Automob Technol 11:48–50
6. Sheng K, Qun Q, Song H (2005) The control method and technology of centralized lubrication system for vehicle chassis. Trans Chin Soc Agric Mach 36(8):26
7. Sheng K, Qian X, Song H (2005) The control method and technology of centralized lubrication system for vehicle chassis. Trans Chin Soc Agric Mach 36(8):41–43
8. Ding X (1995) In heavy automotive applications of centralized lubrication system for vehicle chassis. Heavy Car 5(54):10–15
9. Ma X (2009) Centralized lubricating for vehicle chassis in domestic market analysis. Urban Veh 7:35

Chapter 55
Study on Southeast and Inshore Area Environmental Worthiness of a Certain Equipment

Hongyan Ou, Zhibing Pang, Shuai Mu, Honglei Li, Quanliang Yin and Gongyin Zhang

Abstract Combining influence of "heat, high wet, and high salt fog" environment of southeast and inshore area to a certain equipment operational performance, this paper inquired into the environment characteristics of the southeast and inshore area, and this paper analyzed the concrete influence of the southeast and inshore area environment to a certain equipment thoroughly. Combining the actual circumstance of the southeast and inshore area, this paper put forward a series of protection counterplans to raise environmental worthiness of the certain equipment. The related study and achievement of this paper had a lot of aggressive theory meaning and practice value to strengthening combat efficiency of the certain equipment.

Keywords Certain equipment · Southeast and inshore area · Environmental worthiness · Study

55.1 Introduction

The word of environmental worthiness at the earliest was exited in "Environment Engineering Consideration and Laboratory Test" for American military standard MIL-STD-810F in 2000. It was defined as the capability that equipment, subsystems, or components realize the full set of predetermined function in the expected environment. "General Requirements on Environment Engineering for Equipment" for GJB4239 is supplement for the definition for Environmental worthiness, which is "equipment (products) realization on all of its predetermined function, performance and (or) capability that can't be damaged with the action of

H. Ou (✉) · Z. Pang · S. Mu · H. Li · Q. Yin · G. Zhang
Air Defense Forces Academy, Zhengzhou, China
e-mail: fkb_mmes@163.com

S. Long and B. S. Dhillon (eds.), *Proceedings of the 13th International Conference on Man-Machine-Environment System Engineering*, Lecture Notes in Electrical Engineering 259, DOI: 10.1007/978-3-642-38968-9_55,

various of environment that could be expected to encounter in its lifetime is one of the important quality characteristics of the equipment" [1].

Environmental worthiness is not the only one important characteristic quality but also the important part of the military operation for the weapon system, which directly impact the generation, preservation, and improvement in the combat effectiveness. Certain equipment is the advanced weapon system in our country. It presently is installed in the southeast and inshore area, the important air defense weapon for the carries-to-filed air defense, adjoint-air defense, and point air defense. With the characteristic of "high temperature, high humidity, and high salt fog" environment of southeast and inshore area, the affect on the fight application of a certain equipment caused by is more obvious. Therefore, it is very necessary to start the study on the environmental worthiness of this equipment in southeast and inshore area.

55.2 Environment Characteristic of Southeast and Inshore Area

The southeast and inshore area belongs to the subtropics and tropical marine climate; the air temperature is high, the humidity is high, and the density of salt fog is high [2]. Aiming at the characteristic of high temperature, high humidity, and high salt fog, this article concretely analyzes it from three aspects.

55.2.1 Analysis on the Temperature Characteristic

The average air temperature of the southeast and inshore area is 20–30 °C that is higher in the south area and lower in the north area, and the extreme maximum reached is to be 39.7 °C, and the surface temperature of black objects reached is to be 70 °C at noon. Due to the influences of the marine climate, the characteristics of the air temperature changes in the southeast and inshore area are mainly as follows: First, the daily range is smaller than that of inland area of same latitude. Second, the annual range is smaller than that of same latitude area. Third, the monthly average air temperature is high. Taking any area of Fujian as an example (see Table 55.1), the average air temperature of July is 28.1 °C and the annual average air temperature is 20.7 °C [2].

55.2.2 Analysis on the Humidity Characteristic

The duration time of the cold and warm air exchange in the southeast and inshore area is rather long, having rather long rain season. It is plum rain season from

Table 55.1 Monthly average air temperature in certain place in Fukien and relative humidity circumstance

Month	1	2	3	4	5	6	7	8	9	10	11	12	Average value
Temperature (°C)	12.5	12.4	14.8	19.1	22.9	26.0	28.1	28.0	26.5	23.2	19.3	14.2	20.7
Humidity (%)	74	78	81	82	84	86	82	81	77	69	69	70	78

February to April which always is rainy, and it is typhoon and heavy rain season from May to September with high humidity. Meanwhile, due to closed to the sea, the degree of humidity in this place is often more higher than the inland place. Take any area of Fujian as an example in the same way (see Table 55.1), even in the arid winter for the inland area (December, January, and February), the relative humidity of atmosphere in the southeast and inshore area still maintains at 70 % and above. We can see from this that the humidity environment of this area has the characteristics of high humidity, long action time, and no obvious distinguish for the moist and arid seasons.

55.2.3 Analysis on Salt Fog Characteristics

The density of the salt fog of the southeast and inshore area is more higher than the inland place because the salt molecule of the seawater has been taken into the air and taken to the land by the monsoon and typhoon. The salt content in air is concerned with the distance from the shore. Take the chlorine content of atmosphere of a certain base of Hainan Island as an example: It is 0.38 mg/m^3 when the distance from the shore is 0.5–0.8 m, 0.157 mg/m^3 when the distance is 20 m, 0.071 mg/m^3 when the distance is 200 m, and 0.04 mg/m^3 when the distance is 800 m. It means that the more it is close to the ocean, the higher is the density of salt fog in atmosphere [3].

55.3 Affect that Environment of the Southeast and Inshore Area on a Certain Equipment

55.3.1 Affect that High Air Temperature on this Equipment

The moving speed of the molecule inside the object increases with the increase in the temperature. The kinetic energy increase in the molecule will cause the object to expand, and then, the status and the chemical physical characteristics change accordingly. With the temperature rise, the enlarged material characteristics or parameters include the gas volume under the constant pressure, volume of solid and liquid, sound velocity and particle velocity in gas, gas conductivity, electromagnetic energy emission, vapor pressure of liquid, diffusivity of material molecule, metallic resistance rate, semiconductor conductivity, electron heat emission, solid malleability, chemical reaction. The high temperature not only changes the physical properties and measurements of the materials of this equipment, but also often causes the properties of the components, parts, and even the whole equipment temporarily or perennially to reduce or to damage [4].

In concretely, when he temperature is high, the typical problems would appear as follow: First, the material and mechanical properties change. Second, the expanding inconformity of different materials causes the parts mutually seizure or loose. Third, the lubricant viscosity reduction and lubricant runoff cause the lubrication capability of the joint parts to reduce. Fourth, the material measurements change in all directions or with directionality. Fifth, the sealed cap, gasket, sealed bearing, and axle cause the transformation, occlusion, or invalidation, and then, it causes the mechanical failure or complete damages. Sixth, the outer covers, sealed rings, and gaskets are damaged or transformed. Seventh, the resistance value of the fixed resistance changes. Eighth, the temperature gradient difference or the expansion level difference of different materials cause the electronic circuit and stability change. Ninth, the transformer and the electromechanical components are overheated. Tenth, the operation life shortens. Eleventh, the composite materials deflate. Eleventh, the composite materials may send out some air which is difficult to smell. Thirteenth, the insulating property of the insulation materials reduces and the rubber products become sticky and burn-in. Fourteenth, the solid grain or charging separates. Fifteenth, the explosives and propellants accelerated the combustion. These above-mentioned influences are the results of the combined action of the chemical effect and physical effect.

The high temperature may be caused due to the above-mentioned influences, and many materials of the equipment require a high temperature in practice. The main failure modes of the components of this equipment are indicated as Table 55.2.

55.3.2 Affect that High Humidity on this Equipment

When the degree of humidity is high, this equipment would appear some typical problems as follow [2]: First, the high humidity in the air may cause the adhesion of the water film to this equipment surface that naked eyes cannot detect, then the water film interacts with the acid gas in the air which have the acid property to corrode the metal surface. The increased water content of the equipment causes the wood, paper, textiles, fiberboards, and hydrophilic plastics to deteriorate. Material of the equipment may expand and lose the mechanical strength which would change the property of electrical apparatus. Second, it causes the equipment surface fogging; meanwhile, the moisture absorption in surface may change the

Table 55.2 The heat loses efficacy mode to the type of equipment spare part

Component	Failure mode
Dielectric medium	Dielectric medium leakage increased, the life was shorten; current leakage increased, the resistance was changed
Capacitor	Series capacitor increased
Circuit interrupter	Function disable
Electromagnetic clutch	Coil exceeds the hot spot
Coil	Transformation, melt, instability, and dielectric properties change
Insulator	Epoxy resin spalling, ferrite is peeled off, leakage
Gyroscope	Be drifting
Soldered joint	The strength lost
Magnetron	The life shorten
Engine	Leakage, outage, lubricant deteriorated
Potentiometer	Noise increased; moment, linearity, and resistance changed; insulation resistance under the high temperature reduced
Resistor	Resistance increased, outage and leakage
Servo mechanism	Oil spilled, grease decomposed, high outage voltage
Switch	Contacts oxidized
Thermistor	Outage and leakage increased
Transformer	Dielectric medium properties degraded, outage, leakage, the hot spot abnormal
Semiconductor	Leakage current increased, gain changed, leakage increased, and outage

equipment properties (especially is the insulation property). If the chemical absorption phenomena happen, it is impossible to return to the original status even if it is dried for a long time. Third, the condensation and evaporation alternately appeared impel the metal to corrode rapidly and promotes the organic coating (coating, paint, enamel, nitrocellulose lacquer, etc.) to be damaged. Fourth, the optical instruments are destroyed. When the temperature is low and the degree of humidity is high or the the degree of humidity chages violently, the aqueous vapor may condense on the glassesand permeate into the sealed components. When the temperature raises, there is partial aqueous vapor left in the component, the moisture promotes the microorganism to grow, and it accelerates the destructive effect of the sealed components. Fifth, it corrodes the welding spot of the electronic equipment and precise instrument components and causes the circuit to open or its electrical properties will be changed and then causes the properties of equipment and instrument to disable. Sixth, explosive and propellants absorb the moisture which causes the properties to decrease. Seventh, it has influence on the other properties of the materials, such as reduction in physical strength, decrease in lubrication property, change in electrical insulation and thermal insulation property, delaminate composite materials, change in elasticity or plasticity, reduction in hygroscopic material properties.

In concretely, the influencing effects and mechanism of the humidity to this equipment are indicated in Table 55.3.

Table 55.3 Humidity effect and mechanism

Materials	Influence	Effect of failure
Metal	Slight oxidation and corrosion	Outer electron of metal is stable which do not have drastic chemical reaction if there is no strong acid
	Oxidation and corrosion	Water film adhere to metal surface with acid, oxygen, and salt to aggravate the oxidation
	Serious oxidation and corrosion	The metals with different PD produce the electrochemical action in together
Plastic	Expansion, shrinkage, and embrittlement	Nylon and cellulose acetate fiber transform when encounter water, accelerate the plasticizer lost
	Mechanical strength reduced	Acid dissolved in some keto-water film to catalyze the hydrolysis
	Dielectric constant and power factor low	Perssad plastic which consisted of carbon would hydrolyze and absord moisture
Rubber	Deformation, degradation, polymerization, and resistance	Other acid–base materials dissolved into water film cause the physicochemical properties of nature rubber worse
Coating	Seriously destroy the vanish and pyroxylin lacquer coating	Water penetrates into the base and causes air bubble between coating and base to make the coating to transform and cracking to fall down
Optics	Seriously affect on the light-admitting quality	Moisture condensed on the optical glasses to change the optics property

55.3.3 Affect that High Salt Fog on this Equipment

The salt fog is the moisture consisted of the small visible water drops which contain large amount of sea salt and suspend in the air closed to the earth surface. There normally is salt fog exited in the sky of the sea and inshore area [2].

High salt fog can easily cause bad affects to the equipment which mainly as following three aspects:

First is the corrosion effect. The electrochemical reaction causes the equipment corrosion and mildew. It accelerates the stress corrosion, causes the metal to be rusted, and also makes the salt in water ionize to become the acid–base solution corrosion.

Second is the electricity effect. The insulation decreased and causes the air bubbles to the protected coating due to the electrolyzation. Salt deposition causes the conducting layer to damage the electrical equipments. The insulating materials and metals corrode to impact the electrical property and cause or accelerate the insulating materials and metals corrosion to produce the conducting layer.

Third is the physical effect. The sealed rubber aging causes the sealing property to decrease and the properties of optics, precision machineries, and electrical devices to reduce. The mechanical parts block, bond, or deadlock with the movable parts of assembly. The electrolysis causes the coating blistering and drop. The strong wind with sand dust and salt granular can cause the materials protection layer and coating layer attrition and accelerate the corrosion progress.

55.4 The Protection Countermeasures to Enhance the Environmental Worthiness of Certain Equipment in Southeast and Inshore Area

To adapting to the environment changes, reducing the affects of bad environment and improving the application for this equipment, it is very necessary to take some appropriate protecting measures. According to the environment characteristics of the southeast and inshore area, the protecting measures will promote the combatting efficiency of this equipment [5].

55.4.1 Protection Measures on High Temperature

For the protection measures on high temperature, we should do as the following five aspects: First, build the shade-shed for this equipment and wear the sun-care covers or petites for the tires with chariot loaded and the exposed parts of cables and so on. Second, take full advantage of all ventilation and heat dissipation equipments to dissipate heat and decrease the temperature for the chariot. It should appropriately power off to dissipate heat when power is on for long time. In normal condition, it should power off 15–20 min in each 2 h powered on. If the temperature inside chariot is higherthan the environmental temperature, we should extract the air 15–20 min by the draft fan after power off. Third, try best to arrange the energization checking to be done in the morning, in the evening, or at night and keep the windows open to make the inside air and outside air to cross and cooling during the operation. Fourth, do not contact the parts surface with high smooth finished or silvered with the sweaty hands, to prevent the spots or corrosion and impact the properties. Fifth, oil or change oil in time to all kinds of the lubrication parts.

55.4.2 Protection Measures on High Humidity

For the protection measures on high humidity, we should do the following five aspects: First, often check the seal situation of this equipment, to prevent the rain and moisture to come into. In sunny day, we can open the door, windows, and cabinet door and pull out each combination and unlock the missile covers, for ventilation, sunning, and moisture driving. Second, put the moisture proof sand bags in the square cabin combination and often check them and exchange or dry them in time. Third, to enhancing the capability of driving the moisture, we should first warm and ventilate the chariot 15–20 min in rainy or high humidity day after powering on the equipment, and then switch on with high voltage. In addition to fighting situation, we should try best to reducing the get-on times and closing the door and windows. Fourth, choose the position in the area without ponding and low humidity. Put one layer of sand on the surface and dredge the drainage channels to make the environment to dry in the best way. Fifth, check the electrical components and routes in time to making sure if some part of this equipment mildew, corrode, deteriorated or not.

55.4.3 Protection Measures on High Salt Fog

For the protection measures on high salt fog, we should do the following four aspects: First, strictly carry out the maintenance system of this equipment. Often clean the equipment to prevent the salt deposit. Second, in non-job time, it must cover the covering well. Third, it must coat the protection coating on the exposed metal. Fourth, check the rubber and plastic parts in time and exchange the aging parts.

55.5 Conclusion

In short, it has the outstanding characteristics of "high temperature, high humidity, and high salt fog" in the southeast and inshore area, which has obvious influences to the fight efficiency of a certain equipment. To enhance the study on the environmental worthiness questions of this equipment in the southeast and inshore area, it not only positively promotes this weapon to exert the technology properties of the system war, but also strongly improves the fighting capacity of our air defense force. This article mainly qualitatively analyzes it. In the follow-up researches, it should strengthen the quantitative researches, trying best to make the research is more convincing and more meaningful.

References

1. Qi J, Li X, Liu C (2004) The fungi protection of the weapon material technique. Equip Environ Eng 1(2):69–72
2. Liao G, Wu G, Su S (2004) Characteristics and influence of south china sea weather to equipment. In: Proceeding of equipment and environmental engineering, vol 1(2), pp 69–72

3. Wang X, Zhu Y, Wang L (2004) Study on fungi protection of weapon equipment and experiment technique. In: Proceeding of equipment and environmental engineering, pp 270–276
4. Yu G, Zhao L (2007) Military equipment. National Defense University Press, Beijing
5. Pang Z (1999) Air defense forces man-machine-environment system engineering. Zhenzhou Air Defense Forces Academy, China, pp 91

Part VII
Research on the Overall Performance of Man-Machine-Environment System

Chapter 56
Human Reliability Analysis of Traffic Safety

Tianya Zhang and Jingxiong Wang

Abstract Human factors leading to traffic accidents account for more than 90 % of the traffic accidents causes. To improve the safety reliability of all participants in transportation system is the key method to raise the level of road traffic safety and reduce property loss. This paper explains the reason why human errors are inevitability by Murphy's Law. The psychological field theory is applied to improve the theory of planned behavior (TPB), in order to make the TPB more suitable with the driver's behavior characteristics. And the method of the human reliability analysis (HRA) is introduced in this paper to analyze the safety reliabilities of pedestrians, non-motor vehicles, and also traffic managers.

Keywords Traffic safety · HRA · TPB · Traffic participants

56.1 Introduction

Traffic development has brought great convenience for humans, promoted the political, economic, and cultural exchanges, but also brings negative effects, such as traffic accidents, environmental pollution, and traffic congestion. Since the invention of the first vehicle in the world, the number of people killed in road accidents has amounted to more than 30,000,000. According to the WHO (World Health Organization) report in 2013, 1,240,000 people died in traffic accidents around the world every year. Either the number of traffic accidents or motor vehicle amount is characterized by sharp growth trend. If there is no security behind the high-speed economic development, the economic development is obviously contrary to human-oriented concept. Figure 56.1 shows the number of

T. Zhang (✉) · J. Wang
Department of Traffic Management Engineering, People's Public Security University of China, Beijing 100038, China
e-mail: zty_dionysus@sina.com

S. Long and B. S. Dhillon (eds.), *Proceedings of the 13th International Conference on Man-Machine-Environment System Engineering*, Lecture Notes in Electrical Engineering 259, DOI: 10.1007/978-3-642-38968-9_56,

accidents and deaths, injuries, and economic loss since the founding of China. The research of traffic safety theory revealing the laws of traffic activities is of great importance to the whole society's development.

56.2 The Human Factors in Road Traffic Accidents

Road safety is a branch of safety research, different from other industrial areas. Road traffic system is complex dynamic system composed of people, vehicles, road, environment, and management. Although the reasons for traffic accidents are various, human factors are the most important in a number of factors. Human plays a leading and decisive role in traffic system. The safety and reliability of the human are the fundamental guarantee of road traffic safety. Treat and Sabey after a deep research on a large number of traffic accidents came to the conclusions as follows: the road reasons account for 28–34 %, human factors account for 93–94 %, the car reasons account for 8–12 % [1]. According to China's traffic accident statistics, illegal activities of all traffic participants account for 90 % of the causes of road accidents.

56.2.1 The Inevitability of Human Errors: Murphy's Law and Traffic Safety

Murphy's Law revealing a common phenomenon both in social and in natural was put forward by Edward Murphy in the United States Air Force experiments. Its extreme expression is: If bad things may happen, no matter how small possibility it is, it will always occur and cause the maximum damage. The accident can be regarded as a small probability event.

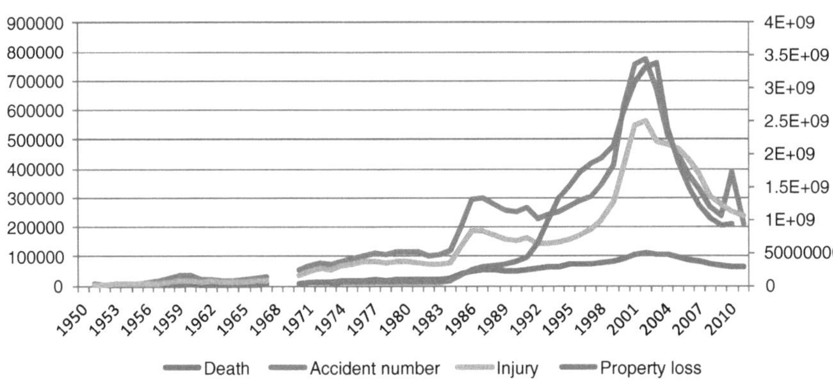

Fig. 56.1 The statistics of the traffic accident, the number of death, injury, and property loss in China for the year 1950–2011

In the randomized trial, set event A the probability for ε.
A_k means A occurs in the kth experiment, let $P(A_k) = \varepsilon$, $P(\overline{A_k}) = 1 - \varepsilon$;
In previous k times mutual independent tests:

$$p(\overline{A_1 A_2 \ldots A_K}) = P(\overline{A_1})P(\overline{A_2})\ldots P(\overline{A_K}) = (1-\varepsilon)^n$$
$$p_n = 1 - P(\overline{A_1 A_2 \ldots A_K}) = 1 - (1-\varepsilon)^n$$

No matter how small ε is, when $n \to \infty$, then $P_n \to 1$.

The above illustrates that small probability events will happen sooner or later. Murphy's Law tells us that the prone to make mistakes is human inborn weakness, no matter how advanced the science and technology are. And the higher the means to solve problems, the more serious the trouble we have to face with. Due to errors in the transport system are inevitable, the reliability level of human factors determines the level of traffic safety.

56.2.2 Human Reliability Analysis

Human reliability analysis (HRA) originated in the 1950s. The theoretical basis of HRA is human factor engineering, system analysis, cognitive science, probability and statistics, behavioral science, and many other disciplines, and the analysis of the quality and quantity for human factors is the core content. HRA's research goal is to analyze, predict, reduce, and prevent human errors. HRA methods can be used in the retrospective analysis and predictive analysis. Through the retrospective analysis, we can find the causes of the accidents and eliminate human errors from the sources to prevent human mistakes from occurring repeatedly, which benefits from the knowledge, experience, and data collection. With predictive analysis, inspectors detect possible defects or weaknesses that easily induce accidents in the environment, predict the possibility of human errors, identify the human errors contributing to systemic risks, improve the design and organization of the system by adopting proper strategies, as well as prevent the possible mistakes through strengthening training on personnel and organization. Therefore, the possible human errors on any system can be analyzed within the framework of the HRA, in practice, which means that we can apply HRA into almost all processes as long as people involved in [2–4].

56.3 Drivers' Safety Reliability Analysis

Drivers compared to the non-motor vehicles and pedestrians are the dominant group in road traffic activities, and transport system reliability analysis is mainly concentrated in drivers. Driver safety reliability factors include physiological and psychological factors, driving skills, and law-abiding consciousness. Driver's physiological and psychological reliabilities are the core factors affecting drivers' reliability. The driver's physiological and psychological reliability determines the reliability of the driver. The reliability of the driver's psychological influencing factors can be summarized as the psychological quality (psychological adaptability and stress capability), traits of character, and emotional factors. Reliability factors in driving skills, including driving knowledge, driving experience, and driving stress ability, determine the emergency driving ability when drivers are faced with unexpected situations. In addition, the driver's legal concept and consciousness also plays a big role. Most driving accidents are related to the unsafe behavior of the driver, and the driver's unsafe behavior is largely due to their insufficient safety awareness as well as the weak sense of traffic laws and regulations [5].

56.3.1 Theory of Planned Behavior [6]

The analysis of the drivers' behavior patterns is the premise to enhance the reliability of drivers. The theory of planned behavior (TPB) is widely used in the analysis of drivers' behavior model. TPB is proposed by Fishbein and Ajezn on the basis of theory of reasoned action (TRA) in 1988 and developed in 1991. TPB is the most famous social psychological attitude-behavior theory. The theory puts the attitude as the predictive value for behavior and constructs a theoretical framework between attitude and behavior, which can be used to describe and explain behavior and attitude controlled completely independently. Structural equation model diagram of TPB is shown in Fig. 56.2.

The four variables of TPB include behavior and attitude, subjective norms, perceived behavior control and behavioral intention. Attitude is the center part of the planned behavior theory, divided into attitude toward behavior and attitude to the subjective matter, which can predict the behavior intention. Subjective norms refer to the individual decision on whether to perform a specific behavior when perceiving social pressure, which is the impact of social factors. Perceived

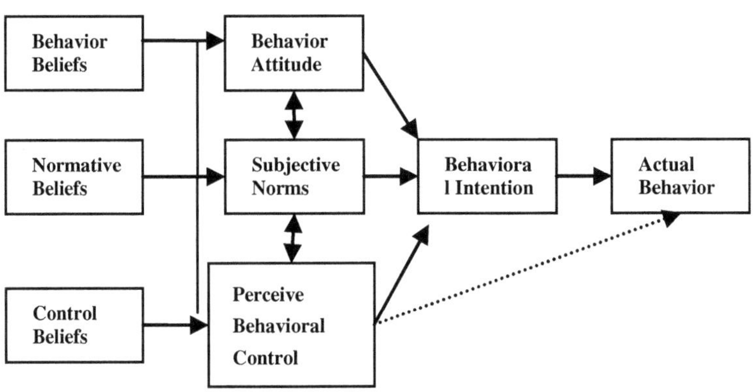

Fig. 56.2 The structure model diagram of the TPB

behavior control is individual perception to the degree of difficulty to perform a certain behavior, reflecting an individual perception promoting or hindering the execution and the will power to perform the behavior.

The theory is that a person's behavioral intention jointly decided by the three variables of behavioral attitude, subjective norms, and perceived behavioral control. Sometimes a single variable plays a decisive role, sometimes two variables play a decisive role, and sometimes combined effects of the three variables play the decisive role. But the traditional TPB in driving behavior analysis focused only on internal factors and not included the vehicles, road conditions, and other external factors. And the measurement variables are only for the driver itself, without considering the interaction of drivers and other traffic participants. So, we put forward the improvement of TPB to make it more in line with the characteristics of driving behavior.

56.3.2 Optimization of Driver's TPB Model

Most of the time, the driver's behavior is not an autonomous decision-making process, but to a large extent influenced by social factors and environment. Therefore, environment variables need to be taken into account. This paper introduces the concept of psychological field into the TPB and establishes a new driving behavior analysis model.

The conception of field in the nature means a mixture of energy. Through the field, during the interaction processes, things receive and transmit their energy. Psychological field is formed between the psychological energy and environment. Psychology field theory is from the basic concepts of physics in energy field, used to describe the psychological effects among objects out of contact with each other. The psychology field theory assumes that people are in psychological field and their psychological activities are affected by all perceived factors of a person, designed to explore people's perception and cognition of its surroundings. Objective environment and the social factors of the traffic system are the important factors in driver's psychological field.

The psychological field formula is $B = f(P \cdot E)$.

f is the function of individual characteristics and environmental, B indicates the personal behavior, P indicates the personal traits, and E represents the environment.

The influencing factors of psychological field consist of natural factors (including the space and time), social factors (population number and structure), and the individual factors. Optimized TPB model is described in the following structure diagram (Fig. 56.3).

Through analysis of the driver's attitude and behavior within the framework of the TPB, we can not only explain and predict the driver's dangerous behavior, but

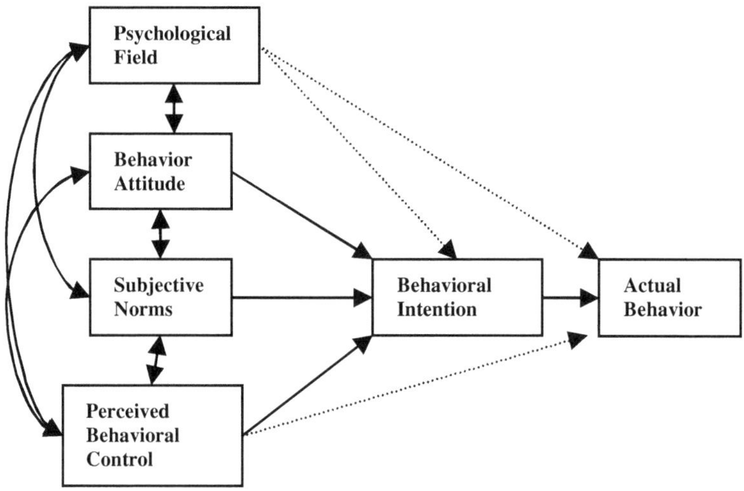

Fig. 56.3 The improved structural equation model of TPB

also intervene the driver's wrongdoings. Improved TPB is not only able to analyze the formation of behavior, attitude, subjective norms, and perceived behavioral control beliefs, but also analysis the driver's interaction with the surrounding environment. Improved TPB makes the analysis of traffic accidents more suitable with the characteristics of fuzzy and randomness, and more adaptive to the continuing changes on the spatial and temporal of the objective environment.

People living in groups influenced by social factors more significantly. Social factors are in the state of flux, which includes population sexual ratio, age structure, and group levels. Besides the social factors and environment, the driver's features differ in thousands of aspects. All the interactions among the three variables make different drivers in traffic activities show different behaviors. The introduction of psychological field enriches connotation of the TPB and improves the explanatory power of driver reliability analysis.

56.4 Pedestrians and Non-motor Vehicles Safety Reliability Analysis

Pedestrians and non-motor vehicles are also the important parts of transport system. About 5 % of the traffic accidents are caused by pedestrians and non-motor vehicles. Pedestrians can be divided into different groups by their age and sex. As a result of the differences of their age and sex, there have been great discrepancies in the perception of risk, information processing, attitudes, and behaviors to avoid dangers. The first step of the process on traffic safety reliability of pedestrians and non-motor vehicles is the perception of risks. If one cannot aware the danger, or cannot realize the possibility of an accident, the accident would be inevitable. Then, the next step is to take the measures if individuals perceive the dangers. It depends on the attitude of pedestrians or non-motor vehicles (such as awareness of traffic safety law and responsibility for their own and others' lives) as well as psychological characteristics, just as someone tend to takes chances, someone likes to take risks, and someone is more cautious. The third step to take actions depends on the ability to avoid dangers, including sport ability, response capability, weight, brain and response characteristics and experiences when facing danger. The ability to avoid danger does not guarantee the safety, since the traffic accidents can sometimes be occasional. The purpose on reliability analysis of pedestrians and non-motor vehicles is to get constructive analysis results and to put forward suggestions for improvement means according to various groups.

56.5 Traffic Managers Safety Reliability Analysis

Management behavior of traffic managers exists in all aspects of the transportation system. Traffic management objective is to guarantee the traffic safety and organize the transportation elements, direct and coordinate transportation activities, coordinate the driver, the vehicle, and the environment. The traffic manager's safety attitude, knowledge, attention, emotional stability, social compassion, legal concepts, self-control, and communication skills concern the transportation system problems to be reasonable tacked with. For example, in the face of unexpected traffic condition or poor traffic environment, whether there can be a reasonable response so as to prevent accidents from happening. Their supervision and punishment of violations directly affect the attitudes and behaviors of traffic participants. If the punishment is too light, it does not arouse the attention of the offender, and heavy punishment causing conflicts between offenders and managers would fail to achieve the goal of education. In addition, traffic managers, as the administrators and leaders of all traffic participants, should carry out reasonable organization and operation on traffic elements such as people, traffic safety facilities, information, and data of accidents. They also should establish a rational management mechanism, make an development of traffic safety regulations, and take results summary. Improving safety reliability of traffic managers is the effective way of realizing traffic safety ultimately.

56.6 Conclusions and Recommendations

Reliability analysis of human factors is an emerging science which is based on behavioral science, cognitive science, information processing and systems analysis, probability statistic, and other theories. The study on human reliability is widely used in nuclear industry, aerospace industry, petrochemical industry, safety of navigation, mining accident, and electric power industry. The measurements of HRA are rarely put into the traffic safety area [7]. Moreover, HRA is involved in many cultural factors, such as scientific literacy, moral level, psychological qualities, attitude in work, habits of thinking and style of work, and safety culture. All the factors above make database on human factors lack of universality. People from different areas have various cognitive structures and different countries data are not portable, making the situation of human reliability being even more complex. It will be a significant contribution to improve the safety and reliability of participants, vehicles, road, and environment in road traffic system if we apply human reliability assessment methods, models of human reliability on the traffic safety research.

References

1. Liu J, Zhang G, Ma Y (2011) Design theory and method of highway traffic safety. Science Press, Beijing
2. Xie H, Sun Z, Li X, et al (2007) Commentary of typical human reliability analysis methods. J Nat Univ Def Technol 29(2):101–107
3. Liao K, Zhang L, Huang X (2006) Research progress on theory of human error. J Mater Sci Technol 16(7):45–50
4. Xiao G, Chen B (2001) Study on the mechanism and reliability of human error. J Mater Sci Technol 11(1):22–26
5. Wang W (1998) Development of reliability analysis of man-machine system. J Syst Eng 13(1):30–45
6. Wen TD, Guang RJ (2008) A review of the theory of planned behavior. Adv Psychol Sci 16:315–320
7. Wang F, Wu C, Wang F, et al (2009) Research progress of human reliability during 1998-2008. Sci Technol Rev 87–94

Chapter 57
Design of Virtual Training System Based on 3D Interactive Graphics

Xiuqing Zhu, Fuchao Hu and Quanchao Hou

Abstract **Objective** In the long-duration spaceflight, a flexible training device is needed for astronauts on board. The computer-aided training based on 3D interactive graphics is a solution. **Methods** Constructing models of virtual objects, importing them into Virtools to realize interaction scheme. Combining text, graphic, and animation, to present training stuff vividly, trainee can interact with the models and drill virtually procedures of operations. **Results** The training system for the extra-vehicular activity procedure and the complex capsules was realized, and applied to personnel training. **Conclusion** The training system affords an effective method for astronaut training on board and also adapts to complement the training devices on ground.

Keywords 3D interactive graphics · Virtual training · Computer-aided training · Spaceflight

57.1 Introduction

With increasingly heavy spaceflight tasks and gradually extended on-orbit duration, there are some new challenges brought out over the training of astronauts. Training is required by all experimental tasks and device maintenance. However, there are a lot of training subjects but with limited training hours and training

This Study is Funded by China Space Medico-engineering Proactive Research Program (2010SY5413004).

X. Zhu (✉) · F. Hu · Q. Hou
National Key Laboratory of Human Factors Engineering, China Astronaut Research and Training Center, 100094 Beijing, China
e-mail: zz_xx_qq@126.com

S. Long and B. S. Dhillon (eds.), *Proceedings of the 13th International Conference on Man-Machine-Environment System Engineering*, Lecture Notes in Electrical Engineering 259, DOI: 10.1007/978-3-642-38968-9_57,

devices. So a flexible convenient training approach shall be developed for astronauts. On the another hand, the maintenance of training memory and refresh training before mission also bring out urgent demand on flexible convenient training methods and approaches. The computer-aided training based on 3D interactive graphics has drawn extensive attention and realized wide applications in major spaceflight countries and organizations for its simple hardware platform, vivid training stuff presented, and prominent training performance at anywhere and anytime including on-orbit training [1–3].

57.2 Applications Review

57.2.1 NASA's Applications

The Virtual Reality Lab of Johnson Space Center, NASA, developed a 3D graphic software package–DOUG [4, 5]. The package started on-orbit applications since STS-102 task in March, 2001, used by astronauts for robot arm training on space shuttle and international space station, extra-vehicular activity task planning and drilling, extra-vehicular simplified rescue device (SAFER) training, etc. DOUG shares uniform model database with ground simulator. To adapt to the graphic capacity of on-orbit notebook computer, the model has been processed with different detail accuracies. Relative low detail accuracy may be chosen to maintain 3D graphic at appropriate detail degree and refresh rate in display. DOUG software may be used to receive actually measured data of each joint of robot arm and accordingly generate virtual views of each monitor camera. So it can be used for virtual training without real operation load.

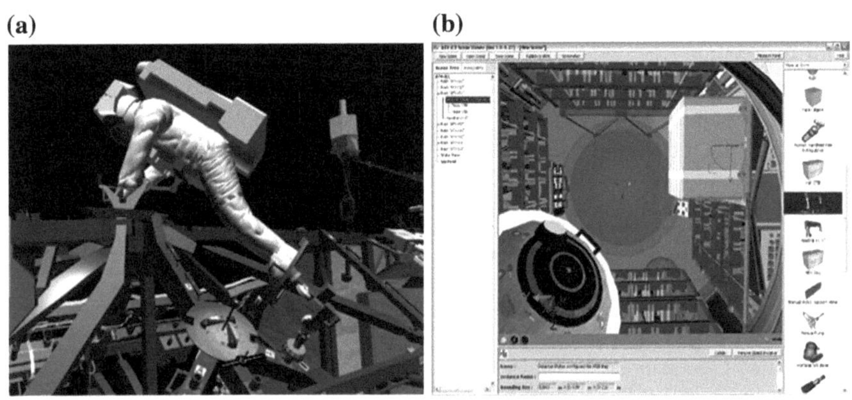

Fig. 57.1 virtual scene from NASA DOUG software package (**a**), ESA 3D interactive graphics training system (**b**)

57.2.2 ESA's Applications

European space administration (ESA) applies 3D interactive design tool Corton-a3D in innovative design of spaceflight manual. A traditional spaceflight task operation manual contains pounds of literary materials, involving various languages and versions and the use of abbreviations. So, some issues of inconsistence or difficulty in understanding appear inevitably. With CAD data and documentary materials for spaceflight device engineering design, ESA designed an electronic manual integrating text, graphics, and 3D interactive animation to greatly enhance efficient learning and accurate understanding in the more visual and more vivid presentation. By using 3D interactive graphics, ESA designed ATV training system aiming at Europe's automated transfer vehicle (ATV) which is able to realize training for visualization and operation maintenance of ATV cargo storage devices and conduct effective planning and maintenance for cargo storage to enhance utilization efficiency of the hold [6]. Such small-sized computer-aided devices may be not only used in ground training, but also applied on on-orbit just-in-time training. Its carrier is portable computer or other small-sized mobile computer devices, so it can be used in training at anywhere and anytime so as to greatly reduce training cost and enhance training efficiency (Fig. 57.1 a, b).

57.3 System Design and Realization

57.3.1 Purpose

According to experience of America, Russia, and Europe in manned spaceflight training field, astronaut training needs training tools of different levels in different forms, involving training devices at different levels from operation manual, computer-aided training systems, single-item system training devices to comprehensive task simulator, etc. [1–3].

Generally, a large-sized training device needs high development cost, long development period, limited training capacity, special requirements on training support and even with certain risk factors. To expand training resources, more assisting training tools shall be developed. In this way, astronaut can conduct primary training with these assisting tools and then use more complicated training devices after certain proficiency is reached so as to greatly enhance utilization efficiency of devices. Today international spaceflight cooperation is getting closer. Astronauts need remote training sometimes and the on-orbit astronauts in international space station even have to conduct on-orbit training. All these bring out demand on small-sized, portable training devices.

Computer-aided training device based on 3D interactive graphics, due to its portability, flexibility, multimedia content presentation, conformity to human's character of cognition and long-duration memory keeping, plays an important role in training applications in the industry.

57.3.2 Methods

57.3.2.1 Design Thought

By constructing spacecraft operation control device model and building up 3D space virtual environment for work scene, student can click device in virtual environment with a mouse and get knowledge about functions and operating methods of device through text popped-up and phonic interpretation and basic experiences of interactive operation. System also provides specific operation contents of each step in learning course and some training contents like acquiring instrumental data required to be monitored simultaneously. Software 3DS MAX is used to build up 3D graphic model of the device and Virtools is used as graphic interactive scheme software. Web browser is used as running platform and 3D graphic interactive contents are played by plug-in on browser.

57.3.2.2 Design Platform and Tools

The system operation software platform is Windows XP. Training software is running on IExplorer. The system is able to operate the contents of 3D interactive graphics just with a playing plug-in, without requiring any other additional software or hardware support. This makes release and application of the software easy.

Main tool software applications:

(a) Image editing and processing software Photoshop: used in processing of image materials like texture and letter.
(b) Audit processing software Cool Edit: used in the processing of interpretation, like recording, editing, and cutting.
(c) 3D modeling software 3DS MAX: used for design and production of 3D graphic model.
(d) 3D graphic interaction (model- driven) design software Virtools [7]: it envelops a huge volume of routine instruction blocks in the form of script prebuilt block. These instruction blocks can be used to realize programming easily and enhance design efficiency. With unique visualized programming method suitable for extensive user groups, it has become one of the international mainstream 3D development platforms.

57.3.2.3 Scene Modeling

Training system adopts 3D graphic model. 3D modeling of spacesuit and devices in complex capsules are the fundamental works of system development. To ensure the reality degree of 3D model of training system, a lot of works in dimensional measurement and texture collection and processing of the real devices were completed. Some models were converted from the original engineering design drawings. On the premise of ensuring display effect, the redundant facets have been deleted or reduced to meet the requirement on the speed of display of 3D graphic scene.

Polygonal modeling method is adopted to establish some models like spacecraft, extra-vehicular spacesuit, and devices in complex capsules. The extra-vehicular spacesuit model is composed of models of main body of suit, control device, lockup mechanism, etc. The models of devices in air lock capsules include suit control device, extra-vehicular activity supporting device, capsule pressure meter, release valve block, charge valve block. The photographs of real product after processed are adopted as texture map and then edited by UWV Map coordinates to make the coordinates of the UVW Map consistent with those of model map. Experiment shows that such production process can realize satisfying effect.

57.3.2.4 Interactive Design and Scene Management

The interactive technology of 3D model is the key to the system design. Currently, representative tools include Vega and Virtools. Virtools is used in this system. The software is embedded with hundreds of BBs (Behavior Building Blocks) which can be endowed on appropriate objects or virtual roles in 3D graphic scene in the form of flow chart, and the prioritization of processing of BBs is determined to realize visualized interactive script design and through editing gradually form a set of complete 3D interactive graphics training system.

The development flow of applying Virtools in interactive design includes some stages: model import, scene deployment, interactive design, debugging, and output.

(a) Model import: 3DS Max-built model needs importing 3D model to Virtools with special conversion software. After modeling, NMO file format will be saved on the hard disk. Choose resource/import file of menu instruction in Virtools and find out NMO file output. In this way, the model is imported to Virtools successfully.
(b) Scene deployment: conduct scene layout according to demand to make scene layout conforming to simulated real scene and add some scene elements like light sources and camera. This involves following three contents: (1) object combination: combine individual objects together and place them in a proper position at reasonable angle in an appropriate proportion. (2) Light source setting: appropriate light source setting creates reasonable contrast of light and shadow with an appropriate sense of depth. (3) Camera setting: set a single viewpoint according to demand or set several cameras for switch from to different viewpoints; besides, angle of the view field of camera can be set as well.
(c) Interactive design: to conduct interactive design after completing model import and scene deployment. Virtools has over 500 BBs built, and the arrangement and connection among these BBs can form many different interactive setting through combination. Virtools' interactive design is

relatively logical. It converts program writing into a vivid flow chart and each key instructive sentence into BBs; many jacks set in each block including input jack, output jack, input parameter and output parameter. These BBs are the most fundamental elements constituting the function of an object. 3D objects can interact with each other only if different BBs are connected together reasonably.

(d) Debugging: the visualized editing environment of Virtools not only makes program design easier, but also enables user to fast debug. It supports real-time output. After modification of interactive design, user can preview modification effect immediately so as to greatly enhance design efficiency.

(e) Output: it is okay to save the production contents in the form of Web page file in the Web page release mode. During running, the learning contents produced can be presented via Web browser in a common computer platform after installing a browser plug-in 3D LifePlayer specially used for Virtools.

Applications of Virtools in program development greatly enhance efficiency of 3D graphic scene management and interactive design and make training system design easier and faster.

57.3.2.5 Roaming Design

Set a keyboard panel to control movements of six degrees of freedom including horizontal movement and rotation in the field of vision of screen. Manipulate camera to freely roam in complex capsules with corresponding control keys. Set the devices on the capsule walls and those inside the capsules as fixed obstacles, detect the collision with obstacles in real time during roaming of camera, and avoid unreal situation of penetrating obstacles.

Roaming inside complex capsules enables trainee to know about layout of devices inside and become familiar with work environment in the complex capsules and reduce occupying hours of simulator mock-up or replace the mock-up when it is unavailable.

57.3.2.6 Device Teaching

The design realized two device-teaching plans. According to a plan, in a virtual scene, mouse will directly click device and camera automatically moves to preset appropriate location, while name and literary introduction of the device will be shown on the display with phonic interpretation played. In another plan, set a device list and then search the device name on the list. In the case that only device name is available, in order to fast easily find out specific device, main device checklist is set. Click corresponding name of device on the checklist, the camera will automatically move to preset location and name and literary introduction of

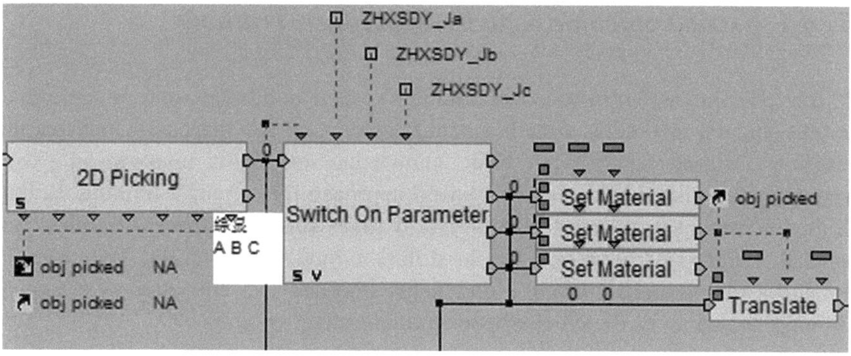

Fig. 57.2 Scheme of virtual buttons interaction in Virtools

device will be shown with phonic interpretation played. The checklist menu out of use will be concealed automatically.

Trainee can learn about the positions and functions of devices in the capsule in a short period and fast localize devices through the training.

57.3.2.7 Virtual Operation

Physical training devices may be delivered late, expensive, or easy to wear and with limited service life. In virtual training system, the physical objects are substituted by virtual objects so as to realize virtual human-machine interaction, provide means for trainee to know about operation method of device and feedback information, and also create an effective approach for astronaut to realize on-orbit refresh and review of operation procedure.

With virtual model and simple interactive tool, the trainee can learn about use procedure and steps of operation device, like operation of instruments, on/off of valves, open/close procedure of port lid. Figure 57.2 shows part of design block of virtual buttons interaction.

57.4 Results

For the extra-vehicular activity procedure training and training of complex capsules, the computer-aided training system based on 3D interactive graphics technology has been designed so as to realize the training functions like familiarity of work environment, device teaching and virtual operation; besides, this system has been used in astronaut training and task experiment group personnel support.

57.4.1 Extra-Vehicular Activity Procedure Training

Extra-vehicular activity procedure training system builds up some models, like extra-vehicular spacesuit, airlock module device, and the spacecraft, realizes the teaching training for relevant basic knowledge and skills involved in extra-vehicular and also provides new technical approach for astronaut training. Before complex simulator is launched into use, it saves time in astronaut training and enables training to be conducted in different modes and levels. Therefore, it enhances the training efficiency and helps improve use efficiency of complex training device so as to achieve better training effect (Fig. 57.3).

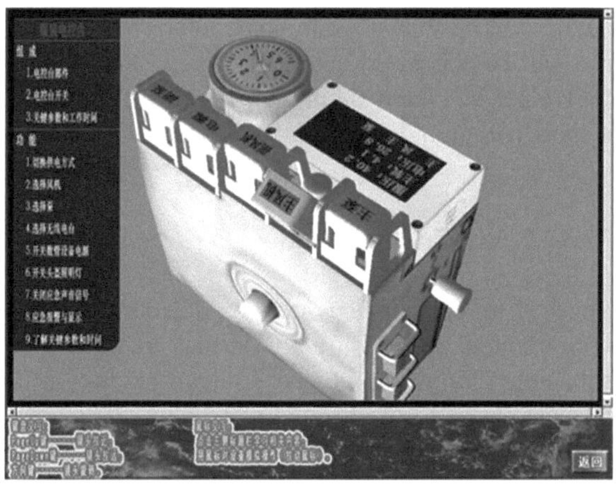

Fig. 57.3 A frame of training EVA spacesuit operation

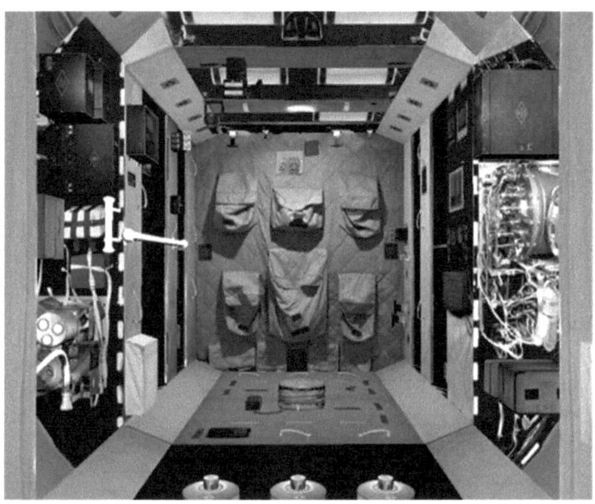

Fig. 57.4 A virtual scene of complex capsules

57.4.2 Training of Complex Capsules

In the training of complex capsules, the 3D complex model was constructed, covering Tiangong I and spacecraft orbital module. The design realized teaching of cabin roaming, device installation positioning and functions and virtual operation trainings like panel simulation operation, valve simulation operation and switch and port lid simulation operation. Besides providing application of astronaut training, the system is also applied in the field of flight control task so as to provide powerful assisting tools for task presentation and planning by personnel (Fig. 57.4).

57.5 Conclusion

With easy flexible applications, low-cost, on-orbit training and some other advantages, portable-computer-aided training system based on 3D interactive graphics has extensive perspective in future long-duration spaceflight tasks. Besides, it can also be used as advance training tool before using of ground large training device and refresh and review tool after training so as to enrich training approaches.

3D interactive graphics system may directly input CAD data of engineering design product and simultaneously develop product maintenance and training system to make product fast applied in training. To enhance efficiency of training system design, the channels of communication between product design institution and training system design department shall be further smoothed out to allow CAD product data to be directly converted into 3D model database of training system. At the same time, a virtual training, operation, and maintenance system shall be created by uniform 3D data to greatly reduce development cost while ensuring the consistency and precision of the system.

References

1. Astronaut Selection and Training (1997) NASA information summaries
2. Cornelissen F, Neerincx MA, Smets N, et al (2012) Gamification for astronaut training. In: The 12th international conference on space operations, Stockholm, Sweden
3. Gancet J, Chintamani K, Letier P, et al (2012) Force feedback and immersive technologies suit (FITS): an advanced concept for facility-less astronaut training. In: International symposium on artificial intelligence, robotics and automation in space, Turin, Italy
4. Aoki H, Oman CM, Buckland DA et al (2008) Desktop-VR system for preflight 3D navigation training. Acta Astronaut 63:841–847
5. Johnson Space Center (2004) Implementation plan [EB/OL] 2003. http://jsc.nasa.gov
6. European Space Agency Employs Cortona3D's three-dimensional, interactive animation to speed training [EB/OL]. http://www.cortona3D.com
7. Dassault System. 3DVIA Virtools 4.0 Online Reference

Chapter 58
Operation Measurement in Deploying Phase of a Certain Chariot Based on Stopwatch Time Study

Zhibing Pang and Tao Li

Abstract Based on human-machine operating optimization analysis of a certain type of chariot, using stopwatch time studies, by selecting qualified staff, scientific plotting out operating unit, and calculating the number of observations, and carrying out time data acquisition and analysis in accordance with the operating procedures, and eliminating outlier data, and combining the Westinghouse appraisal coefficient to determine the time standard and so on, operation measurement quantitative scientific and accuracy of a certain type of chariot deploying phase. So as to providing a reference for the formulation of a certain type of chariot operation courses appraisal standards, its core methods also have guiding significance for the study of related comprehend.

Keywords Stopwatch time studies · Deployment phase · Operation measurement

58.1 Introduction

With high-tech progress, a certain type of weapon system is used to equip army gradually. However, there are three issues affecting the generation of battle effectiveness of this weapon system: no operator selection standard, no standard operation tutorial, and no accurate assessment standard. To solve out these issues, based on the optimization of main chariot operation flow of this weapon system,

---

Z. Pang (✉) · T. Li
Air Defense Forces Academy, Zhengzhou, China
e-mail: fkb_mmes@163.com

T. Li
Unit 66440 of PLA, Shijiazhuang, China

S. Long and B. S. Dhillon (eds.), *Proceedings of the 13th International Conference on Man-Machine-Environment System Engineering*, Lecture Notes in Electrical Engineering 259, DOI: 10.1007/978-3-642-38968-9_58,

this paper conducts quantitative study on how to conduct operation measurement in chariot deployment phase with stopwatch time method.

"Stopwatch time study" is also called direct time study-intensive sampling. It is a method of conducting direct successive observation on the operation of operator with stopwatch or electronic timer in a certain period, recording operating time and other operation-related parameters together and determining standard time required by operator to complete a certain operation in combination with pre-set allowance. Measurement of standard operating time is for the purpose of better scientifically using time and managing to wipe off ineffective time. It has two immediate objectives: first, finding ineffective time hidden in current operation method and managing to wipe off; secondly, making time standard for operator assessment. Both the above important objectives have very important value and significant influence on enhancement of operation efficiency.

58.2 Steps for Stopwatch Time Study

58.2.1 Select Qualified Operator

The top priority is to select qualified operator as study object before making a standard time. The selected operator must have necessary skills and knowledge, high physical quality and teamwork spirit; besides, his operation level shall be medium proficiency to ensure the practicability and typicality of acquired time for measurement [1, 2].

58.2.2 Plot out Operating Units

To plot out operating units is to divide human-machine operation into several operating units for measurement and analysis. Stopwatch time is measured and recorded in the unit of operating unit and the sum of time of all individual units equals to whole operation time. However, when measuring with a stopwatch, if the time of each unit is too short, the measurement and record will be impossible. To easy measurement by stopwatch, the operation is divided into appropriate units generally. According to the actual conditions of deployment operation of a certain type of chariot, the unit division shall follow two principles [3]: first, the boundary between operating units shall be clear and every unit shall have identifiable starting point and end point. Sometimes to make identification easy, take the moment when an operating unit ends and another starts as the boundary; second, the duration of each unit shall be appropriate. To process data with stopwatch time method, the operating units with duration shorter than 0.04 min each shall be merged to ensure the time of each operating unit is longer than 0.04 min.

58.2.3 Determine Number of Observation

Stopwatch time study is a process of sampling observation. To acquire scientific time standard, adequate capacity of sample is required. The bigger the sample scale is, the more accurate the result is; however, the bigger the sample scale is, the more time and energy consumed is. Therefore, it is especially important to scientifically determine the number of observation. Formula (58.1) may be used to calculate number of observation when error scope [4] is ±5 % and reliability is 95 %:

$$N = \left(\frac{40\sqrt{n\sum_{i=1}^{n} X_i^2 - \left(\sum_{i=1}^{n} X_i\right)^2}}{\sum_{i=1}^{n} X_i} \right)^2 \tag{58.1}$$

In formula (58.1), N presents required number of observation in stopwatch time study; n presents actual number of observation; X presents actual time reading on stopwatch. The calculated number of observation is 20 in this stopwatch time study.

58.2.4 Minute

Due to relatively clear boundary of operating units of deployment operation of certain type of chariot, to directly record operating time of each operating unit [5], return-to-zero method among stopwatch time methods is adopted for time measurement. This method means, during observation process, stop the stopwatch and see the reading at the end of each operating unit and then immediately fast return the pointer on the stopwatch back to zero. Restart the watch at the beginning of the next operating unit. The point ending previous operating unit is the point starting this unit, and therefore, it is a simple easy way to record duration of each unit.

58.2.5 Weeding Abnormal Data

Abnormal value refers to the numerical value of duration of certain unit beyond normal range due to influence of exterior factors. For different reasons, it is inevitable to see some abnormal values in time measurement data. Generally, abnormal values are wiped off by treble standard deviation method. The calculation method is as below:

Assume that, in n times of measurement, the measured duration of the same operating unit is $x_1, x_2, \ldots, x_n$ respectively, the average value is

$$\bar{x} = \frac{1}{n}\sum_{i=1}^{n} x_i \tag{58.2}$$

The standard deviation is

$$\sigma = \sqrt{\frac{\sum_{i=1}^{n}(x_i - \bar{x})^2}{n}} \tag{58.3}$$

Accordingly, we get the control upper limit of normal value UCL $= \bar{x} + 3\sigma$ and control lower limit LCL $= \bar{x} - 3\sigma$. So, measured data shall be numerical value within $\bar{x} \pm 3\sigma$. If the value is beyond this range, it shall be deemed as an abnormal value.

58.2.6 Appraisal Coefficient

Westinghouse appraisal method is adopted in this research [6]. The proficiency, effort, and cooperative skill of operator are taken as the main appraisal factors to measure operation conditions of the operator. Every appraisal factor is further classified into six degrees: perfect, outstanding, good, pass, common, and bad. Considering the actual demand of measurement of measured weaponry operation, this research just considers three factors: outstanding, good, and pass. Table 58.1 is the factors listed in Westinghouse appraisal method.

In Table 58.1, proficiency is the reflection of degree of grasping a given operation method; effort is the preference of operator in enhancing operation efficiency subjectively; cooperation skill is the degree of influence of cooperation among operators on operation time.

Table 58.1 Rating grade and appraisal coefficient

Proficiency			Effort			Cooperation skill		
Grade	Code	Coefficient K	Grade	Code	Coefficient K	Grade	Code	Coefficient K
Outstanding	$A1$	−0.11	Outstanding	$B2$	−0.10	Outstanding	$C1$	−0.03
Good	$A2$	−0.06	Good	$B2$	−0.05	Good	$C2$	−0.01
Pass	$A3$	0	Pass	$B3$	0	Pass	$C3$	0

58.2.7 Determine Time Standard

The appraisal standard for human-machine operation in deployment phase of certain type of chariot is acquirable in combination with human-machine operation experimental data and stopwatch study for deployment of certain type of chariot. In experimental test phase, as operation level of selected operators is medium, we may take the average of the whole operating time of operator as “pass” time. Through “pass” time standard, we can draw time standard for “outstanding” and “good”

$$T_{良} = T_{及} \times (1 + K_{A2} + K_{B2} + K_{C2}) = 0.88T_{及} \tag{58.4}$$

$$T_{优} = T_{及} \times (1 + K_{A1} + K_{B1} + K_{C1}) = 0.76T_{及} \tag{58.5}$$

58.3 Application of Stopwatch Time Study in Deployment Operation of Certain Type of Chariot

58.3.1 Division of Operating Units in Deployment Operation

According to principle of operating unit division in stopwatch time study, the optimized operation of each operator in deployment phase may be divided into several operating units.

The operation of operator 1 may be divided into following 7 operating units: (1) erect hanging ladder of port; (2) open port and take a hammer; (3) connect overhead antenna pole on the roof of the cabin; (4) take hammer away; (5) wait for boot-up of oil engine; (6) electrify device; (7) open display of console.

The operation of operator 2 may be divided into the following 9 operating units: (1) take stow-wood of case at right side; (2) place stow-wood at right rear side and run to the place for piling; (3) cooperate in piling; (4) take stow-wood of case at left side; (5) place stow-woods at left rear side and left front side; (6) open hydraulic control box and wait for electrifying device; (7) level vehicle (8) erect feed system; and (9) erect hanging ladder of shelter and go on post.

The operation of operator 3 may be divided into the following 7 operating units: (1) take pile; (2) take hammer and run to place for piling; (3) cooperate in piling; (4) put hammer back and place the hanging ladder of shelter on the ground; (5) claim to the roof of the cabin; (6) install overhead antenna pole; and (7) take off roof of the cabin and go on post.

The operation of operator 4 may be divided into the following 9 operating units: (1) erect hanging ladder of power plant; (2) place stow-wood at right front side; (3) open panel of power plant; (4) wait for completion of grounding; (5) boot-up of oil engine; (6) place hanging ladder of power plant on the ground; (7) Rise antenna; (8) erect hanging ladder of power plant; and (9) enter cab.

For the purpose of statistics and analysis, the operation acts of each operator may be numbered in a predefined order. For instance, the first move of operator 1, erecting port handing ladder, is 11 here, and the second move, opening the port and getting the sledgehammer, is 12, the first move of operator 2, taking a stow-wood of case at right side, is 21. The rest may be deduced by analogy.

58.3.2 Time Measurement and Data Processing

Researcher conducted 20 times of data collection for optimized operation in deployment phase of measured weaponry with return-to-zero method. Besides, the researcher also calculated the control upper limit of the measurement data of each operating unit of operator, UCL $= \bar{x} + 3\sigma$, and control lower limit, LCL $= \bar{x} - 3\sigma$, with SPSS software and then wipe off data out of the range of $\bar{x} \pm 3\sigma$; at last, the researcher calculated the average of the remaining data of each operating unit. The result is shown in Table 58.2. The dark background means abnormal data.

58.3.3 Determine Operating Time Standard of Deployment Phase

According to the Table 58.2, the “pass” time of optimized operation of operator 1 in deployment phase of certain type of chariot is $T_{及} = 83$ s; the “pass” time of operator 2 is $T_{及} = 236$ s; the “pass” time of operator 3 is $T_{及} = 69$ s; the “pass” time of operator 4 is $T_{及} = 272$ s; and the “pass” time of optimized operation in deployment phase is $T_{及} = 272$ s.

According to the formulas (58.4) and (58.5), “good” time of operator 1 in deployment phase of certain type of chariot is $T_{良} = 73$ s; “Outstanding” time is $T_{优} = 63$ s. In this way, we can deduce the “good” time and “outstanding” time of operators 2–4 and the figures after optimization in deployment phase, as shown in Table 58.3.

Table 58.2 Time studies in the deployment-phase optimization

Operating unit	Operating time (s)																			
	1	2	3	4	5	6	7	8	9	10	11	12	13	14	15	16	17	18	19	20
11	9	8.6	8.2	7.1	7.4	6.9	7	7.4	7.8	7.6	7.9	8.3	8.4	7.9	8.6	8.4	7.2	7.3	7.6	7.9
12	5.2	5.6	4.7	5.1	4.6	4.4	5	4.6	4.5	5	5.2	5.6	4.7	5.1	4.6	4.4	5	4.6	4.5	5
13	19.8	17.3	19.5	19.2	17.6	18.1	17.6	17.3	17.9	18	17.6	18.3	18.9	19	18.6	*24.6*	18.4	18.6	18.7	18.4
14	3.7	3	3.5	3.5	3	3.4	2.6	2.9	3.4	3.4	3	2.9	3.7	4	4.3	3.7	3.8	3.8	3.1	3.6
15	42.3	31.8	33.7	33.1	40.5	36.5	35.8	31.1	33.6	38.4	44.2	35	38.5	30.4	30.5	35.2	42.4	31.8	34.5	37.3
16	6.2	5.3	6	5.9	5.6	4.8	4.8	4.6	4.9	5.4	5.6	5.8	5.7	5.3	6	6.1	5.9	5.8	5.9	5.3
17	7.8	7.2	7.8	7.2	6.2	6	6.4	6.6	7.3	7.4	7	7.9	6.7	7.2	6.4	7.4	7.4	6.9	7.5	7
21	11.9	12	12.5	11.8	12.1	10.7	10.8	11	12.8	13.4	12.9	14.5	10.9	13.8	12.9	14	13.7	10.2	10	11.2
22	4.7	5.1	4	4.3	3.9	3.8	4	3.8	3.9	4.5	5	4.8	4.2	4.6	4.4	3.8	3.8	4	4.6	4.2
23	10.1	10.3	10.1	10.5	10.1	9.9	10	9.8	10	11.4	*14.1*	11.8	11.8	9.5	10.2	9.8	11.4	10.6	11.7	10.6
24	17.9	16.3	20.6	19.2	17.4	20	16.8	18.2	19.7	22.6	21.4	19.4	19	18.3	18.7	17.2	16	16.8	17.2	17.2
25	8.2	7.8	7.2	6.9	6.3	7.8	7.4	6.8	7.6	7.8	7.5	7.5	8.2	8.1	6.9	7	6.4	8.9	7.4	7.1
26	34.4	20.1	21.2	21.2	28.9	21.9	23.8	17.3	18.1	18.1	24.2	17.9	26	18.1	20.9	26.7	31	21.6	23.9	27
27	75	74.8	71	73	71.5	70	70.2	70.5	72	73.6	71.5	70.9	74	75	69	70.3	74	*83*	73.8	73.9
28	78.5	77	77.3	75.2	76	74.3	74	75.7	79.4	80	80.4	76.7	79.1	74.6	76	74.3	78	77.1	74.8	73.9
29	12.3	11.2	10.7	9.7	10.4	9.7	9.7	9.5	10.8	11.4	10.6	12.7	12.7	11.9	10.7	13.2	12.8	11.9	12.4	11.6
31	3.2	3	3.4	3.1	2.6	2.6	2.8	2.5	2.8	3.4	3.6	3.1	3.8	4	2.9	2.7	3.6	3.4	3.4	3
32	12.7	13	13.1	11.4	11.4	10.8	10.2	11	12.4	11.6	12.8	13.2	11.8	10.6	11.9	11.9	12.4	10.3	10.8	11.4
33	10.1	10.3	10.1	10.5	10.1	9.9	10	9.8	10	11.4	*14.1*	11.8	11.8	9.5	10.2	9.8	11.4	10.6	11.7	10.6
34	5	4.2	4.4	4	3.9	4	3.6	3.7	4.6	4.7	4.2	5.3	3.4	3.9	4	4.6	4.9	3.7	3.9	4.2
35	9.2	9	9.5	8.7	8.5	7.2	7.7	7.4	7.6	8	8.2	7.9	9.1	10.3	9.2	8.1	7.9	8.6	9.5	7.8
36	22.6	20.1	19	17.9	19.3	17.5	18	17.6	18.2	18.6	23	18.4	22.6	18.3	21.7	22.4	21	19.3	22.8	19.5
37	9.8	9.5	10.5	10.7	11.3	9.6	10.2	9.7	10.7	11.6	12.3	11.4	13.9	12.1	14	11.6	9.7	10.5	11.1	10.2
41	8	7.5	8	7.9	6.8	7	6.5	6.7	7.3	7.9	8.2	7.1	9	6.4	7.6	8.4	7	6.3	8.4	7.9
42	5.6	4.9	5.2	5.6	4.8	4.3	4.5	4.3	4.9	5.6	4.2	6.1	5.7	4.9	4.2	5.7	5.2	4.8	5	4.6

(continued)

Table 58.2 (continued)

Operating unit	Operating time (s)																			
	1	2	3	4	5	6	7	8	9	10	11	12	13	14	15	16	17	18	19	20
43	4.4	4.2	4.5	4.6	3.8	3.3	3.7	3.5	4.6	3.9	5.2	5.8	4.9	3.1	3.8	4.7	5.3	3.9	3.7	4.2
44	8	9.7	8.9	6.9	8.7	8.7	8.3	8.8	8.4	9	11.4	9.1	7.8	9.7	9.4	5.6	9.9	9.3	8.8	8.3
45	55	40	43	43	49	46	45	39	42	46	49	42	47	43	43	48	49	42	43	47
46	4.4	3.8	4.5	3.6	3.8	3.6	3.3	3.4	3.8	4.6	3.4	5.3	5.1	4.8	3.5	3.7	3.9	4.6	5	4.3
47	202	186	193	185	197	175	172	186	190	176	184	183	198	196	188	179	192	186	190	183
48	4.3	3.5	4.9	4.3	3.2	3	3.4	3	3.6	2.9	4.7	3.7	3.7	4.7	3.8	3.2	4.9	3.4	4.9	4.6
49	6.2	6.1	7.1	6.6	6.3	6.2	6.9	6.6	6.9	7.3	7.3	6	9	7.1	6.9	7.2	6.4	5.9	6	6.4

Table 58.3 Operating time standard after optimization in deployment phase

	Pass	Good	Outstanding
Operator 1	83	73	63
Operator 2	236	208	179
Operator 3	69	61	52
Operator 4	272	239	207
Optimized operation	272	239	207

58.4 Conclusion

There are meanings of this research. First, further optimizing the operation flow of a model of tank, remove effective hours during operation process, shorten training hours, alleviate operation load and realize the optimized combination of human and machine; secondary, provide reference for the making of assessment standard for operation subject of a model of chariot to enhance efficiency and quality of exercise. The core method has same guiding meaning for other researches in relevant fields. However, there is certain subjectivity in both the selection of tested persons and determination of selection parameters during research so as to cause deviation of time standard determined. Therefore, in the follow-up works, the objective analysis on above factors shall be stressed to get more applicable time standard.

References

1. Guo X, Guo D (1991) Outline of ordnance operation. Tianjing EST Translation press, Tianjing
2. Pang Z (1999) Air defense forces man-machine-environment system engineering. Air Defense Forces Academy Press, Zhengzhou
3. Guo F, Li S (2003) Research on work measurement technology application. Ind Eng 05:57–60
4. Yi S, Jiang Z, Chen Y (2011) Work study and human factors engineering. Tsinghua University Press, Beijing
5. Wang Y (2009) Study on stopwatch method to develop standard working hours. Inner Mongolia Sci Technol Econ 05:130–131
6. Yi S, Guo F (2005) Basic industrial engineering. Machine Press, Beijing

Chapter 59
Research of Five Kilometers Armed Off-Road Training Methods

Ming Kong, Shuai Mu, Zhibing Pang, Min Chen, Honglei Li and Hongyan Ou

Abstract With the increasingly complex battlefield environment in the future and the more complete of single-soldier combat equipment, load-bearing ability needs a higher degree for the soldiers. But as an effective means of cultivating this ability, 5 km armed off-road needs breakthrough on training methods. Aiming at this problem, guided by the man-machine environment system engineering theory, based on the principle from related disciplines, we made a positive trial on 5 km armed off-road. In conclusion, 5 km armed off-road should follow the principle of step by step, proceed from actual combat, combine aerobic and anaerobic, in line with the comprehensive diversified direction, training rigorously and accurately, thus scientifically and reasonably improving body quality and making training program, and finally reduce the randomness and blindness of the training.

Keywords 5 Kms armed off-road · Experiment research · Training methods

59.1 Introduction

Five kilometers armed off-road training is one of the required training items extensively carried out by armies in all countries in the world. It can not only develop physical function of soldiers, but also boost comprehensive physical development and cultivate brave firm willpower [1]. With increasingly complicated campaign environment in future battles, and gradually improved and complete single-soldier combat equipment, the war has higher and higher requirement on loaded implementation of combat tasks of soldiers [2]. Current 5 km armed off-road teaching training has no any significant breakthrough in our military schools.

M. Kong (✉) · S. Mu · Z. Pang · M. Chen · H. Li · H. Ou
Air Defense Forces Academy, Zhengzhou, China
e-mail: fkb_mmes@163.com

S. Long and B. S. Dhillon (eds.), *Proceedings of the 13th International Conference on Man-Machine-Environment System Engineering*, Lecture Notes in Electrical Engineering 259, DOI: 10.1007/978-3-642-38968-9_59,

Therefore, to solve out above problems, under the instruction of modern teaching theory, starting from fundamentals of relevant subjects, some attempts about training method of 5 km armed off-road training system program are made in this research.

59.2 Purpose of Research

Apply modern PE thought and theoretic outcomes in the 5 km armed off-road training to enhance effect and quality of training. Explore a new simple, easy road suitable for characters of teaching training of military school with better effect to provide theoretic basis for teaching and training of 5 km armed off-road training in the army.

59.3 Object and Method of Research

59.3.1 Object of Research

In this experiment, 75 21–25-year-old male students in a military school were selected as test objects. All of them are in good health with basic physical fitness test meeting the standard, but they have never contacted with armed off-road training.

59.3.2 Site and Facilities

The site setting conforms to the requirements on 5 km armed off-road training field. The clothes include winter camouflage combat uniform, combat hat, camouflage rubber shoes and belts outside the clothes. The armed soldier carries equipment including one 95 type automatic rifle, one setoff cartridge belt, four empty clips, one set of gas mask, four hand grenades, one satchel (with washing tools inside) and one water kettle (full of water).

59.3.3 Research Methods

Comparative analysis method and mathematical statistic method.

59.4 Design of Training Method

59.4.1 Thought of New Training Method Design

Currently, the training method of large physical load and long distance is adopted in traditional long distance running items like 5 km unarmed running training. This method just develops the aerobic metabolism of body in the absence of training in technique and other organ functions and coordination of anaerobic metabolism. Such training method lacks of systematic conception, ignoring influence of human's other functions, so as to limit the further enhancement of aerobic metabolism to a certain extent. We designed 5 km armed off-road training method based on the opinion of systematics: to enhance aerobic metabolism of human body, besides development of function of aerobic metabolism itself, we should also stress enhancement of other functions to make them adapt to aerobic metabolism. The latest research shows aerobic metabolism and anaerobic metabolism always promotes each other [3]. Accordingly, our training method is to organically integration of endurance, speed, and strength trainings and combine aerobic metabolism with anaerobic metabolism equivalently to form a new training procedure, avoid the malpractice of staying on surface, and ensure students' physical quality is enhanced as a whole. Besides, the opinion of control theory penetrating into the whole training ensures the rationalization of training intensity and training volume, avoids high intensity lasting, and prevents training—caused injury occurring.

59.4.2 Training Contents

A time of training includes preparation activity, main training contents, and arrangement activities. The preparation activity is prior to official training contents, lasting for about 10–15 min, composed of by jogging, setting-up exercises, and stretching.

The main training contents include speed, strength, and endurance trainings. Speed diathesis training contains blow—step walk, run, jump, squat to stand up, 100, 10 × 5 m shuttle run and some other items; strength training include push-up and extension training, sit-up and extension training, pull-up and extension training, upside down, squat leg lift some other items; the selected items for the two quality training have low requirement on-site equipment, technical difficulty, easy to be carried out. But we have changed the items of our endurance training and mainly adopt follow two items.

Firstly, ability-based group training. Ability group training means grouping according to ability of students and arranging different training contents and load. The training intensity of each group is enough to bring certain training effect for the whole group and every student. For instance, when arranging run training, the standard is time instead of distance. In this way, the group with outstanding ability can run longer than the group with common ability so as to realize appropriate training intensity and enhance ability of each student.

Secondly, intermittent repeat training method. In run exercise, students can complete certain number of working set (run) in time limit; working set is followed by a recovery set (walk). For instance, complete a 30 s of sprint and then recover in a 60 s of walk. Take above process as a set of exercise and arrange repeating several sets of exercise as the case may be. Through repeated exercise alternating between high intensity and low intensity, the muscles are stimulated repeatedly to strengthen the fatigue resistance of relevant muscles and simultaneously enhance functions of aerobic metabolism and anaerobic metabolism. The ratio of duration of work set to duration of recovery set in such intermittent run training is 1:2. Based on this principle, extension is better. For instance, set two rounds of run and one round of walk as a set and require student to complete five sets.

Arrangement activity follows main training content, lasting for about 10–15 min. The arrangement activity is composed of relaxing run (walk), stretch exercise, and soothing massage.

59.4.3 Training Volume

The whole training period is 4 weeks, 5 days per week, 60 min per day. Three times of strength training, two times of speed training, and two endurance training are arranged through alternate combination.

Training intensity arrangement is according to the state of the max CO_2 ingestion and the max cardiac output. The best heart rate measurement is a very simple method:

Best heart rate value (time/min) = (max heart rate−quiet heart rate) 70 %+quiet heart rate

Week 1 is adaption week, and the heart rate shall be controlled within 80–85 % of the best heart rate, 85–95 % in Week 2 and 95–100 % in Weeks 3–4.

59.5 Test Results

This experiment includes three tests: unarmed 5 km test and armed 5 km test prior to training, and armed 5 km test after a 4-week training. The contrast of test results is shown in Table 59.1.

Table 59.1 Classification on 5 km running

Statistic result	Unarmed before experiment	Armed before experiment	Armed after experiment
Average achievement	22:17	26:59	25:48
Pass rate	88.5 %	64.2 %	92.8 %

59.6 Test Results

59.6.1 "Step by Step" is the Training Principle of 5 km Armed Off-road Training

Step-by-step training helps body adapt to pressure of training. In case the intensity or duration at early training period is too much or increased too fast to follow the principle of step by step, the physical ability and status of the trainees cannot adapt to demand of training or enter effective recovery. This may cause overtraining or injury. Such situation is common in physical training in our army. Due to particularity of the army, the training always faces high standard, heavy task, and tight schedule and has to meet high requirement in a short period. That causes training organizer keeps stressing large intensity, high load, and long duration. Such spirit is good, but it should not be encouraged in the angle of view of scientific training because such training method is very easy to cause sports injury. If trainees as the main body of the battle effectiveness of a troop are injured in training, the battle effectiveness of the whole troop will be affected inevitably [4].

In this experiment, our training plan reflected this point. The training is in three phases: initial phase, development phase, and maintenance phase. The training in initial phase involves muscle endurance exercise with small intensity and cardiopulmonary endurance exercise at a medium level to control injury by generating the least muscle ache. The training in development phase is to enable trainees to obviously improve fitness level by all-round enhancement of training stimulation. The training in maintenance phase is to maintain effect of cardiopulmonary endurance, muscle strength, and muscle durance training.

59.6.2 Scientificity and Reasonability is the Training Rule of 5 km Armed Off-road Training

Current physical training in our army has not met the requirement of scientificity and reasonability, and there is no unified standard for volume and intensity of many trainings. Organizer tends to make standard based on his experience. Such training is unsystematic. Besides, during organizer of physical training always stresses number of time, duration, and other data which are easy to be quantified and ignores the quantity of act completed. Actually, the quality of technical act is important just like the load and times of repetition of each set. It can not only enhance sports skills and ability, but also reduce possibility of injury caused by error in training. The duration, contents, volume, and intensity of training every day have been calculated in this experiment under on-site instruction of many teachers to ensure strictness and scientificity of training. The test result shows that a precise training plan and strict implementation are the guarantee for efficiency and effect of training.

59.6.3 Start from Actual Combat is the Realistic Requirement which must be Make Clear in 5 km Armed Off-road Training

Start from actual combat is the core of 5 km armed off-road training. There are specific requirements on the clothing, equipment, site setting, and all other respects during training. Among them, weather condition has not been included in "Basic Requirements", but it has big influence on achievements of trainees. Our experiment was conducted in November–December, the coldest time in China's north area, with temperature around 0°. We also encountered bad windy, fogy, and snowy days during training. In the principle of start from actual combat, we did not change our training plan for these difficulties and organized outdoor training as usual. Just for exercises and psychological preparation in daily training, the trainees still reflected high ability to meet emergency and exerted normal level even in the final assessment which was in a snowy day. Therefore, we shall consciously cultivate trainee's adaptability in various weather conditions. Battlefield is not a drilling field, and any combat mission cannot be postponed for weather reason. Soldier must have strong adaptability and is able to maintain high will of fight and confidence in any adverse weathers. Training in different weathers can fully develop psychological endurance of trainee. In this way, trainee can calmly implement combat mission in real battlefield in any adverse weather.

59.6.4 Combination of Anaerobic Metabolism and Aerobic Metabolism is an Effective Approach of 5 km Armed Off-road Training

Besides aerobic endurance training, the experimental group added a large volume of anaerobic training to make aerobic training and anaerobic training in proportion, supplementing each other. The result shows that the training effect is remarkable. Authors believe that such effect shall owe to a mechanism: First of all, anaerobic endurance training with appropriate intensity can enhance level of function of human body in full in an all-round way, boost autonomic nervous system and motor nervous system early harmonious unified, reduce a large accumulation of oxygen lack in aerobic sports, postpone occurrence of the limit, and even span the limit so as to speed up the second breathing transition. That is crucial for high-intensity endurance run items like 5 km armed off-road training. Secondly, anaerobic metabolism training facilitates the accumulation of product of glycolysis of muscle glycogen; long-endurance training can strengthen ability of body to resist the acid; after training, the aerobic oxidation of lactic acid can help effectively recover and excess recover to increase energy reserve of body. In addition, anaerobic training is significant for overcoming adverse psychological phenomena and cultivating fortitudinous willpower. That is just one of necessary diatheses for a soldier on the battlefield [5].

59.6.5 All-round Diversified Balanced Development is the Right Direction of 5 km Armed Off-road Training

No matter from realistic demand of implementing combat task or from effect of 5 km armed off-road training, any of three diatheses, strength, speed and endurance, is indispensable. Among selected experimental objects, several cannot maintain balance among the three diatheses: some are good at endurance and 5 km unarmed run is within 15 min, but their strength diathesis is too bad; some are outstanding in speed diathesis, but with relatively weak endurance; in 5 km armed off-road training item, these people who cannot maintain balance of three diatheses are not satisfying prior to training. After comprehensive training of 4 weeks, all reach the standards. This shows that the comprehensive diversified training adopted is effective. Through diversified training approaches, the trainee keeps accepting new stimulation so as to convert passive training into positive training, mobilize initial of training and double effect of training; 5 km armed off-road training shall take comprehensive training of above three items of quality as the goal and cannot just develop a single item.

References

1. PLA Army Military Training and Examining Outline. PLA Press (2008)
2. Wen-sheng W (2009) The systematic teaching procedure for the military 5 km cross-country in battle gear. J PLA Inst Phys Edu 28(3) Guangzhou
3. Bao-lei Y (2011) Study on the improvement of 5-kilometer cross-country armed exercise performance through training under damp-heat environment. J PLA Inst Phys Edu 30(1), Guangzhou
4. Si-qin P (2004) Study on developing the long- distance runners' capability by double anaerobic exercises. J PLA Inst Phys Edu 23(1), Guangzhou
5. Li J (2001) Study on psychology training method in army. J PLA Sports Acad

Chapter 60
Study on Improving Operating Performance of Human Factors

Cheng Jin, Zhibing Pang, Hua Li, Zhiwen Yuan and Jiang Wu

Abstract The weaponry job performance is an important indicator of the combat effectiveness to the human-machine system. The article starts from the concept of operating performance and human factors analyzes the relationship between job performance and human factors. As the "person" of the weapons and equipment the operator is a major factor in the human-machine combination, emphatic reliability study of the physiological and psychological level should be to improve the operating performance. The article analyzes the physiological level that includes strengthening peacetime exercise, to improve the physical function; pay attention to the protection of organ, to adhere to the regular medical checkups; regularing daily life, to keep them in good condition. This article also analyzes the psychological level that includes: deeping the psychological education, to strengthen the psychological qualities; and universing the psychological counseling, to improve self-adjustment; the scientific training plan is very important, as the same time relieving the mental fatigue. The article proposes to improve the human-machine combined with operating-performance-related countermeasures, the purpose is to guide people during the human-machine combination, emphasis on human physiological and psychological research applications. Certaining reference value for the further study of operator selection and job performance and significance.

Keywords Task performance · Human factor · Physiological · Psychological

C. Jin (✉) · Z. Pang · H. Li · Z. Yuan · J. Wu
Air Defense Forces Academy, Zhengzhou, China
e-mail: fkb_mmes@163.com

S. Long and B. S. Dhillon (eds.), *Proceedings of the 13th International Conference on Man-Machine-Environment System Engineering*, Lecture Notes in Electrical Engineering 259, DOI: 10.1007/978-3-642-38968-9_60,

60.1 Foreword

Task performance is the sum of the describable work behavior and measurable work outcome of individual (group) in organization in a specific period and work effect achievable by the individual (group) in future specific period which is predicted by organization through instruction of improvement and perfection in combination with the quality and ability of the individual (group) in the past work. The influence of human on task performance is crucial. The influencing factors for human may be classified into two types: physiological factors and psychological factors. In both respects, we bring out specific requirements on selection of operator in different conditions and attach full importance to enhancement of physiological functions and psychological diathesis of the operator. If we need further enhancement of task performance of human-machine operation, we must enhance research at both layers to lift task performance of human in an all-round way.

60.2 Task Performance and Human Factors

60.2.1 Basic Conception

60.2.1.1 Task Performance

Literally performance means achievement and efficiency. Currently, there are three understandings on performance: first, deem performance as result; the definition on the result-oriented performance tends to associate performance with task completion conditions, objective achievement conditions, result, output, etc. definition: the record of the output of special job functions or activities in specific period; secondly, see performance as behavior and process. Definition: the measurable thing associated with organizational objective which is achieved by people; thirdly, believe that performance is the sum of above both and that performance not only contains process of behavior, but also stresses result of behavior [1].

60.2.1.2 Human Factors

Human factors science is a subject combining knowledge and methods of psychology, physiology, anatomy, and engineering, researching interaction between human and machine and environment, how to consider work efficiency in work, life and rest, health, safety, and comfort of human, etc. it is to discuss the new field of interrelation among human, work, and life with main function of research how to design a new environment most suitable for human's life and work. In most of the countries in Europe and America include human factors science in the category of industrial engineering study. Through decades of development, this science may play an important role in industrialization and improvement of life quality of public. In nature, the objectives of human factors research include "make things adaptable to human" and "how to make things adaptable to human" [2].

60.2.2 Factors Influencing Task Performance

The factors influencing task performance are in many ways. In the angle of view of MMESE, the factors involve three respects: (1) human factors, (2) Weaponry factors, and (3) Environmental factors.

60.2.2.1 Human Factors

The influence of human factors is mainly reflected in the physiological and psychological diathesis, operation method, operation skills, and proficiency of operator. Human factors are dominating. Performance is the behaviors with assessable factors and these behaviors have positive (negative) influence on efficiency of individual or organization; the state of human plays a positive (negative) role in realizing organizational goal and has influence on the final performance. Therefore, human factors have crucial influence on task performance and shall have priority in analysis.

60.2.2.2 Machine Factors

Weaponry factors are fixed factors. After equipment has established, its performance and tactic technical conditions have been shaped. The reliability, environmental adaptability, and basic combat performance of weaponry are inherent factors and relatively stable factors in capacity in basic combat conditions. Therefore, the influence of weaponry on task performance is relatively fixed.

60.2.2.3 Environmental Factors

Human and machine operate and work in a certain environment and many natural processes in the environment are uncontrollable factors like combat zone, combat season, natural environment, artificial environment, and social environment all which may cause great influence on weaponry's task performance. Generally, environmental factors change frequently with high randomicity. Therefore, environment is also an important factor influencing task performance.

Among above three factors, human is dynamic changeable. Training can enhance ability of human and improve factors adjusting and maintaining ability.

Therefore, the in-depth research on influence of human on task performance is significant.

60.2.3 Influence of Human on Task Performance

Task performance is a comprehensive indicator measuring operation level of operator, influenced by the number of time of operation, proficiency, operation environment, and other factors. Human task performance research is an integral part of ergonomic research. There are many factors influencing task performance. Among these factors, human, weaponry operator and user, for inherent characteristics and attribute, causes the complexity and randomicity of operation result with huge influence on task performance. In the view of systematic engineering, in human-machine-environment relation, human is a dominant factor. Researcher finds, everyone is qualitatively different in physiological characteristics and psychological behaviors and quantitatively distinct in growth speed and development level. Such difference is caused by inherent factors and acquired factors together. The existence of such difference has direct influence on the task performance and work efficiency. Therefore, enhancement of qualitative and quantitative research on psychological and psychological factors of human can directly instruct how to enhance task performance of human-machine combination.

60.3 Countermeasures

Enhancement of task performance shall start from solving out interrelation among factors on the basis of the systematic analysis on three factors [3]: human, machine, and environment. Among these factors, human is the priority of human-machine-environment system design. No matter in equipment, tool and operation environment design, or in arrangement of operation volume and method, the physiological and psychological characteristics of human must be taken into account. Human factors research is a systematic engineering. Human factors influence task performance in many ways. In the view of enhancement of task performance, the two levels "physiological factors" and "psychological factors" must be put on the priority.

60.3.1 Physiological Factors

The influence of physiological factors of human on task performance is dominant factors. These factors are reflected by body, behaviors, route, and skills of operator in the form of standardization, accuracy, and fastness with high visibility, observability, measurability, and controllability. At physiological layer, following three conditions must be satisfied when taking countermeasures of enhancing task performance: (1) enhancing daily exercise and improve physiological functions; (2) paying attention to organ protection and insisting in periodic physical examination; and (3) maintaining regular daily life and good state.

60.3.1.1 Enhancing Daily Exercise and Improve Physiological Functions

"Health is the fundamentality of our body" and any job depends on well health. Due to influences of many factors, today, some applicable weaponries have special requirements on operator no matter in fitness, skill, or intelligence. Some weaponries have requirements in single respect and some in many respects. But the requirements on physical diathesis are the most fundamental at any time. A modern weaponry operator should not only conform to basic physical conditions of a soldier, but also enhance exercise of relevant parts and rise relevant parameter standards according to specific conditions and requirements. The operator should enhance physical exercise and effective improve functions of body to better adapt to high-intensity complicated operation environment and ensure the fastness, accuracy, and efficiency while intensively studying business. He should also extend the duration of stable operation to the largest extent to endure the enhancement of performance of human-machine weaponry operation [4].

60.3.1.2 Paying Attention to Organ Protection and Insisting in Periodic Physical Examination

There are different requirements on physiological functions of human in special operation environments. For instance, the requirement on the eyesight of people responsible air situation search for on the observer post is far higher than that for normal people. Observer must see the object first of all. In addition, there are specified requirements on the extent and precision of the observation. Some equipment and work environment also bring special requirements for operator like cardiopulmonary functions and anti-dizziness ability. It shows the operators of different weaponries must meet different requirements on sensory organs of the individuals according to features of the posts. Therefore, operators of different types must pay attention to the protection and appropriate exercise of key organs according to actual conditions to ensure them always in good state. Besides, we should periodically organize physical examination, pay attention to examination for key parts, and take scientific test and protection approaches to ensure organs in good state in a long period; establish the concept "health means battle effectiveness" and conduct division of responsibility to identify and eliminate signs and phenomena of injury of body functions by work in real time and maintain operation performance at a high level duly.

60.3.1.3 Maintaining Regular Daily Life and Good State

Work and rest system in army helps us maintain a regular daily life and enhance military skills and operation level. As modern weaponry operators in army, we shall not only grasp military and business skills well, but also adjust our physical state and enhance physical diathesis and immunity by regular daily work and rest system. Meanwhile, we must refuse corrosion of various bad habits on our physiological and psychological health. "Don't suppose small good and not to do, and small bad to try it" is never an empty lip service. All common gastric illnesses in army start from small bad habits like ignoring dietary. Once a war breaks out, the regular life will be interrupted. The key to whether body can adapt to high-intensity combat in a short period and whether people can maintain efficient task performance is physical diathesis accumulated daily. Sometimes just a little bit, more insistence is the key to the final victory. Therefore, a regular daily life and a good state are significant for task performance of human-machine combination [5].

60.3.2 Psychological Factors

The influence of psychological factors of human on task performance is a hidden factor, unobservable, unpredictable, and non-valuable. That poses difficulty for identifying, analyzing, and settling the problems. It is crucial to enhance adaption training, self-adjustment, and scientific leading. Psychologically, task performance shall be enhanced in following three ways: first of all, deepening psychological education and intensifying psychological diathesis; second, popularizing psychological consultation and completing self-adjustment; third, alleviating physiological fatigue by a scientific training plan.

60.3.2.1 Deepen Psychological Education and Intensify Psychological Diathesis

Today, more and more people realize psychological health is the real health and a psychologically healthy person can bring out the biggest potential better and thus make more achievements. In particular, a psychologically healthy operator can take more advantages of human being and enhance task performance. Therefore, psychological diathesis is a precondition for a soldier to realize efficiency in military and political diatheses and operation equipment. With mature application of psychological warfare in modern wars, psychological education has been included in military education system step by step. Today, all basic units of the army have full-time psychological consultants to regularly organize psychological education course. Through psychological education, our soldiers can effectively grasp fundament of psychology according to demand of military life, know how to analyze psychological phenomena and grasp methods of maintaining psychological health and solving out common psychological issues. It is significant to intensify psychological diathesis of operator and enhance ability to cope with complicated operation issues and unexpected events. More importantly, it helps operator easier deal with baldness and carelessness which may appear during operation so as to reduce occurrence of error of operation and realization and ensure efficiency of task performance.

60.3.2.2 Popularize Psychological Consultation and Complete Self-adjustment

Psychological consultation is still a fresh thing in China. In basic units in our army, most of political instructors have responsibility for psychological consultation. With gradual progress of society and step-by-step enhancement of modernization level in army, the characteristics of relatively low level, diversification, and hierarchy of psychological health of people become more remarkable. The operators long working in the environment of noise and vibration tend to feel anxiety, fidget, and other psychological discomforts. Some methods like conversation, observation, and test in psychological consultation can help analyze internal reason, make a reasonable self-adjustment measure and postpone occurrence of fatigue and psychological discomforts so as to improve duration of operation and endurance of operator. All basic units of army shall organize periodic psychological test for operators of different types of weaponries to grasp their psychological endurance and self-adjustment ability. In this way, they can effectively ensure psychological health of operators and help them maintain a good state during operation. It is also important to help operators solve out psychological issues, cure psychological diseases, and maintain a sound psychological diathesis according to their personalities and behavioral differences. On the one hand, these units should build up and complete psychological education mechanism; on another hand, they may invite psychological experts to classroom or symposia in order to import psychological knowledge in the brain of operators by subtle influence and enhance their ability of psychological adjustment. Sound ability of psychological self-adjustment and health psychological diathesis is the key to reliable human-machine combination and must draw high attention of all commanders at all levels.

60.3.2.3 Scientific Training Plan and Alleviate Psychological Fatigue

After making weaponry in place in a troop, we will seek for the best point of combination of human and weapon. In the circumstance that weaponries have determined, operators shall take the initiatives to adapt to the weaponry. However, we cannot organize training blindly, and we shall also complete various training tasks within the scope of physical and psychological safety of operators. In special operation environment, the operator is easy to feel physiological fatigue and lower the operation efficiency. In order to change this result, on the one hand, operator must get appropriate psychological dredge and adjustment, on the another hand, the units of all levels shall arrange training plan scientifically. Scientific reasonable arrangement of training plan is the only way to enable operator grasp operation skills in proper sequence, enhance reliability of combination between human and machine, and postpone physiological fatigue and shorten recovery period. That is an important guarantee to maintain long duration and efficiency of human-machine combination.

60.4 Conclusion

The human factors in the human-machine operating performance are completed and uncertain. Due to physiological and psychological properties, human is easy to be affected by many factors outside so as to impact change of operating performance [6]. At physiological and psychological levels of the operators, this paper discusses the relevant measures to enhance human-machine operating performance to instruct people to pay attention to the applications of physiological and psychological research outcomes. No matter the commanders or the operators, all of them shall make clear the relation between physiological and psychological factors in human-machine operating performance, conduct training, management, adjustment, and evaluation in a scientific way and keep enhancing reliability and reducing blindness.

References

1. Sun L (2001) Human factors engineering. China science and technology Press
2. Zhu Z, Ge L, Zhang Z (2000) Engineering psychology. People's Education Press, pp 1–4
3. Pang Z (1999) Air defense forces man-machine-environment system engineering. Zhengzhou Air Defense Forces Academy, p 91
4. Benli X (2006) Special operations soldiers mental health education guide. New Times press
5. Tang C, Pang Z, Zhao H (2011) Study on the influence of stamina to the performance of multiple person operating one machine. In: Proceedings of the 11th conference on man-machine-environment system engineering, vol 10, pp 332–335
6. Li T, Li H, Pang Z (2012) Experiment and study on stamina of multiple person operating multiple machine operation. In: Proceedings of the 11th conference on man-machine-environment system engineering, pp 236–239

Chapter 61
Practice of Virtual Reality Case Teaching Using in the Military Training Based on Virtools

Chenhui Li, Runfeng Hou, Zhibing Pang, Kehua Zou, Xiaofei Zhai and Shili Chu

Abstract In order to promoting case teaching effect of military theory, enhancing consciousness of resisting training, the military academy teaching should expand to method of virtual reality case teaching. Virtual reality case teaching not only makes use of various 3D model and alternant engines but also evolving various military tactics conditions for virtual scene. By assembling virtual scene and material model, the military academy students can carry on diverse and nonlinear step-by-step drill and analysis for military theory and equipment operation, and the military academy students can promote study effect of theory knowledge and equipment operating skill.

Keywords Virtual reality · Virtual reality case teaching · Military theory · Military teaching

With fast development of advanced simulation technologies like virtual reality, the application of new-type teaching and training mode of "enhancing hardware with software, replacing real scene with virtual environment and combining virtual environment in reality" in resisting training in military virtual environment, provides a new approach for satisfying demands in many respects like school teaching, military fighting and commanding training and equipment security training. Virtual reality case teaching is based on virtual reality technology and fighting simulation. Different from traditional teaching, by stimulating various sensatory organs, it makes student addicted into the course and thus accomplish the goal of fast learning, real experience, and enhanced skills.

C. Li (✉) · R. Hou · Z. Pang · K. Zou · X. Zhai · S. Chu
Air Defense Forces Academy, Zhengzhou, China
e-mail: fkb_mmes@163.com

S. Long and B. S. Dhillon (eds.), *Proceedings of the 13th International Conference on Man-Machine-Environment System Engineering*, Lecture Notes in Electrical Engineering 259, DOI: 10.1007/978-3-642-38968-9_61,

61.1 Introduction

The form of modern IT-based war changes extremely fast and the gap between different war forms is becoming narrow. To enhance capacity of army to modern war, it is necessary to enhance consciousness of resisting combat training and create a teaching training environment with tactic scene. Currently, academy is changing the mode of teaching and training in order to realize a new-type teaching method "learning modern war in lab and researching war mode in future development." Change of demand of modern war causes fast development of military technology. New types of weaponries appear one after another with enhanced complexity, better speed, high price, and limited output. Aiming at the status, to convert new technologies and new theories into battle effectiveness as early as possible and provide services for army, during teaching process, we must insist in "combination of theory and practice, integration of virtual technology and practice" and organize diversified nonlinear military specialty resisting training method.

In the traditional military theory and equipment operation teaching, some factors like nonobjective of theory and limitation equipment cause that separate of theory and practice, isolated of equipment operation, unrealizable joint operations training and adaptability tactical training. During weaponry learning and operation, students can enhance their basic weaponry operation ability by listening theoretic interpretation. However, due to impossible creation of tactic training environment, students have weak adaptability to weaponry operation. Therefore, a vivid weaponry operation training environment with virtual reality case teaching method has been built to enable students to accept weaponry adaptability operation training in corresponding tactic background during theoretic learning. Meanwhile, the distributive multi-unit integration grouped training and nonlinear deduction of teaching case have been developed with computer network technology so as to provide a brand-new technical mode for integrated united combat training based on information system in academy. Virtual reality case teaching adopts human-machine interactive functions of simulation and virtual reality technology. In comparison with traditional teaching training, this method can better adapt to requirements of development of weaponry teaching and training in the new situations.

61.2 Virtual Reality Case Teaching Method

During virtual teaching process, virtualization of the real weaponry and creation of a virtual tactic scene can break the limitation of traditional theoretic teaching and enable students to conduct theoretic learning and operation training based on their own characters. With virtual teaching system, teacher can present some event courses and phenomena which are complicated, abstract, and inappropriate to be observed directly so as to show teaching contents in an all-round way, in multi-dimensions. Aiming at real environment which may exist in teaching, corresponding virtual scene is developed to realize visualized virtualized 3D reappearance of tactic scene in order to solve out the problems involved in learning and research, satisfy demand of class teaching, enhance effect of combat training, and shorten period of real weaponry teaching. Therefore, virtual reality case teaching method created by combination between combat simulation technology and virtual reality technology can help academy thoroughly change traditional military theory research method and provide conditions and opportunity for reform and development of modern military theory teaching.

Virtual reality case teaching [1] is a new-type military theory teaching method brought and gradually enriched by scholars including Prof. Hu Xiaofeng from National Defense University. This is a case teaching form which is aiming to realize theory and weaponry on the basis of computer war simulation system; through providing more vivid case environment and space–time, the system virtualizes whole process of various actual or possible cases and events in the past, present, and future and supports multi-variable, multi-nodes virtual play in the cases to achieve the goal of learning and research of military theory. Virtual reality case teaching is a special military case teaching form, similar with other military case teaching methods. It is characterized by subjectivity, practicability, and interactivity. But it also has remarkable differences from common case teaching.

First of all, support of computer war simulation basic environment. War simulation basic environment is the foundation for realization of virtual case. The teaching is to create virtual reoccurrence of researched case with 3D virtual reality technology and adopt multivariate nonlinear deduction and evaluation method instead of traditional linear presentation form. By war simulation system providing vivid war environmental and space–time scene [2], the teaching method presents the realistic condition of the case event to the largest extent while conducting multi-dimensional deduction and evaluation on the case in virtual scene to make case evolvement course dynamic changing.

Secondly, case virtualization of military theory and weaponry operation. Most of the cases used in traditional teaching had actually happened in the past. Through discussion and analysis on these cases, student can achieve the objective of theoretic learning. But the cases are more diverse in virtual reality case teaching. They may be the real cases in the past which are represented by virtual simulation environment to present war in history; they may also be the stimulation of future. Through proactive simulation, integrate all impossible situation in the future like new combat theories, action patterns and weaponries in the virtual battlefield environment; observe development course of war by simulation; and analyze and evaluate the matters involved in the research in order to learn about war in the future. Besides, some events which never exist or happen in reality may be virtualized to learn and research diversified military actions. For instance, by assumptive analysis and research on a specific case of battle, change the course and result of the battle in order to make research on decision making. Therefore, it is valuable for use to conclude experience, realize nature of things and further deepen understanding and realization on uncertainty of the war.

Thirdly, diverse human machine interactive methods. The interaction in traditional case teaching is mainly reflected on the discussion and mutual inspiration, while the interaction in virtual reality case teaching may be in many ways: in terms of form of practice, virtual reality case teaching may involve independent research of individual, homework completed by group discussion, confrontation, and deduction among multiple groups, etc. to adapt to demands of different case teachings. With regard to analysis method, make static analysis on actual conditions of case or make dynamic deduction by changing some conditions to know about different results of case in different situations. As to realization method, make faster-than-real-time simulation by computer based acceleration in order to speed up simulation and shorten simulation duration, or conduct quasi-real-time simulation to research details of some events at a lower speed ,and organize human-machine counteraction to deem computer as rival in research, or multi-person counteraction in order to complete the deduction of the whole case by multiple students playing different roles.

Traditional case teaching is divided into six segments: exercise, assignment, individual learning, group discussion, conclusion, and achievement evaluation. Virtual case teaching is the adjustment and sublimation based on these steps according to its inherent characteristics. With virtual environment as a basic teaching platform, the whole teaching course is in a war simulation environment so as to form seamless joint of whole practice environment, simulation system, and teaching activity. Virtual reality case teaching may be divided into four phases: teaching preparation, learning organization, research on playing, and conclusion. Therefore, no matter virtual case or case virtualization, no matter future virtualization or reality virtualization, all help achieve the goal of innovation of military teaching, enable case teaching of military theory and weaponry to extend to deduction and innovation from simple conclusion and complete experiment and test in actual course of simulation.

61.3 Design and Development of Virtual Case Teaching System

Virtual reality case teaching system is classified in two major parts in general. One part is virtual war simulation basic environment and another part is military specialty training deduction system. In the part of virtual war simulation basic environment, conduct parameterization design for strategic factors like geography, weather, human culture and society and build up corresponding virtual basic environment in order to support teaching and training of military specialty cases, complete multivariate nonlinear deduction and evaluation analysis, etc. [3]. Military specialty training deduction system is a system customized according to demand of teaching of military specialties basically with high pertinence. It helps learn and analyze certain specialty or problems in certain field such as certain air target identification training system. The principle of the system is conducting special theoretic knowledge in combination with specific case in a virtual tactic background and thus obtaining corresponding feedback about operation so as to keep accumulating experiment and achieve the objective of grasping corresponding knowledge and skills.

Now we will introduce design and development method of military specialty training deduction system in detail with certain air target identification training system. The teaching system involves contents like basic knowledge, weaponry operation, target identification, and training assessment. Student can know well and grasp theoretic knowledge about air target identification through all these contents and enhance skill of air target identification by training and operation in corresponding tactic background.

Virtual reality simulation engine Virtools used in this system is a set of non-immersive virtual reality development tools developed by France's DASSAULT. With complete functions, visual development interface and human-machine interactive interface and programming environment of flow chart type, the tools are extensively used in the development of military virtual training system. In the graphic interfaces, developer drags behavior building blocks (BB) to the target and finally builds up multi-functional interactive application. Virtools contains 682 BBs so far and user can also edit and combine in their own ways. By Virtools' script language or Visual C++ and other tools, user can conduct self-defined development and form a BB system with certain function step by step.

Virtual reality case teaching based on Virtools platform and resisting training system include not only basic action and behavior of components (functional simulation), but also work logics among components (performance simulation). However, Virtools has no ability of building up 3D model, so generally other 3D modeling tools (like 3DS Max) are used in realizing 3D models like weaponry and landform models [4, 5].

This system erects tactic scene and production of equipment model on this basis and then convert these models with assisting file plug in of Virtools and finally import in Virtools platform. In the next step, according to relevant requirements of case teaching, after parameter-based editing of tactic conditions, natural conditions, and other conditions, position model and apply embedded BBs mechanically, for instance, it realizes flight path actions of single or multiple planes with plane models and functions like rotation, moving and zoom in/out of camera. Figure 61.1 shows BB's message transmission flow.

Now, we make a simple introduction by taking the example of MouseRotate. First of all, define speed of Entity3D variable and vector variable to save displacement data for saving model and mouse and create management object MB and rendering object RC.

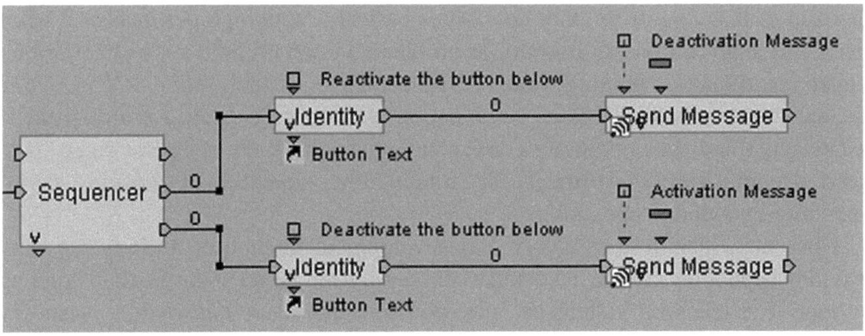

Fig. 61.1 The message function block structures

```
InputManager MB = InputManager.Cast (
  bc.GetManagerByGuid (GetInputManagerGuid ()));
  RenderContext RC = bc.GetPlayerRenderContext ();
  //Initiating variable when click with a mouse
  IntersectionDesc desc;
  rcx.Pick (x, y, desc); //mouse placement coordinates
  obj = desc.Object; //rotate model with mouse displacement
  MB.GetMouseRelativePosition (speed);
  speed.x* = RSpeed; //unit rotation angle
  Float delta = bc.GetDeltaTime ()/1000.0;
  if (axis ==1) {//when rotating around Y axis
  obj.Rotate (1.0, 0.0, 0.0, 0.0, obj));
  obj.Rotate (0.0, 1.0, 0.0, speed.x*delta, obj);
  obj.Rotate (0.0, 0.0, 1.0, 0.0, obj);}
```

When implementing air target identification virtual reality case teaching, in order to achieve the goal of virtual environment simulation, add a part of military task conditions, landform and weather blocks to virtual scene to create tactic scene of corresponding case in combination with 3D model of air target. Then, add operation functions of 3D models of devices like optical detection devices to provide student with sense of scene operation during theoretic knowledge learning, as shown in Fig. 61.2.

The interfaces of the system mainly include initial interfaces and main interfaces. Development of initial interfaces includes operation buttons, operation instruction menu; the development of main interface puts priority on interactive button, mainly realized by push button, wait message, wave play, show, and hide. Mouse is able to change image of button when it is within button scope and can realize functions like show, hiding, trigger, and inhibit by click.

The interactive menu is developed mainly with corresponding BBs, like mouse wait, 2D picking, sequencer, hide, show, active script, deactivate script, set as active camera, switch on parameter. If many menu item parameters are saved in switch on parameter, a part of menu may be hidden or shown by click and

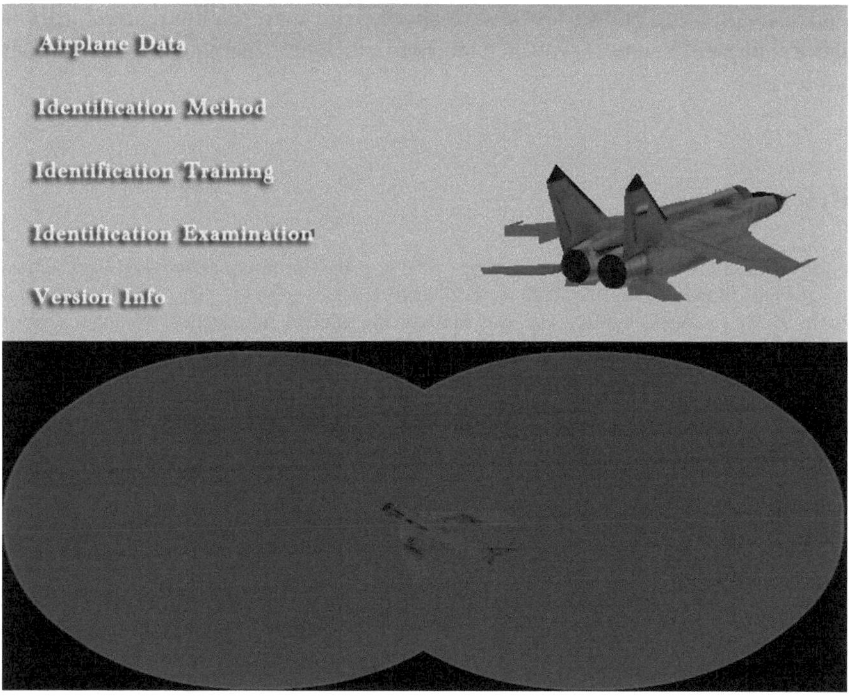

Fig. 61.2 An air target identification virtual case system

corresponding events (roaming, free track and identification operation) and others may be triggered.

After system development, make a function model to enter virtual case system. The system saved as CMO format is unrecognized, therefore release it as an EXE file, and then transfer and operate in the form of module.

61.4 Conclusion

Virtools-based virtual reality case teaching system helps students to visually three-dimensionally understand military theoretic knowledge before virtual tactic scene, grasp basic operation steps and flows of equipment, intensify resisting training consciousness by different case background set in the scene and deepen realization on specialty knowledge. Therefore, it contributes to the enhancement of teaching training effect based on tactic background. Creation of corresponding military case scenes corresponding to different specialties with a virtual reality software system allows students to conduct virtualized diversified nonlinear learning and grouped training prior to actual weaponry operation and tactic training to reduce the operation accident and weaponry loss during military training. It also can provide students with a repeatable, low-cost, high-efficient, short-duration virtual learning and training environment, enhance teaching efficiency, learning effect, and operation skills.

References

1. Hu X, Si G (2005) Virtual case teaching—The new path of information-based ages military teaching. J Equip Command Tech Acad 2005.10
2. Hu X, Si G, Wu L (2004) The war imitates the general introduction. National Defense University Publisher, Peking, pp 16–19
3. Wang P, Zhu Y et al (2010) Drive system to imitate the true platform design according to antiaircraft of HLA and Virtools. Mod Electron Tech (18):176–179
4. Mingkun L (2005) The treasure of 3D game designing. Electronics Sound and Record Image in Sichuan Publisher, Chengdu
5. Zuo T (2012) Design and application of educational software based on Virtools. Softw Ind Eng 4:40–43

Chapter 62
Predicting the Trend of Flight Safety State by According to Variation of Flight System State

Binchun Liu and Hui Wang

Abstract Correctly analyzing and predicting the trend of flight system safe state will do significant meaning to achieving the goal "zero accident" ultimately. Definitely, flight safety is a magnificent system and involving multiple factors. And the uni-effective method to analyze the complex flight safety system is to adopting synthetical integrating method qualitatively and quantitatively. This article advanced a fuzzy relation synthesis theory which can predict the safety state of flight system by synthetic evaluating the accumulated statistical data of one flight system qualitatively and quantitatively. The new method is manipulation adopted and can be high accuracy.

Keywords Flight system · Flight system safety state · Fuzzy relation synthesis theory · Synthetic analysis

62.1 Introduction

Fight system is a system to complete specific flight task. Evaluation and prediction on flight system is one of the very important methods to ensure flight safety. In this way, flight safety system composed of four subsystems—human-machine-environment–management—is taken as research object to research influence of its state change on occurrence of flight accident at different levels. The purpose of the system is to prevent and reduce occurrence of accident and make system in a state

B. Liu (✉) · H. Wang
Air Force Aviation University, Aviation Theory Department, Changchun, 130022 Jilin, China
e-mail: huangtsg@126.com

S. Long and B. S. Dhillon (eds.), *Proceedings of the 13th International Conference on Man-Machine-Environment System Engineering*, Lecture Notes in Electrical Engineering 259, DOI: 10.1007/978-3-642-38968-9_62,

of "zero accident". Generally speaking, there are three major types of flight safety prediction technologies [1]:

1. Qualitative analysis prediction method;
2. Causality prediction method;
3. Time series analysis prediction method. At present, there are some important air accident prediction methods home and abroad, including: exponential smoothing [2, 3] grey prediction method, neural network [4] and support vector machine (SVM) [5], etc. all these methods are mainly used to predict the percentage of accident per 10,000 h, percentage of accident sign per 10,000 h (or 1,000 h) and others. But there is no any real cases of predicting scope of flight system safety status and then pertinently taking preventive measures. Whereas, the relation among influencing factors in flight system is complicated and uncertain, this paper brings out a research method-by historic statistic data of researched flight system, build up membership function of each influencing factor and make quantitative and qualitative analyses, respectively, by fuzzy relation synthesis theory on the premise of relatively stable system in order to predict trend of flight safety status.

62.2 Evaluation on Flight System Safety Status

62.2.1 Main Factors Influencing Safety of Flight System

Flight system is a system to complete specific flight task, composed of three subsystems: human, machine, and environment. However, from the prospective of fight system safety evaluation, there should be another important subsystem inside the flight system-management subsystem.

Among human factors, some factors influences flight quality and flight safety, mainly including: political quality, including responsibility sense of individual for flight, work attitude, state of family, and social relations and law-abiding conditions; knowledge conditions, including knowledge structure, education degree, and specialty knowledge level; business capabilities, including flight technology and special skills, quality and ability to cope with special situation; physical quality, including psychological state and physical health state.

"Machine" in flight system covers the aircraft itself and all the equipment on it. In "machine" factors, there are two factors influencing flight safety including aircraft design and quality of production and maintenance support.

"Environment" involved in flight system mainly refers to the natural environment and support environment. Natural environment includes atmosphere, and geographical and ecological environment; support environment includes technical and logistical support environment.

Safety management subsystem involves: safety guidelines, management-related regulations, systems, standard procedures, safety management institutions, safety management in daily business, technical training, flight safety education, and Safety education.

62.2.2 Synthetic Evaluation on for Major Influencing Factors: "Human-Machine-Environment–Management"

To sum up, the secondary model adopting multihierarchical comprehensive evaluation method is appropriate. The researchers make evaluation on the comprehensive effect of each of four factors "human-machine-environment–management" with the model determined by main factor $M(\wedge, \vee)$[6]. Divide any of four appraisal objects "human-machine-environment–management" into m grades (number of grade divided according to requirements of precision), i.e. $V = \{V_1, V_2, \ldots V_k \ldots, V_m\}$. In the influencing factor set of the four appraisal objectives $U_{\mathrm{ij}} = \{u_1, u_2, \ldots u_k \ldots, u_n\}$, subscript i means appraisal objects (human, machine, environment, and management); j means rating grade ($j = 1, 2$); n means number of influencing factors. The single-factor appraisal subset $u_k = (r_{k1}, r_{k2}, \ldots r_{\mathrm{kk}} \ldots, r_{\mathrm{km}})$, the aggregation of appraisal subsets of n influencing factors of appraisal objectives constitutes an overall appraisal matrix R_{ij}:

$$R_{\mathrm{ij}} = \begin{bmatrix} r_{11} & r_{12} & \ldots & r_{1k} & \ldots & r_{1m} \\ \ldots & \ldots & \ldots & \ldots & \ldots & \ldots \\ r_{k1} & r_{k2} & \ldots & r_{\mathrm{kk}} & \ldots & r_{\mathrm{km}} \\ \ldots & \ldots & \ldots & \ldots & \ldots & \ldots \\ r_{n1} & r_{n2} & \ldots & r_{\mathrm{nk}} & \ldots & r_{\mathrm{nm}} \end{bmatrix}$$

For instance: the secondary grade of environment involves two influencing factors and the aggregation of appraisal subsets of influencing factors constitutes an appraisal matrix $R_{\text{环}2}$.

$$R_{E2} = \begin{bmatrix} r_{11} & r_{12} & \ldots & r_{1k} & \ldots & r_{1m} \\ r_{21} & r_{22} & \ldots & r_{2k} & \ldots & r_{2m} \end{bmatrix}$$

The membership function value of each factor in above appraisal matrix is determined expert appraisal in combination with linear interpolation undetermined coefficient method. Each influencing factor has different importance on appraisal object, presented with fuzzy subset $\underset{\sim}{A}$. $\underset{\sim}{A} = (a_1, a_2, \ldots a_k \ldots, a_n)$, $(0 \le a_k \le 1, \quad k = 1, 2, \ldots, n)$; as weight coefficient, $\sum_{k=1}^{n} a_k = 1$. To avoid limitation on realization of expert, importance coefficient of factor is determined by analytic hierarchy process.

According to fuzzy transform principle, appraisal set $\underset{\sim}{B} = \underset{\sim}{A} \circ \underset{\sim}{R}$; according to maximum membership principle, the grades of four appraisal objects-human, machine, environment, and management are determined. In this way, the status of flight system is determined.

Prediction on safety status of certain unit. The air accident are classified into grade 1, grade 2, grade 3, and air accident sign [7], correspondingly, flight safety status R' is divided into five intervals R'_1, R'_2, R'_3, R'_4, R'_5:

1. R'_1 presents more than two grade-1 or 1 air accidents, meaning safety status is "very bad";
2. R'_2 presents more than one grade-3 or 2 air accident, meaning safety status is "bad";
3. R'_3 presents the sign of more than three air accidents, meaning safety status is "serious safety risk";
4. R'_4 presents the sign of no more than three air accidents, meaning safety status is "safety risk";
5. R'_5 presents no graded air accident or sign of air accident, meaning safety status is "safety";

Take the statistics data of the unit in 5 years from 2001 to 2005 as sample materials and then with foresaid method, create a comprehensive appraisal V in four rating grades for status of the four subsystems—human, machine, environment, and management-in each quarter during the period:

$$V = \{\text{Great, good, common, bad}\}$$

The aggregation of appraisal subsets of n influencing factors of appraisal objectives constitutes an overall appraisal matrix R_{ij}:

$$R_{\text{ij}} = \begin{bmatrix} r_{11} & r_{12} & r_{13} & r_{14} \\ \cdots & \cdots & \cdots & \cdots \\ r_{k1} & r_{k2} & r_{k3} & r_{k4} \\ \cdots & \cdots & \cdots & \cdots \\ r_{n1} & r_{n2} & r_{n3} & r_{n4} \end{bmatrix}$$

List evaluation conclusion and statistic results in Table 62.1.

62.3 Weight of Factor Interval and Weight of Different Safety Status in Each Interval

Mass statistics data show, the trend of flight system safety status is closely related to states of four subsystems in flight system-human, machine, environment, and management.

Table 62.1 Synthetic evaluation of four factors in each statistical quarter and according safety status

Quarter	01	01	01	01	02	02	02	02	03	03	03	03	04	04	04	04	05	05	05	05
	(1)	(2)	(3)	(4)	(1)	(2)	(3)	(4)	(1)	(2)	(3)	(4)	(1)	(2)	(3)	(4)	(1)	(2)	(3)	(4)
Human	△	△	▲	▲	△	○	▲	▲	○	▲	○	○	▲	○	○	○	○	●	●	●
Machine	△	▲	△	▲	△	△	▲	△	▲	○	○	○	○	○	○	○	○	●	○	●
Environment	▲	△	△	▲	△	▲	▲	△	▲	○	△	○	○	○	●	●	○	○	●	○
Tube	△	△	▲	△	▲	▲	○	▲	○	○	○	○	○	○	●	●	○	●	●	●
Quarter	01	01	01	02	02	02	02	03	03	03	03	04	04	04	04	05	05	05	05	06
	(2)	(3)	(4)	(1)	(2)	(3)	(4)	(1)	(2)	(3)	(4)	(1)	(2)	(3)	(4)	(1)	(2)	(3)	(4)	(1)
Safety status	■	■	■	☆	★	☆	□	■	☆	■	■	■	☆	☆	□	■	■	■	□	□

In Table 62.1: △—bad; ▲—common; ○—good; ●—great; ★★—very bad safety status; ★—bad safety status; ☆—serious risk; ■—risk; □—safety

Table 62.2 Factor interval weight and safety status of membership in corresponding factor interval

Factor interval	Interval weight	Interval object weight	Interval weight	Interval object weight
Bad	$\alpha_{\text{人}} = 3/20 = 0.15$	$M_{R'\text{人}1} = 0$ $M_{R'\text{人}2} = 1/3$ $M_{R'\text{人}3} = 0$ $M_{R'\text{人}4} = 2/3$ $M_{R'\text{人}5} = 0$	$\alpha_{\text{机}} = 5/20 = 0.25$	$M_{R'\text{机}1} = 0$ $M_{R'\text{机}2} = 1/5$ $M_{R'\text{机}3} = 1/5$ $M_{R'\text{机}4} = 3/5$ $M_{R'\text{机}5} = 0$
Common	$\alpha_{\text{人}} = 6/20 = 0.3$	$M_{R'\text{人}1} = 0$ $M_{R'\text{人}2} = 0$ $M_{R'\text{人}3} = 2/6$ $M_{R'\text{人}4} = 3/6$ $M_{R'\text{人}5} = 1/6$	$\alpha_{\text{机}} = 4/20 = 0.2$	$M_{R'\text{机}1} = 0$ $M_{R'\text{机}2} = 0$ $M_{R'\text{机}3} = 1/5$ $M_{R'\text{机}4} = 0$ $M_{R'\text{机}5} = 0$
Good	$\alpha_{\text{人}} = 8/20 = 0.4$	$M_{R'\text{人}1} = 0$ $M_{R'\text{人}2} = 0$ $M_{R'\text{人}3} = 3/8$ $M_{R'\text{人}4} = 4/8$ $M_{R'\text{人}5} = 1/8$	$\alpha_{\text{机}} = 9/20 = 0.045$	$M_{R'\text{机}1} = 0$ $M_{R'\text{机}2} = 1/5$ $M_{R'\text{机}3} = 2/9$ $M_{R'\text{机}4} = 5/9$ $M_{R'\text{机}5} = 2/9$
Great	$\alpha_{\text{人}} = 3/20 = 0.15$	$M_{R'\text{人}1} = 0$ $M_{R'\text{人}2} = 0$ $M_{R'\text{人}3} = 0$ $M_{R'\text{人}4} = 1/3$ $M_{R'\text{人}5} = 2/3$	$\alpha_{\text{机}} = 2/20 = 0.1$	$M_{R'\text{机}1} = 0$ $M_{R'\text{机}2} = 0$ $M_{R'\text{机}3} = 0$ $M_{R'\text{机}4} = 1/2$ $M_{R'\text{机}5} = 1/2$
Bad	$\alpha_{\text{环}} = 5/20 = 0.25$	$M_{R\text{环}1} = 0$ $M_{R\text{环}2} = 1/5$ $M_{R\text{环}3} = 0$ $M_{R\text{环}4} = 4/5$ $M_{R\text{环}5} = 0$	$\alpha_{\text{管}} = 3/20 = 0.15$	$M_{R'\text{管}1} = 0$ $M_{R'\text{管}2} = 0$ $M_{R'\text{管}3} = 1/3$ $M_{R'\text{管}4} = 2/3$ $M_{R'\text{管}5} = 0$
Common	$\alpha_{\text{环}} = 5/20 = 0.25$	$M_{R\text{环}1} = 0$ $M_{R\text{环}2} = 0$ $M_{R\text{环}3} = 3/5$ $M_{R\text{环}4} = 1/5$ $M_{R\text{环}5} = 1/5$	$\alpha_{\text{管}} = 4/20 = 0.2$	$M_{R'\text{管}1} = 0$ $M_{R'\text{管}2} = 1/4$ $M_{R'\text{管}3} = 1/4$ $M_{R'\text{管}4} = 2/4$ $M_{R'\text{管}5} = 0$
Good	$\alpha_{\text{环}} = 7/20 = 0.35$	$M_{R\text{环}1} = 0$ $M_{R\text{环}2} = 0$ $M_{R\text{环}3} = 2/7$ $M_{R\text{环}4} = 4/7$ $M_{R\text{环}5} = 1/7$	$\alpha_{\text{管}} = 8/20 = 0.4$	$M_{R'\text{管}1} = 0$ $M_{R'\text{管}2} = 0$ $M_{R'\text{管}3} = 3/8$ $M_{R'\text{管}4} = 4/8$ $M_{R'\text{管}5} = 1/8$
Great	$\alpha_{\text{环}} = 3/20 = 0.15$	$M_{R\text{环}1} = 0$ $M_{R\text{环}2} = 0$ $M_{R\text{环}3} = 2/7$ $M_{R\text{环}4} = 4/7$ $M_{R\text{环}5} = 1/7$	$\alpha_{\text{管}} = 5/20 = 0.25$	$M_{R'\text{管}1} = 0$ $M_{R'\text{管}2} = 0$ $M_{R'\text{管}3} = 0$ $M_{R'\text{管}4} = 2/5$ $M_{R'\text{管}5} = 3/5$

The change of relation among subsystems is uncertain, so fuzzy relation between factor (system) and object (safety status) is determined by establishing the membership function of subsystems.

In Table 62.1, the ratios of number of samples in each factor interval to total number of samples is the weights $a_{\text{人}}$, $a_{\text{机}}$, $a_{\text{环}}$, $a_{\text{管}}$ is the weights of samples in each factor interval to total samples respectively.

In each factor interval, the ratio of number of samples in each safety status R'_1, R'_2, R'_3, R'_4, R'_5 to number of samples in corresponding factor interval, $M_{R'1}$ $M_{R'2}$, $M_{R'3}$, $M_{R'4}$, $M_{R'5}$, is the membership of each safety status in the interval. See Table 62.2.

62.4 Synthetic Estimation of Multi-Factor Air Accidents

In fact fuzzy relation $\underset{\sim}{Q}$ between factors is a fuzzy subset of direct product:

$$\text{Man} \times \text{machine} \times \text{environment} \times \text{management} = C$$

Fuzzy relation $\underset{\sim}{S}$ between factor and safety status is fuzzy subset of $C \times R'$. So safety trend may be predicted with fuzzy relation synthetic evaluation method.

For instance, the result of comprehensive assessment on human, machine, environment, and environment of the unit in the first half of 2011 is: good, good, great, and great. According to result of search on Table 62.2, the weights of human, machine, environment, and management are $\alpha_{人} = 0.4$, $\alpha_{机} = 0.45$, $\alpha_{环} = 0.15$, $\alpha_{管} = 0.25$ respectively. Therefore the fuzzy set $\underset{\sim}{Q}$ is:

$$\underset{\sim}{Q} = (0.4, 0.45, 0.15, 0.25)$$

Similarly, the weight off each object (safety status) interval can be found in Table 62.2, and accordingly fuzzy relation $\underset{\sim}{R} = (r_{\text{ij}})$. The data are as below:

$r_{11} = M_{R'人1} = 0,\ r_{12} = M_{R'人2} = 0,\ r_{13} = M_{R'人3} = 3/8$
$r_{14} = M_{R'人4} = 1/2,\ r_{15} = M_{R'人5} = 0$

Among machine-related factors:

$r_{21} = M_{R'机1} = 0,\ r_{22} = M_{R'机2} = 0,\ r_{23} = M_{R'机3} = 2/9$
$r_{24} = M_{R'机4} = 5/9,\ r_{25} = M_{R'人5} = 2/9$

Among environmental factors:

$r_{31} = M_{R'环1} = 0,\ r_{32} = M_{R'环2} = 0,\ r_{33} = M_{R'环3} = 0$
$r_{34} = M_{R'环4} = 1/3,\ r_{35} = M_{R'环5} = 2/3$

Among management-related factors:

$r_{41} = M_{R'管1} = 0,\ r_{42} = M_{R'管2} = 0,\ r_{43} = M_{R'管3} = 0$
$r_{44} = M_{R'管4} = 2/5,\ r_{45} = M_{R'管5} = 3/5$

$$\underset{\sim}{R} = \begin{bmatrix} 0 & 0 & 3/8 & 1/2 & 0 \\ 0 & 0 & 2/9 & 5/9 & 2/9 \\ 0 & 0 & 0 & 1/3 & 2/3 \\ 0 & 0 & 0 & 2/5 & 3/5 \end{bmatrix}$$

If $\underset{\sim}{Q} = (0.4, 0.45, 0.15, 0.25)$ is taken as input vector and $\underset{\sim}{R}$ is fuzzy convertor, output $\underset{\sim}{B}$ is:

$$\underset{\sim}{B} = \underset{\sim}{Q} \circ \underset{\sim}{R}$$

$$\underset{\sim}{B} = (0.4, 0.45, 0.15, 0.25) \circ \begin{bmatrix} 0 & 0 & 3/8 & 1/2 & 0 \\ 0 & 0 & 2/9 & 5/9 & 2/9 \\ 0 & 0 & 0 & 1/3 & 2/3 \\ 0 & 0 & 0 & 2/5 & 3/5 \end{bmatrix}$$

According to maximum-minimum rule, the result is:

$$\underset{\sim}{B} = (0, 0, 3/8, 0.45, 0.25)$$

where, 0.45 is the weight of safety risk interval, bigger than both of 0, 3/8 and 0.25, therefore it suggests that there is safety risk for safety situation in the second half of 2011. That is consistent with actual conditions.

62.5 Conclusion

1. Flight safety status is controlled by many factors and featuring fuzzy judgment, so it is practical to judge the state of four influencing factors "human-machine-environment–management" with multilayer group fuzzy judgment method in fuzzy mathematics and predict safety status with fuzzy synthetic fundament in fuzzy mathematics.
2. Above method is a kind of analysis and prediction method by predicting flight safety status (object) changes with change of influencing factors in the future, within the category of causality analysis and prediction. On the premise of relatively stable system, multifactor flight safety status is predicted with fuzzy synthetic principle in combination with previous statistics and calculation data through comprehensive evaluation on four influencing factors "human-machine-environment–management" in previous flight quarter. The accuracy rate of predication by this method is above 90 %, meeting requirements on safety management and safety—related decision making.
3. Due to complexity of things and diversity of realizations of people, when determining importance coefficient with analytic hierarchy process, the judgment matrix built by people may has no consistency, while judgment matrix is the main basis to calculate sorting weight vector. If the consistency of judgment matrix is relatively low, the calculated character vector shows relatively big deviation as sorting weight vector and thus may cause error of sorting and failure of prediction. Therefore, to ensure general consistency of matrix, the consistency test must be conducted.

Acknowledgments Here we'd like to thank the teachers for their relevant opinions and selfless support during preparation of this paper.

References

1. Xu B (2005) Introduction to flight safety evaluation. Blue Sky Press, Beijing
2. Wang Y, Li N (2007) Study on forecast of incidents resulting from crew. J Civ Aviat Univ China 25(1):25–28
3. Du Y (2007) Forecast analysis of flight accident-rate and flight incident-rate. J Civ Aviat Univ China 25(6):9–10
4. Gan X, Zang H, Cheng Y (2008) An adaptive fuzzy neural network prediction method on aircraft accident. J Syst Sci Math Sci 28(4):425–433
5. Li D, Xu H, Hu L (2009) The prediction of flight accident rate based on support vector machine. Math Pract Theory 39(8):124–128
6. Lu H (2003) Research of military aircraft accident. National Defense Industry Press, Beijing
7. Yang L, Gao Y (2001) The principle and application of fuzzy mathematics. South China Science and Technology Press, Guangzhou

Part VIII
Theory and Application Research

Chapter 63
Scientific and Technical Informatics of Manned Space Based on Xuesen Qian's Systematic Theory

Hong Xu, Hong Liang, Jianhui Li, Peng Zhou, Zhijun Xiao, Jing Qiang, Jinsong Meng, Haipeng Sun, Yongsheng Wang, Jiutong Zhu, Fangzhou Kong, Jiayuan Zhang and Xiaoyu Zuo

Abstract Xuesen Qian's academic thinking in system science has deep effects on the subject construction of Scientific and Technical Informatics (STI) for manned space. This paper introduced the establishment and development of this subject as well as its contribution during the foundation of a new advanced space science branch—Space Medico-Engineering Discipline. The subject's basic theories, framework (including study on scientific and technical information, construction of informative resource, audio and image technology, documents and archives, magazine editing and printing), and its application in the manned space project were discussed and analyzed. To construct the knowledge service system and integrated service system of manned space STI and to strengthen the quality management and quality control of informatics services and products, these ideas and proposals for the sustainable development of the subject were put forward at the end.

Keywords Manned space · System science · Scientific and technical informatics · Subject construction · Space medico-engineering discipline

63.1 Introduction

Under instruction of Qian Xuesen's system science theory, manned space Scientific and Technical Informatics (STI) has made notable progress, following the step of China manned space. Especially in the past two decades of implementation of China manned space engineering, the informatics has been playing a role of assuring and supporting the accomplishment of tasks of manned space project and maintaining healthy sustainable development of construction of manned space STI.

H. Xu (✉) · H. Liang · J. Li · P. Zhou · Z. Xiao ·
J. Qiang · J. Meng · H. Sun · Y. Wang ·
J. Zhu · F. Kong · J. Zhang · X. Zuo
China Astronaut Research and Training Center, Beijing 100094, China
e-mail: xuhong_2255@126.com

S. Long and B. S. Dhillon (eds.), *Proceedings of the 13th International Conference on Man-Machine-Environment System Engineering*, Lecture Notes in Electrical Engineering 259, DOI: 10.1007/978-3-642-38968-9_63,

63.2 Formation of Manned Space STI

In April 1968, a Space Medico-Engineering Research Team was established. This team is composed of some scientific and technical elites dispatched by three research institutions including Academy of Military Medical Science. Based on this team, an information research group was set up by several researchers. In April 1973, an information research department facing four disciplines was founded. In the condition of blocked information channel and short resources, all information researchers worked together to compile and published "References of International Space Medicine" and "References of International Space Medicine and Engineering-Specialty Materials" in turn.

Both of works drew attention of Qian Xuesen, famous scientist, CAS member and vice director of COSTIND. After reading, Qian modified the works with red ink and then sent to the information research department with a holograph attached. In the letter [1], Qian said, the work was "inspiring."

After receiving Qian's letter, through repeated supplementations and modifications, the information researchers completed a theme report "Significances of Manned Space." Through review and modification, this report was finalized and submitted to the central government in the beginning of 1970. "China's Manned Spacecraft Theme Symposium" held in Beijing soon after with the support of the CPC Central Committee and the State Council. On July 14, 1970, Chairman Mao approved this report about development of manned space in China after reading and endorsing. That is the "714" project, starting the R&D of "Shuguang" manned spacecraft mission [2].

After start-up of national "863" program, manned space project was put on the table once again with the purpose of constructing a space station. In the beginning of conceptual argument, there was a large divergence and extremely furious argumentation on whether or not and how China should develop manned space. A big disagreement also existed in the matter how manned space should start up.

Based on the collection and reading of a huge volume of international and domestic relevant materials and data, information researchers wrote some argumentation reports like "Basic Approach to Development of China's Manned Space," "Roles of Human in Space," and "Environmental Control and Life Support System of Manned Spacecraft" to clearly bring out that development of manned space and breakthrough of manned space technology are the only right approach to the development of manned space understanding in China. Therefore, this report played a very important role in the establishment and start-up of manned space project [3].

With regard to the roles of human in space, STI research believes that the roles of human shall be correctly understood in theory in the conceptual research stage of the manned space program. The roles of human shall be realized with engineering design in R&D stage. The roles of human shall be guaranteed by technical measures in spacecraft implementation stage [4].

So far, two decades of practice in China manned space have fully proven that the thought and direction of China's manned space STI research are totally right. The intelligence information is timely, accurate, and reliable.

63.3 Construction of Manned Space STI

For the first time, Qian Xuesen's speech on the National Thinking Science Symposium in 1984 pointed out that informatics was an applied science of thinking science and outlined the layered structure of informatics. He said that the key to STI was the research on its scientific and technical issues. To sum up, informatics is an applied science which develops the information-related works into a theoretic systematical discipline to make STI work as an effective organizational structure system [5].

Qian's dissertations provide us with not only a clear profile of informatics, but also the position of STI in whole modern scientific and technical system. No doubt it has a huge influence on the development of manned space STI.

Just with the guidance and influence of theory and thought of Qian Xuesen's systematic science, manned space STI has formed and become mature as a part of space medico-engineering discipline. In the construction and development of space, medico-engineering discipline and model-based tasks of manned space project play an important role in technical information support and guarantee.

63.3.1 Manned Space STI System

Taking the goal of space medico-engineering discipline construction, based on clear task demands and application of methods, systems and approaches of systemic science theory, thinking science theory, fundamental research of applied informatics and methods of applied IT and information resource science, through understanding and grasp cross and integration of space medico-engineering technology, a manned space STI system with clear features was gradually founded in order to ensure systematic, complete, accurate, reliable, and effective IT demand in manned space missions (Fig. 63.1).

63.3.2 Application of Manned Space STI System

63.3.2.1 STI Research

Information research leads to the preparation of research [3]. Manned space STI research means that closely coordinate the making of manned space project and

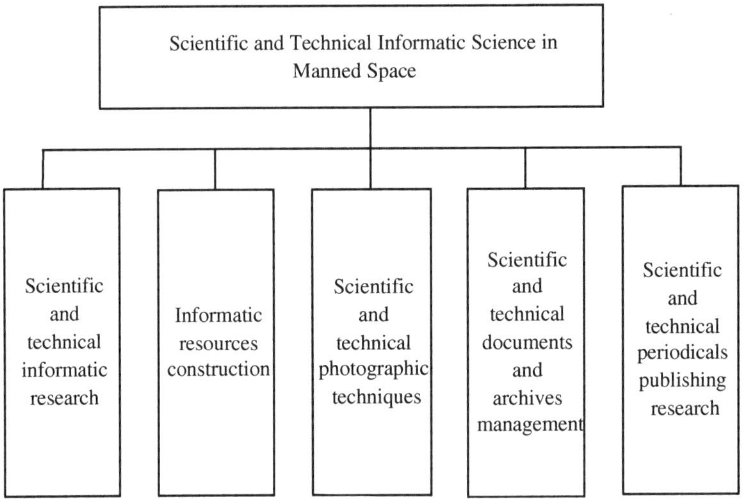

Fig. 63.1 Systematic framework of STI (STI) for manned space

research on development report, carrying out comprehensively proactive and innovation information research and accordingly making pertinent reliable strategic information research analysis report.

In manned space project, some strategic information researches have been started in time like space station, scientific and technical application research, key technical information research of space laboratory medical engineering, medical guarantee research for overseas astronauts in long-duration flight, space–time biological theoretic and practical information research.

Aiming at specific demands and problems of astronauts in manned space project, the researchers have carried out information research and management of manned space model-based service in time, medico-engineering key technical research on mid- and long-term overseas manned space, extra-vehicular activity medical monitor and medical guarantee information research, extra-vehicular activity and large-sized ground simulation device key technical information research, lunar space suit equipment-related technical information research, etc.

Manned space STI dynamic tracking mainly involves tracking and reporting researches on new changes, new dynamics and new situations in manned space development to realize the accuracy, scientificity, timeliness, and readability of dynamic tracking report. Especially during implementation of space tasks by Shenzhou 1 to Shenzhou 10, tracking reports have been made on international and domestic media in time accurately. Accordingly, special issues and monographs of "Manned Space Express" and "Manned Space Information" have been compiled and published in time, playing an important role in information guide.

63.3.2.2 STI Resource Construction Research

According to demand of manned space research, special information researches and information research platform constructing were fulfilled, including manned space STI resource construction, space medicine database, MMESE fact-type database ''Space Medico-engineering Subject Heading List'' and other. The establishment of all these database and resources greatly develops and exploits library information resources and powerfully supports in-depth development of research.

In the respects of collection, processing, and sorting of STI resources, a huge volume of fundamental works have been done to coordinate researches: completed collection, exhibition, and compilation of various title catalogs and table of contents such as "Library Journal Catalogue" and "Library Life Support System Reference Catalogue" in the special researches like astronaut life security system; explored library resources in many ways, recommended library resources, and developed and exploited their value to enable them to play effective roles in research works; established Chinese and Foreign Language Library Database, introduced information resources including Aerospace Database, EI, NTIS, and MEDLINE and Chinese language academic journal databases A, C, E, and I, constructed network terminal service area, carried out information resource search, inquiry, and consultation service by network platform, in-house Chinese and foreign language journal information resource search service, and provided researchers with networking digitalized information resources.

63.3.2.3 STI Acoustic and Image Technical Research

During implementation of manned space project for the respect of scientific and technical acoustic image collection, researchers have made whole-process, all-round acoustic image collection tracking and record for launching fields, landing fields, flight control fields and astronaut selection and training, medical monitor and medical security, space suit R&D and test, research of products for spacecraft ECLSS, research and production of space foods, space ergonomics research, bioelectronic devices and large-sized ground experimental, etc. Based on these efforts, the STI department has accumulated true, complete reliable acoustic image information for manned space project and edited a group of scientific and technical acoustic image products and obtained satisfying outcomes in scientific and technical acoustic image. They collected and edited authentic objective historic records and evidences for scientific research and experiment. The STI plays an important role in promoting the spirit of manned space, publicizing culture of manned space, expanding influence of Astronaut Research and Training Center, showing the elegant demeanor of China's astronauts.

63.3.2.4 STI Documentation Research

After manned space project was started up, by taking experiment of management in this field home and abroad, in combination with demand of model-based task assumed, the STI department conducts ISO quality management for manned space products. In past two decades, the STI department has prepared operation instruction and level-3 documents compliant to model-based tasks STI quality management in combination with products researched in model-based tasks on the site of XQC's quality system certification audit, update audit and other audits and kept refreshing, modifying and completing them so as to make technical documents and scientific and technical documents under closed-loop control and management mode for the respect of "document control" factor. In all past XQC's audits on quality system and edition update, the STI department obtained satisfying achievements without any "fail" item.

With success of manned space missions, the STI department keeps rising its level in the integration of technical documents and scientific and technical documents and automation construction and management. The STI department has established scientific and technical documents, technical drawings, software product database so as to ensure standardized, routinized control and management of technical documents, scientific and technical documents and software products. The STI department completed the research on the space medico-engineering scientific and technical document management system, research on model-based task technical document computer management system and research and production of design drawing management, realized integrated management in all links of technical document and design (process) drawing, and achieved the secondary and full-text information management functions of technical documents, drawings, scientific and technical documents, and multimedia information.

63.3.2.5 STI Information Editing and Publication Research

"Space Medicine and Medical Engineering" is a scientific and technical journal. It has been included by "EI," "CSA," "MEDLINE," and "Aerospace Database" successively in America and over 10 important domestic databases like "Chinese Science Citation Database" (CSCD) and "China Core Journal Database." The journal has won a number of big awards like "National First Scientific and Technical Journal Award" and "Prominent Medical Journal Award" and entered "Chinese Center Journals Catalogue." The journal plays an important role as an academic platform in order to push for the development of China's manned space, communicating new technologies and new knowledge of manned space science and technology and boosting discipline construction and development and drive growth of talents in this field. Besides, the journal puts priority on the subjects in the discipline direction of the field and has set "Special Dissertations" column, directly facing services of model-based tasks of manned space project. The number and academic level of the articles are rising in the course of operation. So far, it has presented research academic outcomes and papers of space medico-engineering to the society and the whole world so as to boost academic communication and cooperation home and abroad and contributed to the construction and development of space medico-engineering discipline. It is an important bridge and link to introduce more outcomes and talents of space medico-engineering discipline in international manned space domain.

"Astronauts" is a popular science journal directly facing training, work, and life of China's astronauts. The journey was popular with adolescents all over the country with extensive influence home and abroad for timely tracking and reporting state of astronauts in manned space missions, presenting spiritual appearance and image of astronauts to all China people and carrying out "Astronaut Experience Camp" activities among adolescents in China.

63.4 Thinking About Sustainable Development of Manned Space STI

During decades of development of manned space STI, guided by the theory and thought of Qian Xuesen's systematic science and informatics, a relatively scientific system has been formed under the efforts of an STI team whose members are selflessly devoted themselves to space undertaking to pursue breakthroughs through close cooperation. Now, STI focus on manned space frontier technology and pursuit for a more perfect manned spaceflight and has become a necessary way to healthy sustainable development of this discipline.

63.4.1 Satisfy User's Demands to Manned Space STI Knowledge

Manned space STI shall involve discipline-oriented knowledge service facing follow-up task of manned space. Information resource platform construction shall face key research laboratories. The feature information knowledge service shall face tasks of research experiment and demands of researchers.

63.4.2 Building up an Integration Service System of Manned Space STI

According to demand of follow-up tasks of manned space, different information sources and different types of technical information should be integrated into systematic information system. STI should satisfy information-related demands of manned space project tasks and researchers at all levels, for instance, establishing information research dynamic database, basic database, research outcome bank, specialty knowledge bank, feature resource bank.

63.4.3 Intensify Quality Control of Manned Space STI

Facing follow-up tasks of manned space project, STI should enhance organization and planning of informatics research and put emphasis on the research project establishment and theme selection, opening argumentation, mid-term examination, closing acceptance and outcome validation at the beginning of the research task. STI should intensify quality of construction of informatics research and enhance construction and implementation from quality standard and operation system. Furthermore, STI should conduct standardized routinized quality control and management in the mechanisms like information demand, information feedback, coordinative research guarantee, fast response, task dispatch, interest distribution.

With progress of China manned space project, manned space STI will face new tasks and new challenges. Therefore, we shall enhance efficient STI service quality around manned space project and space medico-engineering innovation. STI should fully play the important role of informatics in supporting strategic decision-making, research activities, and knowledge communication and make all effort to become a new force to ensure the information for development of manned space project.

References

1. Collection of Qian Xuesern's letters (2007). National Defence Industry Press, Beijing
2. Ye Y (2009) All about Qian Xuesen. Shanghai Jiaotong University Press, Shanghai
3. Symposium of Quan Xuesen's academic thinking and researches (2011). National Defence Industry Press, Beijing
4. Zhang R (1991) The fundamental research on space medico-engineering. National Defence Industry Press, Beijing
5. Wang W (2007) Academic thinking of Qian Xuesen. Sichuan Scientific and Technical Press, Chengdu

Chapter 64
Review of Man-Machine-Environment System Engineering in Maritime Technology

Song Ding, Duanfeng Han and Boshi Zhang

Abstract To help reduce the potential for human error in ship accidents, it is important to effectively integrate man-machine-environment system engineering principles into ship design so that systems encompass human capabilities and limitations, while increasing system availability/performance, and personnel satisfaction. The importance of the human element in maritime safety is increasingly being recognized by the shipping and offshore communities. In this paper, the author is going to make an introduction of the application of the human factors in maritime technology in recent years and talk about some topics about the failure of situational awareness which are controversial and need to be handled.

Keywords Human factors · Maritime technology · Human error · Situation awareness · HFACS · Man–machine–environment system engineering

64.1 Man-Machine-Environment System Engineering

MMESE is a comprehensive boundary technical science born in China in 1981 under the instruction of great scientist Qian Xuesen. In past decades, more and more better products have been designed and produced in many engineering fields with the help of MMESE. Marine system is a closely interconnected, moderately interactive system and also an engineering activity in which MMESE must be taken into consideration to enhance safety, additional driving, and comfort.

Since 2000, MMESE applications in ship and warship field draw more and more attentions. Marine organizations and classification societies began to make standards or guiding documents about man-machine-environment system [1]:

S. Ding (✉) · D. Han · B. Zhang
College of Shipbuilding Engineering, Harbin Engineering University, Harbin, China
e-mail: dingsongsm@126.com

S. Long and B. S. Dhillon (eds.), *Proceedings of the 13th International Conference on Man-Machine-Environment System Engineering*, Lecture Notes in Electrical Engineering 259, DOI: 10.1007/978-3-642-38968-9_64,

- IMO's Guidelines on Ergonomic Criteria for Bridge Equipment and Layout (MSC 73/Circ982).
- ABS Guidance Notes on the Application of Ergonomics to Marine System.
- ABS Guidance Notes on Crew Habitability on Ships.
- ASTM's Standard Practice for Human Engineering Program Requirements for Ships and Maritime Systems, Equipment, and Facilities (F-1337, 1991).

Among them, "ABS Guidance Notes on the Application of Ergonomics to Marine System" (Republished in 2004) released by ABS in 1998 is extensively accepted.

64.2 MMESE of Ships and Warships

ABS human-machine model gathers up the important factors influencing safety and efficiency of system, shown as Fig. 64.1. These factors include ship design and layout, operation environment factor, organization and management related to ship manipulation and crew manipulating the ship. The ignorance of any of above factors will have adverse impact on safety, production, and work efficiency [2].

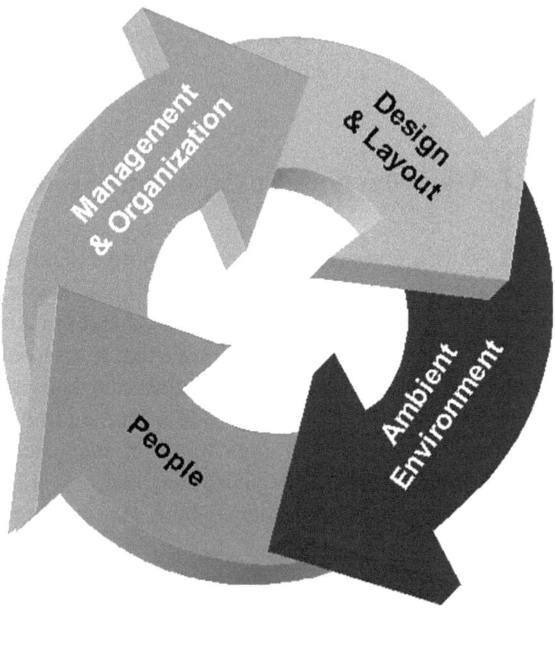

Fig. 64.1 Model of man-machine-environment system

64.2.1 Design and Layout

Design and layout is the basis of man-machine-environment system model, mainly involving integration of human and interfaces of equipment, system, software, and hardware, including control system, presentation system, alarm system, computer workstation, tagging system, stairs, workspace arrangement.

Integrate the psychological and physiological limitations and demands of crew with hardware and software design and layout in order to make the product design meeting the needs and habits of users. Workspace design shall include design of each unit in the workspace, layout of whole area and its influence on the crew.

Besides, cultural and regional differences of crew shall be taken into account in the design like cultural differences which lead to different understandings on the meaning of the colors, influence of heavy suit on the size of human body model in cold weather, potential physical differences (e.g., high/short, male/female, North American/Asian ethnic group), etc. in order to enable the equipment design and the overall layout of the work environment suitable for majority of people [3].

If above factors are not taken seriously enough, the probability of human error will increase greatly, while the damage caused by non-friendly design cannot be repaired by additional training, maintenance and detailed instruction [4].

64.2.2 Environmental Factors

Environmental factors include vibration, noise, indoor climate environment, and lighting system. This part mainly discusses ship habitability and occupational health problems associated with these environmental factors. Non-conforming working environment cannot only reduce work efficiency, but also increase stress and fatigue. High noise, for example, may lead to communication disorders and high temperature may accelerate fatigue. Ambient factors also involve friendly design of living place, because a good environment can effectively relieve fatigue.

64.2.3 Human Factors

Remarkable increase in complexity and information volume of modern navigation equipment brings out higher requirements on vigilance, skills, and experience of the people during operation. Therefore, selection of appropriate qualified crew is very important for safe navigation of a ship. Following factors shall be included in the scope of crew assessment:

- Basic knowledge of individual, knowledge level, and ability as a general rule.
- Specific ship manipulation skills.

- Physical qualities and characteristics, like stature, shoulder breath and height of eyes.
- Physical endurance, like its ability to resist fatigue, visual accuracy, tolerance, acute/chronic diseases.
- Psychological characteristics, like stress tolerance and risk tolerance.

In the whole system, what causes the finally different result is human's behavior usually. Therefore, selection of appropriate crew is crucial for the safety and economy of the navigation. The incompetent crew may not only lower the efficiency of the whole system, but also increase probability of accident.

64.2.4 Management Factor

A reasonable management system can effectively reduce accident risks and shall be implemented in the whole life circle, involving:

- Uniformly implementing management standard without discriminated treatment.
- Clear, straightforward interpretation of management system.
- Safety system and man-machine-environment engineering which shall be operated with sufficient financial support and integrated into the whole system.
- Making reasonable schedule to reduce fatigue of crew.
- Rewarding employees who comply with company's safety procedures.
- Perfect schedule maintenance mechanism.

64.3 Situation Awareness

With progress of science and technology, reduction in personnel quota of crew and alleviation of physical load of crew, the automation level of marine equipment becomes higher and the role of crew gradually turns to "monitor" to "implementer" and intervener just in necessary (such as unattended machinery space, ship monitor, fire alarm, and safety monitor). High-speed rotation and mass information cause crew to lack understanding on current operation status of the ship. Accordingly, crew cannot judge what is wrong first time in emergency and even does not know current status of the ship. Such problems are called Situation Awareness (SA) error generally [5].

Among marine accidents, nearly 80 % are related to human errors [6]; among many types of human errors, SA error is the most common in human error. Figure 64.2 shows the proportion of each sub-factor in human factors in accident reports [7].

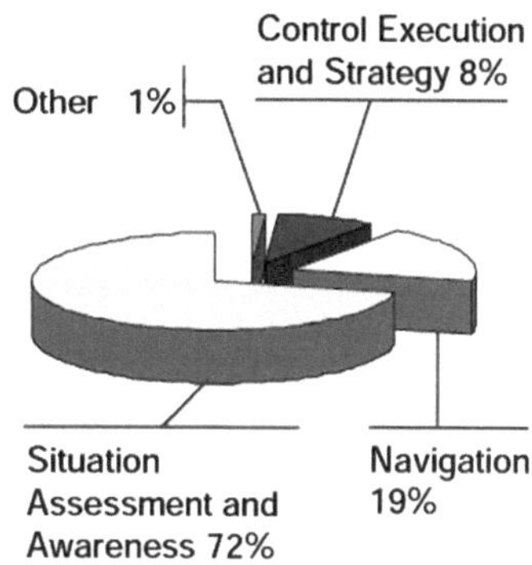

Fig. 64.2 Human errors in ship accident (USCG data)

Table 64.1 Endsley's modified SA error taxonomy

Level 1: Failure to identify information, such as
• Unavailable information
• Information difficult to be distinguished or identified
• No information monitored or observed
• Wrong understanding on information
Level 2: Failure to treat or understand information, such as
• Lacking information model
• Using inaccurate information model
• Over depending on wrong information
• Forgetting relevant contents
Level 3: Failure to judge future action of system, such as
• Wrongly estimating current trend of system

With regard to SA, Endsley has established classification model of failure of SA [8] as shown in Fig. 64.1. Currently, most of the researches on SA focus on navigation field and extend to medical field to some extent [9]. However, marine system is a closely interconnected, moderately interactive system, SA is also very important. There is almost no any research on SA of marine system [10] (Table 64.1).

64.4 Human Factors Analysis and Classification System

These years the problems about marine safety draw extensive attentions in the world. Accident occurrence rate is still high though level of modernization and automation of marine equipment. However, there is not a unified method to analyze human factors which account for the biggest proportion in factors causing marine accidents. Besides, it is difficult to collect, classify, and quantify human factors data [11].

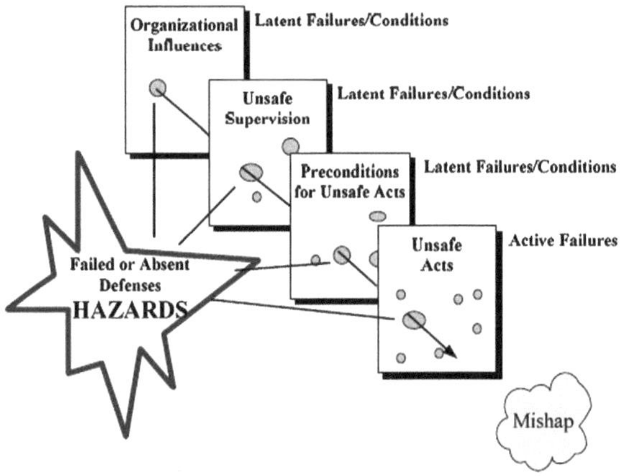

Fig. 64.3 The Swiss cheese model

Human Factors Analysis and Classification System (HFACS) was applied in the analysis on spaceflight accident at the very beginning to rebuild the roles of human in various accidents based on Swiss Cheese Model theoretically, shown in Fig. 64.3. James Reason 1990 is the first one describing HFACS's structure in his "Human Error" in 1990. The structure classifies the accident factors into four levels, among them "acts" is direct factor and "preconditions," "supervision," and "organizational influences" are potential factors. The human factors behind the accident may be classified based on HFACS [12].

There is different among analysis methods for this structure. Metin Celik calculated the weight of each factor in HFACS structure with fuzzy analytic hierarchy process [12] and then analyzed the importance of each factor with quantification process based on expert system; based on HFACS, Shanghai Maritime University conducted classification and statistics for the reports of 866 accidents and acquired the proportion of each hierarchy factor in HFACS [11].

There is certain difficulty for human factors classification because involved person tends to hide the truth to avoid liability and protect himself. So far, there is not an extensively acceptable human factors analysis method in the world.

64.5 Conclusion

To sum up, for China urgently requiring enhancing level and profit in ship field, it is time to conduct a big volume of systematic researches on MMESE application in ship design to assist manufacturer to enhance the level of ship design. Meanwhile as the chief factor of marine accident, marine safety researchers shall take full consideration of "human factors" in ship design.

References

1. Card JC, Baker CC, McSweeney KP, McCafferty DB (2005) Human factors in classification and certification. Trans Soc Naval Arch Mar Eng 99:129
2. American Bureau of Shipping (1998) Guidance notes on the application of ergonomics to marine system. American Bureau of Shipping, New York
3. Chen Z Human factor—New direction of ship design
4. McSweeney KP (2002) Revision of the ABS guidance notes on the application of ergonomics to marine system. Technology conference, Houston, Texas USA, 6–9 May 2002
5. Rothblum AM (2000) Human error and marine safety. National Safety Council Congress and Expo, Orlando
6. O'Neil, MA (2001) The human element in shipping. Biennial Symposium of Seafarers International Research Centre, Cardiff, Wales, 29 June 2001
7. Baker CC (2002) Analysis of coast guard marine safety management system (MSMS) data. Technical report. The American Bureau of Shipping, Houston, Texas
8. Grech M, Horberry T, Smith A (2007) Human error in maritime operation: analysis of accident reports using the Leximancer tool. The 4th Annual Meeting of the Human Factors and Ergonomics Society, Baltimore, USA
9. Endsley MR (1996) Situation awareness measurement in test and evaluation. In: O'Brien TGC, Samuel G (ed) Handbook of human factors testing and evaluation. Lawrence Erlbaum Associates, Hillsdale, NJ, USA, pp 159–180
10. Grech M, Horbeery T (2002) Human error in maritime operations: situation awareness and accident reports. The 5th international workshop on human error, safety and systems development, Newcastle, Australia
11. Xi YT, Fang QG, Chen WJ, Hu SP (2009) Case-based HFACS for collecting, classifying and analyzing human errors in marine accidents. In: Proceedings of the 2009 IEEE international conference on industrial engineering and engineering management, Hongkong, pp 2148–2153
12. Celik M (2009) Analytical HFACS for investing human errors in shipping accidents. Accid Anal Prev 41:66–75

Chapter 65
Application of Man-Machine-Environment System Engineering to Create a New World of Changqing Oilfield

Jiajun Sun and Ye Ren

Abstract A small oil field has been unknown in the past 30 years, why in the last 10 years it created an amazing miracle in China and even in the global oil industry? To answer this question scientifically, you have to have a scientific theory. In terms of depth and breadth of this problem, undoubtedly the most scientific theory is systems engineering theory; and from the perspective of oilfield production growth referred to on this issue, the most appropriate theory is man-machine-environment system engineering (MMESE) theory. Changqing oilfield applied MMESE to develop low-permeability reservoirs with extraordinary results, proving the value of MMESE.

Keywords Application · MMESE · Changqing oilfield

65.1 Tremendous Yield Squeezed from Low-permeability Oilfield

Ranging across the vast northern China in the middle (N35–40, E106–111) locates an extremely significant area with unique history, and infertile earth, however, supremely abundant underground resources. To find it, it is easy to first find a twisted line in the northern China on a map of whatever size or scale, that is, Yellow River. The Yellow River flows northward and then turns to the east, and suddenly goes downwards, shaping a big Chinese character of "Ji," of which the enclosed area is just the place we are looking for. It lies in the middle of Ordos

J. Sun (✉) · Y. Ren
Petrochina Changqing Oilfield Company Supervision Department, Xian 710021, China
e-mail: sunxuejun716@sina.com

Y. Ren
e-mail: renye_cq@petrochina.com.cn

S. Long and B. S. Dhillon (eds.), *Proceedings of the 13th International Conference on Man-Machine-Environment System Engineering*, Lecture Notes in Electrical Engineering 259, DOI: 10.1007/978-3-642-38968-9_65,

Basin, to the west of Gansu and Ningxia provinces, to the north of Inner Mongolian Plateau, to the east of Shanjin Grand Canyon, and to the south of Weihe River Plain. To give it a popular and simply name, name it Shanbei. An overall impression of Shanbei can be concluded into four colors, i.e., Red—revolution history, Black—petroleum and coal, Yellow—ecological issues, and Green—returning farmland to forestry [1]. There is a state-owned enterprise with a mission of petroleum and gas extraction on this land. Its name is Changqing Oilfield. "Speaking of oilfield, Chinese people all know Daqing Oilfield. But Changqing Oilfield is rarely known. Nevertheless, it has made a miracle surprising China and even the world" [2]. Since 2001, the new yield and reserves, two key indexes in the oil industry, of Changqing Oilfield have remained the first in China for 12 consecutive years. At least 5 million tons of petroleum and gas equivalents was increased of annual output in the recent five years, which amounts to contribute a medium-sized oilfield, like Dagang or Zhongyuan Oilfield. What is the most amazing is that Changqing Oilfield is an old oilfield which has been extracted for 43 years, and known by the world thanks to its "low pressure, low permeability, and low abundance." Particularly the low permeability, Changqing Oilfield is the most famous for this characteristic in the world. Its oil-bearing reservoir is known as the "knife grinder" without any void spaces. In the 1980s, the US energy consultation authority conducted an investigation of the Changqing Ansai Oilfield and reached a conclusion that this oilfield is a typical marginal field without any development value. Speaking of its inferior geological conditions, the core point is low permeability, which is also the key reason why the oil industry does not regard it as a promising oilfield. Layman might not understand "low permeability." Experts explain it with a metaphor that it is easy to squeeze out the oil if it is in a sponge; however, if it is in a brick, it is hard to squeeze. Comparing with the sponge, brick is low permeable. "Millidarcy" is the basic unit for permeability in the oil industry. The smaller the value is, the lower the permeability is. Oilfield with permeability less than 50 millidarcy is regarded as the low-permeability oilfield according to the global standard, while 70 % of Changqing Oilfield has permeability less than 1 millidarcy. With such a low permeability, Changqing is of course considered as an unpromising oilfield. Changqing Oilfield has always faced an awkward situation of low yield with much reservoir since 1970. The annual oil and gas output in the 1990s lingers about 1.4 million tons. In fact, Changqing Oilfield has abundant oil and gas resources. According to the general investigation of national geological resources, Ordos Basin boasts 8.588 billion tons of oil and 1.07 trillion cubic meters of gas. Changqing people have diligently persisted in squeezing out the oil and gas resources for the past 43 years. Finally, they succeeded. Year 2012 saw a yield over 45 million tons in Changqing Oilfield and 2013 will reach the target of 50 million. The western "Daqing Oilfield" is emerging and rising from the Ordos Basin with attention from the whole world. Chairman Xi Jinping, formerly the vice chairman in 2009, had a field trip to Changqing Oilfield; and in 2012 Spring Festival, Premier Wen Jiabao paid a New Year call to the workers at Changqing Oilfield.

65.2 Application of MMESE in Changqing Oilfield

Why has this unpromising Changqing Oilfield over 30 years made such a tremendous wonder in the recent 10 years? To answer this question scientifically, a theory is a must. From the perspective of breadth and depth, the most scientific theory is systems engineering, while considering the actual yield increase of this oilfield, it is the man-machine-environment systems engineering theory. Any oilfield can be defined as a large-scale MME system. The overall function is achieved by optimization of the three factors, namely man, machine, and environment. For a specific oilfield, the factor "man" refers to the workers with a mission to exploit the underground oil and gas. They accumulate various experiences from the practices in oil and gas exploration and development, mixed with success and failure, faults and correctness, as well as pain and pleasure. The most valuable part of these experiences becomes the operational knowledge that can be transformed into instruments, tools, machines, and equipments through technical smelting and materialization. With these devices, oil wells and gas wells are produced. All these are referred to as the factor "machine." As all the machines have their own distinct, specific and local applications and situations that constitutes the specific environment of the oilfield. Among the many environmental facets of oilfield, the most crucial factor is the property and condition of the oil and gas hidden in the thousands of meters deep underground. For Changqing Oilfield, its environmental factor is low permeability. It is extremely difficult to investigate this environmental factor hidden in the depth. Black box is the only way to understand it gradually. Fortunately, MMESE offers the technical staff a scientific method of correctly understanding the properties of hydrocarbon reservoir. There are various reasons and factors for the impressive performance of Changqing Oilfield which, nevertheless, can be concluded to one point, i.e., Changqing people employ machines suitable for low-permeability oilfield based on their thorough understanding of low permeability, and deliver the human operations though all kinds of ground equipments and down hole tools to the low permeable hydrocarbon reservoir, and then change this reservoir as intended. All this is MMESE applied in Changqing Oilfield. A comprehensive boundary science—man-machine-environment systems engineering—was born in China under the guidance of famous scientist Qian Xuesen in 1981. Academician Qian highly regarded this new science and wrote to Han Longshen on October 22, 1993 that you have created an important modern science in the socialist China. MMESE employs systematic scientific theory and engineering method to handle and optimize the relation of man, machine, and environment [3]. This new sciences studies human property, machine property, environmental property, man-machine relation, man–environment relation, machine-environment relation, and the overall performance of man-machine-environment system [4]. The basic theories of MMESE are control theory, model theory, and optimum theory. Control theory contributes to break the boundary between the living and non-living through some general concepts and terms, like system, information, and feedback and offers the unified perspective and standard for researches of the three totally different and distinct objects, namely the man, machine, and environment. Model theory provides a complete set of mathematical analysis tools for the MMESE study. Evidently, MMESE requires not only qualitative analysis, but quantitative representation of the motion rules of the whole system. For this reason, proper mathematical model is introduced for different objects, and objective rules in the real world are elaborated via the mathematical language through steps of modeling, parameter identification, simulation, and verification. Optimum theory aims to seek the best or some better solutions among all the combinations of man-machine-environment systems, which is the essence of MMESE. It is a mathematical method [5]. Although MMESE is an emerging boundary science, it has been applied widely in various industries, for example, in the exploitation of low-permeability reservoir at Changqing Oilfield.

65.3 How does MMESE Meet Changqing Oilfield?

People who are familiar with MMESE may have such a question that why MMESE, as a marginal science refined from the experiences of aerospace scientists during their researches and experiments of various spacecrafts as well as with a profound national defense and military background and glorious traditions of the liberation army, was applied in the Changqing Oilfield which is only an oil exploitation enterprise on the land comparing with the aerospace researches? To answer this question, it is necessary to start from 43 years ago. Satellite Dong Fang Hong I transmitted the sound of China to the space for the first time on April 24, 1970 at Jiuquan, Gansu Province. From then, dreams of China Aerospace came true one by one, including manned space flight. 43 years have seen China become a globally recognized leading country in aerospace industry. Scientists have refined their precious experience into theories, one of which is MMESE. At the same year, on August 7, 1970, industrial oil flow is found at the No. 3 well of Huachi County, Gansu Province. 49 days later, another industrial oil flow is seen at the No. 1 well of Qingyang County, Gansu Province. Started from this, Changqing Oilfield has grown up from an unknown small oilfield to the No. 1 in China after 43 years. Many exploitation experiences are refined into theories. For example, academician Hu Wen who has worked in Changqing Oilfield for 34 years published the book *Introduction to Low Permeability Oilfield*. Both the aerospace and oil exploration course started at the same time and went through similar courses. Isn't that the fate? Moreover, people who are familiar with the history of Changqing Oilfield know that this oilfield is closely related with the liberation army. The division No. 57 of army No. 19 was totally converted into the No. 1 division of oil engineering as required by Chairman Mao. Most soldiers have participated in the construction of Changqing Oilfield. Preliminarily, this troop was led by Lanzhou Military Base delegated by the State Council and Central Military Commission. These soldiers took off the military uniform and took part in the construction of Changqing Oilfield. With such profound relation with the liberation army, oil exploration industry is closely related to the national defense and military background of MMESE. Isn't that another fate? Most oil wells and gas wells drilled in the recent ten years at Changqing Oilfield are directional wells, some are horizontal wells. Determination of direction is the core technique. It actually originated from the aerospace technology because determination of direction is also required in the aerospace industry. In addition, domestic oil industry employs the so-called HSE safety management system from this century. This system, namely an integrated management method of human health, machine safety, and environment protection, came from the Royal Dutch Shell Company. People might associate this HSE system with MMESE established by our aerospace scientists. From the inherent relation between production and safety, MMESE is the root while HSE derives from the root. Actually, MMESE overmatches HSE from the mathematization of theoretical roots, familiarity with the domestic conditions, and logical preciseness between different parts. MMESE works better than HSE at Changqing Oilfield based on the actual conditions, because MMESE is a domestic theory although it is converted from the military industry while HSE is an exotic theory that does not accord with the domestic conditions severely and requires localization. From the above analysis, that Changqing Oilfield applies MMESE consciously to solve its low-permeability problem is not groundless, but natural integration due to the native conditions, profound origin, and countless connections. It is worthy to declare that MMESE is not applied in the exploitation of low-permeability reservoir overnight, but following three phases. First of all, the attitude transforms from spontaneity to awareness, i.e., the concept of man-machine-environment system was established among the technicians. Secondly, the knowledge rises from perceptual to conceptual, i.e., technicians started to think about the issue systematically, taking all the three factors, man, machine, and environment into consideration. Lastly, methodology of researching the relations of the three factors changed from static to static, from doctrinal to dialectical, from one-sided to all-sided, from qualitative to quantitative, thus making the foundation for parameters discrimination and system modeling. Particularly, to point out that Changqing Oilfield, a previously unknown oilfield could strategically accept and absorb the advance technology so quickly, MMESE, from the aerospace industry had strong relations to three great minds. World famous scientist Qian Xuesen offered his enormous attention and meticulous guidance, namely his system engineering theory published in his book *Technology of Organization and Management—System Engineering* (Sep. 27, 1978, Wen Hui Bao, Shanghai, by Qian Xuesen, Xu Guozhi, Wang Shouyun). Although Mr. Qian had not directly focused or guided the Changqing Oilfield, his theory had been regarded as canons and read for hundreds of times by the technicians and managers. The second great mind is Professor Long Shenzhao, who established the science of MMESE as an aerospace medicine engineering expert, held the annual MMESE symposiums for 12 consecutive years. His papers spread this discipline around the world and are learned by the technicians of Changqing Oilfield. The third leader is academician Hu Wenrui, the former general manager of Changqing Oilfield in 1999, who vigorously advocated and promoted the re-understanding of Ordos, low permeability, and ourselves. His leadership brought Changqing Oilfield out of the predicament into a bright new world. The origin of his leadership is the very MMESE.

65.4 MMESE Technology to Create a New World

65.4.1 Crushing and Sand-Filling Technology to Improve Seepage of Low-permeability Reservoir

Low-permeability reservoir features not only low output or non-output of oil well due to the ground pressure, but also constrained oil in the reservoir due to the undergrown fissure and bad connectivity of stratums. To improve this situation, seepage of the reservoir needs to be changed manually, i.e., making artificial fissure by high-pressure water and filling sand into the fissure immediately to prevent crack closing, which is the so-called revolutionary crushing improvement technology. All the oil wells of Changqing Oilfield had to be crushed and transformed so as to produce oil. A series of small-scale (sand filling less than 20 m^3), medium-scale (sand filling between 25 and 30 m^3), and large-scale (sand filling more than 50 m^3) tests were carried out on hundreds of wells, and eventually, the classic method was screened out. Be noted that crushing and sand-filling technology is a typical MMESE, in which man is the conductor, designer, and evaluator of the crushing proposal as well as their knowledge, skills, and experiences; machine is the crushing machines, well tube, underground tools, and various auxiliary equipment; environment is the crushing liquid, crushing sand, and targeted reservoir.

65.4.2 Pre-water-injection Technology to Increase Pressure on Low-permeability Reservoir

When drilling the ground to the targeted stratum, oil will automatically flow into the well if the stratum has high pressure, good fluidity, and regular fissure. Then, the oil will flow to the ground with the help of pumps continuously. However, as to the low-permeability reservoir, the preliminary output fails due to the low stratum pressure and energy. Pre-water-injection technology refers to injection of water 3 or 6 months ahead of well drilling so as to maintain the original stratum pressure at 110 or 120 %. The ground equipment will inject water that matches the underground fluids into the stratum to increase the pressure manually. This revolutionary technology successfully solved the two global challenges in low-permeability reservoir exploration, one is low pressure and energy of the low-permeability reservoir, and the other is the continuously declining output of the low-permeability oilfield. This technology made dozens of undeveloped petroliferous areas into small oilfields with an annual yield over 300,000 tons. The two typical examples are Jiyuan Oilfield and Xifeng Oilfield. The former had gone through five times of ups and downs without any oil. This technology made the sixth time succeed greatly. It has become a medium-sized oilfield, yielding over 3 million tons in 2012. Similarly, the latter drilled out oil with the assistance of the technology at the fourth time. It made an output of 1.5 million tons in 2012. Finally, to elaborate that pre-water-injection is also a typical man-machine-environment system, in which man is the conductor, designer and evaluator of the crushing proposal as well as their knowledge, skills, and experiences; machine is the crushing machines, well tube, underground tools, and various auxiliary equipment; environment is the water source, water quality, and targeted reservoir.

65.4.3 Cluster Well to Save Great Amount of Land

Cluster well means that a number of oil wells are drilled with small spacing and different depth on a small area. The large cluster well is constituted with vertical wells, small-shift wells, small cluster wells, and medium cluster wells. Technologies, including drilling, investigation, crushing, testing, extracting, are transformed to support the cluster well technology, because cluster well can save great amount of land and pre-drilling costs and simplify the ground procedures. The recent twenty years have seen various cluster wells of 4,410 sets (20,250 wells). Investment was saved 32 million RMB and land resources saved 80,880 mu (1mu = 666.7 m^2) [6].

65.4.4 Horizontal Well to Increase Oil Drainage Area

It is undoubtedly that inclined well has more oil drainage area than that of the vertical well at the oil reservoir. How about the horizontal well? If other factors remain unchanged, the larger the oil drainage area is, the higher the output is. With the largest oil drainage area comparing with the vertical and inclined well, horizontal well can have its oil drainage area increase along with its extension in the horizontal direction; however, the area of vertical and inclined well are fixed. Horizontal well is actually a well inclined close to 90 degrees, penetrating the oil reservoir hundreds or even thousands of meters. Drilling of horizontal well is no longer an issue thanks to the perfection of directional device, direction technology and drilling equipment, plus the transformed crushing, testing, and extracting technology. The amount of horizontal wells on the Changqing Oilfield increases annually in recent three years. In 2012, 21 wells are drilled and put into production at Ji Yuan Oilfield, each well yielding 14.3 tons per day, 8 times of that of the aligning vertical wells. 635 wells in total are drilled on Changqing Oilfield in 2012. There are approximately 1,000 wells on the oilfield. Comparing with the vertical well, horizontal well can, with other factors remaining unchanged, output 3 times oil. It is another revolutionary technology.

65.4.5 Ansai Mode

Ansai Oilfield is the first low-permeability oilfield that is developed successfully and extensively in China. Eight pieces of technologies are created, namely medium-scale crushing, pre-water-injection, large-scale cluster well, optimized perforation, optimized extraction, optimized well network, optimized dynamic monitoring, optimized ground procedures. Ansai Oilfield yielded more than 1 million tons in 1997 and 3 million tons in 2008. Ground investment decreased greatly from the original 52 to 36 % in the total investment.

65.4.6 Jingan Mode

Jianan Mode copied the Ansai Mode, plus technologies of double-tube unheated closed transportation, pressurization, regional oil transfer, oil–gas mixing transportation and ring-network water injection. The output of Jianan Oilfield surges from 1 million in 2001 to 3 million in 2007.

65.4.7 Digital Archive Mode

Information center selects and digitalizes all the important documents of each project under production annually, including engineering design, construction scheme, operation procedures, installation description, drawings, summaries, original records. These documents are stored in the central data base for end-user logging on the server to get information. New projects or future project can obtain information from this data base, so as to achieve standardized design, modular construction, digitalized management, and market-based operation. In addition, repeated working is decreased, design period shortened, labor costs saved greatly thanks to the circulation of information resources. One thing to notify is that digital archive is an information system, also a typical man-machine-environment system, in which man is the producer, manager, and user of the archive, machine is the computer and network, and environment is the effective information in the archive.

65.5 Conclusions

Development of low-permeability oilfield is not only a technical and economic issue, but also a strategic issue of energy security. Due to the increasing development of oil and gas resources, the percentage of low-permeability reservoir rises. It is discovered that the amount of low-permeability reservoir takes up 60 % of the total unexploited oil reserves. It is a long-term issue confronted by China to exploit the low-permeability reservoir with high speed and amount. Changqing Oilfield applies MMESE and sets a good example for the domestic and even global oil industry in the exploitation of low-permeability reservoir.

References

1. Jiajun S (2010) Standing on high ground in northern Shaanxi, Shaanxi Daily, January 18, 2010 the northern Shaanxi observed column
2. Yongping R, Wei W (2010) "grindstone" take the oil "the People's Daily, June 5, 2010 front-page
3. Monograph (2004) "man-machine-environment system engineering theory and application foundation" according to Huang Ruisheng, Longsheng Science Press
4. Shengzhao L (1993) Man–machine–environmental systems engineering theory and its significance in the development of the productive forces in the literature sources–machine-environment system engineering research progress. Science and Technology Press vol 1
5. Chen X, Shengzhao L (1985) The man-machine-environment systems engineering (learning) Intro J Nat 8(1)
6. Wenrui H, Zhai G (2010) Ordos Basin oil and gas exploration and development practice and sustainable development. China Engineering 12, 5

Chapter 66
Application of Man-Machine-Environment System Engineering in Special Forces

Xiangqian Wang and Weiming Deng

Abstract Man-machine-Environment system engineering in the US military is widely used. Man-machine-environment system engineering in our military is increasingly widely used in aerospace, armament research and development design, tanks, submarines, and other, greatly improving the effectiveness of the fight. Based on the Special Forces training and operational characteristics, this paper explores the purpose and use of man-machine-environment system engineering and the content of the study on the application. The aim of this study is to promote integration of Special Forces training and human-machine-environment system engineering and to help Special Forces in training and operational procedures more efficient, safe, and reliable, and ultimately increase the effectiveness of the special operations.

Keywords Man-machine-environment system engineering (MMESE) · Special forces · Fight · Train

66.1 Introduction

Foreign military pays much attention to the human factor in the research and development of weapons and equipments; for example, US Defense department established the human factor technical consulting group in 1976 and US Natick Military Equipment Center had incorporated man-machines engineering into the R&D and optimization of all parts of soldier system [1]. Chinese researchers put forward the concept of man-machine-environment system engineering in general with the guidance of famous scientist Qian Xue Sen in 1981 [2]. Later some

X. Wang (✉) · W. Deng
Institute of Physical Education of PLA, Guangzhou 510502, China
e-mail: wxq_cn@qq.com

S. Long and B. S. Dhillon (eds.), *Proceedings of the 13th International Conference on Man-Machine-Environment System Engineering*, Lecture Notes in Electrical Engineering 259, DOI: 10.1007/978-3-642-38968-9_66,

military research institutes and universities co-established certain organizations to conduct researches on the field of military equipments, weapon development and training of air defense forces, aerospace and aviation, tanks and canons, etc., contributing a lot to Chinese military.

Special operation, as an important combat mode, has been widely adopted in local battles, armed conflicts, and military operations other than war. Comparing with the conventional forces, Special Forces have stronger war power and can be charged with the most special and tough operational missions. They can give the enemy fateful attacks at unexpected time or location. Sometimes one special operation can realize the goal of a battle or war [3]. Special Forces are strictly selected, specially trained, armed with special weapons, and requested to conduct special operations. Facing the tough training and fierce combat, Special Forces are confronted with trouble of injuries, reliability of weapons, and of how to save energy and improve efficiency. It is a crucial issue to study how to improve the efficiency, safety, and reliability during the training and combat of Special Forces. MMESE can effectively solve this issue.

66.2 Objective of Application of MMESE

66.2.1 Prevention of Training Injuries of Special Forces

The extraordinary stamina, skills, and courage of the Special Forces come from the trainings beyond the imagination of normal people. They are thus more likely to get injuries. Injuries and safety issues in training have always troubled the Special Forces, which is also the officers' concern. Researchers are bound to solve this issue and scientifically improve their training effects. MMESE can assist to enhance the training safety and decrease occurrence rate of injuries effectively. For example, occurrence rate of injuries on the lower limbs is high in the long distance armed running training. Fatigue periostitis and fatigue fractures are multiple injuries. Cause morewith improper training arrangements and subject to higher stresson the lower limbs. The stresses are closely related to the running techniques, cushion of shoes, and ground features. Researches about the running touchdown and cushion performance of shoes can effectively solve this problem.

66.2.2 Improvement of Battle Skills and Training Efficiency

Battle skills are the foundation for Special Forces in operations. We found that these special combat skills are described based on experiences instead of scientific theories, which have brought up certain obstacles for the teaching and training. For instance, beginners can master the basic skills in the armed climbing training;

however, in the medium level, they cannot improve anymore and reach the excellence. Although teachers tried many ways in the training, effects are not satisfactory. The reason is lack of scientific research of the skills techniques. Without the scientific guidance training is low efficiency.

The skills techniques are necessary to be studied and optimized for the Special Forces. Technique principles will guide the teaching and training, making training efficiency improved and occurrence rate of injuries lowered, as well as stamina saved thanks to efficient actions.

66.2.3 Save Stamina and Allocate Body Consumptions Rationally

Stamina of Special Forces during operations should be saved and less consumed. Although the Special Forces are trained hard to acquire the stamina required for the special operations, how to save the stamina to the greatest degree during the operations is seldom paid attention to. In actual operations, Special Forces are equipped with heavy weapons and battle for days without any support. Thus, they need to save stamina to the maximum. Researchers must try to explore ways to lower the consumptions and save the stamina. For example, Special Forces normally carry 25 kg deep in the enemy's rear, or even 40 kg. Study shows that different carrying ways and equipment carriers will greatly affect the body consumptions under the same carrying load. Uniform equipment carrier has been used in the military, but the carrier for the Special Forces must consider ergonomics to the greatest extent for the sake of saving stamina. And the constantly changing information and working day and night will bring much pressure and consume a lot during the detection operations like battle television and radar. Therefore, reasonable ergonomics will help to give full play to the soldier and equipment by rational allocation of tasks.

Three ways to save the stamina: The first is to design the weapon ergonomically; the second is to improve the skills; the third is to enhance the comfortability of uniform and weapons.

Researchers must initiatively try to save the stamina for the Special Forces for the sake of better performance in the battles.

66.2.4 Enhance Reliability of Forces

Weapons for Special Forces are more and more with the advancing national strength. Human physiological and psychological characteristics must be taken into consideration during the weapons design, manufacturing, and usage. If not, the weapon will not be operated in an effective, safe, and reliable way or even cause major accidents. From the perspective of man-machine-environment engineering, the weapons will be designed. For the already existing weapons, we can find out the unreasonable design and improve them accordingly. Besides, specific training can also prevent major mistakes.

Operation reliability of the weapons, such as motorized delta-wing aircraft, underwater carrier, and various detection devices, must be improved by studying the operational skills, so as to prevent mistakes in operations.

66.3 Research Content of Application of MMESE in Special Forces

MMESE covers a wide range of fields; however, its basic study is to explore the relations between man, machine, and environment, so that the overall performance can be optimized. It studies seven aspects: characteristics of man, machine, and environment, relation between man and machine, man and environment, machine and environment, and man, machine, and environment system. For the Special Forces, the main researches are as follows.

66.3.1 MMESE Study of Individual Equipment for Special Force

Working performance and ergonomic performance of the Special Force individual equipment are the important safeguard for the individual battle capability and viability. Comprehensive performance of individual equipment can be assessed by the biomechanics and physiology monitor methods.

66.3.1.1 Research of Military Uniform Comfortableness

Unlike civil clothes, military uniform can help soldiers in many ways. And for the Special Forces, high demands are for the uniform due to the extreme battle conditions. It must meet the basic requirements for special operations and have high comfortableness. Thus, researchers must consider the body, uniform, and environment and adopt advanced materials to make the uniform and protect the Special Forces. For example, new insulation materials, thermoregulation materials, and waterproof, moisture permeable fabrics can be used to protect the Special Forces under extreme weathers. These new materials are normally light with strong performance and comfortableness.

66.3.1.2 Research of Military Shoes

Military shoes must not only satisfy the basic requirements, but wear comfortably and protect the feet without hurting the feet; further help the soldiers enhance the mobility. During the Gulf War, the US military forces used the jungle shoes while crossing the desert, which troubled them for a few months because of the additional steel plate at the shoe bottom. An American research shows that the energy consumptions because of the shoes weight correspond to the six times of the same weight carried by the soldier. Therefore, the weight, comfortableness, and adaptability of battle conditions will affect the soldier's performance severely [4]. However, in our Special Forces, the shoes are not considered about the characteristics and needs of special operations and have less categories and functions.

66.3.1.3 Research of Carrying Devices for the Special Force

In modern wars, various battles modes and conditions require more and more equipment and living materials. Thus, carrying device is becoming increasingly important. It must be designed ergonomically to save energy consumption and improve battle efficacy.

66.3.1.4 Researches of Other Individual Equipment

Researches of the fitness, stability, comfortableness, heat dissipation of helmets; MMESE of individual computer, etc.

66.3.2 MMESE Study of Individual Weapon for Special Forces

Individual weapons serve for the specific purposes. Weapons must be designed and developed based on the MMESE suggestions and consider the human limitations, so as to improve the combined performances of human and weapon. Then, advices must be offered during the usage of weapons. Function allocation can be more reasonable between the soldier and weapon and operation of weapons can be more ergonomic. Furthermore, special weapons must be designed for the Special Forces.

66.3.3 Research of Skills for Special Forces

Research of special skills aims to fully utilize the stamina and effectively perform the activities. It is the human factors in MMESE. Besides, correct skills can prevent injuries in training.

66.3.4 MMESE Study of Special Equipment Operation

Study of operation of special equipment aims to find the perfect combination of human and weapon by considering each own characteristics. Thus, the system is more reliable and efficient. For example, the study of operation of motorized delta-wing aircraft, underwater carrier, and various detection devices.

66.4 Conclusion

MMESE has been widely applied in the US Army. We are gradually applying it in the aerospace, R&D of weapons, tanks, and submarines and have improved the battle performances greatly. And for the Special Forces, MMESE will definitely assist them to achieve more reliable, safe, and economical training and battle performances.

References

1. Yan S (2009) Human factor in weapons and equipment. Harbin Institute of Technology Press, Harbin
2. Xin C (1988) Man-machine-environment system engineering. People's Medical Publishing House, Beijing, pp 135–137
3. Lu J (2010) Special operations into the future. National Defense University Press, Beijing, p 275
4. Bensel et al (1976) Technical Report of Natick. The Effects of Tropical and Leather Combat Boots on Lower Extremity Disorders Among US Marine Corps Recruits. 5

Chapter 67
Application of Man-Machine-Environment System Engineering in Teaching

Zhanxi Zhao, Xiaoli Wu, Xiaoyan Wang and Qia Yu

Abstract Based on the principle of man-machine-environment system engineering (MMESE), the paper expounds how to build a good interpersonal relationship, establish the best man-machine relationship and man-machine-environment in teaching. Various strategies such as good introduction class, the relationship and summary of the related knowledge, taking advantage of historical and cultural allusions, and teaching method of random discussion are adopted to enhance the attraction of teaching and improve the quality of teaching. We conclude that during the overall teaching process, morality education should be infiltrated and it is more important than teaching. In other words, we should educate the students in teaching to develop the good habits such as punctuality, preciseness, and self-motivation at any time.

Keywords MMESE · Teaching · Application

67.1 Introduction

MMESE studies characteristics of man, machine, environment, and relationship of man-machine, man-environment, machine-environment, and man-machine-environment system performance [1]. Working efficiency and error rate can reach the best level when man works with the appropriate machine under the proper environment.

MMESE can also be applied in the teaching activity, in which teachers and students are the subjects, machine refers to all the articles in the classroom, including computer, projector, chalks, blackboards, platform, desks, and chairs,

Z. Zhao (✉) · X. Wu · X. Wang · Q. Yu
College of Mechanical and Electrical Engineering, Hohai University,
Changzhou 213022, China
e-mail: zhaozhanxi@sohu.com

S. Long and B. S. Dhillon (eds.), *Proceedings of the 13th International Conference on Man-Machine-Environment System Engineering*, Lecture Notes in Electrical Engineering 259, DOI: 10.1007/978-3-642-38968-9_67,

environment consists of the natural environment of temperature, illumination, noises, etc. and interpersonal environment between teachers and students.

This paper aims to improve the teaching quality by application of MMESE.

Psychology believes that interest is the emotional expression in need of information. Based on the previous wonderful experiences, the interest generates and makes people intended to understand new things actively and enduringly. Interest is indispensible for motivation [2]. Interest will motivate the student to initiatively obtain knowledge and broaden the horizon, as well as overcome obstacles and difficulties.

67.2 Importance of Interpersonal Relationship

Interpersonal relationship guarantees the classroom discipline and teaching effects.

"Man" is the key of MMESE. It is the prerequisite to keep harmonious relationship for carrying out activities. Thus, it is significant to motivate the students and create good interactive relationship between teachers and students. As the students most have smart phones with wireless network, the teacher should switch off the phone with the students so as to prevent interference.

The teacher can take advantage of an excerpt from *One prospers in worries and hardships, and perishes in ease and comfort* by Mencius, "Therefore, when God decides to assign great commission to someone, he will pain his heart, exhaust his body, starve his stomach, impoverish his life and discourage what he does, so as to strengthen his mind and will and enhance his capability." And teacher can modify this sentence into "Therefore, when God decides to assign great commission to someone, he will switch off his cell phone, take his PAD, pull his cables, shut off his WLAN…." This method might ask the students to switch off their smart phones in a humorous way. And good teaching environment and interpersonal relationship can be created for the sake of teaching effects.

67.3 Importance of Machines

In broad sense, machine refers to all the articles used. Take laser pen as an example to analyze the man-machine relationship.

Laser pen can remote control the power points so that teachers can leave the platform and go into the students. As Fig. 67.1 shows, laser pen normally has three buttons, namely up, down, and signal.

The downside is that bags for laser pen are mostly soft materials as Fig. 67.2 shows. When putting the laser pen into a soft bag after the class, it will be pressed or rubbed and switch on automatically. However, people cannot easily notice this

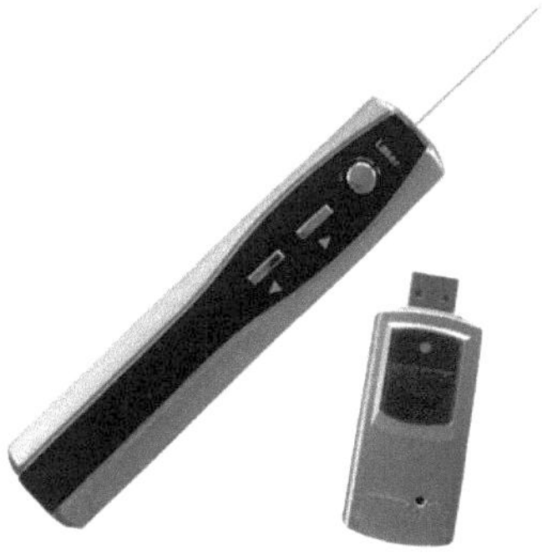

Fig. 67.1 Laser pen

and they will find that the laser pen does not work because of out of battery next day. If changing the soft bag to hard bag, this problem can be solved. Also, a manual switch can be added on the laser pen to prevent misoperation.

67.4 Importance of Environment

People should work in the best environment so that they will not get tired easily or make mistakes. Environment in teaching classes include temperature, humidity, illumination, noises, and air quality.

Take illumination as the example, there are many lights and switches in the classroom. According to the consistency principle of display and control [3], most switches and lights are one-to-one correspondence or have certain logical relations. But some classrooms do not consider the consistency principle due to negligence and much time and electricity are wasted.

In the design of illumination over the platform, there is normally only one switch to control two or three lights, as Fig. 67.3 shows. However, the projector screen does not need lights and blackboard needs lights. So, the two cannot use simultaneously, making much waste and inconvenience. The reasonable design should set no lights over the projector screen or each light can be controlled separately.

Fig. 67.2 Laser pen bag

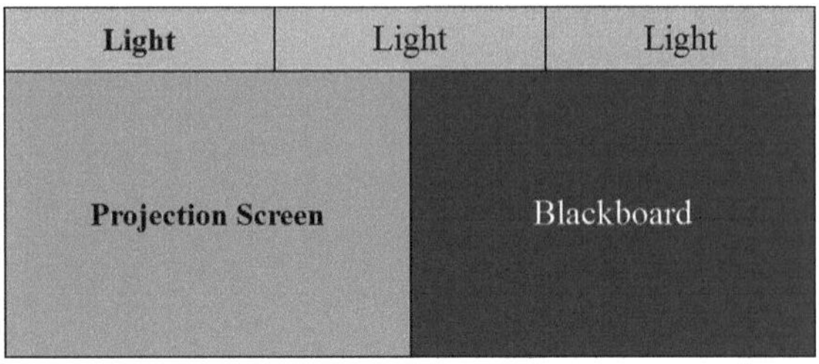

Fig. 67.3 Relative position between the classroom blackboard/projection screen and dome light

67.5 Importance of Introduction Class

Certain relations exist between different disciplines or within one discipline. The introduction of a course should consider how to connect different disciplines and engineering practices. From the perspective of system engineering, the first class of a course is crucial if regarding the course as a system.

As a western saying goes that well begun is half done, it is of great significance to interest the students at the first class.

The introduction class should involve related courses. Any information does not exist alone. There are internal rules and relations. The primary courses of mechanics major connect with each other closely as shown in Fig. 67.4, including engineering drawings, engineering mechanics, mechanical principles, mechanical design, engineering materials, and materials processing technology.

Teacher should review the previous classes and explain the different courses for the students to understand the interrelations and prevent negligence on certain courses.

While motivating the students in the introduction class, the teacher should not involve many obscure concepts which might intimidate the students.

Fig. 67.4 Mechanics major primary courses in progressive order

Course	Role of professional	Instructions
Engineering graphics	Engineering language	Must have knowledge
↓	↓	↓
Engineering mechanics	Stress analysis	Instructions for use
↓	↓	↓
Mechanical principles	Method of functionality	In principle feasible
↓	↓	↓
Mechanical design	Means of implementation	The design method
↓	↓	↓
Engineering materials	variety	Appropriate material selection
↓	↓	↓
Material forming technology	Hot and cold forming	Blank to the parts

67.6 Taking Advantage of Allusions to Motivate the Students

Humanistic environment will also arouse the students' interest for learning.

Interests will motivate the people to learn more about the world, as also called the cognitive interests, while the interest for scientific knowledge is called thirst for knowledge [4]. People with the two kinds of interests will learn with great diligence and motivation and satisfy themselves during the process. Higher level interest will also generate with the increasing knowledge. Therefore, cognitive interest is the core impetus for learning.

Teacher can adopt the historic materials and scientific development to stir up the passion, national pride, and historical responsibility of the students. China not only has the four inventions, namely papermaking, printing, powder, and compass, but also achieve greatly in metal process and usage. It was a leap from iron ore to cast iron in the iron-making industry. It was in the fourteenth century that Europe started iron making, about 1900 years later than China [5].

With a long history, Zhenjiang City, Jiangsu, is one of the birthplaces of Wu Culture. Recorded history has over 3000 years. Many important historic events happed here, such as Sun Quan established his nation, Chen Yi set up the anti-Japanese base in Maoshan, Zhenjiang, which is one of the five mountain bases. In addition, Zhenjiang boasts many great minds, including Xie Linyun, Li Bai, Su Shi, Wang Anshi, and Fan Zhongyan.

Zhenjiang, adjoining the mountains (Jin Mountain, Jiao Mountain, Beigu Mountain) and waters (Chang Jiang River), features unique beauty and landscape. The three famous mountains are the treasure of southern architecture with the landscape of river, mountain, temples, caves, springs, and forests.

Beigu Mountain is known by its Ganlu Temple for the allusion of Liu Bei marriage. There is an iron tower, formerly nine levels, now only four levels left due to wars and tsunami. The three mountains face the Chang Jiang River, which is described by one poem, "The Chang Jiang River is like an ink stone. Take the Jin Mountain and Jiao Mountain as the ink to grind. The iron tower serves as the pen and it writes on the blue sky a few line."

Arouse the students' interest for learning by introduction of history, humanity, and geography and combination with the courses, so as to improve the quality.

67.7 Random Discussion

Human cognitive ability characterizes that the brain will keep running while being stimulated continuously. The teaching practices indicate that interactive teaching method will stimulate the students greatly.

The teacher should get to the classroom 10 or 15 min early and ask the students to get to the classroom before the first bell and sign in person. Students who come after the bell will be deemed as being late for class. This will create tensions in the students mind and help them develop good habits of punctuality and self-motivation.

The teacher should switch off the phones together with the students when standing up before the class, so as to prevent interference and create good environment.

Sometimes, the homework can be arranged in advance before the contents are taught. This aims to push the students to prepare lessons before class and improve classroom efficiency when having questions.

67.8 Conclusions

Teaching is a systematic engineering. Various aspects of man, machine, and environment should be considered to create a proper teaching environment.

There is no fixed rule of teaching. As soon as the basic principles, processes, applications, and structures of members are explained clearly, and the students have deep impressions, the basic objective of teaching has been achieved. Based on this, if engineering practices beyond the book can be supplemented, the initiative of students will be motivated and teaching will be more effective.

During the overall teaching process, morality education should be infiltrated and it is more important than teaching. In other words, we should educate the students in teaching to develop the good habits such as punctuality, preciseness, and self-motivation at any time.

Acknowledgments The author thanks the mechanical majors of the "12th five-year plan" universities of Jiangsu Province for funding this research.

References

1. Chen X, Yuan X (2001) The man-machine environment system engineering physiological basis. Beijing University of aeronautics and astronautics press, Beijing
2. Zhang Z (1984) Psychology. People's Education Press, Beijing
3. Ding Y (2011) Ergonomics (Version 4). Beijing Institute of Technology Press, Beijing
4. Xie F (2005) From the teaching practice of "engineering materials" stimulate students' interest in learning. J Jiangsu Polytech Inst 6(4):86–88
5. Yang K (2005) Iron smelting technology in ancient China history. Shanghai People's Publishing House, Shanghai

Chapter 68
Construction of Assessment Indicator System of Tourism Destination Safety Based on the Man-Machine-Environment System Engineering

Jingfeng Luo

Abstract Tourism destination is a complex system. The system is made up of four elements: man, machine, environment, and management. Tourism destination safety has attracted much attention in recent years. Based on the analysis of the impact factors of tourism destination safety, in the perspective of the man-machine-environment system engineering. The assessment indicator system of tourism destination safety based on man-machine-environment-management was tried to be constructed, which included three layers and 23 specific indicators. The assessment indicator system can function as a good way to reflect safety situation of tourism destination. It provided a new model system for tourism safety assessment, so as to provide decision-making bases for the tourism safety management.

Keywords Man-machine-environment system engineering · Tourism destination · Safety assessment · Indicator system

68.1 Introduction

The release of "Outline of national tourism and relaxation" (2013–2020 year) symbolized that Chinese tourism development have entered one new golden historical opportunity period. How to effectively respond and resolve the tourism safety problem which is outstanding day by day already has been the key problem

J. Luo (✉)
College of Tourism, Huaqiao University, Quanzhou 362021, China
e-mail: luojingfeng2006@126.com

J. Luo
Center for Tourism Safety and Security Research of China Tourism Academy, Quanzhou 362021, China

S. Long and B. S. Dhillon (eds.), *Proceedings of the 13th International Conference on Man-Machine-Environment System Engineering*, Lecture Notes in Electrical Engineering 259, DOI: 10.1007/978-3-642-38968-9_68,

if it possible to really carry out "People first, service for people's livelihood, safety first, green consumption and try the best to promote the healthy, civilized, and environmentally friendly tourism and relaxation idea". The assessment of the tourism destination safety is the effective way and means to quantize degree of safety of tourism destination system, recognizing the hazard of tourism destination system and preventing the accidents. Its critical problem is how to construct the assessment indicator system of tourism destination safety. At present, the research on construction of assessment indicator system of tourism destination safety has gained certain achievements [1–6], but it still has many insufficiency: (1) considered factors of indicated system which has been constructed are not comprehensive and impact the assessment result, then it cannot reflect truly the safety situation of tourism destination and (2) the research is scattered, not systematic, not deep going, and it is still in the stage of primary searching. Therefore, based on the idea of man-machine-environment-management system engineering [7], combined with basic principle of system safety assessment, construct the assessment indicator system of tourism destination safety based on man-machine-environment management, expecting to provide one more objective, scientific, reliable assessment model system for tourism destination safety assessment.

68.2 Analysis of Influencing Factors for Tourism Destination Safety

The tourism system safety is the ultimate goal. The key to achieve this goal is tourism safety system. Tourism destination safety is one complicated, opened system, which consisted factors of man (tourists, tourism employees, residents of tourism destination, etc.), machine (the related equipments and facilities in tourism destination), environment (natural environment, social environment, economic environment, etc.), management (various rules of management, laws, and regulations), etc. in together. Therefore, it can systematically analyze and recognize the related risk factors that impact tourism destination safety, according to the idea of man-machine-environment system engineering.

68.2.1 Human Factor

According to the accident causation theory, people's non-safety actions are the direct reason to cause the accident. Therefore, when analyzing the influencing factors for tourism safety, the primary task is to recognize the human factors in tourism safety system. Human factor is the main incentive of the lot of tourist safety in accident. The people in tourism safety system include tourists, tourism employees (tourism industry managers, tour guides, leaders, other people engaged in tourism-related jobs in tourism destination), and residents in tourism destination. The human factors influencing tourism safety mainly are tourists' physical conditions, safety perception of tourists, tourism experiences of tourists, familiarity on the environment by tourists, tourism employees' professional qualification, safety perception of tourism employees, tourism perception, and attitude of residents of tourism destination.

The tourists' physical condition is the premise possible to complete the tourism action safe, successful and satisfied. It includes the following: the physical condition before the tourism (such as physical self-control, diseases, physical reserves, and so on) and the physical condition in the trip (such as inadaptation in physiology, diseases, physical power overdraw, recovery in time, and so on).

The safety perception degree of tourists high or low will directly influence the probability of tourists' self-safety problem happening. The safety perception of tourists is the understanding, attention, and reaction to the safety problems by tourists. Different tourists have different understandings on tourism safety due to their individual differences (such as gender, age, education background, family, life experiences, and so on). The main route for tourists to get safety perception is through personal experiences and news media. The safety attention by tourists is higher, his safety awareness level is higher, the safety perception degree is better, and his control capability facing to risk accidents is stronger.

The tourism experience of tourists is the knowledge and skills accumulated in many times tourism practical experiences by tourists, which positively prompt to realize the tourism safety. It concretely can be summarized as follows: the adaptive capacity in unfamiliar environment, the correct communicative skills with strangers, if possible take the necessary travel drugs according to personal physical status, weather conditions, and scenic regions characteristics; choose the available time, season, location, and travel routs; the correct decision on tourism difficulty and tourism safety; and the emergency processing capability for the various emergencies in the trip and so on.

The familiarity of tourists is the natural, social, and economic environment data collection and familiarity of the destination, the understanding on the trip routs in tourism area, the understanding and communication skills on the local language, culture, religion, custom, and so on.

The tourism employees' professional qualification is the basic condition for his job, also is the premise and guarantee to realize the safe tourism and prevent the accidents. The tourism employees' professional qualification should be embodied as corresponding qualification certificates, which include the certification for tourism industry managers, tour guide (temporary tour guide) certification, leader certification, certification for employees of tourist hotels, certification for employees of tourist transportation industry, certification for employees of tourist entertainment industry, and certification for employees in other related industries to tourism, and so on. Seriously carry out the professional qualification and grade standards for tourism employees, and making strict supervision and audit system is the most basic link to eliminate the risks and realize the essential safety.

Safety cognition of tourism practitioner includes two aspects: safety awareness education and safety management of tourism practitioner; safety awareness training and safety management cognition of tourism practitioner are effective approaches for guaranteeing tourism safety and reducing occurrence rate of safety accident. The ignorance and carelessness are other important reasons for tourist safety in accident. The purposes of safety cognition of tourism practitioner are as follows: firstly, to improve tourism practitioner's safety awareness and safety precaution abilities of tourism practitioners and secondly, to constitute safety guidance education service system for tourists of the tourism practitioners. This is profoundly significant for realizing inherent safety of tourism.

As main bearer of tourism impact, tourism cognition and attitude of the locals in the destination are other non-negligible important factors that impact tourism safety. Generally, tourism cognition and attitude of the locals in the destination are expressed with local's hospitality degrees. Local's hospitality degrees refer to the local's attitude to the tourists, reflecting locals' support to the development of tourism. If local's hospitality degrees are high and their culture development level is high, then the development of local tourism is smooth and host–guest conflict probability between the local in the destination and the tourist is low, and safety issue is decreased; oppositely, if the development of local tourism is blocked, host–guest conflict probability between the locals in the destination and the tourist is high, and safety issue is increased.

68.2.2 Machine Factor

Similarly, in accordance with accident causation theory, unsafe status of matter (machine) is a direct cause for the accident. It is involved in traffic facilities, recreational facilities, fire-fighting emergency facilities, and other facilities in tourism safety system. The safety of lives and properties of the tourists in destination can be guaranteed only if the facilities and equipments are constantly under good status.

Besides airplane, train, bus, and tourism boat, mainstream tourism traffic tools also include non-mainstream traffic tool, little train, protective sightseeing vehicle, water motor yacht, sailboat, skiing board, manpower tricycle, sliding rod, carriage and sled. Traffic tools, traffic facilities and equipments are main causes for various safe accidents in destination, strict management and control are necessary approach for reducing accident risk, safe and reliable operation.

Recreational facilities can be categorized into recreational facilities on water (e.g., bath field, swimming pool, water amusement park, and yacht) and recreational facilities on land (amusement center, tourist car, fire balloon, skiing field, and children amusement park). Improper construction, improper use, lack of maintenance, and overload operation always cause the occurrence of accidents. Regular safety check to these recreational facilities and equipments is an effective approach to discover dangerous and hazardous factors in time and reduce casualty, especially for safety check to various highly risky recreational facilities and equipments.

Fire-fighting emergency facilities are the fundamental conditions and are necessary to guarantee safety during accidents, which includes the following: fire automation warning system, automatic extinguishing system, fire hydrant system, smoke proof and emission system, various portable extinguishing equipments, emergency broadcast and emergency lighting, safe evacuation facilities, communication facilities, traffic facilities, and medical first-aid facilities.

Other facilities, e.g., safety monitoring facilities that can monitor the capacity of the resort on real-time basis, containing various safety logo systems in the resort with various safety warning signs.

68.2.3 Environmental Factor

Environment is the carrier for the development of tourism; environmental factor is crucial for tourism safety, directly impacting the tourist's choice of destination. Tourism destination environment includes natural environment, social environment, and economic environment.

Natural environment refers to preserving level of ecological environment, meteorological and climate characteristics, and natural disaster in the tourism destination. Beautiful ecological environment is an important tourism resource to attract the tourists, as well as premise for the region to develop tourism. Currently, ecological environments in various tourism destinations are being destroyed, which have been seriously impacting healthy and sustainable development of local tourism. Tourism climate and meteorological condition have scientific guidance significance for correctly locating the types of natural disasters and their occurrence and development laws. Natural disaster is tourism safety problem caused by the factor of force majeure including typhoon, earthquake, mud-rock flow, landslide, flood, rainstorm, and tornado, as well as one kind of common form causing tourism accident. Conducting real-time monitoring and warning to natural disaster are effective approaches to reduce loss and improve safe level of tourism.

Social environment refers to political factor, social culture difference, and social public safety status in the tourism destination. Political factor is the leading reason; there is no tourism without stable political situation. Good policy environment and system environment can exert positive results of various factors at maximum and promote healthy and rapid development of tourism industry in the tourism destination. When the tourists are conducting tourism activities in another place, it is easy to generate the phenomenon including fraud and crime due to conflict with the locals in the destination, impacted by different social culture. Therefore, the differentiation of social culture is also a non-negligible factor. Social public safety refers to the problems jeopardizing social safety and stability including terrorist event, urban fire, plague and epidemic disease, group violence, political riot, financial crisis turmoil, food safety, cyber safety, and key traffic accidents, which will cause major threat to personal safety of the tourist.

Good economic environment is the foundation and pillar for the development of tourism. Generally, the more developed the economy in the tourism destination is, the higher is the opening-up, the richer is the tourism products, the more complete is the tourism service facilities, the higher is the tourism safety management level, and the stronger is the attraction to the tourist, as a result, the tourism will enjoy rapid and healthy development.

68.2.4 Management Factor

Management factor is the fundamental guarantee for realizing the safety of tourism system. Management factors of tourism safety include production organization of tourism safety, production responsibility system of tourism safety, production management regulations and systems of tourism safety, configuration of full-time safety personnel, promotion and education of tourism safety, supervision and check of tourism safety, warning mechanism of tourism safety, laws and regulations of tourism safety, tourism insurance system, and tourism emergency rescue system.

Safety production organizations in tourism enterprise and public institutions are embodied in the form of Safety Production Committee. Safety Production Committee comprises of leading person in charge, person in charge of safety production, other persons in charge involved, and persons in charge of various function departments, which is the highest leadership organization responsible for safety production management in the unit.

Tourism safety production responsibility system defines detailed rules for safe production responsibilities for various departments and various personnel in specific position, which is the guarantee for implementing tourism safety.

Tourism safety production management regulations and systems are important components of various management regulations and systems in the enterprises and public institutes. Constituting complete various management regulations and systems for tourism safety production are the most important fundamental work for tourism involved enterprises and public institutions to achieve good safety production. Safety production management regulations and systems are the results obtained by the people learning about objective laws during the process of production and operation repeatedly, even experienced conclusion exchanged with fresh blood and lives.

Under direct leadership of persons in charge of safety production from tourism enterprises and public institutions, full-time (part-time) personnel in charge of safety are responsible for tourism safety production in the unit under authorized range.

The targets of tourism safety promotion and education are mainly tourists, the locals in the destination, and various tourism practitioners; its main contents include relevant laws and regulations, risk identification, precautious measures involved in tourism, emergent measures of breaking events, and analysis of accident case Official media is mainly responsible for safety promotion education, assisted by public and commercial media, combining various training classes, in order to form a kind of stable tourism safety culture atmosphere.

Relevant department of the people's government at or above the county level shall be responsible for subordinates to tourism enterprises in tourism safety supervision and inspection. Tourism safety supervision and inspection mechanism can promote the tourism enterprises before the accident to eliminate hazard.

As important component of tourism safety guarantee system, tourism safety warning mechanism warns the tourists and tourism enterprises through issuing the information on tourism safety and exerts enormous promoting action for strengthening tourism safety awareness and improving tourism safety precaution and control abilities.

Tourism safety laws and regulations are foundations and premises for conducting all the work involved of tourism safety. Currently, the laws and regulations about tourism in our country are not complete, which needs to be further developed and completed.

Tourism insurance system is the guarantee for dealing with the work after tourist safety in accident and safeguarding legal rights and interests of the tourists. Whether providing the tourists with liability insurance of tourism agency in accordance with "Provisions of Tourism Agency Buying Liability Insurance of Tourism Agency" or not and whether recommending the tourists with personal insurance are indices for measuring tourism agency to implement and perform tourism safety insurance system.

Tourism emergency rescue mechanism is effective measure and approach to reduce the loss of tourism accident; tourism emergency rescue work can be summarized as tourism emergency rescue command system, tourism safety emergency rescue team establishment, tourism safety emergency preplan preparation, and tourism safety emergency rescue training and drilling. Establishment of emergency rescue systems and teams of various governments, tourism industry associations, tourism enterprises (tourism agencies, tourism restaurants and tourism resorts), preparation and drilling status of preplan, general mutual linkage status is the key for conducing emergency rescue and reducing accident loss in time and effectively after tourism safety in accident.

68.3 Assessment Indicator System of Tourism Destination Safety

On the basis of comprehensive analysis on factors influencing tourism security such as people, machine, environment, and management, we built hierarchical structure model of assessment indicator system of tourism destination safety as

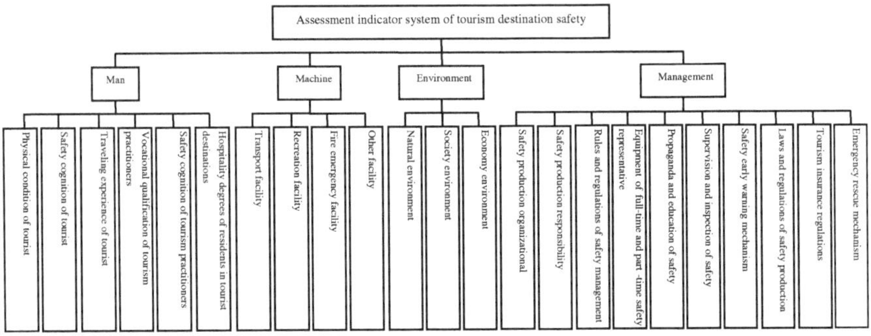

Fig. 68.1 Assessment indicator system of tourism destination safety

shown in Fig. 68.1 in accordance with the scientific nature, measurability, systemic, and conciseness principle, judging by "Tourism Law of the People's Republic of China (draft)", "Interim Measures for Tourism Safety Management", "Implementing Rules of Interim Measures for Tourism Safety Management", "Travel Agency Regulations", "Implementing Rules for the Regulation on Travel Agencies", "Travel Insurance Travel Agency Liability Insurance Regulation", "Regulation on Administration of Guiders", Implementing Measures for Administration of Guiders", "Trial Procedures for Major Travel Security Incident Report System", "Trial Implementation Measures for the Processing Procedures of Major Tourism Security Incident", "Safety and Service Quality of Amusement Park" (GB/T 16767-1997), and a series of related laws, rules, regulations, and measures of tourist safety.

68.4 Conclusion

It is necessary to comprehensively consider the factors affecting the safety of tourism system from system point of view in the whole process of building a tourist safety assessment indicator system as the requirement for comprehensive characteristics of tourism safety. The assessment indicator system of tourism destination safety based on human-machine-environment management can objectively reflect the safety situation in tourist destination, which can provide model for the next tourism safety work in combination with qualitative and quantitative assessment mode.

Acknowledgments Subsidized information: Subsidized project by scientific research fund of Huaqiao University (13SKBS114).

References

1. Zhen X (2009) Generality of tourism safety and security. Chinese Tourism Press, Beijing, pp 1–296
2. Xi J, Liu H, Qi X et al (2007) The risk assessment model of tourism destination-taking ten adventure tourism lines as a case study. J Mountain Sci 25(3):370–375
3. Cui X (2005) Fuzzy safety classification for tourism regions. J Saf Sci Tech 1(1):64–67
4. Zhao H, Wang M (2006) Study on risk system of tourism security. China Saf Sci J 16(1):17–21
5. Lin X, Li J (2004) Drifting tourism and its major latent. J Saf Environ 4(2):63–65
6. Wang L, Yu J (2010) Evaluation indicator system of tourism safety in the urban tourism destination. Saf Environ Eng 17(2):85–89
7. Long S, Huang D, Chen D et al (2004) Theory of man-machine-environment system engineering and its application base. Science Press, Beijing, pp 3–19

Chapter 69
Integrated Analysis of System Reliability and Safety by Man-Machine-Environment System Engineering

Baiqiao Huang and Jun Zhang

Abstract To solve the problem that the efficiency of the traditional system reliability and safety analysis is not high, and the problem that the human factor analysis is usually ignored, an integrated analysis method of reliability and safety based on man-machine-environment system engineering (MMESE) is proposed. This method integrates the system reliability, safety, environment adaptability, and human factor analysis and considers the human factor as the important element of the analysis; thus, the efficiency of the analysis is evidently improved. Finally, this method is applied in the reliability and safety analysis of a shipborne fueling station.

Keywords MMESE · Reliability analysis · Safety analysis · Human factor analysis · Environment adaptability analysis · Object failure mode analysis

69.1 Introduction

It is an effective way to improve the quality of weaponry by conducting analysis and design of reliability and safety during the R&D process of weaponry, and this approach has gained much attention. In current practices of analysis and design, there exists a prominent problem, namely the contradiction between micro-perspective of reliability and macro-perspective of safety. Reliability analysis for the product only considers the product itself, while the factors of environment and human are seldom considered. For safety analysis, although the above factors are all included, the factors incurring severe consequences are considered only, instead of systematic and overall analysis. In regard of the two analyses, two problems are

B. Huang (✉) · J. Zhang
System Engineering Research Institute of China State Shipbuilding Corporation, Beijing 100094, China
e-mail: sunshinefly@dse.buaa.edu.cn

S. Long and B. S. Dhillon (eds.), *Proceedings of the 13th International Conference on Man-Machine-Environment System Engineering*, Lecture Notes in Electrical Engineering 259, DOI: 10.1007/978-3-642-38968-9_69,

obvious. The first problem is the exclusion of the reliability analysis and safety analysis, i.e., some repeated works are done, and for example, efficiency is decreased if FMEA is conducted in both analyses. The second problem is the absence of comprehensive analysis for the environment and human factors [1]. Concerning these problems, this paper put forward an integrated analytical method for reliability and safety from the MMESE perspective. FMEA is carried out for the MMESE. Factors including reliability, safety, environment adaptability, and human are all covered. Thus, the efficiency of analysis is improved. And the human factor is regarded as the crucial part. Object-FMA [2, 3] is introduced to analyze each object comprehensively. This approach is applied for the analysis of reliability and safety of a shipborne fueling station, so as to verify its feasibility and effectiveness.

69.2 Macro-perspective of Reliability by MMESE

69.2.1 Micro-perspective of Reliability and Macro-perspective of Safety

Traditional definition of reliability is that the product completes its required function in specified conditions and period. The specified conditions consist of the environmental condition, service condition, and maintenance condition. The product itself is considered only while the environment and human factors are excluded as the assumed conditions. Environment factor is considered in the environmental engineering and human factor in the human factors engineering. This is a local, isolated perspective, instead of a systematic one. Environment and human factors will affect the function realization of the product, particularly the human factor. According to the analysis of civil aviation accidents by ICAO, around half of the accidents are caused by human [4]. Therefore, this paper defines the traditional reliability as the "micro-perspective of reliability."

The traditional definition of safety is the ability to prevent accidents [4]. Causes of accidents include not only the product itself, but also hazardous factors in the environment and operation factors of human. Thus, the traditional safety analysis corresponds to the MMESE. This paper defines it as the "macro-perspective of safety."

69.2.2 Analysis of Reliability and Safety from Macro-perspective

Reliability focuses on the ability of product to realize the expected functions. It is the main content analyzed by the traditional reliability to consider the product itself firstly. However, before the full automation of products, human are necessary to operate and monitor them. Without the participation of human, whatever perfect functions cannot be realized. Skills and mental and physical state of the operators will affect the output. Thus, human factor is the indispensible part for the realization of product function. Moreover, product and human are both in the environment, which will influence the product state and the physical and mental state of human, too. Environment is also an important factor. In short, the reliability of product should include the whole system of man, machine, and environment, which is called the "macro-perspective of reliability."

The macro-perspective of reliability not only considers the original product itself, but also the environment and human factors. And correspondingly, analyses regarding the failure of product, hazards in the environment, and operation failures of human, which might cause severe consequences, are the contents of safety analysis. Thus, this macro-perspective of reliability solves the problem of exclusion between the two analyses and low efficiency. Meanwhile, this new approach can better consider the interrelations between the three factors, making the analysis more thorough.

69.3 Analytical Method for Reliability and Safety by MMESE

FMEA is the most used method for reliability analysis. Suggestions for improvement are brought up from the analysis of the potential failure mode and influences. This paper adopts FMEA to analyze the man-machine-environment for the product, covering the traditional reliability, environmental adaptability, safety, and human factor. In order to better analyze the potential failure mode, this paper introduces a systematic method, namely Object-FMA. This method is object oriented. Steps of the integrated analysis of reliability and safety by MMESE are as follows.

69.3.1 Set Up MMESE Model

Man, machine, and environment are interrelated as shown in Fig. 69.1. The operator receives the information and feedback from the machine and processes them. Then, the brain will lead the body to operate the machine. The machine will function as per the input and output the expected products. The activities of machine and human will affect the microenvironment, which will affect the machine and human state, too. In the actual analysis, the MMESE model should be firstly set up for the product object according to the actual conditions.

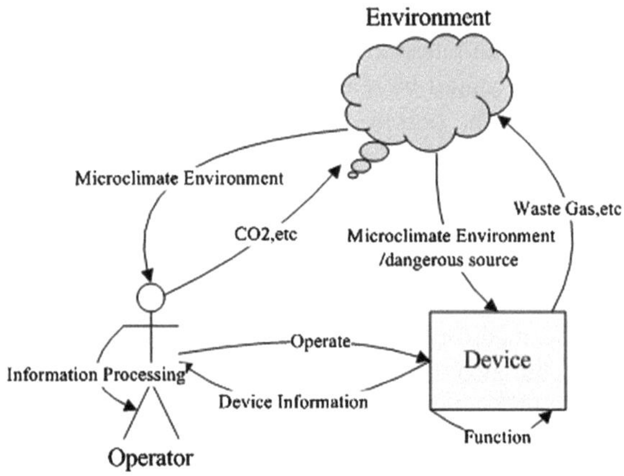

Fig. 69.1 Information interactive of man-machine-environment system

69.3.2 Failure Model Effectiveness Analysis of Man, Machine, and Environment

Failure model effectiveness analysis will be conducted for the man, machine, and environment, respectively.

FMEA analyzes the object with properties and methods. The properties and methods of object must satisfy certain constraint conditions. Violation of these conditions is deemed as the failure mode. Thus, there are three steps of FMA: (1) analysis of properties and methods of the object; (2) analysis of the constraint conditions for the properties and methods; (3) deem the violation of constrain conditions as the failure mode of the object. Object-FMA accords with the way human recognizes the world. And the analysis can go deep into the object and find the internal rules by analyzing the properties and methods.

FMEA for object "human". In the man-machine-environment system, the operator sees or hears the conditions and output from the machine, processes the information, and then leads to body to operate the machine. During the information process, operator will be affected by his/her mental state, including the character, ability, motive, motion, and will [5]. In addition, the operator is also affected by the physical state. The "methods" for the man are the operations and vary from the actual conditions. Failure mode analysis is conducted at the three steps in Table 69.1. The contents in Table 69.1 can be added or deleted according to the actual conditions.

FMEA for object "environment". It is easier to extract the factors from the environment and determine the constraint conditions by regarding the environment as an object. Environment factor includes two categories. The first category is the microclimate environment, namely the environment in which human and machine work. And considering the constraint conditions, it must be appropriate for the human and machine working without any influence on the mental and physical state, and without damage to the machine. The other category is the source of

Table 69.1 "Man" failure mode analysis

Object	Property/method			Constraints	Failure mode
Human	Property	Receive information	Vision	Read information and recognize color correctly	Fail to read information
					Fail to read information
					Fail to distinguish
			Auditory	Hear and distinguish the sound category correctly	Fail to hear
					Fail to distinguish sound
		Process information (mental state)	Character	Meet the working requirement	Impatient, timid
			Ability	Meet the working requirement	Unskilled
					Fail to meet the requirement
			Motive	Positive	Overactive or negative
			Emotion	Stable	Overexcited
			Will	Good will	Undetermined, afraid of difficulties
		Operation	Physical state	Height, weight, arm length, strength meet the working requirement	Unqualified
			Mental state	Meet the working requirement	Sick state for long term or temporarily
	Method	Specific operations		Meet the working requirement	Misoperation

hazard. GJB/Z99 can be referred to for the hazard sources. The "methods" for the environment are considering the extreme weather or geological disasters. The results of analysis of environment object by the FMEA are as shown in Table 69.2.

FMEA for object "machine". It is the same as the traditional way. First of all, disassemble the machine and get the product tree. Then, analyze each part in the product tree by the Object-FMA.

69.3.3 FMEA of MMESE

Similar to the traditional FMEA analysis of reliability, for the FMEA of man-machine-environment system, analysis table must be filled in, including the reason, influences, grade, and measures. The slight difference lies in the assessment of failure mode which should include not only the influence to the function, but also

Table 69.2 "Environment" failure mode analysis (hint)

Object	Properties/methods			Constraints	Failure mode
Environment	Properties	Microclimate environment	Temperature, humidity, salinity, wind speed, luminance, etc	Comfortable for human, no damage to machine	……
		Source of hazard	Hazard sources specified in GJB/Z99, heat, pressure, radiation, hazardous gas, etc	Avoid hazard or without severe results	……
	Methods	Typhoon, rainstorm, earthquake, etc		Without affecting human and machine	……

damage to the machine and operator (for the sake of safety). In the FMEA for "human," human factor is included and operational and support hazard analysis, as well as some occupational health analysis is also considered. In the FMEA for "machine," traditional analysis for reliability and hazard analysis of system/subsystem for safety are considered. In the FMEA for "environment," environment adaptability analysis and initial hazard analysis, as well as some occupational hazard analysis, are conducted. The coverage of our new approach of the traditional reliability and safety analysis is shown in Table 69.3.

69.4 Application

This approach is applied in the reliability and safety analysis for a shipborne fueling station. For the "human" factor, as the operators all meet the working requirements, the only mental state of "ability" is considered in the FMEA. And for the "methods," namely the operations of humanusing the "machine". The procedures in the product manual are followed, particularly paying attention to the special operations. For the "environment" factor, much attention is paid to the temperature, salinity (corrosion to machine), oxygen density in the microenvironment, as well as the flammable gases, pressure, static electricity, fire, explosion of the hazard sources; for the "methods," rainstorm, hurricane, and lightning are

Table 69.3 The coverage of our approach of traditional reliability and safety analysis

FMEA category	Reliability	Safety	Environment adaptability	Human factor
Man		Covered		Covered
Machine	Covered	Covered		
Environment		Covered	Covered	

considered. For the "machine" factor, disassemble the station into parts, such as filter, relief valve, oil gun, flammable gas density detector, and pressure detector. Object-FMA is adopted firstly to analyze the failure mode and conduct FMEA for all parts. The results show that this approach is more comprehensive comparing with the traditional method, especially for the "environment" (working condition improvement suggestion) and "human" (improvement of operational procedures) factors. Furthermore, the FMEA analysis covers four items which largely raises the efficiency and proves the feasibility and effectiveness of this new approach.

69.5 Conclusions

This paper provides an integrated analytical method of reliability and safety by man-machine-environment system engineering. This method conducts FMEA for the man-machine-environment system of the product, covering not only the content of traditional reliability analysis, but also the safety analysis, environment adaptability, and human factor. Thus, the analytical efficiency is improved entirely, and human factor is regarded as the important part. During the analysis, Object-FMA method is adopted to analyze the failure mode. This method focuses on the object, which accords with the thinking pattern of human, and offers more comprehensive analysis results.

References

1. Long Z (1993) Human-machine-environment system engineering theory. and its significance in productivity development. Progress in Human-Machine-Environment System Engineering Research, vol 01, Beijing Science and Technology Press, Beijing, pp 2–13
2. Huang B, Zhang H, Lu M et al (2010) Object-FMA based software code inspection approach. J Beijing Univ Aeronaut Astronautics 36(12):1473–1479
3. Huang B (2011) Study on integrated application of software FMEA and software developing process. Beijing University of Aeronautics and Astronautics, pp 62–72
4. Committee of National Defense Science and Technology (1990) GJB900-1990 General program for system safety. Committee of National Defense Science and Technology Standard Publishing Department, Beijing
5. Liu W, Feng S (2009) Modern human-machine-environment system engineering. Beijing University of Aeronautics and Astronautics Press

Chapter 70
Analysis of DoD 2011–2012 New HE General Standards

Hong Liang, Jianhui Li, Guohua Jiang, Zhijun Xiao, Ruiying Cao, Yongjie Ma, Yumei Zhang, Ping Liao and Mengjie Sun

Abstract This paper summarily analyzes the two new human engineering (HE) general standards released by DoD (Department of Defense, USA) in 2011–2012. There are four parts in this paper. First part is introduction. Second part introduces the development history of MIL-STD-46855 series from 1968 to 2011 and features as well as contents of MIL-STD-46855A (2011) including preface, main body (focus on the general and detailed requirements of this standard), and appendix. Third part reveals the updates of MIL-STD-1472 series from 1989 to 2012 and composition as well as main updates of MIL-STD-1472G (2012) including scope, applicable documents, general and detailed requirements, notes, and appendix A and B. Forth part briefly compares the relationship, similarities and differences, and functions of MIL-STD-46855A (DoD Standard practice: Human engineering requirements for military systems, equipment, and facilities) and MIL-STD-1472G (DoD Design criteria standard: Human engineering).

Keywords Human engineering · Human factors engineering · Military system · Standard · DoD

70.1 Introduction

Along with the development of the time and improvement of technology, in order to demonstrate USA's leading position in technology standards and military advantage in weapon equipment, in May 2011 and January 2012, DoD separately released two newly amended human engineering standards, MIL-STD-46855A (DoD Standard practice: Human engineering requirements for military systems,

H. Liang (✉) · J. Li · G. Jiang · Z. Xiao · R. Cao · Y. Ma · Y. Zhang · P. Liao · M. Sun
China Astronaut Research and Training Center, Beijing 100094, China
e-mail: 1302715079@qq.com

S. Long and B. S. Dhillon (eds.), *Proceedings of the 13th International Conference on Man-Machine-Environment System Engineering*, Lecture Notes in Electrical Engineering 259, DOI: 10.1007/978-3-642-38968-9_70,

equipment, and facilities 2011) and MIL-STD-1472G (DoD Design criteria standard: Human engineering 2012) [1, 2]. This paper preliminarily analyzes MIL-STD-46855A (2011)and MIL-STD-1472G (2012), in order to explore the sustainable development mode of US military standards and to sever the leaping development research of our country military man-machine-environment system engineering standards.

70.2 Brief History and Highlights of MIL-STD-46855A (2011)

70.2.1 Brief History of MIL-STD-46855 Series

In June 1968, under the leadership of DoD, the first US Military human engineering standard MIL-H-46855 *Human Engineering Requirements for Military Systems, Equipment and Facilities* was instituted. After that, military human engineering criteria standards that reflect the time needs and replace the old versions have been amended one after another, in which, 5 were released in hand book form, and 2 were released in standard form. They are MIL-H-46855A released in May 1972, MIL-H-46855B released in January 1979, MIL-HDBK-46855 released in May 1994, MIL-HDBK-46855 released in January 1996, MIL-HDBK-46855A released in May 1999, and MIL-STD-46855A released in May 2011, twelve years later [1, 3].

In the standards system managed by DoD, the mandatory function of H or HDBK standard is weaker than that of STD standard. The HDBK standard is quasi-standard, while the STD standard is formal standard. The switch between H or HDBK and STD mainly depends on the maturity degree and the acceptable as well as enforceable degree of the standard.

70.2.2 Highlights of MIL-STD-46855A (2011)

MIL-STD-46855A establishes and defines the requirements for applying human engineering to the design, development, and acquisition of military systems, equipment, and facilities. These requirements include the work to be accomplished in conducting a human engineering effort integrated with the total system engineering and development effort. These requirements are the basis for including human engineering in proposals; system, equipment, software and associated user interfaces, and facility analysis, design and test; and documentation and reporting. This standard military human engineering requirements with over 40 years' condensed military human engineering practical experiences, in accordance with combination of principle and innovation, referring to the construction of

MIL-STD-46855 [4], have majorly amended and perfected into two chapters, the standard's "general requirements" and "detailed requirements." The new standard is divided into three parts, preface, main body, and appendix, and the main body includes six chapters.

70.2.2.1 Preface

In order to adapt to all phases of the development and application of military system, equipment and facilities, and to make it convenient for users to reasonably apply the standard requirements, this standard regulated overall and general mission analysis, design, testing, and related requirements. Since this standard is top-level standard, it has not taken in detailed human engineering requirements. Every specific military program human engineering standard requirement will be regulated by other specified standard document.

70.2.2.2 Main Body

The main body includes six chapters: scope, applicable documents, definitions [5], general requirements, detailed requirements, and notes. In which, general requirements and detailed requirements have largely changed and almost been rewritten.

General requirements are divided into 9 sections.

Section 1 specifies that the scope and nature of standard work is to effectively integrate the human factor into system engineering developed by military system, equipment and facilities, and to run through the whole process of the system engineering's analysis, design and development, and testing and evaluation. Meanwhile, it explicitly points out that there are five ways to realize this fundamental requirement.

Section 2 is new added content. It specifies that system engineering development should apply human engineering standard at the preliminary stage, and the human engineering activity should run through the whole system engineering development process [6]. It emphasizes that where there is human and system interface, there is human engineering problem to be solved.

Section 3 is human engineering program planning. It requires that human engineering activity have to form document, such as a human engineering program plan (HEPP).

Section 4 is risk management. It requires that the standard contents of ergonomics and human engineering should involve identification, analysis of potential technology, cost, and planning risk, to eliminate the risk or to reduce it to an acceptable level.

Section 5 is requirement on synthesis analysis.

Section 6 is requirement on program trial and coordination. It requires that human engineering program shall be integrated into system integration program and management program, and that human engineering program shall intercoordinate with RAM (reliability, available, maintainability), system security, durability and vulnerability, equipment engineering, integrated logistic support, and other human factor in line (including biomedical functions, life support, personnel selection and training), and shall blend in with the whole system plan.

Section 7 is data management. It majorly requires standardize the use of data and the traceability.

Section 8 is the requirement to incorporate the human-machine system engineering sub-contractor and supplier into system management.

Section 9 is the requirement of highlighting task characteristics. Work done according to human engineering standard should be consistent with the standard requirement and should even more comply with task contract requirement instead of fully follow the standard without analysis. Every specific military program human engineering standard requirement will be regulated by other specified standard document.

Detailed requirements are divided into 3 sections.

Section 1 is the analysis requirement of human engineering. This section clearly defines various analysis and its implicit task and requirement, assures appropriate system distribution and coordination function, and makes sure the task performer (operator, maintainer, guarantor) complete the system required work in the range of human capability and limitation. The basic analysis includes human and system performance requirement analysis, task and workload analysis, and alternative operation equipment analysis. The analysis basis is human engineering requirement stands. This section has newly added 23 aspects to be focus on, at the task analysis part.

Section 2 is the design and development requirement of human engineering. This section is rewritten content and requires the input of human engineering requirement and the output of human engineering analysis result to convert to specific construction design features or engineering parameters, in the design and development process. Design of equipment should be in accordance with human and system integration system, MIL-STD-1472 standards and other engineering contract regulated human factor engineering standards. This section has newly added 8 requirements of basic human engineering design and development.

Section 3 is testing and evaluation requirement of human engineering. It requires contractor to set up and implement testing and evaluation program; there are 6 main contents. Eventually, human engineering verification has to be done in the complete system. Performance of operator or maintainer being in normal system work condition, emergency, and abnormal work condition should also be evaluated in the system-level testing eventually. Basic human engineering testing and evaluation requirements contain import of testing and evaluation plan, performance of testing and evaluation regulated by plan, and actualization of malfunction and error analysis.

70.2.2.3 Appendix

This section provides appendix A (selection and clipping of MIL-STD-46855A) that is applicable to human engineering program contract formulation and incremental applying human engineering requirement standard status.

70.3 Brief History and Highlights of MIL-STD-1472G (2012)

70.3.1 Brief History of MIL-STD-1472 Series

Based on 20 years' practice of MIL-H-46855 (1968) *Human Engineering Requirements For Military Systems, Equipment And Facilities*, according to practical demand and the technique improving level of the time, DoD released MIL-STD-1472D *Human Engineering Design Criteria For Military Systems, Equipment And Facilities* in March 1989. Later, after 7 years' application, under the influence of "Large and Comprehensive" amendment guiding ideology, big frame MIL-STD-1472E *Human Engineering Design Criteria Standard* was released in 1996. Three years later, MIL-STD-1472F *Human Engineering Design Criteria Standard* was released in August 1999. Compared with MIL-STD-1472E of 1996, MIL-STD-1472F mainly added contents of MIL-HDBK-759C *Human Engineering Design Guidelines*.

Over 10 years later, along with US military strategy adjustment and military tactic development, the old version MIL-STD-1472F has not been able to adapt to the need of human engineering guarantee under the new circumstances, especially with the fast development of information technology, portable and wearable computer equipment has been or will be normal demand, warfare simulation and real warfare handling have proposed very big challenge for the cognitive competence of military equipment user, maintainer, guarantor and even manager, and has formed great contrast between the fast technology updates and user capability's self-limitation. For this reason, in order to outstand the new times technology improving features in the twenty-first century, DoD organized forces and implemented large extent amendment on the guiding MIL-STD-1472F and released MIL-STD-1472G *Human Engineering Design Criteria Standard for Military Systems, Subsystems, Equipment and Facilities* in January 2012.

70.3.2 Highlights of MIL-STD-1472G (2012)

MIL-STD-1472G establishes general human engineering criteria for design and development of military systems, equipment, and facilities. Its purpose is to present human engineering design criteria, principles, and practices to be applied in the design of systems, equipment, and facilities. This standard has implemented full sorting, review, amendment, and improvement over the old versions, especially in the Detailed Requirement chapter that detailed, strongly operable requirements have been proposed against 15 aspects of military human engineering and 2 very valuable normative appendixes have been given. The new human engineering design criteria standard includes applicable scope, reference documents, terminology, general requirements, detailed requirements, annotations, attaching two mainappendices and an index.

70.3.2.1 Scope

This standard clearly defines as general standard, applicable to the development, and application of military system, sub-system, equipment, and facilities, with the purpose of assuring system task's success. When applying this standard, contract two parties are allowed to reasonably cut out relative clauses and indicate in the contract according to specific task requirement.

70.3.2.2 Applicable Documents

This standard reference document is divided into two categories, one is government documents (including standards, specifications and handbooks released by the federal government, and DoD), the other is non-government organization publications (including standard made by the International Organization for Standardization and standard made by American Industry and Professional Association). This chapter specifies under general condition, in the aspect of reference document sequence, terminologies, general requirements, and detailed requirements of this standard has legal priority.

70.3.2.3 Definitions

This chapter lists 58 acronyms and 28 abbreviations used in this standard, and 23 technical terms.

70.3.2.4 General Requirements

This standard specifies general requirements of 16 aspects. They include design objectives, standardization, off-the-shelf equipment, human engineering design, fail-safe design, simplicity of design, interaction, safety, ruggedness, chemical/biological/radiological/nuclear and high-yield explosive (CBRNE) survivability, electromagnetic pulse hardening, automation, functional use of color, design of aircrew systems, system integration, recycled/recovered or environmentally preferable materials.

70.3.2.5 Detailed Requirements

This standard regulated detailed requirements on 15 aspects. They are controls, visual displays, speech and audio systems, labeling, environment, ground vehicles, warning/hazards and safety, physical accommodation, maintenance accessibility, workspace design, physical environment design, virtual environments/remotely handled system/automated system/telepresence and teleoperation, small systems/ equipment and weapons, peripherals, ship and marine structure valves.

70.3.2.6 Notes

To better understand and use the standard, it provides annotations on 4 aspects. They are intended use, acquisition requirements, 17 subject keywords (aerospace vehicles, anthropometry, controls, control–display integration, design, displays, environment, human factors engineering, human-system integration, labeling, maintainability, safety, remote handling, user-computer interface, valves, vehicles, and workspace), and changes from previous issue.

70.3.2.7 Appendix

This standard has two main appendixes. Appendix A lists 13 service key points on maintainability considerations: scope, design for maintainability, design of tools, design of mechanical equipment, mounting and packaging, electronic modules, stowage of equipment, covers, fasteners, conductors, test and service points, test equipment, batteries. Appendix B gives anthropometric reference values, there are 18 values: standing body dimensions, seated body dimensions, depth and breadth dimensions, circumference and surface dimensions, hand and foot dimensions, head and face dimensions, which for general forces, army air crew and air force pilots.

70.3.2.8 Index

For user's retrieving convenience, this standard has listed 52 index terms directed into over 400 classes.

70.4 Relationship Between MIL-STD-46855A and 1472G

70.4.1 Interdependent Relationship

DoD manages US military standard centralizedly. US military standard system is comparatively numerous and jumbled, it is divided into two categories, one is military general standard, the other is army/navy/air force specific standard. At present, US effective human engineering general standards are the following: (1) MIL-STD-46855A, Department Of Defense Standard Practice: Human Engineering Requirements For Military Systems, Equipment, And Facilities; (2) MIL-STD-1472G, Department Of Defense Design Criteria Standard: Human Engineering; (3) MIL-HDBK-759C Department Of Defense Handbook: Human Engineering Design Guidelines, and (4) MIL-HDBK-1908B, Department Of Defense Handbook: Definitions Of Human Factors Terms [5, 7–10].

As MIL-STD-46855A and MIL-STD-1472G belong to the top-level standards of US military human engineering standard system, MIL-STD-46855A, Department Of Defense Standard Practice: Human Engineering Requirements For Military Systems, Equipment, And Facilities, is the theory guiding standard of the military human engineering construction, while MIL-STD-1472G, Department Of Defense Design Criteria Standard: Human Engineering is the practical guiding standard of military human engineering construction, these two standards are interdependent and indispensable in the research and development of human engineering.

70.4.2 Complementary Relationship

In 1968, DoD first released MIL-H-46855, Human Engineering Requirements for Military Systems, Equipment Facilities, and in 1989 according to practical demand, released MIL-STD-1472G, Military Standard: Human Engineering, Design Criteria for Military Systems, Equipment, And Facilities. This proves that "Engineering requirement first, and design guideline follows" is the developing rule for human engineering. Separately after over 40 years and 20 years practical experience, the fact that "engineering requirement" and "design guideline" have formed new version human engineering general standard MIL-STD-46855A (2011), Department of Defense Standard Practice: Human Engineering Requirements for Military Systems, Equipment and Facilities and MIL-STD-1472G (2012), Department Of Defense Design Criteria Standard: Human Engineering proves that the two standards have very close complementary relationship.

To sum up, MIL-STD-46855A and MIL-STD-1472G two human engineering general standards have three similar and three different functions: same purpose, same object, same applicable scope and different content, different function, different guiding significance. All in all, to satisfy the demand of military human engineering, MIL-STD-46855A (2011), Department of Defense Standard Practice: Human Engineering Requirements for Military Systems, Equipment and Facilities provides direction, and MIL-STD-1472G (2012), Department Of Defense Design Criteria Standard: Human Engineering provides method, both independent and reinforcing mutually.

Acknowledgments This paper has gained passionate guidance from Professor Siguang Jia, senior aerospace medical engineering expert, member of the International Academy of Astronautics. Here is my sincere gratitude.

References

1. MIL-STD-46855A (2011) Department of defense standard practice: human engineering requirements for military systems, equipment, and facilities
2. MIL-STD-1472G (2012) Department of defense design criteria standard: human engineering
3. Chow R, Kobierski B, Curtis C et al. (2006) Applied comparison between hierarchical goal analysis and mission, function and task analysis. In: Proceeding of the human factors and ergonomics society 50th Annual Meeting pp 520–524
4. MIL-STD-46855 (1994) Human engineering requirements for military systems, equipment, and facilities
5. MIL-HDBK-1908B (1999) Definitions of human factors terms
6. McDonald B, Campbell G (1999) Integrating human engineering requirements into the early systems engineering process
7. MIL-HDBK-759C (1995) Handbook for human engineering design guidelines
8. MIL-HDBK-759C (1997) Notice 1
9. MIL-HDBK-759C (1998) Notice 2
10. Standards Library Base Navigation. http:// www. everyspec. com/ library.php

Chapter 71
Establishment of Military Ergonomics

Zhibing Pang, Hui Gui, Quanliang Yin, Honglei Li and Chenhui Li

Abstract Man-machine-environment system engineering is the basic theory for establishing military ergonomics. In this paper, the necessity of establishing military ergonomics is discussed from three aspects: design of weaponry, selection and training of military officers, and combat command management. This paper dissects the relationship between ergonomics and military science; also it puts particular emphasis on the internal connection between ergonomics and science of military command, military equipment, military training, military psychology, and military management. Through analyzing extensive application of ergonomics in military field, significance of military ergonomics is expounded and it is mainly embodied in three aspects as follows: absorbing and integrating the military ergonomics theory, guiding military operations, and promoting the development of ergonomics.

Keywords Ergonomics · Military operations · Military ergonomics · Establishment of discipline

Ergonomics studies the working efficiency by considering various interactive factors such as man, machine, and environment and offers the optimum proposal. It is widely applied in various fields, including the military. The military ergonomics, as a sub-branch, comes into shape based on the practical application from the partial to the whole, from theories to practices, and vise versa repeatedly. With the advancement of weaponry in terms of informatization, digitization, automation, and sophistication, the system of "soldier-weaponry-military environment" has become unprecedentedly complex. Due to the particularities of military system, the study of military ergonomics has more special requirements than the ordinary ergonomics. Therefore, establishment of military ergonomics aims to

Z. Pang (✉) · H. Gui · Q. Yin · H. Li · C. Li
Air Defense Forces Academy, Zhengzhou 450052, China
e-mail: fkb_mmes@163.com

S. Long and B. S. Dhillon (eds.), *Proceedings of the 13th International Conference on Man-Machine-Environment System Engineering*, Lecture Notes in Electrical Engineering 259, DOI: 10.1007/978-3-642-38968-9_71,

research the ergonomics in the military field systematically, educate various military talents for the future, meet the demand of information war, and enhance the performance of the forces.

71.1 Military as an Important Field for Application of Ergonomics

Ergonomics studies the man, machine, and the environment, while the military officers conduct operations in the battlefield with weaponry, which constitutes the typical "man-machine-environment" system. Thus, ergonomics is naturally applied in the military when it was born and develops with the advancement of weaponry technology. During the World War II, some countries developed new weaponry in real earnest. As they neglected the study and training of operators for these advanced weaponry, issues emerged. The decision makers and engineers started to pay high attention to this problem and ergonomics was born. Ergonomics is applied in the following three aspects of military field.

71.1.1 Military Weaponry as the First Field for Application of Military Ergonomics

Weaponry, as the "machine" of ergonomics, is the foundation for fighting capacity. Introduction of ergonomics into the design of weaponry will directly determine the performance of the weaponry function. With the increasingly digitalized, informationalized, and precise weaponry, operators will receive and process much more information and shoulder heavier duties. Thus, the soldiers are demanded higher. However, there must be limits for the human. For this reason, a gap exists between the sophistication and precision of weaponry and the capabilities of the operators. If the gap is neglected in the design of weaponry, its performance will be degraded, or even fail to achieve its intended function. Therefore, ergonomics applied in the design of weaponry is the primary step.

71.1.2 Selection and Training of Personnel as the Crucial Field for Application of Military Ergonomics

With the increasing complexity of weaponry, the "soldier-weaponry-military environment" system becomes more complicated due to the interactive relations between various factors. The factor "man" is no longer the single "soldier" but a military group or a few groups. Meanwhile, it requires higher for the skills of the operators. For example, whether an antiaircraft gun company can shoot down the enemy plane or not depends on the coordination and cooperation of dozens or hundreds of operators, including radar searching, tracking, shooting command, gun filling and launching operation, shooting observation and correction, etc. Any failure by any operator might have severe consequences on the whole battle. Therefore, it is getting difficult to select and train the operators for different operations, because the relations between operators and weaponry, between military groups, and between individuals are extremely complex. Thus, selection and training of personnel becomes the crucial part for improvement of efficiency.

71.1.3 Combat Command Management as the Key Field for Application of Military Ergonomics

The information war features human command of the information system. The computer raises the human information transfer ability in multiples, namely with the speed, precision, and reliability that man cannot achieve. Application of computers in the combat command management brings out a typical ergonomics problem, i.e., the optimum coordination of computer and human. Therefore, combat command management is the key field for the military ergonomics.

71.2 Relation Between Ergonomics and Military Science

Military science studies the nature and rules of wars and taking them as the guidance for the preparation and implementation of wars [1]. Its essence is the science of researching the optimum combination of military officers with weaponry in the specific battlefield. Weaponry, as the material foundation for battles, is vital for the war [2]. However, only the operator can use the weaponry in the specific environment to realize its function. Therefore, the objective of military science is to improve the performance of "soldier-weaponry-military environment" system, which coincides with that of the ergonomics. Many sub-branches of the both are closely interrelated.

71.2.1 Relation with Science of Command

Science of military command studies the theories, methods, and rules of military organization and command. Taking the combat command as the research object, it focuses on the exploration of command factors, principles, and art so that the military command system is established. In addition, it covers the assessment of command methods and performance, as well as the characteristics and trend for the modern military command.

Research results of ergonomics can be used as the theoretical references and basic data for the battle group scale and structure, and establishment of combat command system. Meanwhile, the research results of the science of military command, like computer analog data and drill data, can provide the basic support for the analysis and research of ergonomics.

71.2.2 Relation with Science of Military Equipment

Science of military equipment studies the characteristics and rules of development and management of military equipment [2]. Taking the development and application of the whole military equipment system as the object, it combines the military science and technology organically. It discovers the inherent contradiction between development and management of military equipment, as well as the primary features and objective rules, from the macro and systematic perspective. It offers the theoretical bases for the development and management of military equipment.

The systematic theory and methods of ergonomics must be introduced for the work such as argument, R&D, procurement, and maintenance of military equipment. Only the military equipment which is designed based on the human psychology, physiology and capability can realize the designed and intended performance to the largest extent. For this reason, ergonomics is the foundation of the science of military equipment.

71.2.3 Relation with Science of Military Training

Science of military training studies the rules of military training and guidance. It aims to answer and solve all theoretical questions about the military training, so as to guide the practices [3]. Training is to improve the skills of using weapons of the soldiers, and then, combat capacity enhanced.

The combat capacity must achieve the optimum condition of "soldier-weaponry-military environment" system. Thus, qualified operators must be selected and trained scientifically so that they can operate and maintain the system. On the other hand, quantitative researches on the detailed procedures, actions, and time allocation must be conducted so as to design the reasonable actions and procedures for the operators. This is the problem to be solved by ergonomics.

71.2.4 Relation with Military Psychology

Military psychology studies the generation and development of psychological state of the military operation subjects. It aims to understand the soldiers' psychological conditions, predict the changes, adjust the state, and retain a healthy and active mental state [4].

The physiological and psychological characteristics of the operators, influences of emotional changes on the operations, etc., are the important content of ergonomics. Therefore, military psychology can provide theoretical foundation and methods for the ergonomics.

71.2.5 Relation with Military Management

Science of military management studies the rules of military management, integrating social science, natural science, and technology science [5]. Based on the objective rules of military combats and construction, it allocates and uses various resources effectively to achieve better performance of the troops by the actions of making decisions, planning, organization, coordination, control, and incentive.

Good management brings benefits and fighting capacity. Military system comprises of man, machine, and environment. Efficiency of working, training and assessment relies on the scientific and efficient management. In short, military ergonomics is closely related and interactive with the military management.

71.3 Significance of Establishment of Military Ergonomics

Military ergonomics belongs to the scope of ergonomics undoubtedly although it studies the special object. Firstly, it is not complete for the ergonomics without military ergonomics. "Soldier-weaponry-military environment" system is a typical "man-machine-environment" system. Both have the same objective of achieving higher performance. Secondly, the particularities of military ergonomics make it irreplaceable by the ordinary one, including the rigorousness of military operations, severe military environment, serious consequences of military operations, and the complexity of the military operations, etc. For this reason, military ergonomics is the indispensable part of the ergonomics. It is of great significance to conduct researches on the military ergonomics.

71.3.1 Absorb and Integrate the Military Ergonomics Theory

Establishment of military ergonomics will absorb and integrate the theories. It is necessary to establish the discipline scientifically and systematically. This has two aspects of effects. On the one hand, the absorption of military ergonomics theories, namely all the scientific and rational theories in this field, will be absorbed by the military ergonomics. On the other hand, the military ergonomics theories is integrated. As a discipline focusing on the whole military field, its theory must be a complete system. It must categorize and arrange all the theories logically and systematically.

71.3.2 Guide Military Operations

Establishment of military ergonomics will guide the military operations. It is the basic goal of military ergonomics. Three aspects can be guided: firstly, operator-oriented design of weaponry; secondly, scientific selection of soldiers and optimized military training; thirdly, systematic and standardized command [6].

71.3.3 Promote the Development of Ergonomics

Establishment of military ergonomics benefits the ergonomics for the in-depth application in the military field and expedites the research process in military field. As a sub-branch of ergonomics, military ergonomics will enrich the whole system and promote the development of ergonomics. It is the inevitable result of application of ergonomics in the military field after so many years of development.

Military ergonomics is the consequent outcome of ergonomics. Establishment of the military ergonomics not only promotes the further study of ergonomics in the military field, but also facilitates the overall structure of the ergonomics. In conclusion, it is urgent to establish the military ergonomics and carry out extensive studies.

References

1. Li Y, Li Q (2005) Military science tutorial. Ocean University of China Press, Qingdao
2. Gaoda Y, Lusheng Z (2007) Military equipment. National Defense University Press, Beijing
3. Wu Q (2003) Military training. Military Science Press, Beijing
4. Kennedy CH, He L, Gao X (2007) (translation) Military psychology. East China Normal University Press, Shanghai, China
5. Wang J, Fang N (2000) Army Management tutorial. Military Science Press, Beijing
6. Pang Z (1999) Air defense forces man-machine-environment system engineering. Zhenzhou Air Defense Forces Academy, p 91

Erratum to: Research on the Background Noise of Ordinary Classrooms

Jie Li, Kang Jiang and Jian-Zhong Guo

Erratum to: Chapter 24 in: S. Long and B. S. Dhillon (ed.), *Proceedings of the 13th International Conference on Man-Machine-Environment System Engineering*, DOI 10.1007/978-3-642-38968-9_24

Artwork of Fig. 24.4a was published wrongly. It should be displayed as below:

The online version of the original chapters can be found under DOI 10.1007/978-3-642-38968-9_24

J. Li (✉) · K. Jiang · J.-Z. Guo
Capital University of Economics and Business, Beijing 100070, China
e-mail: lijie_jerry@126.com

S. Long and B. S. Dhillon (eds.), *Proceedings of the 13th International Conference on Man-Machine-Environment System Engineering*, Lecture Notes in Electrical Engineering 259, DOI: 10.1007/978-3-642-38968-9_72,

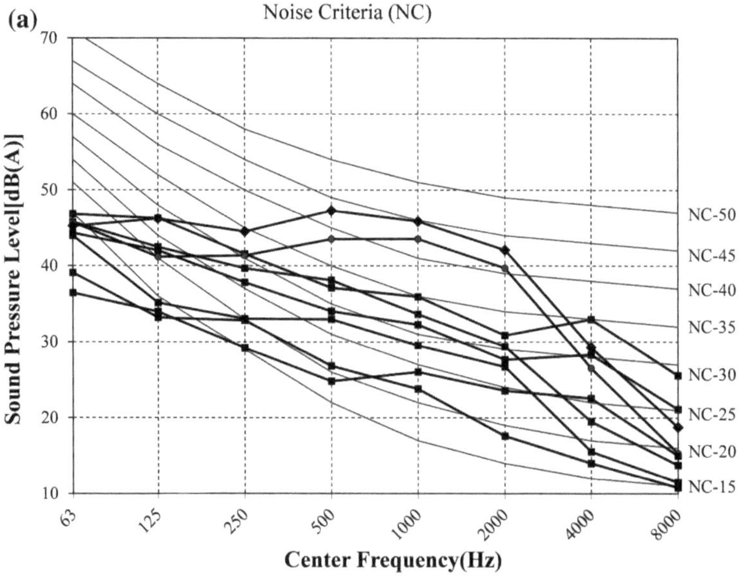

Fig. 24.4 The location of the old and new classrooms octave band sound pressure level in the NC curve **a** The NC value of the old classroom, **b** The NC value of the new classroom